Fisheries and Aquatic Ecology

Edited by Halsey Parkinson

SYRAWOOD
PUBLISHING HOUSE
New York

Published by Syrawood Publishing House,
750 Third Avenue, 9th Floor,
New York, NY 10017, USA
www.syrawoodpublishinghouse.com

Fisheries and Aquatic Ecology
Edited by Halsey Parkinson

International Standard Book Number: 978-1-68286-723-5 (Hardback)

Cataloging-in-Publication Data

Fisheries and aquatic ecology / edited by Halsey Parkinson.
     p. cm.
Includes bibliographical references and index.
ISBN 978-1-68286-723-5
1. Fisheries. 2. Aquatic ecology. 3. Fishery sciences. 4. Aquaculture.
I. Parkinson, Halsey.
SH331 .F57 2019
639.2--dc23

TABLE OF CONTENTS

PREFACE

It is often said that books are a boon to mankind. They document every progress and pass on the knowledge from one generation to the other. They play a crucial role in our lives. Thus I was both excited and nervous while editing this book. I was pleased by the thought of being able to make a mark but I was also nervous to do it right because the future of students depends upon it. Hence, I took a few months to research further into the discipline, revise my knowledge and also explore some more aspects. Post this process, I begun with the editing of this book.

Fisheries science, a branch of aquatic ecology, undertakes the scientific study of fisheries. Fisheries are harvested for commercial purposes. The fish harvested can be marine, freshwater, wild or farmed. The oceans and seas contribute nearly 90% of the total fishery catches. An understanding of aquatic ecology is essential for continued sustenance of fisheries. Some of the factors that are crucial for the sustainability of aquatic ecosystems are water depth, nutrient levels, salinity, temperature, flow, etc. Any activity that results in a change in any of these parameters results in a stress on the aquatic ecosystem. This book covers in detail some existing theories and innovative concepts revolving around fisheries and aquatic ecology. Different approaches, methodologies and advanced studies on these domains have been included in this book. It is meant to assist students and experts who want to broaden their knowledge in these fields.

I thank my publisher with all my heart for considering me worthy of this unparalleled opportunity and for showing unwavering faith in my skills. I would also like to thank the editorial team who worked closely with me at every step and contributed immensely towards the successful completion of this book. Last but not the least, I wish to thank my friends and colleagues for their support.

Editor

Isolation and in vitro culture of primary cell populations derived from ovarian tissues of the rockfish, *Sebastes schlegeli*

Jun Hyung Ryu[1], Hak Jun Kim[2], Seung Seob Bae[3], Choon Goo Jung[3] and Seung Pyo Gong[1,4*]

Abstract

This study was conducted to identify the general conditions for the isolation and in vitro culture of ovary-derived cells in rockfish (*Sebastes schlegeli*). The effects of three different enzymes on cell retrieval from ovarian tissues were evaluated first, and then the ovary-dissociated cells were cultured under various culture conditions, with varying basal media and culture temperatures, addition of growth factors, and/or culture types. We found that collagenase type I treatment was effective for cell isolation from ovarian tissues. From a total of 42 trials to evaluate the effects of basal media and culture temperatures on cell culture of ovary-dissociated cells, we observed that Leibovitz's L15 medium was more supportive than Dulbecco's modified Eagle's medium for culture, and the cells could grow at all three temperatures tested, 15, 20, and 25 °C, at least up to passage 2. However, growth factor addition did not improve cell growth. Introduction of suspension culture after monolayer culture expanded the culture period significantly more than did monolayer culture alone. Our results may provide a basis for developing an in vitro system for *S. schlegeli* germline cell culture, which will ultimately lead to improvement of the species.

Background

Establishment of an in vitro system can provide a useful tool for biotechnology applications as well as various research areas (Lakra et al., 2011; Smagghe et al., 2009; Smith, 2006; Stacey, 2012). Particularly, in vitro utilization of germline cells that possess greater developmental potential can offer a novel way of deriving superior animals through grafting this in vitro system onto existing genetic breeding and transgenic technology (Gong et al., 2008; Shin et al., 2008; Kim et al., 2009; Kim and Izpisua Belmonte, 2011). In fish, however, a limited number of studies has been conducted to establish in vitro systems using germline cells, and most of those have been restricted to small fish models (Hong et al., 2004; Kawasaki et al., 2012; Nóbrega et al., 2010; Wong and Collodi, 2013; Wong et al., 2013). Thus, there has been little research regarding farmed fish species (Lacerda et al., 2010).

Rockfish (*Sebastes schlegeli*) is a major farmed marine finfish, representing the second largest production of fish cultured in Korea (Jayasinghe et al., 2015). Thus, attempts to develop an in vitro system for culturing germline cells followed by improvement of the breed are important due to its commercial scale as a valuable food resource. In this study, as a first step towards the long-term goal of establishing a germline cell culture system in *S. schlegeli*, we first conducted primary cell cultures derived from *S. schlegeli* ovarian tissues to establish general guidelines for *S. schlegeli* cell culture by determining optimal culture conditions. We first compared the effects of three different enzymes on cell retrieval from ovarian tissues for efficient cell isolation and subsequently cultured the ovary-dissociated cells under various conditions to evaluate the effects of basal media, culture temperature, growth factors, and culture type on in vitro culture.

Methods
Fish

Rockfish (*S. schlegeli*) were purchased from a local market from July to October. In total, 17 fish were used for this

* Correspondence: gongsp@pknu.ac.kr
[1]Department of Fisheries Biology, Pukyong National University, Busan 608-737, Korea
[4]Laboratory of Cell Biotechnology, Department of Marine Biomaterials and Aquaculture, College of Fisheries Science, Pukyong National University, Busan 608-737, Korea
Full list of author information is available at the end of the article

study, and the average weight and body length were 534.17 ± 64.46 g and 32.24 ± 1.71 cm, respectively. All procedures for animal management, euthanasia and surgery were complied with the guidelines of Institutional Animal Care and Use Committee (IACUC) of Pukyong National University and the ethical guidelines published by International Council for Laboratory Animal Science (ICLAS).

Ovarian tissue collection and cell isolation

To collect ovarian tissues, healthy female rockfish were sterilized using 70 % ethanol (SK Chemicals, Sungnam, Korea) for 5 min. The ovarian tissues were removed from the bodies using sterilized surgical equipment and washed five times in Ca^{2+}/Mg^{2+}-free Dulbecco's phosphate-buffered saline (DPBS; Gibco, Grand Island, NY, USA). To evaluate the effects of different enzymes on cell isolation, 200.14 ± 0.11 mg ovarian tissue fragments were placed in 35-mm Petri dishes (SPL life Sciences, Pocheon, Korea) filled with different digestive solutions: DPBS supplemented with 0.05 % trypsin-EDTA (Gibco), 500 U/mL collagenase type I (Worthington Biochemical Corp., Lakewood Township, NJ, USA), and 500 U/mL collagenase type IV (Worthington Biochemical Corp.). Concentrations of each enzyme were determined on the basis of our previous study (Kim et al., 2014). Two collagenase solutions were prepared in Ca^{2+}/Mg^{2+}-containing DPBS, whereas the others were prepared in Ca^{2+}/Mg^{2+}-free DPBS. Then, the tissue fragments were chopped using a surgical blade and incubated for 30 min at 25 °C with periodic pipetting using wide-pore tips every 10 min. After digestion, tissue derivatives were filtered through a 40-μm cell strainer (BD Falcon, San Jose, CA, USA), and the cells were collected by centrifugation (400 g, 4 min). The cells were then treated with red blood cell (RBC) lysis buffer (155 mM NH_4Cl, 10 mM $KHCO_3$, and 0.1 mM $Na_2\cdot EDTA$) for 5 min at 4 °C to remove RBCs and then collected by centrifugation (400 g, 4 min) after inactivation of the RBC lysis buffer by adding two volumes of Ca^{2+}/Mg^{2+}-free DPBS. Cell counting was conducted using a hemocytometer (Marienfeld, Lauda-Königshofen, Germany) after trypan blue (Gibco) staining. Cell viability was calculated as the number of live cells/number of total cells × 100. For cell culture, the cells were isolated from whole ovaries under the same protocol as above, but RBS lysis buffer was not used due to its detrimental effects on cell culture.

Culture media and supplements

Two basal culture media, Leibovitz's L15 medium (L15; Gibco) and Dulbecco's modified Eagle's medium (DMEM; Gibco) containing 25 mM HEPES, were used in this study. For cell culture, both media were supplemented with 20 % (v/v) fetal bovine serum (Gibco) and a 1 % (v/v) mixed solution of penicillin and streptomycin (Gibco). According to the treatment groups, 10 ng/mL basic fibroblast growth factor (bFGF; Gibco), 25 ng/mL epidermal growth factor (EGF; Sigma-Aldrich, St. Louis, MO, USA), 1000 units/mL leukemia inhibitory factor (LIF; Millipore, Billerica, MA, USA), 1 % (v/v) fish serum from rainbow trout (FS; Caisson Laboratories, Smithfield, UT, USA), and/or 50 μg/mL medaka embryo extract (MEE) were added to the culture media. MEE was extracted as described previously (Lee et al., 2013).

Cell culture

The isolated cells were seeded in 12-well tissue culture plates (BD Falcon) coated with 0.1 % gelatin (Sigma-Aldrich). A high cell density, 1.28 ± 0.35 (mean ± standard deviation) × 10^6 cells/cm^2, was used for each culture, and the cells were cultured at 15, 20, or 25 °C in air. On day 3 after the initial seeding, cells were washed twice with Ca^{2+}/Mg^{2+}-free DPBS, and the culture medium was refreshed. Thereafter, half of the medium was replaced every 2 or 3 days. When the cells reached 80–90 % confluency, subculturing of the cells was conducted. The cells were washed twice with Ca^{2+}/Mg^{2+}-free DPBS and detached using 0.05 % trypsin-EDTA. After collecting the cells by centrifugation (400 g, 4 min), they were suspended in culture medium and seeded on 0.1 % gelatin-coated tissue culture plates by splitting at a 1:2 ratio. For suspension culture, the collected cells were suspended in L15 and cultured on 35 or 60 mm Petri dishes (SPL life Sciences) at a density of 3.5 × 10^4 cells/cm^2. To exchange the medium in suspension cultures, cells and aggregates were collected by centrifugation (400 g, 4 min), and the aggregates were dissociated by treating with 0.05 % trypsin-EDTA for 3 min. After collecting the cells by centrifugation (400 g, 4 min), the cells were suspended in L15 and cultured on 35 or 60 mm Petri dishes. Medium exchange was conducted every 6 or 7 days. Primary cell attachment and cell morphology were examined visually under an inverted microscope (TS100-F, Nikon, Tokyo, Japan).

Statistical analysis

The SAS (SAS Institute, Cary, NC, USA) software was used to analyze the effect of each treatment. If a significant main effect was detected by analysis of variance, treatments were analyzed subsequently by the least squares method or Duncan's method. Differences among treatments were considered significant at a P value < 0.05.

Results & discussion
Effects of different enzyme solutions on cell isolation

To develop an effective cell isolation method, we evaluated the effects of three different enzymes on cell retrieval and viability in the isolation procedure. As shown in Fig. 1, high levels of cell retrieval were achieved in the

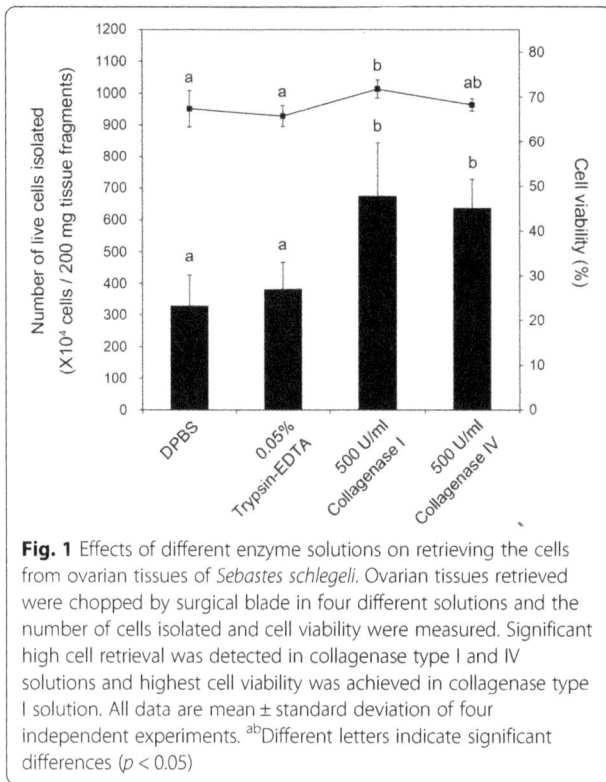

Fig. 1 Effects of different enzyme solutions on retrieving the cells from ovarian tissues of *Sebastes schlegeli*. Ovarian tissues retrieved were chopped by surgical blade in four different solutions and the number of cells isolated and cell viability were measured. Significant high cell retrieval was detected in collagenase type I and IV solutions and highest cell viability was achieved in collagenase type I solution. All data are mean ± standard deviation of four independent experiments. [ab]Different letters indicate significant differences (*p* < 0.05)

Cell isolation by tissue dissociation is a major step in cell culture, and many factors can affect tissue dissociation procedures. Of these factors, the choice of enzyme is critical to maximize the viable cell yield, and it can usually be determined by the type of tissue subjected to dissociation, because different tissues have different extracellular matrix compositions. However, due to the variability in the extracellular matrix according to the physiological status of the animal species and type of tissue, enzyme selection needs to be conducted empirically. Trypsin is a commonly used enzyme in tissue disaggregation, because it is effective on many cells and tissues. Similarly, the use of crude collagenase preparations has been found suitable for many mammalian cell cultures because of their effectiveness at breaking intercellular tissue matrices via collagenolytic and proteolytic activities (Freshney, 2010). Among the crude collagenase preparations evaluated, collagenase type I, which contains the original balance among collagenase, caseinase, clostripain, and tryptic activities (Can and Karahuseyinoglu, 2007), is widely used for the dissociation of many different kinds of tissues, including ovarian tissues in several mammalian species, such as the mouse (Gong et al., 2010), rat (Ando et al., 1999), and rabbit (Setrakian et al., 1993). Similar results were derived from our study, demonstrating suitability for *S. schlegeli* ovarian tissues as well. This provides fundamental information for cell culture of *S. schlegeli*, as well as other fish species, for which tissue dissociation procedures have not been well established. Nevertheless, additional conditions, such as the temperature, incubation time, and enzyme concentration for enzymatic dissociation, remain to be optimized in each case.

two groups treated with collagenase type I and collagenase type IV. The mean ± standard deviation numbers of live cells isolated from 200 mg tissue fragments were $328 \pm 98 \times 10^4$, $381 \pm 86 \times 10^4$, $675 \pm 168 \times 10^4$, and $637 \pm 90 \times 10^4$ in DPBS, 0.05 % trypsin-EDTA, 500 U/mL collagenase type I, and 500 U/mL collagenase type IV, respectively, from four independent experiments (*p* = 0.0019). In terms of cell viability, the cells isolated by collagenase type I treatment showed the highest value: 71.8 ± 2.0 % (*p* = 0.0450). Overall, we found that 500 U/mL collagenase type I treatment was an effective way to retrieve live cells from *S. schlegeli* ovarian tissues.

Effects of media and temperature on the culture of ovary-dissociated cells

To determine the optimal culture conditions for *S. schlegeli* ovary-dissociated cells, we first investigated the effects of basal medium and culture temperature on primary cell attachment and continuous culture. Two different basal

Table 1 Culture outcome of *Sebastes schlegeli* ovary-dissociated cells according to temperature and basal media

Temperature (°C)	Basal media[a]	No. of cell populations tested	No. (%)[b] of cell populations initially attached	No. (%)[b] of cell populations subcultured to			
				Passage 1	Passage 2	Passage 3	Passage 4
15	L15	7	7 (100)	7 (100)	3 (43)	1 (14)	0 (0)
	DMEM	7	7 (100)	3 (43)	1 (14)	0 (0)	0 (0)
20	L15	7	7 (100)	6 (86)	3 (43)	1 (14)	0 (0)
	DMEM	7	7 (100)	3 (43)	0 (0)	0 (0)	0 (0)
25	L15	7	7 (100)	5 (71)	2 (29)	1 (14)	1 (14)
	DMEM	7	7 (100)	3 (43)	0 (0)	0 (0)	0 (0)
	Total	42	42 (100)	27 (64)	9 (21)	3 (7)	1 (2)

[a]For cell culture, basal media were supplemented with 20 % (v/v) fetal bovine serum and 1 % (v/v) mixed solution of penicillin and streptomycin
[b]Percentage of the number of cell populations tested

Table 2 Effects of temperature and basal media on the culture of *Sebastes schlegeli* ovary-dissociated cells

Factors	Group	No. of cell populations tested	No. (%)[a] of cell populations initially attached	No. (%)[a] of cell populations subcultured to			
				Passage 1	Passage 2	Passage 3	Passage 4
Whole replicates	Merged	42	42 (100)	27 (64)	9 (21)	3 (7)	1 (2)
Basal media	L15	21	21 (100)	18 (86)	8 (38)	3 (14)	1 (5)
	DMEM	21	21 (100)	9 (43)	1 (5)	0 (0)	0 (0)
	P-value		-	0.003	0.0076	0.0754	0.3233
Temperature	15	14	14 (100)	10 (71)	4 (29)	1 (7)	0 (0)
(°C)	20	14	14 (100)	9 (64)	3 (21)	1 (7)	0 (0)
	25	14	14 (100)	8 (57)	2 (14)	1 (7)	1 (7)
	P-value		-	0.7475	0.6717	1.0000	0.3771

This table was derived from reallocation of data from Table 1
[a]Percentage of the number of cell populations tested

media, L15 and DMEM, were assessed in cell culture under three different temperatures of 15, 20, and 25 °C. As shown in Table 1, all 42 cell populations cultured showed primary cell attachment regardless of the experimental treatments. Of the cell populations, 64 % (27/42), 21 % (9/42), 7 % (3/42), and 2 % (1/42) grew and survived beyond the first, second, third, and fourth subcultures, respectively, but no cell population reached the fifth subculture. We found that L15 as a basal medium was better than DMEM for cell culture (Table 2). A higher number of cell populations grew to passage 2 when cultured in L15 than in DMEM (86 % vs. 43 % to passage 1, *p* = 0.03, and 38 % vs. 5 % to passage 2, *p* = 0.0076). However, no significant treatment effect on the initial cell culture was detected at the different culture temperatures, suggesting that the cultured cells can grow in the temperature range of 15 to 25 °C, at least up to passage 2. Significant differences in cell growth were observed between the cells

cultured in L15 and DMEM. On the basis of many previous studies that reported successful fish cell culture in an air atmosphere (Abdul Majeed et al., 2013; Kim et al., 2014; Lee et al., 2015), we used an air atmosphere instead of CO_2 gas to culture *S. schlegeli* ovary-dissociated cells. L15 medium was originally designed for use in culture without CO_2, and DMEM was supplemented with HEPES to control the physiological pH (Will et al., 2011). Less support for DMEM for cell growth might be a result of phototoxicity caused by the production of hydrogen peroxide from light-exposed HEPES (Lepe-Zuniga et al., 1987; Zigler et al., 1985). More evidence may be derived from additional experiments using appropriate reactive oxygen species scavengers.

As an important factor controlling the physicochemical culture properties, the optimal culture temperature is largely dependent on the body temperature of the animal from which the cells were

Table 3 Effects of various growth factors on the culture of *Sebastes schlegeli* ovary-dissociated cells

Composition of culture medium	No. of cell populations tested	No. (%)[a] of cell populations initially attached	No. (%)[a] of cell populations subcultured to	
			Passage 1	Passage 2
B	6	6 (100)	6 (100)	2 (33)
B + E	3	3 (100)	3 (100)	1 (33)
B + F	3	3 (100)	3 (100)	1 (33)
B + L	3	3 (100)	3 (100)	1 (33)
B + E + F	3	3 (100)	3 (100)	1 (33)
B + E + L	3	3 (100)	3 (100)	1 (33)
B + F + L	3	3 (100)	3 (100)	1 (33)
B + E + F + L	3	3 (100)	3 (100)	1 (33)
B + FS	3	3 (100)	3 (100)	2 (67)
B + MEE	3	3 (100)	3 (100)	2 (67)
B + FS + MEE	3	3 (100)	3 (100)	2 (67)

B, L15 medium containing 20 % fetal bovine serum and 1 % mixed solution of penicillin and streptomycin; E, 25 ng/ml epidermal growth factor; F, 10 ng/ml basic fibroblast growth factor; L, 1000 units/ml leukemia inhibitory factor; FS, 1 % fish serum; MEE, 50 μg/ml medaka embryo extract
[a]Percentage of the number of cell populations tested

derived (Freshney, 2010). In the case of *S. schlegeli*, a wide range of temperatures was tested, because as a poikilotherm, it can tolerate a wide temperature range (Md Mizanur et al., 2014). Because all of the temperatures tested supported the growth of *S. schlegeli* ovary-dissociated cells to passage 2, we decided to fix the culture temperature for further study at 15 °C due to the tolerance of cultured cells to low relative to higher temperatures (Freshney, 2010). Further examinations targeting stable continuous cell lines may be required to determine the optimal culture temperature.

There may be several reasons why cell growth was limited to passage 1 or 2 in most cases. It may be the original senescence timing of *S. schlegeli* ovary-dissociated cells, or it may be caused by a lack of appropriate signaling molecules stimulating cell growth. Next, we evaluated the effects of various growth factors, including EGF, bFGF, LIF, FS, and MEE, on cell growth in L15 medium and 15 °C

culture temperature. We confirmed again that ovary-dissociated cells can grow and survive at least to passage 1 when cultured under these conditions regardless of added growth factors but observed that none of the growth factors tested induced further growth of the cultured cells (Table 3). These results may indicate that the limitation in cell growth was due to the senescence of cultured cells at passage 1 or 2, although additional signaling molecules that originate from *S. schlegeli* or allied species need to be tested.

Effects of culture types on in vitro maintenance of ovary-dissociated cells

Throughout the culture of *S. schlegeli* ovary-dissociated cells, we observed that all cultured cell populations formed primarily a monolayer in culture (Fig. 2a), but they detached spontaneously from the substratum as the culture progressed, resulting in the removal of a

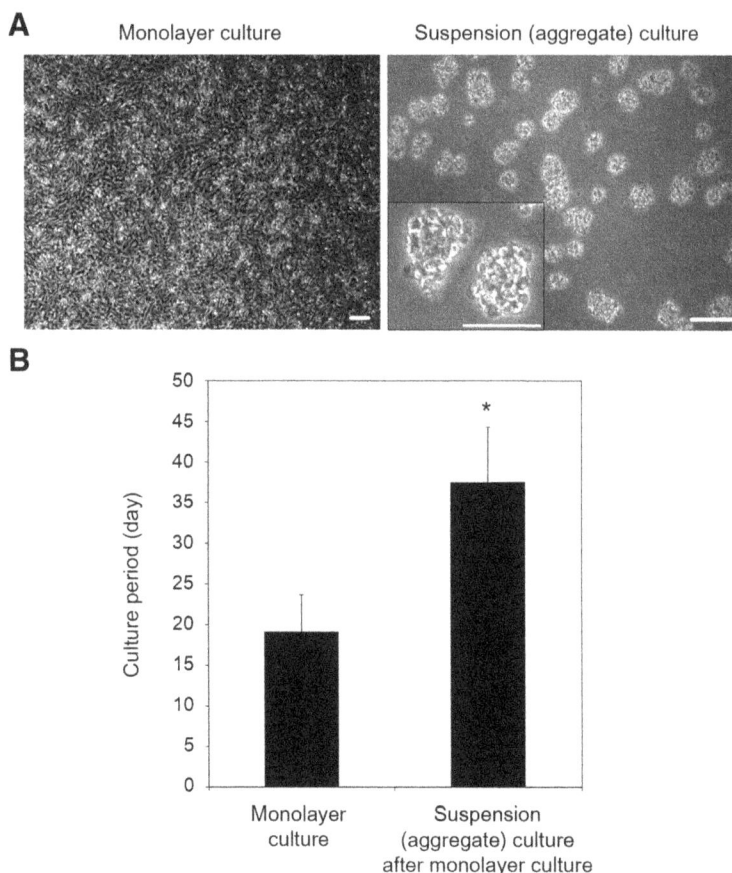

Fig. 2 Effects of culture type on culturing *Sebastes schlegeli* ovary-dissociated cells. Four cell populations derived from monolayer culture were detached on substratum at first or second passage and subsequently cultured as a form of aggregates in suspension manner for each individual. Culture period was measured individually on all cell populations cultured. **a** Morphology of the cells and cell aggregates in monolayer and suspension culture, respectively. Box within the picture of suspension culture shows a magnified image of aggregates. Scale bar = 100 μm and 50 μm in pictures and box, respectively. **b** Comparison of culture period between monolayer culture and suspension (aggregate) culture after monolayer culture. Combination of monolayer and suspension culture induced a statistically significant long-term maintenance of the cell populations compared with the sole monolayer culture. The data are mean ± standard deviation of nine and four independent experiments in monolayer and suspension culture, respectively. Asterisk indicates significant difference, *p* < 0.05

large proportion of the cells during media exchange. Eventually, it brought about a shortening of the culture period in vitro. To increase the culture period of the cells, we attempted a suspension culture after a monolayer culture and compared the total culture period between suspension culture after monolayer culture and monolayer culture alone. Four cell populations in the monolayer culture derived from different individuals were detached artificially from the substratum using 0.05 % trypsin-EDTA treatment at the first or second passage and subsequently cultured in a suspension manner individually. The cells in suspension culture formed cell aggregates spontaneously (Fig. 2a), which were maintained without significant morphological changes. A significant increase in culture period was detected in suspension culture after monolayer culture versus monolayer culture alone (Fig. 2b; 37.5 ± 6.8 vs. 19.1 ± 4.6 days, $p < 0.0001$). Short-term primary cultures have limited application because of a lack of reproducibility and cell homogeneity. Thus, long-term culture or development of continuous cell lines is required for certain biotechnological applications of cultured cells (Bols et al., 1994). Consequently, extension of the culture period by introducing suspension culture is important in that it provides more time to add additional treatments for establishing long-term cultivable cells, such as spontaneous induction of cell immortalization and artificial induction of cell transformation. Additionally, aggregate formation in suspension culture has potential advantages, such as ease of cell maintenance, culture at high cell densities (Kyung et al., 1992), and provision of a three-dimensional microenvironment (Welter et al., 2007). These suggest the potential for further development of this culture system. Additional studies to activate cell proliferation followed by establishment of continuous cell lines are needed.

Conclusions

We report the general conditions for in vitro culture of *S. schlegeli* ovary-dissociated cells. Tissue dissociation and cell isolation can be implemented effectively using collagenase type I, and the cells can be cultured in L15 medium under a range of temperatures. Additionally, a combination of monolayer and suspension culture can extend the culture period significantly. The results of this study provide a basis for establishing an in vitro system for *S. schlegeli* germline cell culture.

Competing interests
The authors declare that they have no competing interests.

Authors' contributions
JHR carried out the experiments. SSB, CGJ, and HJK participated in experimental design and data analysis. SPG conceived and designed the study, analyzed the data, and wrote manuscript. All authors read and approved the final manuscript.

Acknowledgments
This research was supported by National Marine Biodiversity Institute Research Program (2015 M00700) and Basic Science Research Program through the National Research Foundation of Korea (NRF) funded by the Ministry of Education (NRF-2015R1D1A1A01059056).

Author details
[1]Department of Fisheries Biology, Pukyong National University, Busan 608-737, Korea. [2]Department of Chemistry, Pukyong National University, Busan 608-737, Korea. [3]Marine Biodiversity Institute of Korea, Seochun 33662, Korea. [4]Laboratory of Cell Biotechnology, Department of Marine Biomaterials and Aquaculture, College of Fisheries Science, Pukyong National University, Busan 608-737, Korea.

References
Abdul Majeed S, Nambi KS, Taju G, Sundar Raj N, Madan N, Sahul Hameed AS. Establishment and characterization of permanent cell line from gill tissue of *Labeo rohita* (Hamilton) and its application in gene expression and toxicology. Cell Biol Toxicol. 2013;29:59–73.

Ando M, Kol S, Irahara M, Sirois J, Adashi EY. Non-steroidal anti-inflammatory drugs (NSAIDs) block the late, prostanoid-dependent/ceramide-independent component of ovarian IL-1 action: implications for the ovulatory process. Mol Cell Endocrinol. 1999;157:21–30.

Bols NC, Barlian A, Chirina-Trejo M, Caldwell SJ, Goegan P, Lee LEJ. Development of a cell line from primary cultures of rainbow trout, *Oncorhynchus mykiss* (Walbaum), gills. J Fish Dis. 1994;17:601–11.

Can A, Karahuseyinoglu S. Concise review: human umbilical cord stroma with regard to the source of fetus-derived stem cells. Stem Cells. 2007;25:2886–95.

Freshney RI. Culture of Animal Cells: A Manual of Basic Technique and Specialized Applications. 6th ed. Hoboken: Wiley-Blackwell; 2010.

Gong SP, Lee EJ, Lee ST, Kim H, Lee SH, Han HJ, et al. Improved establishment of autologous stem cells derived from preantral follicle culture and oocyte parthenogenesis. Stem Cells Dev. 2008;17:695–712.

Gong SP, Lee ST, Lee EJ, Kim DY, Lee G, Chi SG, et al. Embryonic stem cell-like cells established by culture of adult ovarian cells in mice. Fertil Steril. 2010;93: 2594–601.

Hong Y, Liu T, Zhao H, Xu H, Wang W, Liu R, et al. Establishment of a normal medakafish spermatogonial cell line capable of sperm production *in vitro*. Proc Natl Acad Sci U S A. 2004;101:8011–6.

Jayasinghe JD, Elvitigala DA, Whang I, Nam BH, Lee J. Molecular characterization of two immunity-related acute-phase proteins: haptoglobin and serum amyloid a from black rockfish (*Sebastes schlegeli*). Fish Shellfish Immunol. 2015;45:680–8.

Kawasaki T, Saito K, Sakai C, Shinya M, Sakai N. Production of zebrafish offspring from cultured spermatogonial stem cells. Genes Cells. 2012;17:316–25.

Kim IW, Gong SP, Yoo CR, Choi JH, Kim DY, Lim JM. Derivation of developmentally competent oocytes by the culture of preantral follicles retrieved from adult ovaries: maturation, blastocyst formation, and embryonic stem cell transformation. Fertil Steril. 2009;92:1716–24.

Kim MS, Nam YK, Park C, Kim HW, Ahn J, Lim JM, et al. Establishment condition and characterization of heart-derived cell culture in Siberian sturgeon (*Acipenser baerii*). In Vitro Cell Dev Biol Anim. 2014;50:909–17.

Kim S, Izpisua Belmonte JC. Pluripotency of male germline stem cells. Mol Cells. 2011;32:113–21.

Kyung YS, Peshwa MV, Perusich CM, Hu WS. Dynamics of Aggregate Culture of Mammalian Cells. In: Animal Cell Technology: Basic & Applied Aspects. Netherlands: Springer; 1992. p. 65–9.

Lacerda SM, Batlouni SR, Costa GM, Segatelli TM, Quirino BR, Queiroz BM, et al. A new and fast technique to generate offspring after germ cells transplantation in adult fish: the Nile tilapia (*Oreochromis niloticus*) model. PLoS ONE. 2010;5:e10740.

Lakra WS, Swaminathan TR, Joy KP. Development, characterization, conservation and storage of fish cell lines: a review. Fish Physiol Biochem. 2011;37:1–20.

Lee D, Kim MS, Nam YK, Kim DS, Gong SP. Establishment and characterization of permanent cell lines from *Oryzias dancena* embryos. Fish Aquat Sci. 2013;16: 177–85.

Lee D, Ryu JH, Lee ST, Nam YK, Kim DS, Gong SP. Identification of embryonic stem cell activities in an embryonic cell line derived from marine medaka (*Oryzias dancena*). Fish Physiol Biochem. 2015;41:1569–76.

Isolation and in vitro culture of primary cell populations derived from ovarian tissues...

7

Lepe-Zuniga JL, Zigler Jr JS, Gery I. Toxicity of light-exposed Hepes media. J Immunol Methods. 1987;103:145.

Md Mizanur R, Yun H, Moniruzzaman M, Ferreira F, Kim KW, Bai SC. Effects of feeding rate and water temperature on growth and body composition of juvenile Korean rockfish, *Sebastes schlegeli* (Hilgendorf 1880). Asian-Australas J Anim Sci. 2014;27:690–9.

Nóbrega RH, Greebe CD, van de Kant H, Bogerd J, de França LR, Schulz RW. Spermatogonial stem cell niche and spermatogonial stem cell transplantation in zebrafish. PLoS ONE. 2010;5:e12808.

Setrakian S, Oliveros-Saunders B, Nicosia SV. Growth stimulation of ovarian and extraovarian mesothelial cells by corpus luteum extract. In Vitro Cell Dev Biol Anim. 1993;29A:879–83.

Smith CL. Mammalian cell culture. Curr Protoc Mol Biol. 2006;28:1–28.0.2.

Smagghe G, Goodman CL, Stanley D. Insect cell culture and applications to research and pest management. In Vitro Cell Dev Biol Anim. 2009;45:93–105.

Stacey G. Current developments in cell culture technology. Adv Exp Med Biol. 2012;745:1–13.

Shin SS, Kim TM, Kim SY, Kim TW, Seo HW, Lee SK, et al. Generation of transgenic quail through germ cell-mediated germline transmission. FASEB J. 2008;22: 2435–44.

Welter JF, Solchaga LA, Penick KJ. Simplification of aggregate culture of human mesenchymal stem cells as a chondrogenic screening assay. Biotechniques. 2007;42:732–7.

Will MA, Clark NA, Swain JE. Biological pH buffers in IVF: help or hindrance to success. J Assist Reprod Genet. 2011;28:711–24.

Wong TT, Collodi P. Dorsomorphin promotes survival and germline competence of zebrafish spermatogonial stem cells in culture. PLoS ONE. 2013;8, e71332.

Wong TT, Tesfamichael A, Collodi P. Production of zebrafish offspring from cultured female germline stem cells. PLoS ONE. 2013;8:e62660.

Zigler Jr JS, Lepe-Zuniga JL, Vistica B, Gery I. Analysis of the cytotoxic effects of light-exposed HEPES-containing culture medium. In Vitro Cell Dev Biol. 1985; 21:282–7.

2

Seasonal variation of physicochemical factor and fecal pollution in the Hansan-Geojeman area, Korea

Young Cheol Park[1], Poong Ho Kim[2], Yeoun Joong Jung[1], Ka Jeong Lee[1], Min Seon Kim[1], Kyeong Ri Go[1], Sang Gi Park[1], Soon Jae Kwon[1], Ji Hye Yang[1] and Jong Soo Mok[1*]

Abstract

The seasonal variation of fecal coliforms (FCs) and physicochemical factors was determined in seawaters of the Hansan-Geojeman area, including a designated area for oyster, and in inland pollution sources of its drainage basin. The mean daily loads of FCs in inland pollution sources ranged from 1.2×10^9 to 3.1×10^{11} most probable number (MPN)/day; however, the pollutants could not be reached at the designated area. FC concentrations of seawaters were closely related to season, rainfall, and inland contaminants, however, within the regulation limit of various countries for shellfish. The highest concentrations for chemical oxygen demand (COD) and chlorophyll-a in seawaters were shown in the surface layer during August with high rainfall, whereas the lowest for dissolved oxygen (DO) in the bottom layer of the same month. Therefore, it indicates that the concentrations of FC, COD, DO, and chlorophyll-a of seawaters were closely related to season and rainfall.

Keywords: Fecal coliform, Physicochemical factor, Hansan-Geojeman area, Inland pollutant, Seawater

Background

The Hansan-Geojeman area is located between the Geoje and Hansan islands in Gyeongnam province on the south coast of Korea (Fig. 1). The sea surface area is about 55 km^2, and the longest length of the east-west or south-north is about 10 km. The drainage basin of the Hansan-Geojeman area is 161.8 km^2, and most parts are farmlands and mountains. Approximately 17,550 people lived in this drainage area in 2012. Of the area, 72.3 % (116.9 km^2) was occupied by forestry field and its 18.5 % was cultivated as both of rice paddy and dry paddy (Yoo et al. 2013). The Hansan-Geojeman area has been a designated shellfish-growing area for export since 1974 because it is a major oyster production area in Korea (Ha et al. 2011). According to Statistics Korea (2013), Korea produced 239,779 tons of oysters, the first largest amount of shellfish produced in Korea. In particular, Gyeongnam province, including the Hansan-Geojeman

area, produced the largest amount of oysters in Korea, accounting for about 90 % of oyster products. The products are consumed domestically or exported mainly to the USA, Japan, and the European Union (EU) (Mok et al. 2013, 2014a, 2015). The Korean government has established a memorandum of understanding with the USA and EU, and there are designated shellfish-growing areas for export along the southern coast of Korea that meet the standards set by these countries (Mok et al. 2014b, 2015). In particular, oysters are commonly consumed raw in many cultures, including Korea. Therefore, the sanitary status of seawaters in this area is needed to assess oyster quality both for Korean populations and for consumers in importing countries.

Fecal or chemical contaminations of inland and coastal waters may have a negative impact on shellfish sanitary status (Feldhusen 2000; Dorfman and Sinclair Rosselot 2008) resulting in economic losses due to shellfish bed closures (Rabinovici et al. 2004). The pollutions can also deteriorate the aquatic environment for producing, harvesting, and consuming shellfish. To protect public health, the authorities in various countries, such as Korea, the USA, and EU, have established regulatory limits and

---

* Correspondence: mjs0620@korea.kr
[1]Southeast Sea Fisheries Research Institute, National Fisheries Research & Development Institute, 397-68, Sanyang-iljuro, Sanyang-up, Tongyeong 650-943, Republic of Korea
Full list of author information is available at the end of the article

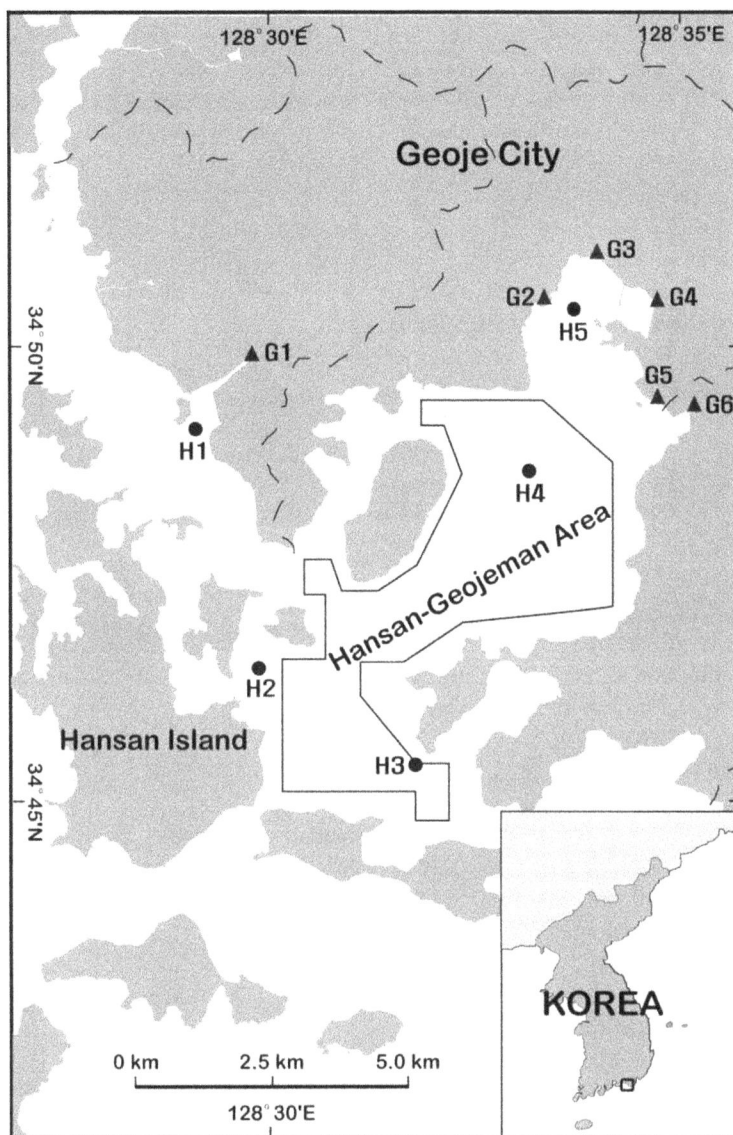

Fig. 1 Sampling locations of inland pollution sources and seawaters from the Hansan-Geojeman area on the southern coast of Korea

monitoring programs using the fecal coliforms (FCs) for bivalves or their growing area (EC 2005; US FDA 2013; KMFDS 2015; MOF 2015). Impact of inland pollution sources on water quality of shellfish-growing areas must also be estimated regularly for identifying pollution sources and protecting the growing area.

In the present study, we determined the concentrations of FCs and physicochemical factors in the seawaters collected from the Hansan-Geojeman area on the southern coast of Korea. In addition, we also attempt to compare the spatial and seasonal variation of FCs and physicochemical factors of major inland pollution sources in the drainage basin of this area. To our knowledge, this is the first report on the comparison and contribution of FCs and physicochemical factors of inland pollution

sources discharged to the Hansan-Geojeman area of Korea.

Methods
Sample collection

The sampling locations for creek water and seawater from the Hansan-Geojeman area on the southern coast of Korea, where a designated shellfish-growing area for export was located, are presented in Fig. 1. Creek water samples were collected at low tide in 2014 from the six fixed sampling stations, classified as major inland pollution sources, in the drainage area of the Hansan-Geojeman area to evaluate the effects of inland pollution sources on water quality of the shellfish-growing area. The samples of surface seawater for fecal pollution-indicative bacteria were collected once a month in 2014 at the five fixed stations in the surveyed area. Seawater samples for physicochemical factors were also collected bimonthly in 2014 from the surface and bottom layers of water using a Niskin water sampler at the same stations for fecal pollution-indicative bacteria. Samples were maintained below 10 °C during transport to the laboratory.

Physicochemical factor analysis

Physicochemical factors included temperature, salinity, pH, dissolved oxygen (DO), dissolved inorganic nutrients, and chemical oxygen demand (COD), as well as a biological factor, chlorophyll-*a* (Chl-*a*). The measurement methods are as follows. Temperature, salinity, pH, and DO values were observed on-site using a YSI 556 multiprobe system (Yellow Springs, OH, USA). They were completely processed on-site before being transported to the laboratory. Dissolved inorganic nutrients, COD, and Chl-*a* were analyzed according to the Marine Environmental Process Exam Standards (MOF 2002). Dissolved inorganic nutrients were analyzed by filtering the sample through a 0.45-μm membrane filter paper (nitrate cellulose), then measuring ammonia nitrogen (NH_3), nitrate nitrogen (NO_3), and nitrite nitrogen (NO_2), and dissolved inorganic phosphate (DIP) using a nutrient auto-analyzer (Quattro four-channel, Seal Analytical, WI, USA). Dissolved inorganic nitrogen (DIN) was expressed as the sum of NH_3, NO_3, and NO_2. Chl-*a* was measured by filtering 500 mL of the sample through a 0.45-μm membrane filter paper, then extracting the color dye with 90 % acetone in a cold darkroom, and using a fluorescence spectrometer (10-AU, Turner Designs, CA, USA). COD of seawater was analyzed with the alkaline permanganate method on non-filtered samples.

Fecal pollution-indicative bacteria analysis

The bacteriological examination of water samples were immediately performed after receiving. The numbers of FC in the samples were examined according to the recommended procedures for the examination of seawater and shellfish (APHA 1970). FC counts were determined by a multiple dilution series using the most probable number (MPN) method. Five tubes were used for each dilution. Lauryl tryptose broth (Difco, Detroit, MI, USA) was used as the presumptive medium. The culture tubes of presumptive positive, in which gas formed within 48 h after inoculation at 35.0 °C, were confirmed for FC using an EC medium (Difco) at 44.5 °C. FC populations were expressed as MPN per 100 mL.

Evaluation of inland pollution source

The flow velocity of discharges from inland pollution sources was measured on-site using a hydrometer

(Flo-Mate 2000, Marsh McBirney, Loveland, CO, USA) for their flow rate, which was calculated using the velocity–area method. Evaluation method of pollution sources is suggested in the US Food and Drug Administration (FDA) in accordance with the following equations (Park et al. 2012; Shim et al. 2012). Thus, it is calculated as the amount of dilution water required that can dilute the FC density to less than 14 MPN/100 mL of the standard level in the seawater based on the US FDA guidance.

Daily load (DL, MPN/day) = FC concentration (MPN/100 mL) × conversion factor (liter to milliliter; 1000 mL/L) × conversion factor (minutes per day; 1440 min/day) × flow rate (L/min).

Dilution water required (DWR, m^3/day) = DL (MPN/day)/ (standard [14 MPN/100 mL] × conversion factor [milliliter to cubic meter; m^3/1,000,000 mL]).

Area required (AR, m^2/day) = DWR (m^3/day)/average depth (AD, m).

Radius of half-circle (m/day) = square root (AR[m^2/day] × 2/3.14).

Statistical analysis

Statistical evaluation was conducted using analysis of variance with the general linear model procedure (SAS version 9.2, SAS Institute, Cary, NC, USA). Duncan's multiple-range test was applied to determine the significance of differences between the concentrations of FCs or physicochemical factors.

Results and discussion

Concentration of FCs in inland pollution sources

The concentrations, daily loads, and diffusion ranges of FC in the major inland pollution sources collected from six fixed stations at the drainage area of the Hansan-Geojeman area in 2014 are shown in Fig. 2. Nonpoint source (NPS) pollution occurs when rainfall, snowmelt water, or irrigation water flows over land, carrying and depositing pollutants into streams, lakes, and coastal waters (Wu and Chen 2013). The level of bacterial indicator may be influenced by rainfall events, and therefore, rainfall is often not considered when monitoring the microbial quality of waters (Lipp et al. 2001; Rose et al. 2001). Therefore, this survey was conducted when no precipitation occurred during a week before the sampling for minimizing NPS inputs into the inland pollution sources.

In the six inland water samples, the mean flow rates at each station ranged from 43 to 4529 L/min; the highest mean value was found at station G6, which is a Sanyang stream in Geoje city (Fig. 2a). Specially, the highest

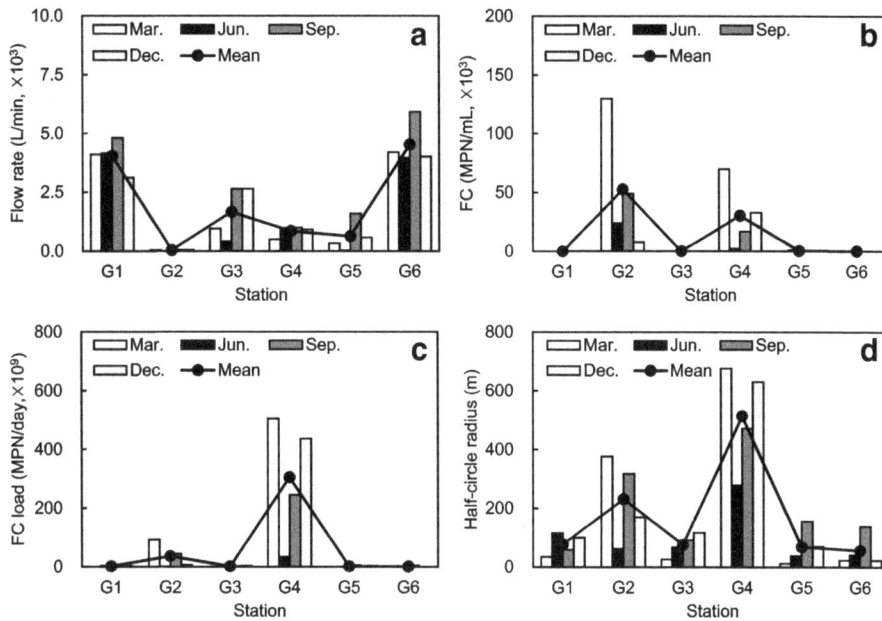

Fig. 2 Spatial and seasonal variation of flow rates (**a**), fecal coliform (FC) concentrations (**b**), daily loads of FC (**c**), and half-circle radii of FC (**d**) of discharges from major inland contamination sources in the drainage area of the Hansan-Geojeman area, Korea

values of flow rate at all the stations were found in September. It is presumed that rainwater was largely stored in reservoirs or streams during the summer season with heavy rainfall and used for rice farming between July and September. FC concentrations of major pollution sources ranged from 2.0 to 130,000 MPN/100 mL; the highest level was found in March at station G2, which is a Neagan stream in Geoje city (Fig. 2b). It is presumed that the stream is surrounded by dry fields and orchards on which fertilizers including animal feces are spread during spring season in Korea.

According to the formula in the "Methods" section, the mean daily loads of FC at each station ranged from 1.2×10^9 to 3.1×10^{11} MPN/day (Fig. 2c). The mean half-circle radii of discharges at each site on the sea area ranged from 56.6 to 514.5 m; the highest radius was found at station G4, which is a Seojeong stream (Fig. 2d), because of the relatively high FC concentration of that, which flows through a relatively densely populated residential area into the sea area. However, the pollutants could be not reached at the boundary line of the designated area because of the buffer zone between shoreline and boundary line of the designated area.

Distributions of FCs and bacteriological water quality

FC concentrations of seawater samples in 2014 from the five stations fixed in the Hansan-Geojeman area, including a designated area for oyster, are summarized in Fig. 3. The monthly values of geometric mean for FC ranged from <1.8 to 46.0 MPN/100 mL; the maximum level was found in August (Fig. 3a). Monthly rainfall of Geoje city

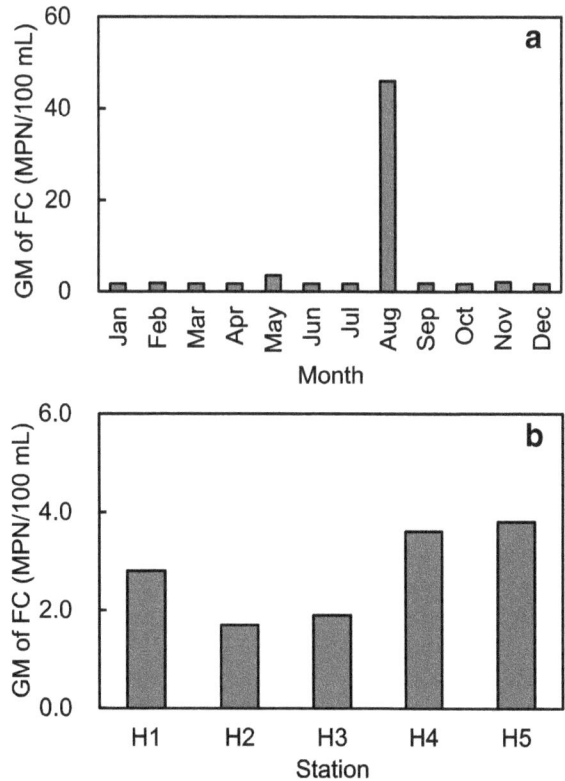

Fig. 3 (**a**, **b**) Spatial and seasonal variation of the geometric mean (GM) of fecal coliform (FC) for seawater samples collected from the Hansan-Geojeman area, Korea, in 2014

in 2014 varied from 13.1 to 589.4 mm with seasonal variation (data not shown). The highest value found in August was about fourfold higher than monthly mean rainfall in 2014. In particular, the samples were collected after approximately 200-mm rainfall in August 2014. Seasonal temperature variation may affect the abundance of bacteria (Rippey 1994). Avila et al. (1989) reported that microbial counts were twofold higher in fall and spring and around eightfold higher in summer than in winter during the monitoring of seawater quality on the southern coast of Spain. Chigbua et al. (2004) also reported that FC levels had a positive relationship with rainfall in Mississippi Sound. This study also shows a similar result that FC concentrations were relatively detected higher in August than in other seasons and affected greatly by rainfall. Therefore, it indicates that FC counts had a seasonal variation and, specially, were detected higher after a heavy rainfall before the sampling day. But, because the oyster harvesting period in Korea is from October to May in the following year, oysters are not fortunately harvested in the summer season with heavy rainfall.

Additionally, the values of geometric mean for FC at each station ranged from <1.8 to 3.8 MPN/100 mL; the highest value was found at site H5 (Fig. 3b), which is close to that of major pollution sources (G2 and G4) with high concentration and daily load of FC (Fig. 2). It was found that the high geometric mean level of FC in August 2014 was caused by FC (920 MPN/100 mL) of a sample collected after rainfall at site H5, which is close to that of waste discharges from the sites G2 and G4 along the shoreline with a populated residential area (Fig. 1). The stations G2 and G4 for pollution source had high concentration and daily load of FC in March (Fig. 2); however, the FC value at site H5 was low in the same month without rainfall before sampling (Fig. 3a). In this study, all stations showed the values far below the regulation limit (14 MPN/100 mL) of geometric mean value for FC set by Korea (MOF 2015), the USA (US FDA 2013), and New Zealand (NZFSA 2006).

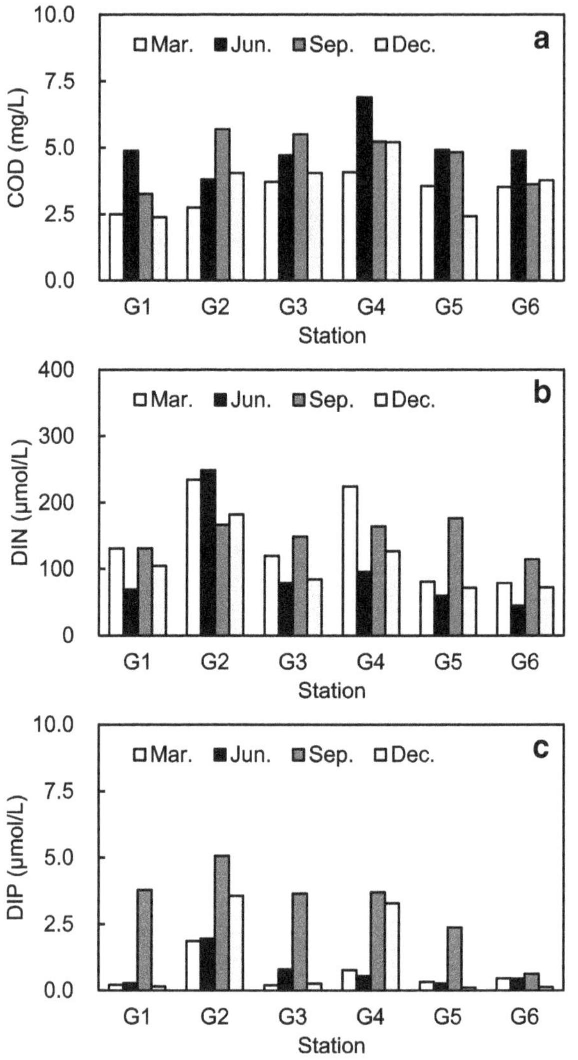

Concentration of physicochemical factors in inland pollution sources

The concentrations and daily loads of COD, DIN, and DIP in the major inland pollution sources, collected quarterly from the six fixed stations at the drainage area of the Hansan-Geojeman area, are shown in Figs. 4 and 5, respectively. This survey was also conducted when no precipitation occurred during a week before the sampling for minimizing NPS inputs into the inland pollution sources. COD concentrations of the pollution sources ranged from 2.4 to 6.9 mg/L; the highest level was found at station G4, which is a Seojeong stream in Geoje city (Fig. 4a), whereas the daily loads of COD at each station

Fig. 4 Spatial and seasonal variation of chemical oxygen demand (COD) (**a**), dissolved inorganic nitrogen (DIN) (**b**), and dissolved inorganic phosphate (DIP) (**c**) of discharges from major inland contamination sources in the drainage area of the Hansan-Geojeman area, Korea

ranged from 1.4×10^5 to 3.1×10^8 mg/day; higher values were found at stations G1 and G6 (Fig. 5a) with a relatively high flow rate (Fig. 2a). In addition, DIN and DIP concentrations of pollution sources ranged from 45 to 249 μmol/L and 0.11 to 5.06 μmol/L, respectively; the highest levels were found at station G2, which is a Naegan stream in Geoje city (Fig. 4b, c); however, the daily loads of DIN and DIP were the lowest at site G2 with relatively much low flow rate (Fig. 5b, c). The concentrations of DIP were relatively higher in September than in other months surveyed in this study. It assumes that artificial fertilizer including phosphate is largely spread on rice fields between July and September in Korea; Geoje city supplied 946 tons of artificial fertilizer including 177 tons of

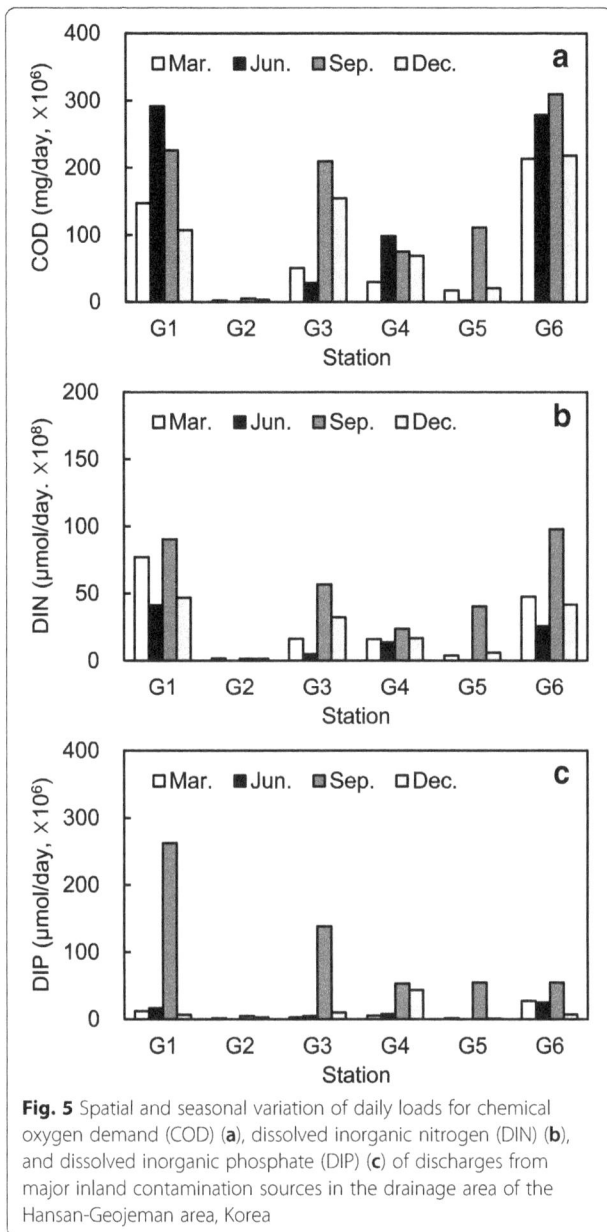

Fig. 5 Spatial and seasonal variation of daily loads for chemical oxygen demand (COD) (**a**), dissolved inorganic nitrogen (DIN) (**b**), and dissolved inorganic phosphate (DIP) (**c**) of discharges from major inland contamination sources in the drainage area of the Hansan-Geojeman area, Korea

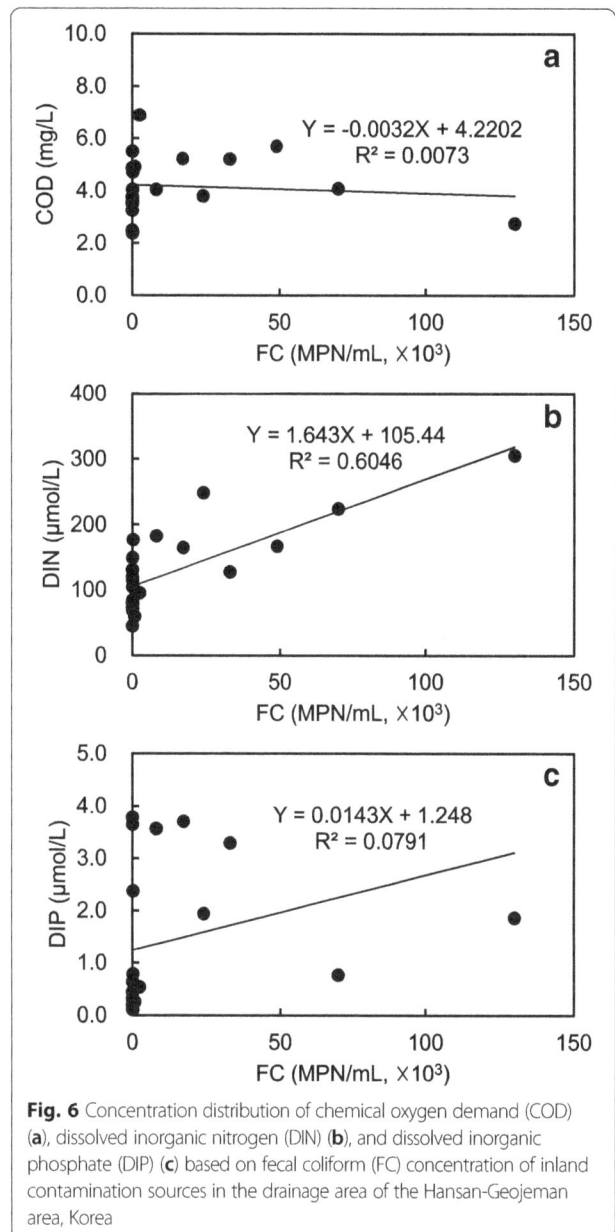

Fig. 6 Concentration distribution of chemical oxygen demand (COD) (**a**), dissolved inorganic nitrogen (DIN) (**b**), and dissolved inorganic phosphate (DIP) (**c**) based on fecal coliform (FC) concentration of inland contamination sources in the drainage area of the Hansan-Geojeman area, Korea

phosphate in 2013 (Geoje city 2014). The daily loads of DIN and DIP were relatively higher in September than in other months because of relatively high flow rates and/or concentrations (Figs. 2a and 4b, c). In particular, the daily load of DIP was very high in September at stations G1 and G3, which are surrounded by a lot of rice fields.

Moreover, a significant relationship was identified between FC and DIN concentrations in the major inland pollution sources ($p < 0.05$); the correlation coefficients (R^2) were 0.60, indicating close correlation with both FC and DIN (Fig. 6). The FC concentrations did not show almost the correlation with COD and DIP concentrations in the inland pollutions.

Distributions of physicochemical factors in seawaters

The concentration distribution of physicochemical factors of seawater samples collected from the surface and bottom layers of the Hansan-Geojeman area is shown in Tables 1 and 2 and Fig. 7. All physicochemical parameters tended to show low difference between the surface and bottom layers during December. Mean water temperatures were within the range of 9.02 to 25.59 °C at the surface layer, and 8.95 to 21.12 °C at the bottom layer, with the largest temperature difference between the layers recorded in August (approximately 4.5 °C). Monthly mean surface salinity varied from 29.47 to 34.15 with seasonal variation. Because of heavy rainfall during wet season in Korea, August had the lowest salinity value of 29.47 at the

Table 1 Values of temperature, salinity, pH, dissolved oxygen (DO), chemical oxygen demand (COD), and chlorophyll-*a* in surface and bottom waters of the Hansan-Geojeman area, Korea, in 2014

Month	Temperature (°C)	Salinity	pH	DO (mg/L)	COD (mg/L)	Chlorophyll-*a* (µg/L)
Surface						
Feb.	9.02 ± 1.04	34.15 ± 0.09	8.22 ± 0.07	10.36 ± 0.58	1.92 ± 0.15	7.51 ± 4.01
Apr.	13.26 ± 0.36	33.84 ± 0.31	8.19 ± 0.04	8.38 ± 0.14	1.67 ± 0.54	1.33 ± 0.37
Jun.	18.52 ± 0.31	33.85 ± 0.04	7.92 ± 0.04	7.95 ± 0.35	1.54 ± 0.29	2.87 ± 1.25
Aug.	25.59 ± 0.71	29.47 ± 2.49	8.32 ± 0.28	6.50 ± 1.71	2.20 ± 0.98	18.03 ± 14.28
Oct.	19.90 ± 0.48	32.73 ± 0.22	8.13 ± 0.02	7.27 ± 0.71	2.31 ± 0.18	9.47 ± 1.87
Dec.	14.68 ± 0.47	33.30 ± 0.30	8.11 ± 0.02	7.56 ± 0.25	1.57 ± 0.23	3.50 ± 2.66
Bottom						
Feb.	8.95 ± 0.96	34.18 ± 0.08	8.23 ± 0.06	10.00 ± 0.51	2.04 ± 0.32	5.64 ± 3.03
Apr.	12.96 ± 0.12	34.06 ± 0.09	8.20 ± 0.04	8.63 ± 0.29	2.10 ± 0.26	2.71 ± 1.76
Jun.	17.04 ± 0.49	34.09 ± 0.05	7.96 ± 0.03	7.85 ± 0.24	1.93 ± 0.41	1.52 ± 0.22
Aug.	21.12 ± 2.28	33.47 ± 0.42	8.04 ± 0.04	4.38 ± 0.28	2.32 ± 0.30	11.09 ± 11.40
Oct.	19.85 ± 0.48	32.84 ± 0.09	8.10 ± 0.02	7.05 ± 0.56	2.49 ± 0.37	10.50 ± 5.25
Dec.	14.69 ± 0.48	33.16 ± 0.21	8.11 ± 0.02	7.38 ± 0.66	1.50 ± 0.39	4.48 ± 3.25

Values are means ± standard deviations

surface layer and the largest difference between the surface/bottom layers, with 4.0. But the differences of water temperature and salinity among the sampling stations were not significant (data not shown).

Besides, COD, which is comparatively sensitive to the inflow of terrestrial contaminants, showed the largest difference at station H1 in August, when there is high rainfall. COD in the surface layer was also highest at station H1

Table 2 Concentrations of dissolved inorganic nutrients in surface and bottom waters of the Hansan-Geojeman area, Korea, in 2014

Month	Concentrations of dissolved inorganic nutrients (µmol/L)				
	NH$_3$	NO$_2$	NO$_3$	DIN	DIP
Surface					
Feb.	1.28 ± 0.31	0.12 ± 0.04	2.48 ± 0.96	3.59 ± 1.24	0.18 ± 0.08
Apr.	1.30 ± 0.56	0.10 ± 0.03	0.91 ± 0.47	2.30 ± 1.02	0.22 ± 0.11
Jun.	1.90 ± 1.18	0.18 ± 0.08	1.53 ± 0.68	3.60 ± 1.91	0.20 ± 0.12
Aug.	1.60 ± 0.78	0.38 ± 0.18	3.89 ± 1.36	5.88 ± 1.84	0.31 ± 0.14
Oct.	5.30 ± 2.00	0.74 ± 0.20	2.06 ± 0.46	8.10 ± 2.60	0.61 ± 0.14
Dec.	1.20 ± 0.97	0.61 ± 0.38	4.27 ± 3.11	6.08 ± 4.45	0.64 ± 0.29
Bottom					
Feb.	1.54 ± 0.17	0.08 ± 0.03	2.36 ± 1.26	4.10 ± 0.93	0.22 ± 0.10
Apr.	1.58 ± 0.44	0.09 ± 0.04	0.93 ± 0.19	2.60 ± 0.62	0.21 ± 0.06
Jun.	2.31 ± 1.04	0.33 ± 0.13	0.68 ± 0.80	3.31 ± 0.99	0.28 ± 0.06
Aug.	3.82 ± 0.61	0.49 ± 0.12	3.52 ± 0.38	7.84 ± 0.61	0.50 ± 0.02
Oct.	6.14 ± 1.33	1.07 ± 0.35	3.92 ± 0.89	11.13 ± 2.27	0.67 ± 0.17
Dec.	1.33 ± 0.69	0.61 ± 0.37	3.87 ± 2.70	5.81 ± 3.55	0.45 ± 0.21

Values are means ± standard deviations, *DIN* dissolved inorganic nitrogen, *DIP* dissolved inorganic phosphate

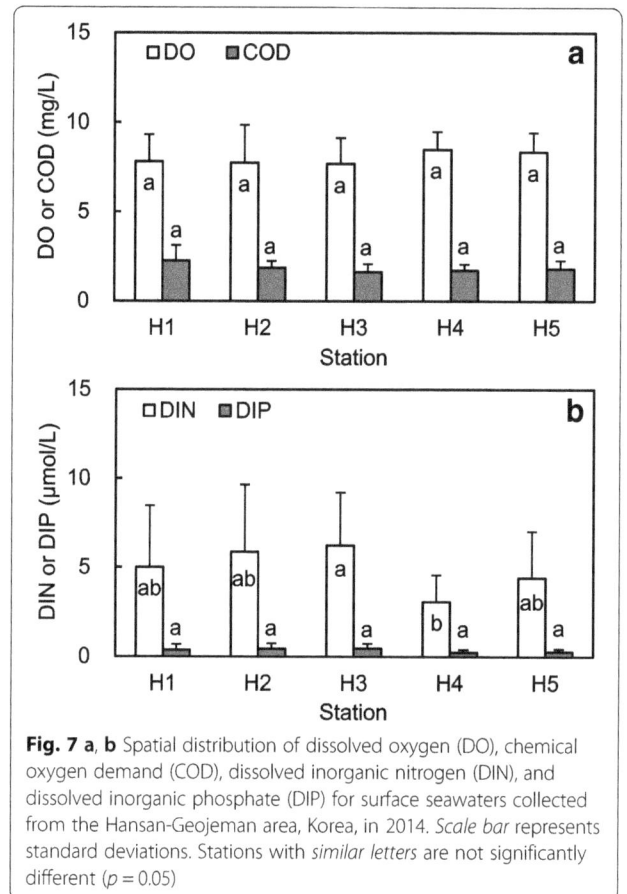

Fig. 7 a, b Spatial distribution of dissolved oxygen (DO), chemical oxygen demand (COD), dissolved inorganic nitrogen (DIN), and dissolved inorganic phosphate (DIP) for surface seawaters collected from the Hansan-Geojeman area, Korea, in 2014. *Scale bar* represents standard deviations. Stations with *similar letters* are not significantly different (*p* = 0.05)

during August, which is close to inland pollution source (G1) with relatively high daily load of COD (Fig. 5a). COD concentrations tended to increase gradually from the middle of the sea area to the coastal regions; however, the differences of those among stations were not significant ($p = 0.05$) (Fig. 7a).

Oxygen-deficient water mass (ODW), which is defined as water containing less than the normal 3 mg/L DO concentration (Pearson and Rosenberg 1978), occurs when the dead organisms and nutrient compounds sink to the bottom and are decomposed by bacteria, using the available dissolved oxygen (Rabalais et al. 2002; Middelburg and Levin 2009). The areas where ODW occurs have mainly semi-enclosed water bodies with poor exchange of water, such as the Hansan-Geojeman area. The Hansan-Geojeman area is representative of aquaculture regions of Korea, with active farming of oysters, fish, sea squirts, etc., and specially has a designated area for oysters. DO mean concentrations showed the largest difference in August, and the lowest concentration of DO was also found in August at the bottom layer (Table 1). The lowest concentration of DO during August showed similar pattern in Jinhae Bay, Korea, reported by Kim et al. (2015). In Jinhae Bay, ODW began in June, which is the heavy rainfall season, and peaked in August, which is the high-temperature rainfall season. It began to weaken in September, at the start of the autumn season, and was already completely dissipated in December. But, it confirms that ODW was not shown in the surface and bottom waters of the Hansan-Geojeman area during all of the survey periods. The Chl-a mean concentration was the highest in August at both layers, while April and June showed the lowest concentration in the surface and bottom layers, respectively (Table 1). Generally, the proliferation of phytoplankton, containing Chl-a, using DO has a direct effect on the fluctuation of DO concentration (Kim et al. 2015). Chl-a distribution showed strong positive mutuality with DO in the summer month of August. Therefore, August showed the highest concentrations for COD and Chl-a in the surface layer, whereas the lowest for DO in the bottom layer. In this study, all values of pH, salinity, and DO were within 7–9, 12–38, and ≥80 %, respectively, of guide or mandatory levels for the quality required of shellfish waters set by the EU (EC 2006), with the exception of DO concentrations in August. The mean concentration of DIN and DIP were the highest in October at the bottom layer, while April and February showed the lowest mean concentration at the surface (Table 2). DIN concentrations at station 3 were significantly higher ($p < 0.05$) than those at station 4; however, the differences among other stations were not

significant (Fig. 7b). The differences of DIP concentrations among stations were not significant ($p = 0.05$).

Conclusions

In this study, the mean daily loads and half-circle radii of FCs at each inland pollution source in the drainage basin of the Hansan-Geojeman area, including a designated area for oysters, ranged from 1.2×10^9 to 3.1×10^{11} MPN/day and 56.6 to 514.5 m, respectively; the highest radius was found at station G4, which is a Seojeong stream with relatively high FC concentration. However, the pollutants could not be reached at the boundary line of the designated area thanks to the existing buffer zone, in which bacteria are diluted and reduced. The geometric mean level for FC of the seawater samples was the maximum in August. Therefore, FC counts had a seasonal variation and, specially, were detected higher after heavy rainfall before the sampling day. The FC concentrations at each station met the criteria of various countries for the approved area for shellfish.

The daily loads of COD of inland pollution sources showed higher values at stations (G1 and G6) with relatively high flow rate. The daily loads of DIN and DIP were relatively higher in September than in other months because of relatively high flow rates and/or concentrations of those. The highest concentrations for COD and Chl-a were shown in the surface layer during August with high rainfall, whereas the lowest for DO in the bottom layer of the same month. Chl-a distribution showed strong positive mutuality with DO in August. But, ODW was not shown in both of the surface and bottom waters during all survey periods. Therefore, this study shows that the concentrations of FC, COD, DO, and Chl-a of seawaters were relatively detected higher in the summer season, with large rainfall, than in other seasons.

Acknowledgements
This work was supported by a grant from the National Fisheries Research and Development Institute of Korea (R2015062).

Authors' contributions
The manuscript is the result of ongoing collaboration and discussion between the authors. YCP carried out the study design and wrote the first draft. JSM added to and revised the manuscript. PHK, YJJ, KJL, MSK, KRG, SGP, SJK, and JHY participated in sampling and analysis. All authors read and approved the final manuscript.

Competing interests
The authors declare that they have no competing interests.

Author details
[1]Southeast Sea Fisheries Research Institute, National Fisheries Research & Development Institute, 397-68, Sanyang-iljuro, Sanyang-up, Tongyeong 650-943, Republic of Korea. [2]Food Safety Research Division, National Fisheries Research & Development Institute, 216, Gijang-haeanro, Gijang-up, Gijang-gun, Busan 619-705, Republic of Korea.

References

APHA (American Public Health Association). Recommended procedures for the examination of seawater and shellfish. 4th ed. Washington D.C., USA: American Public Health Association; 1970. p. 1–47.

Avila MJ, Morinigo MA, Cornax R, Romero P, Borrego JJ. Comparative study of coliform-enumeration media from seawater samples. J Microbiol Meth. 1989; 9:175–93.

Chigbua P, Gordonb S, Strangec T. Influence of inter-annual variations in climatic factors on fecal coliform levels in Mississippi Sound. Water Res. 2004;38:4341–52.

Dorfman M, Sinclair Rosselot K. Testing the waters 2008: a guide to water quality testing at vacation beaches. Washington D.C., USA: Natural Resources Defense Council; 2008.

EC (European Commission). Commission Regulation (EC) No. 2073/2005 of 15 November 2005 on microbiological criteria for foodstuffs. 2005. Accessed 22 Jun 2015, http://eur-lex.europa.eu/LexUriServ/LexUriServ.do?uri=CONSLEG: 2005R2073:20071227:EN:PDF.

EC (European Commission). Directive 2006/113/EC of the European parliament and of the council of 12 December 2006 on the quality required of shellfish waters. 2006. Accessed 2 Sep 2015, http://eur-lex.europa.eu/legal-content/ EN/TXT/?uri=uriserv:OJ.L_.2006.376.01.0014.01.ENG.

Feldhusen F. The role of seafood in bacterial foodborne disease. Microbes Infect. 2000;2:1651–60.

Geoje city. The 19th statistical yearbook of Geoje. 2014. Accessed 01 Jan 2016, http:// ebook.geoje.go.kr/src/viewer/main.php?host=main&site=20150205_095443.

Ha KS, Yoo HD, Shim KB, Kim JH, Lee TS, Kim PH, et al. Evaluation of the influence of inland pollution sources on shellfish growing areas after rainfall events in Geoje bay, Korea. Kor J Fish Aquat Sci. 2011;44:612–21.

Kim YS, Lee YH, Kwon JN, Choi HG. The effect of low oxygen conditions on biogeochemical cycling of nutrients in a shallow seasonally stratified bay in southeast Korea (Jinhae Bay). Mar Pollut Bull. 2015;95:333–41.

KMFDS (Korea Ministry of Food and Drug Safety). Korea food code. 2015. Accessed 22 Jun 2015, http://fse.foodnara.go.kr/residue/RS/jsp/menu_02_01_01.jsp.

Lipp EK, Kurz R, Vincent R, Rodriguez-Palacios C, Farrah SR, Rose JB. The effects of seasonal variability and weather on microbial fecal pollution and enteric pathogens in a subtropical estuary. Estuaries Coast. 2001;24:266–76.

Middelburg JJ, Levin LA. Coastal hypoxia and sediment biogeochemistry. Biogeosciences. 2009;6:1273–93.

MOF (Ministry of Oceans and Fisheries). Marine environment process exam standard. Sejong, Korea: Ministry of Oceans and Fisheries; 2002. p. 43–268.

MOF (Ministry of Oceans and Fisheries). Korean Shellfish Sanitation Program (KSSP). Sejong, Korea: Ministry of Oceans and Fisheries; 2015. p. 1–110.

Mok JS, Song KC, Lee KJ, Kim JH. Variation and profile of paralytic shellfish poisoning toxins in Jinhae bay, Korea. Fish Aquat Sci. 2013;16:137–42.

Mok JS, Kwon JY, Son KT, Choi WS, Kang SR, Ha NY, et al. Contents and risk assessment of heavy metals in marine invertebrates from Korean coastal fish markets. J Food Prot. 2014a;77:1022–30.

Mok JS, Yoo HD, Kim PH, Yoon HD, Park YC, Lee TS, et al. Bioaccumulation of heavy metals in mussel *Mytilus galloprovincialis* in the Changseon area, Korea, and assessment of potential risk to human health. Fish Aquat Sci. 2014b;17:313–8.

Mok JS, Yoo HD, Kim PH, Yoon HD, Park YC, Lee TS, et al. Bioaccumulation of heavy metals in oysters from the southern coast of Korea: assessment of potential risk to human health. Bull Environ Contam Toxicol. 2015;94:749–55.

NZFSA (New Zealand Food Safety Authority). Animal products (specifications for bivalve molluscan shellfish). 2006. Accessed 22 Jun 2015, http://www. foodsafety.govt.nz/elibrary/industry/Animal_Products-Applies_Anyone.pdf.

Park KBW, Jo MR, Kim YK, Lee HJ, Kwon JY, Son KT, et al. Evaluation of the effects of the inland pollution sources after rainfall events on the bacteriological water quality in Narodo area, Korea. Kor J Fish Aquat Sci. 2012;45:414–22.

Pearson TH, Rosenberg R. Macrobenthic succession in relation to organic enrichment and pollution of the marine environment. Oceanogr Mar Biol Annu Rev. 1978;16:229–311.

Rabalais NN, Turner RE, Wiseman JR. Hypoxia in the Gulf of Mexico, a.k.a. "The dead zone". Annu Rev Ecol Syst. 2002;33:235–63.

Rabinovici SJ, Bernknopf RL, Wein AM, Coursey DL, Whitman RL. Economic and health risk trade-offs of swim closures at a Lake Michigan beach. Environ Sci Technol. 2004;38:2737–45.

Rippey SR. Infectious diseases associated with molluscan shellfish consumption. Clin Microbiol Rev. 1994;7:419–25.

Rose JB, Epstein PR, Lipp EK, Sherman BH, Bernard SM, Patz JA. Climate variability and change in the United States: potential impacts on water- and foodborne diseases caused by microbiologic agents. Environ Health Perspect. 2001;109:211–21.

Shim KB, Ha KS, Yoo HD, Lee TS, Kim JH. Impact of pollution sources on the bacteriological water quality in the Yongnam-Gwangdo shellfish growing area of western Jinhae bay, Korea. Kor J Fish Aquat Sci. 2012;45:561–9.

Statistics Korea. Korean Statistical Information Service (KOSIS). 2013. Accessed 20 Jun 2015, http://kosis.kr.

U.S. Food and Drug Administration (FDA). National Shellfish Sanitation Program (NSSP), Guide for the control of molluscan shellfish. 2013. Accessed 20 Jun 2015, http://www.fda.gov/Food/GuidanceRegulation/ FederalStateFoodPrograms/ucm2006754.htm.

Wu Y, Chen J. Investigating the effects of point source and nonpoint source pollution on the water quality of the East River (Dongjiang) in South China. Ecol Indic. 2013;32:294–304.

Yoo HD, Yoon HD, Kim PH, Mok JS, Ha KS, Lee HJ. Sanitary survey of shellfish growing area in Hansan-Geojeman area (2011–2013). Busan, Korea: National Fisheries Research and Development Institute; 2013. p. 1–103.

3

Effect of dietary carbohydrate sources on apparent nutrient digestibility of olive flounder (*Paralichthys olivaceus*) feed

Md Mostafizur Rahman[1], Kyeong-Jun Lee[2] and Sang-Min Lee[1*]

Abstract

Apparent digestibility coefficients (ADCs) of dry matter, crude protein, crude lipid, nitrogen-free extract, and energy in selected carbohydrate sources including wheat flour (WF), α-potato starch (PS), α-corn starch (CS), Na alginate (AL), dextrin (DEX), and carboxymethyl cellulose (CMC) were determined for olive flounder. The olive flounder averaging 150 ± 8.0 g were held in 300-L tanks at a density of 30 fish per tank. Chromic oxide was used as the inert marker. Feces were collected from the flounder by a fecal collector attached to a fish rearing tank. Apparent dry matter and energy digestibilities of flounder fed WF, PS, CS, and DEX diets were significantly higher than those of fish fed AL and CMC diets. Apparent crude protein digestibility coefficients of flounder fed PS and CS diets were significantly higher than those of fish fed AL, DEX, and CMC diets. Apparent crude lipid and nitrogen-free extract digestibility coefficients of flounder fed PS and DEX diets were significantly higher than those of fish fed WF, CS, AL, and CMC diets. The present findings indicate that PS and DEX could be effectively used as dietary carbohydrate energy compared to WF, CS, AL, and CMC for olive flounder.

Keywords: Flounder, Apparent digestibilities, Carbohydrate source

Background

Carbohydrates are valuable ingredient in formulated aquaculture diets (Rosas et al. 2000), and they can spare the use of protein as an energy source (Simon 2009). As the least expensive dietary energy source and good binding agent during the pelleting procedure (Arnesen and Krogdahl 1993; González-Félix et al. 2010), carbohydrate utilization by various species of cultured fish is getting more and more attention to nutritionists and food manufacturers (Niu et al. 2012). Carbohydrate digestion and capacity is variable among fishes where carnivorous fish are less able to utilize them than omnivorous and herbivorous species (Krogdahl et al. 2005). In general, carbohydrate inclusion in carnivorous fish diets is limited to 20 % (NRC 2011). The efficiency of dietary carbohydrate utilization in fish diets has been examined for many species (Peres and Oliva-Teles 2002; Stone 2003; Lee et al. 2003; Wang et al. 2005).

The use of dietary carbohydrate by fish depends on many factors including complexity, type, source, degree, and heat treatment of the carbohydrate; fish species; and environmental condition (NRC 1993; Hemre et al. 2002). However, its efficient utilization is linked to efficient digestibility, and the ability of fish to digest carbohydrate has been reported to be variable between species (Stone et al. 2003). Physiological variations associated with the feeding habits of the species, using the chemical features of the carbohydrate, lead to variable digestibility. This variability demonstrates physiological and functional differences of the gastrointestinal tract and associated organs of fish species (Krogdahl et al. 2005; González-Félix et al. 2010).

González-Félix et al. (2010) reported that carbohydrate digestibility in Florida pompano, *Trachinotus carolinus*, is approximately 50 % emphasizing its restricted availability and therefore confined energy digestibility. Niu et al. (2012) investigated the dry matter and protein digestibility of some carbohydrate feedstuffs in juvenile tiger shrimp, *Penaeus monodon*. Deng et al. (2005) determined apparent digestibility coefficients (ADCs)

* Correspondence: smlee@gwnu.ac.kr
[1]Department of Marine Biotechnology, Gangneung-Wonju National University, 7 Jukheon-gil, Gangneung, Gangwon-do 25457, South Korea
Full list of author information is available at the end of the article

of hydrolyzed potato starch for juvenile white sturgeon. But, no reports are available regarding the ADCs of carbohydrate sources for olive flounder. Information on apparent digestibility of ingredients in flounder diets is needed to improve diet formulation and reduce feed production costs. Therefore, this study was conducted to evaluate the ADC of different carbohydrate sources including wheat flour (WF), α-potato starch (PS), α-corn starch (CS), Na alginate (AL), dextrin (DEX), and carboxymethyl cellulose (CMC) for olive flounder.

Methods

Experimental design and diet preparation

A feeding experiment with three replicates was employed to investigate the effects of dietary carbohydrate source on nutrient digestibilities of olive flounder. Six experimental diets were formulated to contain 20 % of each test ingredient such as WF, PS, CS, AL, DEX, and CMC. Ingredients and chemical composition of the experimental diets are presented in Table 1. Fish and squid liver oils were used as protein and lipid sources. Chromic oxide at a concentration of 0.5 % dry matter was added as an inert marker. The diets were thoroughly mixed with 40 % distilled water, pelleted by a wet pelleting machine, dried at room temperature for 24 h, and stored at −25 °C until use.

Fish and experimental condition

The flounder were obtained from a private hatchery (Jeju Island, Korea) to the Marine and Environmental Research Institute of Jeju National University (Jeju, South Korea). They were acclimated to the laboratory condition for 2 weeks. Afterwards, the experimental fishes (150 ± 8.0 g) were randomly distributed into 300-L cylindrical fiberglass tanks filled with 200 L of water at a density of 30 fish per tank in a flow-through system prior to starting the digestibility test. The fish were hand-fed the experimental diets to visual satiety once a day for 10 days. Filtered seawater was supplied at a flow rate of 3 L/min to each rearing tank. Fish rearing tanks had a sloping bottom leading to a centrally located drainage slot, and the effluent water was directed first over a fecal collection column and then to waste. Photoperiod was maintained on natural condition during the experimental period. The water temperature was maintained at 20.0 ± 0.82 °C.

Fecal collection technique

Three replicate groups of fish were carefully hand-fed the test diets to visual satiety once a day at 10:00 h by the same person during the experimental period. Two hours after feeding, the rearing tanks and collection columns were cleaned by sponges, and uneaten feed and fecal residues were removed. The next day, feces were

Table 1 Ingredients and chemical composition of the experimental diets

	Diets					
	WF	PS	CS	AL	DEX	CMC
Ingredients (%)						
Fish meal	61.0	61.0	61.0	61.0	61.0	61.0
Corn gluten meal	5.0	5.0	5.0	5.0	5.0	5.0
Dehulled soybean meal	5.0	5.0	5.0	5.0	5.0	5.0
Wheat flour	20.0					
α-potato starch		20.0				
α-corn starch			20.0			
Na alginate				20.0		
Dextrin					20.0	
Carboxymethyl cellulose						20.0
Squid liver oil	5.0	5.0	5.0	5.0	5.0	5.0
Vitamin premix[a]	1.2	1.2	1.2	1.2	1.2	1.2
Mineral premix[b]	1.5	1.5	1.5	1.5	1.5	1.5
Vitamin C (50 %)	0.3	0.3	0.3	0.3	0.3	0.3
Vitamin E (25 %)	0.2	0.2	0.2	0.2	0.2	0.2
Choline salt (50 %)	0.3	0.3	0.3	0.3	0.3	0.3
Cr_2O_3	0.5	0.5	0.5	0.5	0.5	0.5
Nutrient content (dry matter basis)						
Crude protein (%)	53.2	49.6	50.7	49.8	49.3	49.4
Crude lipid (%)	7.9	7.5	8.2	9.4	9.2	8.4
Ash (%)	13.8	13.1	13.9	17.8	14.1	17.0
N-free extract (%)[c]	25.1	29.8	27.2	23.0	27.4	25.2
Gross energy (kcal/g diet)	4.2	4.0	4.2	3.7	4.2	4.0

[a]Vitamin premix contained the following amount which were diluted in cellulose (g/kg mix): DL-α-tocopheryl acetate, 18.8; thiamin hydrochloride, 2.7; riboflavin, 9.1; pyridoxine hydrochloride, 1.8; niacin, 36.4; Ca-D-pantothenate, 12.7; myo-inositol, 181.8; D-biotin, 0.27; folic acid (98 %), 0.68; p-aminobenzoic acid, 18.2; menadione, 1.8; retinyl acetate, 0.73; cholecalciferol, 0.003; and cyanocobalamin, 0.003
[b]Mineral premix contained the following ingredients (g/kg mix): $MgSO_4 \cdot 7H_2O$, 80.0; $NaH_2PO_4 \cdot 2H_2O$, 370.0; KCl, 130.0; ferric citrate, 40.0; $ZnSO_4 \cdot 7H_2O$, 20.0; Ca-lactate, 356.5; CuCl, 0.2; $AlCl_3 \cdot 6H_2O$, 0.15; KI, 0.15; $Na_2Se_2O_3$, 0.01; $MnSO_4 \cdot H_2O$, 2.0; and $CoCl_2 \cdot 6H_2O$, 1.0
[c]Calculated = 100 − (crude protein + crude lipid + ash)

collected from the fecal collection columns at 10:00 h for 10 days. The feces were immediately filtered with filter paper (Whatman #1) for 60 min at 4 °C and stored at −75 °C for chemical analyses. Fecal samples from each tank were pooled at the end of the experiment.

Analytical methods

Freeze-dried feed and feces were finely ground using a grinder. Fish scales were removed using a 300-μm sieve before analysis. Crude protein was determined by the Kjeldahl method using an auto Kjeldahl System (Buchi, Flawil, Switzerland). Crude lipid was analyzed with ether extraction in a soxhlet extractor (SER 148, VELP Scientifica, Milan, Italy). Moisture content was determined with a dry oven at 105 °C for 6 h, and the ash content was determined after combustion at 550 °C for 4 h in a muffle furnace. Nitrogen-free extract (NFE) was calculated by difference. Gross energy was analyzed by an adiabatic bomb calorimeter (Parr, USA). Chromic oxide content of the experimental diets and fecal samples were determined by a wet-acid digestion method (Furukawa and Tsukahara 1966). All chemical analyses from each tank were performed in triplicates.

Apparent digestibility coefficients (ADCs) for dry matter, nutrient, and energy of the experimental diets were determined using the following equations:

$$\text{ADC of dry matter (\%)} = (100-(\text{dietary } Cr_2O_3/\text{feces } Cr_2O_3) \times 100),$$

$$\text{ADC of nutrients or energy}$$
$$= \left(1-\frac{\text{dietary } Cr_2O_3}{\text{feces } Cr_2O_3} \times \frac{\text{feces nutrient or energy}}{\text{dietary nutrient or energy}}\right) \times 100.$$

Statistical analysis

All data were subjected to one-way analysis of variance (ANOVA) using SPSS version 20.0 (SPSS Inc., Chicago, IL, USA). Significant differences ($p < 0.05$) among the means were determined using Duncan's multiple range test (Duncan 1955). Correlation of nutrients and energy was assessed using Pearson regression. All data were presented as mean ± SE of three replicate groups.

Results

Apparent nutrient digestibility of flounder fed different dietary carbohydrate sources is presented in Table 2. Apparent dry matter and energy digestibilities of flounder fed WF, PS, CS, and DEX diets were significantly higher than those of fish fed AL and CMC diets. Apparent crude protein digestibility coefficients of flounder fed PS and CS diets were significantly higher than those of fish fed AL, DEX, and CMC diets. Apparent crude lipid and NFE digestibility coefficients of flounder fed PS and DEX diets were significantly higher than those of fish fed WF, CS, AL, and CMC diets.

Discussion

Estimation of ADC values for feedstuffs is an important aspect in screening the nutritive value of feed ingredients and formulation of nutritionally sufficient diets (Irvin and Tabrett 2005). In this study, dry matter digestibilities (65–78 %) of flounder fed the diets containing different carbohydrate sources were higher compared to those (48–63 %) of rockfish fed the diets containing α-potato starch, β-potato starch, β-corn starch, and dextrin (Lee and Pham 2011) but similar to results (68–79 %) for spiny lobster, *Jasus edwardsii*, fed WF, DEX, PS, and CMC (Simon 2009). The low dry matter digestibility of CMC in this study can be attributed to the high content of ash. Lee (2002) reported that dry matter digestibility of rockfish fed the different ingredients appeared to relate to the quantity and chemical composition of the carbohydrate used. It has been found that complexity of the carbohydrate significantly influences its utilization in the fish diet (Jobling 2001; Lee and Pham 2011). Stone et al. (2003) reported that simple carbohydrate has high dry matter digestibility compared to the complex kind.

ADC of protein in the present study ranged from 72 to 90 %. Niu et al. (2012) reported that apparent protein digestibility of wheat starch, sucrose, potato starch, maize starch, and dextrin diets ranged from 81 to 92 % for juvenile tiger shrimp, *P. monodon*. Apparent crude protein digestibility of α-potato starch, β-potato starch, β-corn starch, and dextrin diets ranged from 90 to 95 % for juvenile and grower rockfish (Lee and Pham 2011). Niu et al. (2012) demonstrated that protein digestibility of potato starch diet was significantly higher than that of maize starch and dextrin diets. ADC of protein appeared to have a positive relationship with dry matter (DM) digestibility ($r = 0.90$, $p < 0.01$) of the test diets. In the present study, low crude protein digestibility of CMC diet may be due to the accelerated passage of the digesta from the stomach of olive flounder (Yamamoto and Akiyama 1995).

NFE digestibility exhibited a significant correlation to DM digestibility ($r = 0.76$, $p < 0.01$) of the test diets. Although the actual mechanism responsible for poor carbohydrate utilization in fishes has not been recognized,

Table 2 Apparent nutrient digestibility (%) of olive flounder fed different dietary carbohydrate sources

Diets	Dry matter	Crude protein	Crude lipid	NFE	Gross energy
WF	74.5 ± 0.41^c	89.3 ± 0.17^{cd}	81.4 ± 0.29^c	62.0 ± 0.61^b	78.8 ± 0.35^c
PS	75.4 ± 0.29^c	90.0 ± 0.12^d	89.6 ± 0.12^e	76.7 ± 0.29^c	79.7 ± 0.40^c
CS	74.7 ± 0.80^c	89.9 ± 0.31^d	83.8 ± 0.51^d	59.4 ± 1.31^b	79.8 ± 0.63^c
AL	69.3 ± 0.59^b	78.5 ± 0.42^b	64.1 ± 0.71^a	62.4 ± 0.74^b	67.9 ± 0.62^b
DEX	78.5 ± 3.17^c	87.3 ± 1.86^c	94.3 ± 0.84^f	86.3 ± 2.04^d	87.3 ± 1.86^d
CMC	64.6 ± 0.30^a	72.4 ± 0.27^a	71.6 ± 0.27^b	53.2 ± 0.43^a	62.0 ± 0.33^a

Values (mean ± SE of three replications) in the same column not having a common superscript are significantly different ($p < 0.05$)

inferior digestion of carbohydrate in carnivorous fish might be attributed to the environment in which they evolved (McGoogan and Reigh 1996). Lee and Pham (2011) reported that NFE digestibility of α-potato starch, β-potato starch, β-corn starch, and dextrin ranged from 20 to 56 % for juvenile and grower rockfish. In this study, the apparent NFE digestibility values were 53 to 86 %. The highest NFE digestibility value obtained for DEX was 86 %, and the lowest value observed for CMC was 53 %. This result indicates that differences in NFE digestibility of the diets containing different carbohydrate sources might be due to their indigestible polysaccharide content (Lee and Pham 2011).

Regarding carbohydrate sources, low NFE digestibility of carbohydrates indicates that digestibility of carbohydrates may be attributed to their reduced solubility and to their impediment of α-amylase activity. Generally, the fish has low ability in digestion of carbohydrates due to its specific carbohydrase enzyme (Jobling 2001). The effects of solubility and α-amylase on digestibility of different carbohydrates have been reported in silver perch (Stone et al. 2003). Some researchers suggested that carbohydrates are mainly digested in the interior segment of the digestive tract of fish and depends on their solubility in the fluid of the digestive tract (Fange and Grove 1979; Lovell 1989). There is no available information on NFE digestibility of olive flounder to date. Hence, studies on the topic are necessary to elaborate the carbohydrate utilization of flounder.

ADC of energy ranged from 62 to 80 % in this study. It has been noted that energy ADC of the dietary carbohydrates such as dextrin, gelatinized wheat starch, glucose, raw wheat starch, and raw pea starch seemed to be significantly influenced by carbohydrate kind (Stone et al. 2003). Dextrin has been reported to be a very good carbohydrate energy source and successfully digested by fish (Lee et al. 2003; Stone et al. 2003). Lee and Pham (2011) reported that α-potato starch and dextrin appeared to be effectively digested by rockfish as carbohydrate energy source. ADC of energy appeared to have a positive correlation with DM digestibility ($r = 0.96$, $p < 0.01$) of the test diets. In this study, ADC of energy for DEX was the highest among the ingredients tested. In contrast, ADC of energy was the lowest for CMC among those tested. The poor energy digestibility of CMC in this study may be because of insufficient non-protein energy in feeds.

Conclusion
Among the carbohydrates tested, PS and DEX could be effectively used as dietary carbohydrate energy compared to WF, CS, AL, and CMC for olive flounder.

Acknowledgements
This work was supported by a grant from the National Institute of Fisheries Science (R2016016) in Korea.

Authors' contributions
MMR analyzed the chemical composition and prepared the draft paper. KJL manufactured the feed and conducted the feeding trial. SML designed this study, the feeding system, and the revised paper. All authors read and approved the final manuscript.

Competing interests
The authors declare that they have no competing interests.

Author details
[1]Department of Marine Biotechnology, Gangneung-Wonju National University, 7 Jukheon-gil, Gangneung, Gangwon-do 25457, South Korea. [2]Department of Marine Life Sciences, Jeju National University, Jeju 63243, South Korea.

References
Arnesen P, Krogdahl Å. Crude and pre-extruded products of wheat as nutrients sources in extruded diets for Atlantic salmon (Salmo salar L.) grown in sea water. Aquaculture. 1993;118:105–17.
Deng DF, Hemre GI, Storebakken T, Shiau SY, Hung SSO. Utilization of diets with hydrolyzed potato starch, or glucose by juvenile white sturgeon (Acipenser transmontanus), as affected by Maillard reaction during feed processing. Aquaculture. 2005;248:103–9.
Duncan DB. Multiple-range and multiple F-tests. Biometrics. 1955;11:1–42.
Fange R, Grove D. Digestion. In: Hoar WS, Randall DJ, Bretts JR, editors. Fish physiology. New York: Academic Press; 1979. p. 161–260.
Furukawa A, Tsukahara H. On the acid digestion method for the determination of chromic oxide as an index substance in the study of digestibility of fish feed. Bull Jpn Soc Sci Fish. 1966;32:502–6.
González-Félix ML, Davis DA, Rossi Jr W, Perez-Velazquez M. Evaluation of apparent digestibility coefficient of energy of various vegetable feed ingredients in Florida pompano, Trachinotus carolinus. Aquaculture. 2010;310:240–3.
Hemre GI, Mommsen TP, Krogdahl A. Carbohydrates in fish nutrition: effects on growth, glucose metabolism and hepatic enzymes. Aquac Nutr. 2002;8:175–94.
Irvin SJ, Tabrett SJ. A novel method of collecting fecal samples from spiny lobsters. Aquaculture. 2005;243:269–72.
Jobling M. Feed composition and analysis. In: Houlihan D, Boujard T, Jobling M, editors. Food intake in fish. France: Blackwell Science; 2001.
Krogdahl A, Hamre GI, Mommsen TP. Carbohydrates in fish nutrition: digestion and absorption in postlarval stages. Aquac Nutr. 2005;11:103–22.
Lee SM. Apparent digestibility coefficients of various feed ingredients for juvenile and grower rockfish (Sebastes schlegeli). Aquaculture. 2002;207:79–95.
Lee SM, Pham MA. Effects of carbohydrate and water temperature on nutrient and energy digestibility of juvenile and grower rockfish, Sebastes schlegeli. Asian-Australas J Anim Sci. 2011;24:1615–22.
Lee SM, Kim KD, Lall SP. Utilization of glucose maltose, dextrin and cellulose by juvenile flounder (Paralichthys olivaceus). Aquaculture. 2003;221:427–38.
Lovell T. Digestion and metabolism. In: Lovell T, editor. Nutrition and feeding of fish. New York: Van Nostrand Reinhold; 1989. p. 185–203.
McGoogan BB, Reigh RC. Apparent digestibility of selected ingredients in red drum (Sciaenops ocellatus) diets. Aquaculture. 1996;141:233–44.
National Research Council. Nutrient requirements of fish. Washington: National Research Council of the National Academy Press; 1993. p. 105.
National Research Council. Nutrient requirements of fish and shrimp. Washington: National Research Council of the National Academics; 2011. p. 363.
Niu J, Lin HZ, Jiang SG, Chen X, Wu KC, Tian LX, et al. Effects of seven carbohydrate sources on juvenile Penaeus monodon growth performance, nutrient utilization efficiency and hepatopancreas enzyme activities of 6-phosphogluconate dehydrogenase, hexokinase and amylase. Anim Feed Sci Technol. 2012;174:86–95.

Peres H, Oliva-Teles A. Utilization of raw and gelatinized starch by European sea bass (*Dicentrarchus labrax*) juveniles. Aquaculture. 2002;205:287–99.

Rosas C, Cuzon G, Gaxiola G, Arena L, Lemaire P, Soyez C, et al. Influence of dietary carbohydrate on the metabolism of juvenile *Litopenaeus stylirostris*. J Exp Mar Biol Ecol. 2000;249:181–98.

Simon CJ. The effect of carbohydrate source, inclusion level of gelatinized starch, feed binder and fishmeal particle size on the apparent digestibility of formulated diets for spiny lobster juveniles, *Jasus edwardsii*. Aquaculture. 2009;296:329–36.

Stone DAJ. Dietary carbohydrate utilization by fish. Rev Fish Sci. 2003;11:337–69.

Stone DAJ, Allan GL, Anderson AJ. Carbohydrate utilization by juvenile silver perch, *Bidyanus bidyanus* (Mitchell). II. Digestibility and utilization of starch and its breakdown products. Aquac Res. 2003;34:109–21.

Wang Y, Liu YJ, Tian LX, Du ZY, Wang JT, Wang S, et al. Effects of dietary carbohydrate level on growth and body composition of juvenile tilapia, *Oreochromis niloticusxO.aureus*. Aquac Res. 2005;36:1408–13.

Yamamoto T, Akiyama T. Effect of carboxymethyl cellulose, α-starch and wheat gluten incorporated in diets as binders on growth, feed efficiency, and digestive energy activity of fingerling Japanese flounder. Fish Sci. 1995;61:309–13.

4

Effect of a new phosphorus source, magnesium hydrogen phosphate (MHP) on growth, utilization of phosphorus, and physiological responses in carp *Cyprinus carpio*

Tae-Hyun Yoon[1], Seunggun Won[2], Dong-Hoon Lee[3], Jung-Woo Choi[1], Changsix Ra[1] and Jeong-Dae Kim[1]*

Abstract

Magnesium hydrogen phosphate (MHP, $MgHPO_4$) recovered from swine manure was prepared as an alternative phosphorus (P) source. Conventional P additives, monocalcium phosphate (MCP), dicalcium phosphate (DCP), and tricalcium phosphate (TCP) were compared with the MHP in terms of growth and P availability by juvenile carp *Cyprinus carpio*. A basal diet as a negative control was prepared using practical feed ingredients without P supplementation to which four supplemental P sources were added at the level of 2%. Five groups of 450 fish having mean body weight of 6.5 g following 24 h fasting after 2 weeks of adaptation period were randomly distributed into each of 15 tanks (30 fish/tank). Fish were hand-fed to apparent satiety twice a day for 9 weeks. Fish fed the MHP had weight gain (WG), feed conversion ratio (FCR), protein efficiency ratio (PER), and specific growth rate (SGR) comparable to those fed the MCP. Those values of both the MHP and MCP groups were significantly different ($p < 0.05$) from the other groups. Fish groups fed control and the TCP showed the lowest WG, PER, and SGR and the highest FCR among treatments. No fish were died among treatments during the experimental period. Fish fed control and the TCP showed hematocrit and hemoglobin significantly lower ($p < 0.05$) than fish fed the MHP. The lowest inorganic P (Pi) in plasma was found in the control group. Even though Pi was not significantly different ($p > 0.05$) from other phosphate groups, fish fed the MCP and MHP retained higher P in whole body than the other groups. P availability was determined to be 93.2, 62.4, 6.1, and 98.0% for MCP, DCP, TCP, and MHP, respectively. The present results suggested that the MHP recovered from wastewater stream could be used as an alternative P source in carp diet.

Keywords: *Cyprinus carpio*, Weight gain, Feed conversion ratio, Phosphorus availability, Alternative phosphorus, Magnesium hydrogen phosphate

Background

Phosphorus (P) is a component of organic phosphate in diverse forms, such as nucleotides, phospholipids, coenzymes, deoxyribonucleic acid, and ribonucleic acid. Inorganic phosphates also serve as important buffers to maintain normal pH of intra- and extra-cellular fluids (Zubay 1983). Fish have the dietary requirement of P like other animals for growth. The requirement was reported to range from 0.3 to 0.6% for channel catfish (Wilson et al. 1982), rainbow trout (Rodehutscord and Pfeffer 1995), striped bass (Brown et al. 1993; Dougall et al. 1996), and white fish (Vilema et al. 2002). However, somewhat higher requirement value of 0.7 to 0.9% was estimated for haddock (Roy and Lall 2003), red tilapia (Phromkunthong and Udom 2008), African catfish (Nwanna et al. 2009), red drum (Davis and Robinson 1987), yellow croaker (Ma et al. 2006), and carp (Kim et al. 1998). On the other hand, P discharged into water stream as a major nutrient for

* Correspondence: menzang@gmail.com
[1]College of Animal Life Sciences, Kangwon National University, Chuncheon 24341, Korea
Full list of author information is available at the end of the article

eutrophication is known to play a vital role in promoting algal growth (Beveridge 1984; Auer et al. 1986).

Theoretically, diet for fish should contain all essential nutrients including P above the needs and maintain total P levels as low as its available requirements in order to achieve both maximal growth and minimum P discharge. However, the practical diet contains an excessive P originating from animal and plant sources, which is excreted to the cultural water body due to the low availability by stomachless species like carp (Kim and Ahn 1993). Therefore, soluble P sources like monocalcium phosphate (MCP) and dicalcium phosphate (DCP) are being supplemented to the diet to meet the requirement of P for maximum growth. Based on this point of views, some significant discharge of P is inevitable from fish farming. Kim et al. (1998) reported dietary available P of 0.7% with 2% MCP exerted both maximal growth and minimum P loss in juvenile carp. An adequate combination of low-P protein meals supplemented with 0.5% MCP significantly reduced the P loading on rainbow trout without compromising the growth (Satoh et al. 2003; Hernandez et al. 2004, 2005).

Such an important source for living creatures as one of non-renewable elements in the nature is mostly produced from phosphate rock of which production in the world has constantly increased from 198 million metric tons in 2011 to 210 million metric tons in 2012 according to U.S. Geological Survey (USGS 2013). Shu et al. (2006) expected all reserves of phosphate rock would be depleted by 2090, assuming an annual increase of 1.5% in its demand. Considering this estimation, it is necessary to recover phosphorous from diverse downstream of agricultural and industrial fields. One of resources to retrieve phosphorous could be swine manure which contains high levels of phosphorous and nitrogen. Swine manure becomes a source of pollution on surface waters and induces eutrophication near the site when it is under improper treatment. Thus, the control of wastewater stream must be achieved and struvite precipitation could be an effective way to control phosphorus from wastewater stream with the addition of magnesium (Liu et al. 2011). If this alternative phosphate recovered from swine manure could be effectively employed as an available P source for animals including fish, it would not only substitute for import of several phosphates but also protect our environment through recycling of the waste source.

Cyprinid is the species the most cultured worldwide. In 2012, China produced more than 90% of the world's carp which relies on formulated feed with the exception of filter-feeding species (Cao et al. 2015). Considering the low availability of P in both animal and plant feedstuffs for carp (NRC 1983), the amount of dietary P supplemented through its additive would be substantial. Nevertheless, the potential use of dietary alternative phosphate as an available P source for growth of the fish was not examined until now. This study was therefore carried out to investigate the supplemental effect of various conventional phosphate sources and the alternative (MHP) on growth, feed utilization, plasma inorganic P, whole body P, and the availability of juvenile carp.

Methods

The protocol for the present experiment was approved by the Institutional Animal Care and Use Committee of Kangwon National University, Chuncheon, Republic of Korea. Commercially selling phosphate additives (MCP, BIOFOS®, USA; DCP, SICHUAN MIANJHUSANJIA FEED Co., China; tricalcium phosphate (TCP), FOOD-CHEM, Shandong, China) were obtained from fish feed companies in Korea, and the testing P additive, magnesium hydrogen phosphate (MHP, $MgHPO_4$), was prepared from P-recovery process in swine farm operated by Kangwon National University.

A P-recovery process was a pilot scale with the effective volume of 400 L, and the operational condition was set with hydraulic retention time of 3 h and pH 8 to 9 controlled by aeration (33 L $air/m^3_{reactor}$.min) resulting in CO_2 stripping. Such a condition leads to precipitate crystal called struvite which is composed of magnesium, ammonium, and phosphate in equimolar ratio. In order to provide magnesium source that is a deficient source in swine manure, magnesium chloride ($MgCl_2$) was added to meet Mg to P ratio of approximately 1.2.

The precipitates collected was struvite, which was identified by X-ray diffractometer (XRD) (Rigaku, Model D/Max-2500 V, Japan). The MHP was obtained by removing ammonium-N through incineration of the recovered struvite at 550 °C for 30 min. It was finely ground to use as phosphate additive.

Preparation of diets

The diet was prepared with MCP, DCP, TCP, and MHP at the level of 2% but the control diet did not contain phosphate source of 2% in lieu of cellulose. The major ingredients were constituted with fish meal (25%), soybean meal (40%), wheat flour (27%), fish oil (2%), and soy oil (2%) as a basal diet containing 42.5% protein and 6.5% lipid (Table 1). Prior to diet formulation, chemical composition of fish meal, soybean meal, wheat flour, and four phosphate sources were determined. All the ingredients were weighed following the formula and ground to 100-mesh size by a hammer mill and thoroughly mixed for 10 min using a V-mixer (Hangjin Co., Korea) in order to make a mixture of 500 kg per diet. Then, the mixture was transferred to a twin extruder (Model ATX-2, Fesco Precision Co., Korea) and manufactured to the sinking pellets with two sizes of 1.5 and 2.5 mm, respectively, for feeding during the first 6- and second 3-week

Table 1 Ingredient and chemical composition of the experimental diets[a]

Ingredient (%)	Diet				
	Control	MCP	DCP	TCP	MHP
Fish meal	25.00	25.00	25.00	25.00	25.00
Soybean meal	40.00	40.00	40.00	40.00	40.00
Wheat flour	27.18	27.18	27.18	27.18	27.18
Soya oil	2.00	2.00	2.00	2.00	2.00
Fish oil	2.00	2.00	2.00	2.00	2.00
Vit. mix[a]	0.70	0.70	0.70	0.70	0.70
Min. mix[b]	0.30	0.30	0.30	0.30	0.30
Lysine-HCl	0.30	0.30	0.30	0.30	0.30
DL-methionine	0.20	0.20	0.20	0.20	0.20
Choline-HCl	0.30	0.30	0.30	0.30	0.30
Antioxidant	0.02	0.02	0.02	0.02	0.02
P source	–	2.00	2.00	2.00	2.00
Cellulose	2.00	–	–	–	–
Total	100.00	100.00	100.00	100.00	100.00
Composition (%, DM)					
C. protein	42.27	42.85	42.20	42.70	42.52
C. fat	6.90	6.67	6.62	6.65	6.65
C. ash	8.41	9.38	9.77	10.33	9.63
Ca	1.74	1.93	2.16	2.30	1.75
P	1.20	1.57	1.53	1.51	1.53
Av. P[c]	0.32	0.66	0.52	0.34	0.64

MCP, monocalcium phosphate, DCP dicalcium phosphate, TCP tricalcium phosphate, MHP magnesium hydrogen phosphate
[a]Vitamin added to supply the following (per kg diet): vitamin A, 4000 IU; vitamin D_3, 800 IU; vitamin E, 150 IU; vitamin K_3, 20 mg; thiamine HCl, 25 mg; riboflavin, 50 mg; D-Ca pantothenate, 100 mg; biotin, 1 mg; folic acid, 20 mg; vitamin B_{12}, 0.2 mg; niacin, 200 mg; pyridoxine HCl, 20 mg; ascorbic acid, 500 mg; inositol, 200 mg; BHT, 15 mg; BHA, 15 mg
[b]Mineral added to supply the following (per kg diet): copper sulfate (25.4% Cu), 30.5 mg; zinc sulfate (22.7% Zn), 230 mg; manganous sulfate (32.5% Mn), 100 mg; cobalt chloride (24.8% Co), 20 mg; potassium iodide (76.4% I), 6.5 mg; sodium selenite (45.6% Se), 2.2 mg; sodium fluoride (45.2% F), 8 mg
[c]Available P calculated based on apparent digestibility coefficient values

growth trials, respectively. Operating conditions of extrusion were as follows: feeder speed, 16~18 rpm; conditioner temperature, 80~90 °C; main screw speed, 250~320 rpm; temperature of the second and the third barrel compartment, 105~135 °C; steam heater pressure, 4~6 kgf/cm^2 and temperature of the fourth barrel compartment, 80~90 °C. Extruded pellets were dried at 60 °C for 6 h using a drying oven which resulted in the moderate moisture content of 5 to 8%.

For the measurement of P digestibility, chromic oxide of 1.0% was mixed with an aliquot of 10 kg of each extruded diet fully ground. Then, each diet mixture was added with 20% distilled water, and the mixture was pelletized using a meat chopper and dried for 12 h in a ventilated oven at 60 °C. The diets were stored in a freezer at −20 °C for P digestibility measurement until the growth trials.

Growth trial

Carp fry of 5000 with around body weight of 5 g were kindly provided from Kyeongnam freshwater fish institute and acclimated to the experimental conditions for 3 weeks with a control diet. Following a 24-h fasting, five groups (triplicates/group) of 450 fish of a mean body weight of 6.6 g were randomly allotted to each of 15 tanks (0.4 × 0.6 × 0.36 cm, water volume of 66 L). The feeding experiment lasted 9 weeks during which each diet was hand-fed to apparent satiety twice a day (08:30 and 17:30) at the 4% of body weight in every 6 days per week. A recirculation freshwater system was employed where water temperatures for first, second, and third weeks were maintained at 26 ± 1.2, 22 ± 1.2, and 18 ± 1.5 °C, respectively, and dissolved oxygen 5.5~6.4 mg O_2/L for the whole experimental periods. The flow rate of 5 L/min was constantly set. The extruded pellet of 2.5 mm size was fed for the last feeding of 3 weeks. Fish were bulk-weighed at the beginning of the experiment in every 3 weeks. Daily feed intake (DFI, %/av. body weight/d), weight gain (WG, %), feed conversion ratio (FCR), protein efficiency ratio (PER), specific growth rate (SGR, %), and survival rate (SR, %) were calculated as follows:

$$DFI\ (\%/av.\ body\ wt/d) = feed\ intake\ (g,\ DM)/$$

$$((initial\ wt + final\ wt)/2)/experimental\ days \times 100,$$

$$WG\ (\%) = (final\ weight\ (g) - initial\ weight\ (g)) \times 100/initial\ weight\ (g),$$

$$FCR = feed\ intake\ (g,\ DM)/wet\ weight\ gain\ (g),$$

$$PER = wet\ weight\ gain\ (g)/protein\ intake,$$

$$SGR\ (\%) = (Ln\ final\ weight\ (g) - Ln\ initial\ weight\ (g))/experimental\ days \times 100,$$

and

$$SR\ (\%) = final\ fish\ number/initial\ fish\ number \times 100.$$

Digestibility trial

At the end of the growth trial, digestibility measurement was conducted to calculate the available P of the experimental diets. Following a 24-h fasting, 300 fish (mean body weight, 22.2 g) were randomly distributed into each of five 130 L capacity tanks (60 fish/tank) with a fecal collection column. Following 1 week of feeding, fecal collections were made for three consecutive weeks as described by Kim et al. (2006). Each diet was fed by hand to apparent satiety twice a day (08:30 and 16:30). One hour after final feeding of the day, the drain pipes and fecal collection columns were thoroughly cleaned with a brush to remove feed residues and feces from the system. The settled feces and surrounding water were carefully collected into 250-ml centrifuge bottles each morning (08:00). Apparent digestibility coefficient (ADC) of P in experimental diets was calculated according to Maynard and Loosli (1969). P availability of phosphate additives was calculated by dividing the difference of total P into that of available P between control and each phosphate containing diet:

$$ADC\ (\%) = \left(1 - \frac{ID \times PF}{IF \times PD}\right) \times 100$$

where ID is % indicator in the diet, PF represents % P in the feces, IF indicates % indicator in the feces, and PD is % P in the diet.

P availability of the phosphorus additives was calculated according to the following equation:

$$P\ availability\ (\%) = \frac{APDP\text{-}APCD}{TPDP\text{-}TPCD} \times 100$$

where APDP indicates % available P in the diet containing P source, APCD is % available P in control diet, TPDP shows % total P in the diet containing P source, and TPCD is % total P in control diet.

Sample collection and analysis
At the end of the experimental period, fish were anesthetized with AQUI-S (New Zealand Ltd., Lower Hutt, NZ) and bulk-weighed and counted for calculation of WG, FCR, SGR, PER, and SR. Blood samples were obtained from the caudal vessels with a heparinized syringe from two fish of each tank after fish were starved for 24 h and anesthetized with AQUI-S. Feces collected in the same bottle from each tank for 6 days a week were used as one replicate for the treatment. After collection of three replicate samples of each diet during 3 weeks, fecal samples were lyophilized, finely ground, and frozen at –20 °C until analysis.

Chemical analyses of feed ingredient, diets, and feces were performed by the standard procedure of AOAC (1990) for moisture, crude protein, crude fat, and crude ash. Moisture content was obtained after drying in an oven at 105 °C for 24 h. Crude protein ($N \times 6.25$) was determined by Kjeldahl method after acid digestion. Crude fat was determined by the soxhlet extraction method by using Soxtec system 1046 (Foss, Hoganas, Sweden) and crude ash from incineration in a muffle furnace at 550 °C for 12 h. Chromium in diets and feces for P digestibility measurement was analyzed using a spectrophotometer (Shimadzu, UV-120-12) at a wavelength of 440 nm after perchloric acid digestion (Bolin et al. 1952). Ca in diets and P in diets and the whole body of the final fish (five fish per replicate) were

measured using inductively coupled plasma mass spectrometer (ICP-MS) (Perkin-Elmer, NexION 300D, Waltham, MA, USA) after the pretreatment of test materials following the method from US Environmental protection agency (USEPA 1996). Hematocrit (PCV, %) and hemoglobin (Hb, g/dL) were measured with the same fish (two fish per replicate) by the microhematocrit method (Brown 1980) and the cyanmethemoglobin procedure using Drabkin's solution, respectively. Hb standard prepared from human blood (Sigma Chemical, St. Louis, MO) was employed. Blood plasma from two fish per replicate was obtained after blood centrifugation ($3500 \times g$, 5 min, 4 °C) and stored at –80 °C until inorganic P (Pi) was analyzed. The plasma Pi was measured using a blood chemical analyzer (HITACHI 7600-210, Hitachi High-Technologies Co., Ltd., Japan) with commercial clinical investigation reagent (Clinimate IP, Sekisui medical Co., Ltd., Tokyo, Japan).

Statistical analysis
Data of growth trial (initial and final fish, DFI, WG, FE, PER, SGR, and SR), P availability among both replications and treatments, whole body P, and hematological and serological parameters were analyzed using one-way analysis of variance (ANOVA), and significant differences among treatment means were compared using Duncan's multiple range test (Duncan 1955). Prior to the analysis, homogeneity of variance of all data was verified using Cochran's test (Sokal and Rohlf 1994). All statistical analyses were carried out using the SPSS Version 10 (SPSS 1999). Statistical significance of the differences was determined by a significant level of 5% ($p < 0.05$).

Results
Growth performance, whole body P, and P digestibility
As shown in Table 1, crude protein level of dry diets was determined to be 42.2 (DCP) to 42.9% (MCP) and crude fat to be 6.6 (DCP) to 6.9% (control). At the end of the 9-week growth trial, daily feed intake (DFI, %) per averaged weight of fish ranged from 1.88 (MHP) to 2.17 (control and TCP). Fish fed the MHP showed the highest WG of 278%, which was not significantly different ($p > 0.05$) from that (270%) of fish fed the MCP, while fish fed the control and TCP showed lowest WG among the treatment groups ($p < 0.05$). FCR ranged from 0.99 (MHP) to 1.29 (control). The highest value of PER and SGR were found in fish fed the MCP and MHP, while fish fed the control and TCP showed the lowest level (Table 2). Relative weight gain (RWG) was significantly higher in DCP, MCP, and MHP compared to that in control and TCP (Fig. 1). Fish fed the control showed the lowest P content (0.36%) in whole body, while those fed the MHP the highest (0.46%) after 9-week feeding trial. (Fig. 3). The significant difference in P availability of

Table 2 Growth performance of carp fed diets containing various phosphorus sources for 9 weeks

Parameters	Diet				
	CON	MCP	DCP	TCP	MHP
Initial wt (g/fish)	6.40 ± 0.05^{ns}	6.63 ± 0.07	6.63 ± 0.08	6.60 ± 0.08	6.61 ± 0.12
Final wt (g/fish)	19.19 ± 0.23^d	24.53 ± 0.30^a	22.41 ± 0.28^b	20.14 ± 0.41^c	24.94 ± 0.39^a
DFI (%/av. wt/day)[1]	2.17 ± 0.02^a	1.91 ± 0.02^c	2.01 ± 0.03^b	2.17 ± 0.03^a	1.88 ± 0.03^c
WG (%)[2]	199.87 ± 5.68^c	269.85 ± 2.44^a	237.87 ± 2.76^b	205.19 ± 2.61^c	277.54 ± 12.59^a
FCR[3]	1.29 ± 0.03^a	1.00 ± 0.01^c	1.09 ± 0.01^b	1.28 ± 0.03^a	0.99 ± 0.03^c
PER[4]	1.83 ± 0.04^c	2.34 ± 0.03^a	2.17 ± 0.03^b	1.83 ± 0.04^c	2.39 ± 0.08^a
SGR(%)[5]	1.74 ± 0.03^c	2.08 ± 0.01^a	1.93 ± 0.01^b	1.77 ± 0.01^c	2.11 ± 0.05^a
SR(%)[6]	100 ± 0.00^{ns}	100 ± 0.00	100 ± 0.00	100 ± 0.00	100 ± 0.00

Values (means ± SE of triplicates) with different superscript letters in the same row are significantly different ($p < 0.05$)
ns nonsignificant
[1]Daily feed intake (%/av. wt/day) = dry feed intake (g/fish)/((initial wt + final wt)/2)/experimental days × 100
[2]Weight gain (%) = (final weight (g) − initial weight (g)) × 100/initial weight (g)
[3]Feed conversion ratio = dry feed intake/wet weight gain
[4]Protein efficiency ratio = wet weight gain (g)/protein intake
[5]Specific growth rate (%) = (Ln final weight (g) − Ln initial weight (g))/experimental days × 100
[6]Survival rate (%) = final fish number/initial fish number × 100

repetition was not shown from manure analyses every week. Apparent availability of P in the experimental diets varied from 24.2 to 41.9% for TCP and MHP, respectively. The values of P were found to be 93.2, 62.4, 6.1, and 97.8% for MCP, DCP, TCP, and MHP, respectively (Table 3). The available P level in diets (Table 1) was calculated based on determined P digestibility, which ranged from 0.32 (control) to 0.66% (MCP).

Hematological parameters and plasma inorganic P
Hematological parameters and inorganic P in plasma of fish fed the experimental diets are shown in Figs. 2 and 3, respectively. PCV (%) of fish fed TCP (24.2) and control (25.5) showed a significant difference ($p < 0.05$) from that of fish fed MHP (28.8). Hb (g/dL) of fish ranged

from 7.5 (TCP) to 8.8 (MHP), which were significantly different ($p < 0.05$). Significantly lower P (mg/dL) in plasma was found in fish fed the control (4.3), while the other groups did not show any significant difference ($p > 0.05$) each other.

Discussion
As the environmental risks has been increased by releasing animal fecal wastes including fisheries, the P-recovery from swine manure has been introduced in the form of struvite and widely studied (Yoon et al. 2015). Its application has been only limited as a fertilizer but it is much desirable to apply P recovered to new applications after appropriate manufacturing and securing safety. Magnesium hydrogen phosphate (MHP) was

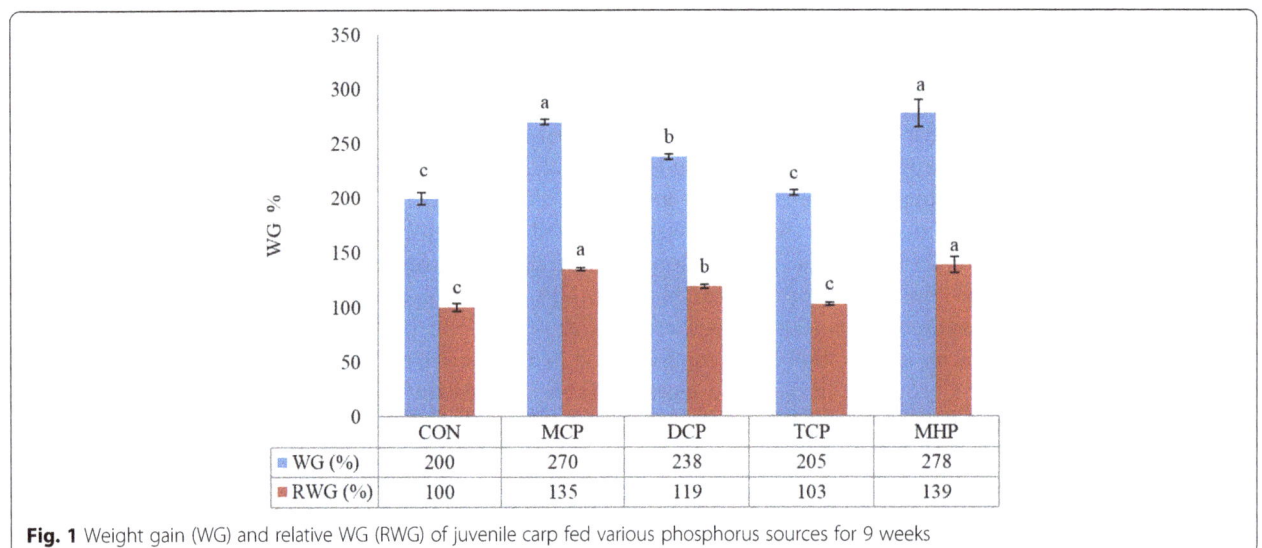

Fig. 1 Weight gain (WG) and relative WG (RWG) of juvenile carp fed various phosphorus sources for 9 weeks

Table 3 Apparent availability of phosphorus in the experimental diets and various phosphate sources

	P availability				
	CON	MCP	DCP	TCP	MHP
Diet	26.57 ± 1.07^c	42.27 ± 0.47^a	34.30 ± 0.46^b	22.37 ± 0.74^d	41.97 ± 0.68^a
Phosphorus sources		93.19 ± 1.96^b	62.42 ± 3.37^c	6.11 ± 2.02^d	97.97 ± 2.06^a

Means ± S.D of three replicates of each group with different superscript letters in the same row are significantly different ($p < 0.05$)

newly manufactured from struvite which was to recycle P from swine manure. It is first attempt to investigate the effect of the MHP as dietary P source on growth and feed utilization of carp. Based on the earlier study (Kim et al. 1998), who found that 2% of MCP in diet for carp improved growth and decreased P loads, the same level of various P sources was incorporated into the experimental diets (Table 1). Fish fed the MHP showed the best WG at the end of growth trial, although it was not significantly different ($p > 0.05$) from that of fish fed the MCP. Same tendencies were found in FCR, PER, and SGR (Table 2). The result suggests that juvenile carp could utilize P from MHP as effectively as MCP. From the results, it was evident that MHP was a good P source competitive with MCP in terms of WG and FCR

in juvenile carp. As shown in Fig. 1, RWG of MCP, DCP, TCP, and MHP to control at the level of 2% in each diet were 135, 119, 103, and 139% on WG, respectively. Such relative differences among various P sources might be due to the difference in availability of P, by which the requirement could be met or not. Available P requirement is known to be 0.6~0.7% (Ogino and Takeda 1976; Kim et al. 1998) for carp. The present study confirmed such a requirement to be optimal.

Ogino et al. (1979) extensively studied the P availability of inorganic P sources and various feed ingredients by fish. They reported that the availability of MCP, DCP, and TCP was 94, 46, and 13 by carp, respectively, using egg albumin-based diet. Though somewhat higher values were obtained for DCP and TCP in the present study, the value for MCP corresponds well with their result.

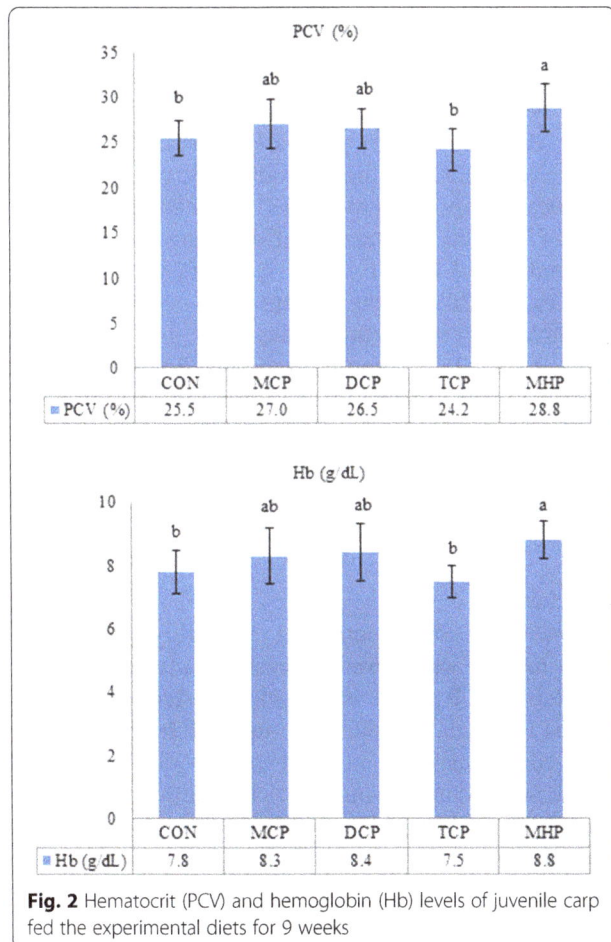

Fig. 2 Hematocrit (PCV) and hemoglobin (Hb) levels of juvenile carp fed the experimental diets for 9 weeks

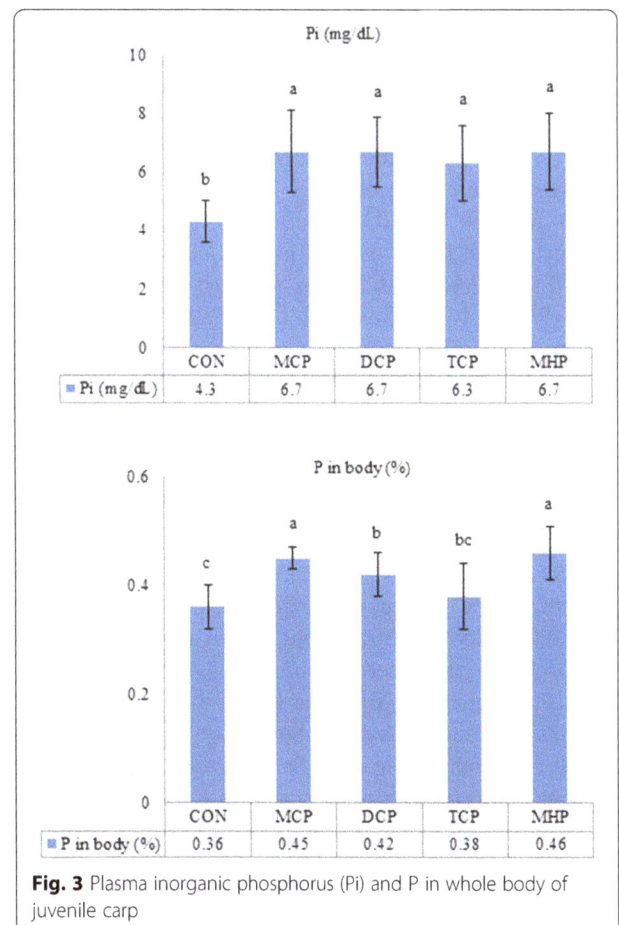

Fig. 3 Plasma inorganic phosphorus (Pi) and P in whole body of juvenile carp

On the other hand, MHP had the higher P availability than MCP, suggesting its potential use for carp. Similar results were reported in the previous study conducted using juvenile catfish (Yoon et al. 2014). They found that MHP (90.9%) had P availability comparable to MCP (88.1%) at 2% incorporation level in diet. Sarker et al. (2009) determined P availability of different P sources by yellowtail using albumin-based diet. They incorporated MCP (2.7%), DCP (3.7%), and TCP (3.6%) into the respective diets and obtained P availabilities of 92.4, 59.2, and 48.8% for MCP, DCP, and TCP, respectively. Lovell (1978) reported that the availability was found to be 94 and 65% for reagent grade MCP and DCP, respectively, in channel catfish. On the other hand, Eya and Lovell (1997) determined the net absorption of P from various P sources using all-plant basal diet in channel catfish. They obtained the values of 81.2, 74.8, and 54.8% for MCP, DCP, and TCP, respectively.

To our knowledge, this is the first time P availability of feed grade MCP, DCP, and TCP as well as MHP was determined for juvenile carp. In the present study, P availabilities of both MCP and DCP were found to be high while that of TCP was very low (Table 3). P availability of MCP seems to be comparable to those reported by Ogino et al. (1979), Lovell (1978), Kim et al. (1997), and Sarker et al. (2009). However, it was higher than those reported by Pimentel-Rodrigues and Oliva-Teles (2007) and Eya and Lovell (1997). The availability of DCP was comparable to those obtained in yellowtail (Sarker et al. 2009) and European sea bass (Pimentel-Rodrigues and Oliva-Teles 2007), while it was lower than those obtained in channel catfish (Lovell 1978), rainbow trout (Ogino et al. 1979), and far eastern channel catfish (Yoon et al. 2014). On the other hand, reported P availability of TCP by stomach fish ranged from 42% (Pimentel-Rodrigues and Oliva-Teles 2007) to 64% (Ogino et al. 1979). In contrast, the value from the present study was as low as that obtained by carp, stomachless species (Ogino et al. 1979). Bioavailability of dietary P is influenced by several factors including chemical form, digestibility of diet, particle size, interaction with other nutrients, feed processing, and water chemistry (Lall 1991). It remains to be explained whether such discrepancies in P availability of various P sources, especially TCP, are due to the differences in species and diet composition employed for the digestibility measurement as well as methodological approach in fecal collection (Kim et al. 1996). On the other hand, P availability of MHP was higher than those of MCP obtained from carp and rainbow trout by Ogino et al. (1979) and from yellowtail by Sarker et al. (2009), suggesting that MHP could be a potential P source for fish.

Hematological and serological parameters are useful in monitoring the physiological status of fish and as indicators of the health of the aquatic environment, although they are not routinely used in fish disease diagnosis (El-Sayed et al.

2007). Hematocrit (PCV, %) provides an indirect measurement of the body's oxygen carrying ability, while Hb (g/dL) a direct measurement of the oxygen carrying capacity of the blood (McClatchey 2002). It was reported that hematological parameters could be influenced by nutritional status (Spannhof et al. 1979), infectious disease (Barham et al. 1980; Iwama et al. 1986), environmental changes (Giles et al. 1984), and stress (Ellsaesser and Clem 1986). The normal ranges of healthy adult Atlantic salmon were reported 44 to 49 and 8.9 to 10.4 for PCV and Hb, respectively (Sandnes et al. 1988). Somewhat lower values for cichlid fish were reported by V´azquez and Guerrero (2007), which were 22.5 to 39.2 for PCV and 5.2 to 8.3 for Hb. Recently, Rahimnejad and Lee (2013) reported 30.7 to 34.3 for PCV and 4.4 to 5.4 for Hb of red sea bream fed various dietary valine levels. Our findings for the parameters are in good agreement with those obtained from tilapia (Hrubec et al. 2000) and striped bass (Hrubec et al. 2001), although there is no available information on the effect of dietary available P on the parameters of fish. In the present study (Fig. 2), PCV and Hb of fish groups fed low available P in diet (control and TCP) were significantly lower than the MHP group ($p < 0.05$). Differently from the present results, Yoon et al. (2014) observed that dietary low P resulted in PCV significantly lower but did not affect Hb. An increase in available P in diet resulted in an increase in plasma Pi (Vielma and Lall 1998; Bureau and Cho 1999; Avila et al. 2000). A clear evidence was observed by Yoon et al. (2014) that plasma Pi increased in fish fed MCP and MHP with higher available P, resulting in the significant improvement in WG by meeting dietary P need. However, any significant difference in plasma Pi was not found in fish groups fed diets with various P sources ($p > 0.05$) in the present study. Nevertheless, P in whole body was greatly affected by dietary available P levels (Fig. 3), suggesting successive P intake through diet maintains plasma Pi constant then deposition in skeletal tissues occurs when P requirement is met. More researches remain to be clarify the difference in plasma Pi level observed in previous (Yoon et al. 2014) and present studies.

Conclusions

In conclusion, the potential use of MHP recovered from swine manure was proven sufficiently to replace MCP as an alternative P source with respect to WG and FCR as well as P availability. "Such a re-use of P from swine manure could positively influence on the development of other useful sources from livestock manure."

Abbreviations
DCP: Dicalcium phosphate; DF: Daily feed intake; FCR: Feed conversion ratio; Hb: Hemoglobin; MCP: Monocalcium phosphate; MHP: Magnesium hydrogen phosphate; P: Phosphorus; PCV: Hematocrit; PER: Protein efficiency ratio; Pi: Inorganic P; RWG: Relative weight gain; SGR: Specific growth rate;

SR: Survival rate; TCP: Tricalcium phosphate; USGS: U.S. Geological Survey; WG: Weight gain; XRD: X-ray diffractometer

Acknowledgements
We are grateful to the funding by Rural Development Administration, Korea. Also, this study was partially supported by 2015 Research Grant from Kangwon National University.

Funding
This study was supported by the grant from Rural Development Administration, Korea (PJ008390) for the design of study, experimental procedures, and interpretation of data derived from this study. In addition, this study was partially supported by 2015 Research Grant from Kangwon National University (No. 520150143) for the experiments in this manuscript.

Authors' contributions
JDK designed this study, led in analyzing data derived from this study, and wrote this manuscript. THY mainly analyzed the data from this study and wrote this manuscript. SW, DHL, JWC, and CR participated in analyzing and writing this manuscript. All authors read and approved the final manuscript.

Competing interests
The authors declare that they have no competing interests.

Author details
[1]College of Animal Life Sciences, Kangwon National University, Chuncheon 24341, Korea. [2]Department of Animal Resources, Daegu University, Gyeongsan 38453, Korea. [3]Gyenoggi Province Maritime and Fisheries Research Institute, 23-2 Sangkwang-Gil, Yangpyeong-Gun 12513, Korea.

References
AOAC. Official Methods of Analysis. Association of Official Analytical Chemists: Arlington, VA, USA;1990. p. 1298.

Auer MT, Kiesser MS, Canale RP. Identification of critical nutrient levels through field verification of models for phosphorus and phytoplankton growth. Can J Fish Aquat Sci. 1986;43:379–88.

Avila EM, Tu H, Basantes S, Ferraris RP. Dietary phosphate regulates intestinal transport and plasma concentrations of phosphate in rainbow trout. J Comp Physiol. 2000;170B:210–09.

Barham WT, Smit GL, Schoonbee HJ. The haematological assessment of bacterial infection in rainbow trout, Salmo gairdneri Richardson. J Fish Biol. 1980;17:275–81.

Beveridge MCM. Cage and pen fish farming: carrying capacity models and environmental impact. FAO Fish Tech Pap. 1984;255:131.

Bolin DW, King RP, Klosrerman EW. A simplified method for the determination of chromic oxide (Cr_2O_3) when used as an inert substance. Science. 1952; 116(3023):634–5.

Brown BA. Routine hematology procedures, Hematology: Principles and Procedures. Philadelphia: Lea and Febiger; 1980. p. 71–112.

Brown ML, Jaramillo Jr F, Gatlin III DM. Dietary phosphorus requirement of juvenile sunshine bass, Morone chrysops × M. saxatilis. Aquaculture. 1993; 113:355–63.

Bureau DP, Cho CY. Phosphorus utilization by rainbow trout (Oncorhynchus mykiss): estimation of dissolved phosphorus waste output. Aquaculture. 1999; 179:127–40.

Cao L, Naylor R, Henriksson P, Leadbitter D, Metian M, Troell M, Zhang W. China's aquaculture and the world's wild fisheries. Science. 2015;347:133–5.

Davis DA, Robinson EH. Dietary phosphorus requirement of juvenile red drum Sciaenops ocellatus. J World Aquacult Soc. 1987;18:129–36.

Dougall DS, Wood III LC, Douglass LW, Soares JH. Dietary phosphorus requirement of juvenile striped bass Morone saxatilis. J World Aquacult Soc. 1996;27:82–91.

Duncan DB. Multiple range and multiple 'F' tests. Biometrics. 1955;11:1–42.

Ellsaesser CF, Clem LW. Haematological and immunological changes in channel catfish stressed by handling and transport. J Fish Biol. 1986;28:511–21.

El-Sayed YS, Saad TT, El-Bahr SM. Acute intoxication of deltamethrin in monosex Nile tilapia, Oreochromis niloticus with special reference to the clinical, biochemical and haematological effects. Environ Toxicol Phamacol. 2007;24:212–7.

Eya JC, Lovell RT. Net absorption of dietary phosphorus from various inorganic sources and effect of fungal phytase on net absorption of plant phosphorus by channel catfish Ictalurus punctatus. J World Aquacult Soc. 1997;28:386–91.

Giles MA, Majewski HS, Hobden B. Osmoregulatory and haematological responses of rainbow trout (Salmo gairdneri) to extended environmental acidification. Can J Fish Aquat Sci. 1984;41:1686–94.

Hernandez A, Satoh S, Kiron V, Watanabe T. Phosphorus retention efficiency in rainbow trout fed diets with low fish meal and alternative protein ingredients. Fish Sci. 2004;70:580–6.

Hernandez A, Satoh S, Kiron V. Effect of monocalcium phosphate supplementation in a low fish meal diet for rainbow trout based on growth, feed utilization, and total phosphorus loading. Fish Sci. 2005;71:817–22.

Hrubec TC, Cardinale JL, Smith SA. Hematology and plasma chemistry reference intervals for cultured tilapia (Oreochromis hybrid). Vet Clin Pathol. 2000;29:7–12.

Hrubec TC, Smith SA, Robertson JL. Age related changes in haematology and chemistry values of hybrid striped bass chrysops Morone saxatilis. Vet Clin Pathol. 2001;30:8–15.

Iwama GK, Greer GL, Randall DJ. Changes in selected haematological parameters in juvenile Chinook salmon subjected to a bacterial challenge and a toxicant. J Fish Biol. 1986;28:563–72.

Kim JD, Ahn KH. Effects of monocalcium phosphate supplementation on phosphorus discharge and growth of carp (Cyprinus carpio) grower. Asian Australas J Anim Sci. 1993;6:521–6.

Kim JD, Kim KS, Song JS, Lee JY, Jeong KS. Optimum level of dietary monocalcium phosphate based on growth and phosphorus excretion of mirror carp, Cyprinus carpio. Aquaculture. 1998;161:337–44.

Kim JD, Kim KS, Song JS, Jeong KS, Woo YB, Choi NJ, Lee JY. Comparison of feces collection methods for determining apparent phosphorus digestibility of feed ingredients in growing mirror carp (Cyprinus carpio). Korean J Anim Nutr Feed. 1996;20:201–6.

Kim JD, Kim KS, Lee SB, Jeong KS. Nutrient and energy digestibilities of various feedstuffs fed to Israeli strain of growing common carp (Cyprinus carpio). J Aquacult. 1997;10:327–34.

Kim JD, Tibbets SM, Milley JE, Lall SP. Effect of the incorporation level of herring meal into test diet on apparent digestibility coefficients for protein and energy by juvenile haddock, Melanogrammus aeglefinus L. Aquaculture. 2006; 258:479–86.

Lall SP. Digestibility, metabolism and excretion of dietary phosphorus in fish. In: Cowey CB, Cho CY, editors. Nutritional Strategies and Aquaculture Wastes. Ontario: University of Guelph; 1991. p. 21–36.

Liu YH, Kumar S, Kwag JH, Kim JH, Kim JD, Ra CS. Recycle of electronically dissolved struvite as an alternative to enhance phosphate and nitrogen recovery from swine wastewater. J Hazard Mater. 2011;195:175–81.

Lovell RT. Dietary phosphorus requirements of channel catfish. Trans Am Fish Soc. 1978;107:617–21.

Ma K, Zhang C, Ai Q, Duan Q, Xu W, Zhang L, Luifu Z, Tan B. Dietary phosphorus requirement of large yellow croaker, Pseudosciaena crocea R. Aquaculture. 2006;251:346–53.

Maynard LA, Loosli JK. Animal nutrition. 6th ed. New York: McGraw-Hill; 1969. p. pp 613.

McClatchey KD. Clinical laboratory medicine. Baltimore: Williams & Wilkins; 2002. p. 833–9.

NRC. Nutrient requirements of warmwater fishes and shellfishes. Washington: National Academy Press; 1983.

Nwanna LC, Adebayo IA, Omitoyin BO. Phosphorus requirements of African catfish, Clarias gariepinus, based on broken-line regression analysis methods. Science Asia. 2009;35:227–33.

Ogino C, Takeda H. Mineral requirements in fish-III calcium and phosphorus requirements in carp. Bull Jap Soc Sci Fish. 1976;42:793–9.

Ogino C, Takeuchi L, Takeda H, Watanabe T. Availability of dietary phosphorus in carp and rainbow trout. Bull Jap Soc Sci Fish. 1979;45:1527–32.

Phromkunthong W, Udom U. Available phosphorus requirement of sex-reversed red tilapia fed all-plant diets. Songklanakarin J Sci Technol. 2008;30:7–16.

Pimentel-Rodrigues A, Oliva-Teles A. Phosphorus availability of inorganic phosphates and fish meal in European sea bass (*Dicentrarchus labrax* L.) juveniles. Aquaculture. 2007;267:300–7.

Rahimnejad S, Lee KJ. Dietary valine requirement of juvenile red sea bream *Pagrus major*. Aquaculture. 2013;416–417:212–8.

Rodehutscord M, Pfeffer E. Requirement for phosphorus in rainbow trout (*Oncorhynchus mykiss*) growing from 50 to 200 g. Water Sci Technol. 1995;31:137–41.

Roy PK, Lall SP. Dietary phosphorus requirement of juvenile haddock (*Melanogrammus aeglefinus* L.). Aquaculture. 2003;221:451–68.

Sandnes K, Lie O, Waagbo R. Normal ranges of some blood chemistry parameters in adult farmed Atlantic salmon, *Salmo salar*. J Fish Biol. 1988;32:129–36.

Sarker PK, Fukada H, Masumoto T. Phosphorus availability from inorganic phosphorus sources in yellowtail (*Seriola quinqueradiata* Termminck and Schlegel). Aquaculture. 2009;289:113–7.

Satoh S, Hernandez A, Tokoro T, Morishita Y, Kiron V, Watanabe T. Comparison of phosphorus retention efficiency between rainbow trout (*Oncorhynchus mykiss*) fed a commercial diet and a low fish meal based diet. Aquaculture. 2003;224:271–82.

Shu L, Schneider P, Jegatheesan V, Johson J. An economical evaluation of phosphorus recovery as struvite from digester supernatant. Bioresour Technol. 2006;97:2211–6.

Sokal RR, Rohlf FJ. Nested analysis of variance, Biometry. New York: Freeman W. H; 1994. p. 272–342.

Spannhof L, Nasev D, Kreutzmann HL. Early recognition of metabolic disturbances in trout (*Salmo gairdneri* Rich.) stocks. Aquaculture. 1979;18:317–23.

SPSS Inc. SPSS Base 10.0 for Windows User's Guide. Chicago: SPSS Inc; 1999.

USEPA (US Environmental protection agency). 1996. Acid digestion of sediments, sludge, and solids, Method 3050B, Revision 2 (December 1996). Available at: https://www.epa.gov/sites/production/files/2015-06/documents/3050b.pdf.

USGS (US Geological survey). 2013. Mineral commodity summaries. Digital Data Series. Available at: http://minerals.usgs.gov/minerals/pubs/mcs/2013/mcs2013.pdf.

V'azquez GR, Guerrero GA. Characterization of blood cells and hematological parameters in *Cichlasoma dimerus* (Teleostei, Perciformes). Tissue Cell. 2007; 39:151–60.

Vilema J, Koskela J, Ruohonen K. Growth, bone mineralization, and heat and low oxygen tolerance in European whitefish (*Coregonnus lavaretus* L.) fed with graded levels of phosphorus. Aquaculture. 2002;212:321–33.

Vielma J, Lall SP. Control of phosphorus homeostasis of Atlantic salmon (*Salmo sala*) in fresh water. Fish Physiol Biochem. 1998;19:83–93.

Wilson RP, Robinson EH, Gatlin III DM, Poe WE. Dietary phosphorus requirement of channel catfish. J Nutr. 1982;112:1197–202.

Yoon TH, Lee DH, Won SG, Ra CS, Kim JD. Effects of dietary supplementation of magnesium hydrogen phosphate ($MgHPO_4$) as an alternative phosphorus source on growth and feed utilization of juvenile far eastern catfish (*Silurus asotus*). Asian Australas J Anim Sci. 2014;27:1141–9.

Yoon TH, Lee DH, Won SG, Ra CS, Kim JD. Optimal incorporation level of dietary alternative phosphate ($MgHPO_4$) and requirement for phosphorus in Juvenile far eastern Catfish (*Silurus asotus*). Asian Australas J Anim Sci. 2015;28:111–9.

Zubay GL. Biochemistry. Reading: MA, Addison-Wesley Publishing Company, Inc.; 1983. p. pp 388.

5

Inhibitory effect of bacteriocin-producing lactic acid bacteria against histamine-forming bacteria isolated from *Myeolchi-jeot*

Eun-Seo Lim

Abstract

The objectives of this study were to identify the histamine-forming bacteria and bacteriocin- producing lactic acid bacteria (LAB) isolated from *Myeolchi-jeot* according to sequence analysis of the 16S rRNA gene, to evaluate the inhibitory effects of the bacteriocin on the growth and histamine accumulation of histamine-forming bacteria, and to assess the physico-chemical properties of the bacteriocin. Based on 16S rRNA gene sequences, histamine-forming bacteria were identified as *Bacillus licheniformis* MCH01, *Serratia marcescens* MCH02, *Staphylococcus xylosus* MCH03, *Aeromonas hydrophila* MCH04, and *Morganella morganii* MCH05. The five LAB strains identified as *Pediococcus acidilactici* MCL11, *Leuconostoc mesenteroides* MCL12, *Enterococcus faecium* MCL13, *Lactobacillus sakei* MCL14, and *Lactobacillus acidophilus* MCL15 were found to produce an antibacterial compound with inhibitory activity against the tested histamine-producing bacteria. The inhibitory activity of these bacteriocins obtained from the five LAB remained stable after incubation at pH 4.0–8.0 and heating for 10 min at 80 °C; however, the bacteriocin activity was destroyed after treatment with papain, pepsin, proteinase K, α-chymotrypsin, or trypsin. Meanwhile, these bacteriocins produced by the tested LAB strains also exhibited histamine-degradation ability. Therefore, these antimicrobial substances may play a role in inhibiting histamine formation in the fermented fish products and preventing seafood-related food-borne disease caused by bacterially generated histamine.

Keywords: Bacteriocin, Histamine, Lactic acid bacteria

Background

Myeolchi-jeot, a traditional Korean salted and fermented seafood, is made of anchovies (*Engraulis japonicas*) and is mostly used as an ingredient in Kimchi. Since proteins are broken down into precursor amino acids of biogenic amines through the action of digestive enzymes and microbes during the fermentation process, it contains relatively high concentrations of biogenic amines (Mah et al., 2002). According to their chemical structure, biogenic amines are classified into aliphatic (putrescine, cadaverine, spermine, spermidine), aromatic (tyramine, phenlethylamine), and heterocyclic (histamine, tryptamine) (Santos, 1996). High doses of biogenic amines such as histamine (>500 mg/kg) and tyramine (100–800 mg/kg) contained mainly in fermented

foods can cause adverse health effects to consumers (Joosten and Nuńez, 1996).

Histamine is the causative agent of scombroid poisoning which is often manifested by a wide variety of symptoms such as rash, urticarial, nausea, vomiting, diarrhea, abdominal cramp, hypotension, localized inflammation, headache, palpitation, and severe respiratory distress (Taylor, 1986). Histamine is a basic nitrogenous compound formed mainly through the decarboxylation of histidine by exogenous decarboxylases released from the many different bacterial species associated with salted and fermented seafoods including *Myeolchi-jeot* (An and Ben-Gigirey, 1998). In *Myeolchi-jeot*, *Bacillus licheniformis* has been known to raise levels of histamine in retail canned anchovies during the storage at an ambient temperature Mah and Hwang (2003). The presence of histidine decarboxylase activity has been described in different microbial groups such as seafood-borne bacteria,

Correspondence: limsm020@tu.ac.kr
Department of Food Science and Nutrition, Tongmyong University, Busan 608-735, Republic of Korea

spoilage microorganisms, and lactic acid bacteria (López-Sabater et al., 1994).

Fortunately, in recent years, some lactic acid bacteria (LAB) have been reported to degrade biogenic amine through the production of amine oxidase enzymes or antimicrobial substances (Garciz-Ruiz et al., 2011; Joosten and Nuñez, 1996). LAB that are usually designated as generally recognized as safe (GRAS) status in foods can also exert a biopreservative effect against other microorganisms as a result of competition for nutrients and/or of the production of antagonistic compounds such as organic acids, diacetyl, acetoin, hydrogen peroxide, antibiotics, and bacteriocins (Schillinger et al., 1996). Among anti-microbial metabolites, bacteriocins are ribosomally synthesized and defined as extracelluarly released peptides or protein molecules produced by specific bacteria during the primary phase of growth, though antibiotics are usually secondary metabolites (Zacharof and Lovitt, 2012). Brillet et al. (2005) has shown that bacteriocin produced by *Carnobacterium divergens* V41 can be used as a biopreservative to inhibit the growth *Listeria monocytogenes* in cold smoked salmon; therefore, bacteriocins can be applied in the seafood industry for extension of shelf life as natural preservatives against pathogens and food spoilage.

In particular, outgrowth of histamine producer *Lactobacillus buchneri* St2A was almost completely inhibited by treatment of bacteriocin-producing enterococci and *Lactococcus lactis* strain as fermentation starters (Joosten and Nuñez, 1996). Tabanelli et al. (2014) reported that bacteriocin forming lactococci strains were able to reduce the growth extent and histamine accumulation of *Streptococcus thermophilus* PRI60. The studies reported previously indicate that the applications of bacteriocins in food industries can extend shelf life of foods, inhibit growth of foodborne pathogens during the manufacture of food, prevent formation of toxic substances by harmful bacteria, ameliorate the economic losses due to food spoilage, and reduce the application of chemical preservatives (Gálvez et al., 2007).

The objectives of this study were to (1) identify the histamine-forming bacteria and bacteriocin-producing LAB isolated from *Myeolchi-jeot* according to sequence analysis of the 16S rRNA gene, (2) evaluate the inhibitory effects of the bacteriocin on the growth and histamine accumulation of histamine-forming bacteria, and (3) assess the physico-chemical properties of the bacteriocin.

Methods
Isolation and identification of histamine-forming bacteria and LAB
Five samples of salted-fermented anchovy were obtained from retail stores in Busan and aseptically collected for these experiments. Each sample (50 g) was diluted with sterile peptone water (450 mL) and homogenized at high speed for 2 min in stomacher. The decimal serial dilutions of the homogenates were then subsequently prepared using a sterile peptone water, and 1.0-mL aliquots of the dilutes were inoculated into brain heart infusion (BHI) agar (BD Difco Co., Sparks, MD, USA) and incubated for 48 h at 37 °C. To isolate histamine-forming bacteria, each strain collected from the plates was subcultured for 48 h at 35 °C in decarboxylating broth contained L-histidine monohydrochloride monohydrate (Sigma-Aldrich, St Louis, MO, USA, 1 g/L) and pyridoxal-5′-phosphate (Sigma-Aldrich, 1 mg/L) according to the method of Bover-Cid and Holzapfel (1999) with minor modifications. The cell culture (0.1 mL) was spread on trypticase soy agar (TSA, BD Difco Co.) supplemented with 2.0% (w/v) L-histidine hydrochloride monohydrate. After incubation under anaerobic conditions for 4 days at 35 °C (Anoxomat system, MART Co., Netherland), the colonies with blue or purple color on the plates were considered as histamine-forming bacteria. Meanwhile, to distinguish lactic acid-producing bacteria from salted-fermented anchovy, the homogenized sample was spread directly onto the surface of MRS (BD Difco Co.) agar plates containing 1% $CaCO_3$. After incubation under aerobic conditions for 48 h at 37 °C, the colonies surrounded by a clear zone were randomly selected from the plates, purified on MRS agar, and examined histamine production as the abovementioned. The identity of histamine-forming isolates was confirmed by using 16S rRNA gene sequence analysis (Chen et al., 2008). Amplification of the isolates was performed with the universal primers UNI-L (5′-AGAGTTT GATCATGGCTCAG-3′) and UNI-R (5′-GTGTGACGG GCGGTGTGTAC-3′). Bacterial cells were cultivated in BHI broth at 37 °C with shaking overnight and centrifuged at 7000×g for 10 min. After washing, the cell pellets were resuspended in PBS (pH 7.0) and lysed by 20% sodium dodecyl sulfate (SDS). After the boiling process for 20 min at 85 °C, the cell debris was removed by centrifugation (13,000×g, 5 min, 4 °C). DNA in the supernatant was precipitated by addition of 70% ethanol and used as template DNA for polymerase chain reaction (PCR). PCR amplification was carried out with a reaction mixture consisting of 10 mM Tris-HCl (pH 8.3), 50 mM KCl, 1.5 mM $MgCl_2$, 20 pmol of each primer, 0.2 mM deoxynucleotide triphosphates, 0.5 U of Taq DNA polymerase (Applied Biosystem, Foster City, CA, USA), and template DNA (10 ng). Cycle conditions were an initial denaturation for 4 min at 94 °C, followed by 35 cycles of 30 s denaturation at 94 °C, 30 s annealing at 55 °C, 1 min primer extension at 72 °C, followed by a final extension for 7 min at 72 °C in a PCR Thermal Cycler (Bio-Rad Laboratories Ltd., Canada). To identify the LAB isolates which do not produce histamine,

the universal primer used for identification of LAB was 27F (5′-AGAGTTTGATCCTGGCTCAG-3′) and 1492RF (5′-GGTTACCTTGTTACGACTT-3′). Thermocycling was carried out using denaturation step at 94 °C for 1 min, annealing at 55 °C for 1 min, and extension at 72 °C for 2 min, for a total of 30 cycles. Amplicons were visualized on a 1.5% agarose gel staining with ethidium bromide to confirm successful amplification and then PCR product was purified using a QIAquick PCR Purification Kit (Qiagen, Valencia, CA, USA). The amplified DNA was directly sequenced with a DNA sequencer (ABI Prism® 3730 Avant Genetic Analyzer, Applied Biosystem) and sequence similarity searching was performed using the National Center for Biotechnology Information (NCBI) Basic Local Align Search Tool (BLAST).

Preparation of bacteriocin solution and determination of bacteriocin activity

To prepare the bacteriocin solution of the LAB strains, the culture extracts of the isolated LAB were obtained after 24-h incubation at 37 °C in MRS broth. The cultures were centrifuged at 7000×g for 10 min at 4 °C. The cell-free supernatant (CFS) was then adjusted with 1 N of NaOH to pH 6.5 to avoid effect of acid, treated with catalase (200 U/mL) for 30 min at 25 °C to remove hydrogen peroxide, and filtered through a 0.22-µm membrane filter (Millipore Corp., USA). Crude bacteriocin was precipitated from the CFS by 50% (w/v) ammonium sulfate and the precipitates were kept overnight at 4 °C with gentle stirring. After centrifugation (at 12,000×g for 30 min at 4 °C), the resulting pellets were dissolved in 20 mM sodium phosphate buffer (pH 6.5) and dialyzed overnight at 4 °C against distilled water using dialysis membrane (molecular weight cutoff, 1,000; Spectrum Labs., Gardena, CA, USA). The antimicrobial activity of the bacteriocin was assayed by microtiter plate assay (Holo et al., 1991), using histamine-forming bacteria as an indicator strain. In brief, each well of the microtiter plate (BD Falcon, Franklin Lakes, NJ, USA) was filled with 800 µL of BHI broth, 100 µL of a twofold serial dilution of the crude bacteriocin solution or PBS (pH 7.0), and 100 µL of cell cultures (1.0×10^5 CFU/mL) of indicator organism grown in BHI broth. The plates were then placed in an aerobic incubator for 12 h at 37 °C. The extent of growth inhibition was measured spectrophotometrically at 600 nm using microplate reader (Bioteck, Inc., Korea). One bacteriocin unit (BU/mL) was arbitrarily defined as the reciprocal of the highest dilution inhibiting the 50% growth of the indicator strain compared to the control.

Inhibitory effect of the bacteriocin on histamine accumulation

Histamine-forming bacteria isolated from the samples were seeded in TSB enriches with L-histidine hydrochloride monohydrate (0.5%) and pyridoxal-HCl (0.0005%) and incubated at 35 °C for 24 h. The cultures (1 mL) were transferred to test tubes containing the same broth (9 mL) and bacteriocin solution (100 and 200 BU/mL) from the LAB isolates and incubated for 24 h at 35 °C. These broth cultures were centrifuged (7000×g, 10 min, 4 °C) and filtered through a 0.22-µm membrane filter. The histamine content in the cultures was performed according to the procedure modified by Eerola et al. (1993) and Mah and Hwang (2003) using high-pressure liquid chromatography (HPLC, Hitachi, Tokyo, Japan). Briefly, 1 mL of the cell cultures or standard histamine solutions was added to 9 mL of 0.4 M perchloric acid (Merck, Darmstadt, Germany) and vigorously mixed. The mixture was then centrifuged at 3000×g for 10 min and the supernatant was filtered through Whatman paper No. 1. The samples (1 mL) were mixed with 2 M sodium hydroxide (200 µL) and saturated sodium bicarbonate solution (300 µL). Subsequently, the samples were added with 2 mL of 10 mg/mL dansyl chloride (Sigma-Aldrich) solution prepared in acetone and the mixture solution was incubated at 40 °C for 45 min. Residual dansyl chloride was removed by addition of 25% ammonium hydroxide (100 µL). After incubation for 30 min at room temperature, the volume of mixture was adjusted to 5 mL with acetonitrile. Finally, the dansyl derivatives were centrifuged at 2500×g for 5 min, and the supernatant was filtered through a 0.22-µm filter. A Nova-Pak C_{18} column (150 mm × 3.9 mm, Waters, Milford, MA, USA) was used for the separation of histamine, and acetonitrile (solvent B) as the mobile phases at the flow rate of 1 mL/min. The sample volume injected was 20 µL, and the eluted sample was monitored at 254 nm. All experiments were performed in triplicate. Data are expressed as means ± standard deviation (SD) and analyzed with SPSS program (ver. 12.0, SPSS Inc., Chicago, IL, USA). Then data comparisons were performed with paired t test and differences were considered statistically significant when P values were <0.05.

Physical and chemical properties of the bacteriocn

The effect of pH on antimicrobial activity of the crude bacteriocin was determined by incubating the crude bacteriocin in various buffers [0.1 M acetate buffer (pH 2.0–5.0), 0.1 M phosphate buffer (pH 6.0–7.0), and Tris HCl (pH 8.0–10.0)] with pH ranging from 2 to 10 for 24 h at 37 °C. To evaluate the heat stability, the crude bacteriocin was heated for 10 min at 80, 100, and 120 °C. Sensitivity of the crude bacteriocin to proteolytic enzymes was examined by incubation of the sample for 1 h at 37 °C after treatment with the following enzymes (1 mg/mL): proteinase K in 50 mM Tris-HCl (pH 7.5), trypsin in 50 mM Tris-HCl (pH 8.0), pepsin in 100 mM

Trish-HCl (pH 3.0), papain in 50 sodium phosphate acetate (pH 5.0), and α-chymotrypsin in 20 mM Tris-HCl (pH 8.0). Following incubation, the enzymes were heat inactivated for 3 min at 100 °C. The residual bacteriocin activity was determined by the microtiter plate assay as described earlier. In addition, the bacteriocin treated under each physical or chemical condition were added to TSB containing L-histidine hydrochloride monohydrate (0.5%) and pyridoxal-HCl (0.0005%) and inoculated with initial population of approximately 1.0×10^5 CFU/mL of histamine-forming bacteria. After incubation for 24 h at 35 °C, histamine levels in the cultures were analyzed using HPLC described above. All the experiments were done in triplicate.

Results and discussion
Isolation and identification of histamine-forming bacteria and bacteriocin-producing LAB
The results of the identification of histamine-forming bacteria and bacteriocin-producing LAB found in *Myeolchi-jeot* are presented in Table 1. Based on 16S rRNA gene sequences, histamine-forming bacteria were identified as *B. licheniformis* MCH01, *Serratia marcescens* MCH02, *Staphylococcus xylosus* MCH03, *Aeromonas hydrophila* MCH04, and *Morganella morganii* MCH05, with 98.0–99.9% similarity to the GenBank database. Meanwhile, 53 LAB strains isolated from *Myeolchi-jeot* were screened for the inhibition activity against histamine-forming bacteria, using a microtiter plate assay. Out of these strains, five strains (9.4%) presented 98.3–100.0% similarity with the 16S rRNA sequences reported for *Pediococcus acidilactici* MCL11, *Leuconostoc mesenteroides* MCL12, *Enterococcus faecium* MCL13, *Lactobacillus sakei* MCL14, and *Lactobacillus acidophilus* MCL15 in the GenBank database.

Several fish species including tuna, mackerel, sardines, and anchovy contain high levels of free histidine in their tissues. These fishes subjected to temperature abuse after the catch and before consumption can cause formation of histamine from histidine by bacterial histidine

decarboylases (Visciano et al., 2012). Histamine-rich foods may cause food intolerance in sensitive individuals and histamine poisoning that is a common seafood-borne disease causing various symptoms such as rash, nausea, vomiting, fever, diarrhea, headache, itching, flushing, and abdominal cramps (Taylor, 1986). *Proteus vulgaris*, *Proteus mirabilis*, *Clostridium perfringens*, *Enterobacter aerogenes*, *Klebsiella pneumonia*, *Hafnia alvei*, and *Vibrio alginolyticus* and enterobacteriaceae have been isolated from fish and described as the predominant histamine-forming bacteria (Shalaby, 1996; Kung et al., 2009). Moreover, there have been several reports describing amino acid decarboxylation activity of different genera, such as *Acinetobacter*, *Aeromonas*, *Bacillus*, *Cedecea*, *Citrobacter*, *Clostridium*, *Escherichia*, *Klebsiella*, *Plesiomonas*, *Proteus*, *Pseudomonas*, *Salmonella*, *Serratia*, *Shigella*, and *Vibrio*, and of some lactic acid bacteria (Kim et al., 2009). Our data are partially consistent with results reported by Guan et al. (2011) and Kung et al. (2009). The commonly isolated genera from *Myeolchi-jeot* were *Microbacterium* sp., *Kocuria* sp., *Vibrio* sp., *Psychrobacter* sp., *Halomonas* sp., *Brevibacillus* sp., *Bacillus* sp., *Enterococcus* sp., *Tetragenococcus* sp., *Weissella* sp., *Lactobacillus* sp., and *Staphylococcus* sp. (Guan et al., 2011). Among bacterial communities in Jeotgal, *M. morganii*, *K. pneumoniae*, and *H. alvei* have been known as the strains causing scombroid poisoning (often called "histamine poisoning") (Kung et al., 2009). In addition, *Staphylococcus epidermidis* obtained during the ripening of anchovies showed a powerful histamine-forming activity (Hernández-Herrero et al. 1999).

Bacteriocin activity of the isolates against histamine-forming bacteria
As shown in Table 2, the five LAB strains isolated from *Myeolchi-jeot* were found to produce an antibacterial compound with inhibitory activity against histamine-producing bacteria such as *B. licheniformis* MCH01, *S. marcescens* MCH02, *S. xylosus* MCH03, *A. hydrophila* MCH04, or *M. moarganii* MCH05. The crude bacteriocin of *P. acidilactici* MCL11 inhibited the growth of *B.*

Table 1 Identification of histamine-forming bacteria and lactic acid bacteria isolated from *Myeolchi-jeot* using 16S rRNA gene sequence analysis

Histamine-forming bacteria				LAB			
Strain	Accession no. related strain in NCBI	Similarity (%)	Identification	Strain	Accession no. related strain in NCBI	Similarity (%)	Identification
MCH01	EF433410	98.6	*Bacillus licheniformis*	MCL11	EF059986	99.9	*Pediococcus acidilactici*
MCH02	EF194094	99.7	*Serratia marcescens*	MCL12	KF673541	100.0	*Leuconostoc mesenteroides*
MCH03	EU266748	99.9	*Staphylococcus xylosus*	MCL13	EU887814	99.2	*Enterococcus faecium*
MCH04	AY538658	99.0	*Aeromonas hydrophila*	MCL14	AB650590	98.3	*Lactobacillus sakei*
MCH05	AB680150	98.0	*Morganella morganii*	MCL15	KT222158	99.5	*Lactobacillus acidophilus*

Table 2 Antibacterial activity of the bacteriocin produced by lactic acid bacteria against histamine-forming bacteria isolated from *Myeolchi-jeot*

Indicator organism	Bacteriocin activity (BU/mL)				
	LAB				
	Pediococcus acidilactici MCL11	*Leuconostoc mesenteroides* MCL12	*Enterococcus faecium* MCL13	*Lactobacillus sakei* MCL14	*Lactobacillus acidophilus* MCL15
Bacillus licheniformis MCH01	256	ND	ND	ND	ND
Serratia marcescens MCH02	ND	128	ND	256	ND
Staphylococcus xylosus MCH03	512	ND	64	ND	ND
Aeromonas hydrophila MCH04	ND	ND	256	ND	64
Morganella morganii MCH05	ND	ND	ND	128	ND

ND not detected

licheniformis MCH01 and *S. xylosus* MCH03. In particular, *P. acidilactici* MCL11 showed the strongest antimicrobial activity (512 BU/mL) against *S. xylosus* MCH03. The antimicrobial compound produced by *L. mesenteroides* MCL12 (128 BU/mL) and *L. sakei* MCL14 (256 BU/mL) strains showed activity against *S. marcescens* MCH02. The crude bacteriocin produced by *E. faecium* MCL13 showed an antimicrobial activity of 64 and 256 BU/mL against *S. xylosus* MCH03 and *A. hydrophila* MCH04, respectively. *L. acidophilus* MCL15 was also found to have a bacteriocin activity against *A. hydrophila* MCH04, which was weaker than that of *E. faecium* MCL13.

Joosten and Nuñez (1996) reported that the bacteriocin-producing enterococci and *L. lactis* strains completely inhibited the outgrowth of histamine producer *L. buchneri* St2A; therefore, no histamine formation was detected in the cheeses made with bacteriocin-producing starters. Our results are in agreement with the findings of previous studies. Gómez-Sala et al. (2015) demonstrated that analysis of 1245 LAB isolates obtained from fish, seafood, and fish products showed that 197 exerted direct antimicrobial activity against 20 spoilage and food-borne pathogenic microorganisms. Furthermore, LAB isolates selected on the basis of their direct antimicrobial activity were identified as *E. faecium*, *E. faecalis*, *Pediococcus pentosaceus*, *Weissella cibaria*, *L. sakei* subsp. *carnosus*, *L. sakei* subsp. *sakei*, *Lactobacillus curvatus*, and *L. mesenteroides* subsp. *cremoris* based on 16S rDNA sequences. *L. mesenteroides* HK4, HK5, and HK11 and *Streptococcus salivarius* HK8 strains isolated from Jeotgal were also chosen by preliminary bacteriocin activity test (Cho and Do, 2006). *L. lactis* subsp. *lactis* VR84 produced nisin Z induced the death of the histamine producing strain *S. thermophilus* PRI60. However, *L. lactis* subsp. *lactis* EG46-produced lacticin 481 did not show a lethal action against PRI60 strain, but were able to reduce its growth extent and histamine accumulation (Tabanelli et al., 2014). Furthermore, the bacteriocin produced by *L. casei* was able to inhibit the activity of the histamine-forming bacteria such as *Pseudomonas* sp.,

Proteus morganii, and *Micrococcus* sp. (Nugrahani et al., 2016). *S. xylosus* no. 0538 obtained from a salted and fermented anchovy (*Myeolchi-jeot*) possessed not only the greater capability to degrade histamine but a detectable ability to degrade tyramine as well. In addition, this strain was also found to produce the bacteriocin-like inhibitory substance(s) and have the highest antimicrobial activity against *B. licheniformis* strains defined as amine producers. *S. xylosus* no. 0538 exhibited significantly greater ability to degrade histamine, degrading histamine to about 62–68% of its initial concentration within 24 h (Mah and Hwang, 2009).

Reduction of histamine accumulation by the bacteriocin of LAB

As shown in Table 3, the five histamine-producing bacteria isolated from *Myeolchi-jeot* had strong ability to produce histamine. Among the tested strains, the highest level of histamine (2869.4 ± 49.0 mg/L) formation was observed for the *M. morganii* MCH05 strain. In addition, the concentration of histamine produced by *S. xylosus* MCH03 (2257 ± 30.7 mg/L) strain was higher than that produced by *B. licheniformis* MCH01 (1699.3 ± 35.6 mg/L), *S. marcescens* MCH02 (1987.2 ± 27.8 mg/L), and *A. hydrophila* MCH04 (1655.5 ± 41.2 mg/L). Meanwhile, the treatment with 100 and 200 BU/mL of the bacteriocin obtained from the tested LAB significantly reduced the histamine content of the five histamine-producing bacteria. After 24 h of incubation in the presence of the bacteriocin (200 BU/mL) of *P. acidilactici* MCL11, histamine contents of *B. licheniformis* MCL01 and *S. xylosus* MCH03 were reduced by 49 and 27%, respectively, as compared to the control (without bacteriocin). The histamine accumulation of *S. xylosus* MCH03 and *A. hydrophilia* MCH04 at 24 h of incubation in the presence of the bacteriocin of *E. faecium* MCL13 was significantly inhibited compared with the control group. The bacteriocin obtained from *L. sakei* MCL14 was effective in degrading histamine produced by *S. marcescens* MCH02 and *M. morganii* MCH05 strains. These bacteriocins produced by *L. mesenteroides* MCL12

Table 3 Inhibitory effect of the bacteriocin from the tested LAB on histamine accumulation of histamine-forming bacteria isolated from *Myeolchi-jeot*

Histamine-forming bacteria	Histamine content (mg/L)											
	Control	Bacteriocin-producing LAB										
		MCL11		MCL12		MCL13		MCL14		MCL15		
		Bacteriocin concentration (BU/mL)										
		100	200	100	200	100	200	100	200	100	200	
Bacillus licheniformis MCH01	1699.3 ± 35.6	1086.9 ± 31.2*	863.2 ± 22.3*	1668.4 ± 45.4	1681.5 ± 24.8	1679.5 ± 28.1	1673.2 ± 15.9	1667.3 ± 40.0	1702.5 ± 12.4	1671.5 ± 23.6	1715.2 ± 39.1	
Serratia marcescens MCH02	1987.2 ± 27.8	1979.1 ± 16.3	1995.8 ± 20.5	1423.5 ± 14.1*	1016.3 ± 10.7*	1964.3 ± 30.5	1990.5 ± 9.9	1619.1 ± 9.8*	1426.3 ± 20.4*	1964.0 ± 13.8	1980.5 ± 34.1	
Staphylococcus xylosus MCH03	2257.4 ± 30.7	1805.8 ± 21.7*	1653.7 ± 10.1*	2234.1 ± 13.2	2294.0 ± 17.5	754.3 ± 12.5*	518.5 ± 16.1*	2219.8 ± 29.4	2273.4 ± 26.0	2281.3 ± 41.2	2270.3 ± 11.6	
Aeromonas hydrophila MCH04	1655.5 ± 41.2	1603.8 ± 30.8	1682.3 ± 25.5	1690.3 ± 33.4	1682.5 ± 26.1	1518.2 ± 17.6*	1305.8 ± 20.4*	1633.9 ± 8.7	1690.5 ± 22.8	909.7 ± 11.4*	825.5 ± 18.8*	
Morganella morganii MCH05	2869.4 ± 49.0	2805.3 ± 10.2	2811.5 ± 20.7	2897.0 ± 36.1	2815.9 ± 7.5	2900.5 ± 16.4	2822.4 ± 13.8	1915.4 ± 30.7*	1688.5 ± 21.2*	2809.9 ± 18.4	2894.6 ± 33.8	

Data are presented as mean ± standard deviation (SD) from three independent experiments

*Significantly differ ($P < 0.05$) from the control group by paired t test

and *L. acidophilus* MCL15 strains also exhibited excellent histamine degradation ability. The histamine degradation ability of these bacteriocins increased in a concentration-dependent manner. The reduction of histamine content by treatment with these bacteriocins may be related to antagonistic activity of these antibacterial substances against histamine-producing bacteria.

Prolific histamine formers in Indian anchovy were identified as *M. morganii*, *P. vulgaris*, and *E. aerogenes*, and produced high histamine content of 104.1–203.0 mg/ 100 mL (Rodtong et al., 2005). Kim et al. (2009) noted that the histamine contents in fish, squid, and shellfish samples remarkably increased up to 36.6–2123.9 mg/kg after 24 h of storage at 25 °C, while the contents began to gradually increase after 2–3 days of storage at 4–10 °C. The dominant microbial group in these samples was enterobacteria throughout the storage period. In our results, the histamine-producing isolates from *Myeolchi-jeot* were identified as *B. licheniformis*, *S. marcescens*, *S. xylosus*, *A. hydrophila*, and *M. morganii* and these strains were capable of producing histamine in the range of 1655.5 to 2869.4 mg/L. Consequently, in raw fish, histamine content is linked to the type of histamine-forming bacteria, the type of seafood, and temperature/time storage conditions (Visciano et al., 2012).

Regarding to the inhibition of histamine formation, Zaman et al. (2011) observed that *Staphylococcus carnosus* FS19 and *Bacillus amyloiquefaciens* FS05 isolated from fish sauce which possess amine oxidase activity were found to be effective in reducing biogenic amine accumulation, and histamine concentration was reduced by 27.7 and 15.4% by FS19 and FS05, respectively, which is in disagreement with our observations that the LAB isolates tested in our study showed histamine degradation capacity by production of the antimicrobial substance such as bacteriocin. The histamine levels produced by *B. licheniformis* MCH01, *S. marcescens* MCH02, *S. xylosus* MCH03, *A. hydrophila* MCH04, and *M. morganii* MCH05 were reduced by 21–77% in the presence of the bacteriocin (200 BU/mL) obtained from the tested LAB.

Effect of enzymes, pH, and temperature on antimicrobial activity of the bacteriocin

These bacteriocins obtained from *P. acidilactici* MCL11 and *L. sakei* MCL14 remained stable after incubation at pH 4.0–8.0. However, no the bacteriocin activity was recorded under extremely acidic condition (pH 2.0). The bacteriocin of *L. mesenteroides* MCL12 and *E. faecium* MCL13 remained stable after incubation for 24 h at pH from 4.0 to 10.0, but not when kept at pH 2.0. The bacteriocin of *L. acidophilus* MCL15 remained active after 24 h of exposure to pH values ranging from 3.0 to 9.0 at 37 °C. The activity of the bacteriocin produced by *P. acidilactici* MCL11, *L. mesenteroides* MCL12, and *E. faecium*

MCL13 remained almost completely after heating for 10 min at 80 °C; however, the bacteriocin activity of these strains was partially destroyed after 10 min at 100 °C. In particular, the bacteriocin produced by *E. faecium* MCL13 was inactivated after 10 min at 120 °C, whereas the inhibitory activity of *L. sakei* MCL14 was 100% stable to heat treatment at 100 °C for up to 10 min, and the antimicrobial activity of *L. acidophilus* MCL15 was not affected by the heat treatment for 10 min at 120 °C. The activity of the bacteriocin produced by *P. acidilactici* MCL11 was destroyed after treatment with papain, pepsin, and proteinase K, but not when treated with α-chymotrypsin and trypsin. However, the treatment with papain and proteinase K had no effect on the activity of the bacteriocin of *L. mesenteroides* MCL12. Treatment of the bacteriocin produced by *E. faecium* MCL13 with papain and trypsin did not result in any activity loss, but the bacteriocin activity partially inactivated by treatment with pepsin, proteinase K, and α-chymotrypsin. Meanwhile, the bacteriocin of *L. sakei* MCL14 was destroyed by all the proteolytic enzymes tested such as papain, pepsin, proteinase K, α-chymotrypsin, and trypsin. Complete inactivation in antimicrobial activity of the bacteriocin produced by *L. acidophilus* MC15 was observed after treatment with α-chymotrypsin and trypsin. The histamine-degrading ability of the bacteriocin inactivated by some physico-chemical treatments was significantly reduced compared to the control group (Table 4).

The bacteriocin activity of *P. acidilactici* MCL11 was not affected by treatment with α-chymotrypsin and trypsin, but was lost after incubation with proteolytic enzymes such as papain, pepsin, and proteinase K. This bacteriocin was stable at up to 80 °C for 10 min and was in the pH range of 4.0–8.0. This is in disagreement with results recorded for pediocin SA-1. Pediocin SA-1 from *P. acidilactici* NRRLB5627 was inhibitory to several food spoilage bacteria and food-borne pathogens such as *Listeria* spp. and found to be very effective against the anaerobic *Clostridium sporogenes* and *Clostridium thiaminolyticum* (Papagianni and Anastasiadou, 2009). This bacteriocin was heat stable for up to 60 min at 121 °C, not impaired even following incubation at 30 °C for 1 week at pH values ranging between 3.0 and 12.0, and found to be resistant to treatment with trypsin, α-chymotrypsin, pepsin, and papain, but not to proteinase K (Anastasiadou et al., 2008).

The bacteriocin produced by *L. mesenteroides* MCL12 showed slight resistance to proteolytic enzymes such as α-chymotrypsin and trypsin. Treatment of this bacteriocin with papain and proteinase K had no effect on activity. The activity of this bacteriocin did not decrease after heat treatment at 80 °C for 10 min and the stability of the antimicrobial activity was observed at pH range of 4.0 to 10.0. Characteristics of the bacteriocin produced by *L. mesenteroides* MCL12 were widely different from

Table 4 Effects of pH, temperature, and proteolytic enzymes on the antibacterial activity and histamine-degrading ability of the bacteriocin from the tested LAB

Treatment		MCL11[a]				MCL12[b]				MCL13[c]				MCL14[d]				MCL15[e]			
		BA (BU/mL)		HC (mg/L)		BA (BU/mL)		HC (mg/L)		BA (BU/mL)		HC (mg/L)		BA (BU/mL)		HC (mg/L)		BA (BU/mL)		HC (mg/L)	
		C	T	C	T	C	T	C	T	C	T	C	T	C	T	C	T	C	T	C	T
pH	2.0	512	ND	1156.4 ± 22.1	2251.4 ± 16.9	128	16	1217.5 ± 30.5	1898.2 ± 16.9	256	128	1236.3 ± 15.4	1476.2 ± 24.2	256	ND	1367.5 ± 11.8	1972.5 ± 13.5	64	32	926.5 ± 37.8	1325.2 ± 11.3
	3.0		128		1715.2 ± 26.5		32		1853.6 ± 15.2		256		1204.3 ± 18.2		64		1707.2 ± 21.3		64		1161.4 ± 19.2
	4.0–8.0		512		1208.2 ± 30.2		128		1198.5 ± 47.1		256		1247.0 ± 40.3		256		1388.2 ± 16.7		64		1185.7 ± 21.2
	9.0		256		1619.5 ± 22.5		128		1219.3 ± 38.2		256		1203.4 ± 32.1		128		1571.4 ± 22.4		64		1201.2 ± 34.2
	10.0		64		1906.1 ± 30.7		128		1199.6 ± 30.8		256		1216.4 ± 25.4		128		1586.1 ± 13.7		32		1334.0 ± 40.2
Heating	80 °C, 10 min		512		1178.9 ± 42.2		128		1203.5 ± 13.2		256		1207.3 ± 19.3		256		1375.1 ± 48.8		64		1194.4 ± 27.1
	100 °C, 10 min		128		1690.5 ± 30.4		64		1688.3 ± 14.5		32		1598.3 ± 15.4		256		1390.1 ± 50.2		64		1203.8 ± 13.6
	120 °C, 10 min		128		1709.2 ± 24.7		64		1674.2 ± 15.8		ND		1708.5 ± 33.6		64		1698.2 ± 20.2		64		1208.9 ± 22.5
Enzyme	Papain		ND		2268.1 ± 22.4		128		1228.5 ± 16.7		256		1200.3 ± 14.2		64		1671.3 ± 16.2		16		1526.3 ± 40.0
	Pepsin		32		1998.1 ± 13.2		ND		2015.0 ± 20.7		32		1602.2 ± 27.1		32		1776.2 ± 13.5		16		1533.7 ± 9.0
	Proteinase K		64		1922.2 ± 25.8		128		1200.4 ± 41.7		64		1579.2 ± 30.5		64		1726.3 ± 40.0		32		1329.2 ± 27.5
	α-Chymotrypsin		512		1047.1 ± 24.1		32		1844.6 ± 23.1		128		1399.5 ± 23.0		64		1715.3 ± 33.0		ND		1676.0 ± 48.2
	Trypsin		512		1188.2 ± 12.6		64		1680.5 ± 22.7		256		1214.1 ± 20.1		128		1599.2 ± 10.7		ND		1673.0 ± 33.1

BA bacteriocin activity, HC histamine content, C bacteriocin before physico-chemical treatment, T bacteriocin after physico-chemical treatment
[a] Activity of P. acidilactici MCL11 against S. xylosus MCH03
[b] Activity of L. mesenteroides MCL12 against S. marcescens MCH02
[c] Activity of E. faecium MCL13 against S. xylosus MCH03
[d] Activity of L. sakei MCL14 against A. hydrophila MCH04
[e] Activity of L. acidophilus MCL15 against M. morganii MCH05

those of the mesentericin Y105. Mesentericin Y105, the bacteriocin from *L. mesenteroides*, had a narrow inhibitory spectrum limited to *Listeria* genus. Neither the Gram-negative and other Gram-positive indicator bacteria nor the related LAB species were inhibited when tested by the well-diffusion assay. This bacteriocin exhibited excellent stability under heating and acidic conditions. However, all of the proteolytic enzymes such as pronase, proteinase K, trypsin, chymotrypsin, and pepsin totally inhibited the antimicrobial activity of this bacteriocin (Héchard et al., 1992).

The bacteriocin activity of *E. faecium* MCL13 was stable the pH range between 3.0 and 10.0. However, a reduction in the activity was observed when the bacteriocin was exposed to 100 °C for 10 min. The bacteriocin activity was not affected by the presence of papain and trypsin. Unlike our results, the bacteriocins produced by the *E. faecium* strain showed a broader spectrum of activity against indicator strains of *Enterococcus* spp., *Listeria* spp., *Clostridium* spp., and *Propionibacterium* spp. This bactercioin was inactivated by α-chymotrypsin, proteinase K, trypsin, pronase, pepsin, and papain, but not by lipase, lysozyme, and catalase. The bacteriocin was heat stable and displayed highest activity at neutral pH (Toit et al., 2000).

Jiang et al. (2012) reported that sakacin LSJ618 produced by the strain *L. sakei* LSJ618 exhibited inhibitory activity against food-spoiling bacteria and food-borne pathogens, including the Gram-positive *L. monocytogenes*, *Staphylococcus aureus*, *Sarcina* spp., *Micrococcus luteus*, and the Gram-negative *Proteus* spp. and *Escherichia coli*, but not against most of the LAB tested. This bactercioin was completely inactivated by pepsin, papain, trypsin, and lipase, was stable between pH 2.0 and 8.0, and was heat resistant (30 min at 121 °C), which is partially in agreement with our observations. The bacteriocin of *L. sakei* MCL14 retained their activity at pH 4.0–8.0 and was thermally stable over a wide temperature range up to 100 °C for 10 min. Also, the bacteriocin activity was unstable after treatment with all the proteolytic enzymes like papain, pepsin, proteinase K, α-chymotrypsin, and trypsin.

The activity of the bacteriocin produced by *L. acidophilus* MCL15 was stable a pH range between 2.0 and 10.0, and remained constant after heating at 80, 100, and 120 °C for 10 min. However, the bacteriocin activity was destroyed or diminished after treatment with proteolytic enzymes such as papain, pepsin, proteinase K, α-chymotrypsin, and trypsin. The same results were recorded for the bacteriocin of *L. acidophilus* IBB 801. Acidophilin 801 obtained from *L. acidophilus* IBB 801 strain displayed a narrow inhibitory spectrum, being active particularly towards closely related lactobacilli and two Gram-negative pathogenic bacteria including *E. coli* Row and *Salmonella panama* 1467, whereas acidolin

and acidophilin produced by *L. acidophilus* strains showed a wide inhibitory spectrum against Gram-positive and Gram-negative bacteria. The antimicrobial activity of acidophilin 801 was insensitive to catalase but sensitive to proteolytic enzymes such as trypsin, proteinase K, and pronase, heat-stable (30 min at 121 °C), and maintained in a wide pH range (pH 3.0–10.0) (Zamfir et al., 1999).

Conclusion

In conclusion, these bacteriocins produced from the LAB isolates (P. *acidilactici* MCL11, *L. mesenteroides* MCL12, *E. faecium* MCL13, *L. sakei* MCL14, and *L. acidophilus* MCL15) may be a useful as a food biopreservative for controlling microbial deterioration, enhancing the hygienic quality, and extending the shelf-life of fish and seafood products. Notably, these antimicrobial substances may play a role in inhibiting histamine formation in the fermented fish products and preventing seafood-related food-borne disease caused by bacterially-generated histamine.

Abbreviations

LAB: lactic acid bacteria; *B. licheniformis*: *Bacillus licheniformis*; *S. marcescens*: *Serratia marcescens*; *S. xylosus*: *Staphylococcus xylosus*; *A. hydrophila*: *Aeromonas hydrophila*; *M. morganii*: *Morganella morganii*; *P. acidilactici*: *Pediococcus acidilactici*; *L. mesenteroides*: *Leuconostoc mesenteroides*; *E. faecium*: *Enterococcus faecium*; *L. sakei*: *Lactobacillus sakei*; *L. acidophilus*: *Lactobacillus acidophilus*; GRAS: generally recognized as safe; CFS: cell-free supernatant; BU: bacteriocin unit; *K. pneumonia*: *Klebsiella pneumonia*; *H. alvei*: *Hafnia alvei*; *P. vulgaris*: *Proteus vulgaris*; *E. aerogenes*: *Enterobacter aerogenes*; *L. buchneri*: *Lactobacillus buchneri*; *L. sakei*: *Lactobacillus sakei*; *L. lactis*: *Lactococcus lactis*; *S. thermophilus*: *Streptococcus thermophilus*; *L. monocytogenes*: *Listeria monocytogenes*; *E. coli*: *Escherichia coli*

Acknowledgements

We acknowledge the technical assistance provided by Lee Jong-Gab Professor.

Funding

This Research was supported by the Tongmyong University Research.

Competing interests

The author declares that he/she has no competing interests.

References

An H, Ben-Gigirey B. Scombrotoxin poisoning. In: Millar I, Gray D, Strachan N, editors. Microbiology of seafoods. London: Chapman & Hall Ltd.; 1998. p. 68–89.

Anastasiadou S, Papagianni M, Filiousis G, Ambrosiadis I, Koidis P. Pediocin SA-1, and antimicrobial peptide from *Pediococcus acidilactici* NRRL B 5627: Production conditions, purification and characterization. Biores Technol. 2008;99:5384–90.

Bover-Cid S, Holzapfel WH. Improved screening procedure for biogenic amine production by lactic acid bacteria. Int J Food Microbiol. 1999;53:33–41.

Brillet A, Pilet MF, Prevost H, Cardinal M, Leroi F. Effect of inoculation of *Carnobacterium divergens* V41, a bio-preservative strain against *Listeria monocytogenes* risk, on the microbiological, chemical and sensory quality of cold-smoked salmon. Int J Food Microbiol. 2005;104:309–24.

Chen HC, Kung HF, Chen WC, Lin WF, Hwang DF, Lee YC, Tsai YH. Determination of histamine and histamine-forming bacteria in tuna dumpling implicated in a food-borne poisoning. Food Chem. 2008;106:612–8.

Cho GS, Do HK. Isolation and identification of lactic acid bacteria isolated from a traditional Jeotgal production in Korea. Ocean Sci J. 2006;41:113–9.

Eerola S, Hinkkanen R, Lindfors E, Hirvi T. Liquid chromatographic determination of biogenic amines in dry sausages. J AOAC Int. 1993;76:575–7.

Gálvez A, Abriouel H, López RL, Omar NB. Bacteriocin-based strategies for food biopreservation. Int J Food Microbiol. 2007;120:51–70.

Garciz-Ruiz A, González-Rompinelli EM, Bartolomé B, Moreno-Arribas V. Potential of wine-associated lactic acid bacteria to degrade biogenic amines. Int J Food Microbiol. 2011;148:115–20.

Gómez-Sala B, Muńoz-Atienza E, Sánchez J, Basanta A, Herranz C, Hernández PE, Cintas LM. Bacteriocin production by lactic acid bacteria isolated from fish, seafood and fish products. Eur Food Res Technol. 2015;241:341–56.

Guan L, Cho KH, Lee JH. Analysis of the cultivable bacterial community in Jeotgal, a Korean salted and fermented seafood, and identification of its dominant bacteria. Food Microbiol. 2011;28:101–13.

Héchard Y, Dérijard B, Letellier F, Cenatiempo Y. Characterization and purification of mesentericin Y105, and anti-*Listeria* bacteriocin from *Leuconostoc mesenteroides*. J Gen Microbiol. 1992;138:2725–31.

Hernández-Herrero MM, Roig-Sagués AX, Rodriguez-Jerez JJ, Mora-Ventura MT. Halotolerant and halophilic histamine-forming bacteria isolated during the ripening of salted anchovies (Engraulis encrasicholus). J Food Protect. 1999;5:509–14.

Holo H, Nilssen O, Nes IF. Lactococcin A, a new bacteriocin from *Lactococcus lactis* subsp. cremoris: isolation and characterization of the protein and its gene. J Bacteriol. 1991;173:3879–87.

Jiang J, Shi B, Zhu D, Cai Q, Chen Y, Li J, Qi K, Zhang M. Characterization of a novel bacteriocin produced by *Lactobacillus sakei* LSJ618 isolated from traditional Chinese fermented radish. Food Control. 2012;23:338–44.

Joosten HM, Nuńez M. Prevention of histamine formation in cheese by bacteriocin-producing lactic acid bacteria. Appl Environ Microbiol. 1996;62:1178–81.

Kim MK, Mah JH, Hwang HJ. Biogenic amine formation and bacterial contribution in fish, squid and shellfish. Food Chem. 2009;116:87–95.

Kung HF, Wang TY, Huang YR, Lin CS, Wu WS, Lin CM, Tsai YH. Isolation and identification of histamine-forming bacteria in tuna sandwiches. Food Control. 2009;20:1013–7.

López-Sabater EI, Rodriguez-Jerez JJ, Hernández-Herrero M, Mora-Ventura MT. Evaluation of histidine decarboxylase activity of bacteria isolated from sardine (Sardina pilchardus) by an enzymatic method. Lett Appl Microbiol. 1994;19:70–5.

Mah JH, Hwang HJ. Effects of food additives on biogenic amine formation in Myeolchi-jeot, a salted and fermented anchovy (Engraulis japonicas). Food Chem. 2003;114:168–73.

Mah JH, Hwang HJ. Inhibition of biogenic amine formation in a salted and fermented anchovy by *Staphylococcus xylosus* as a protective culture. Food Control. 2009;20:796–801.

Mah JH, Han HK, Oh YJ, Kim MG, Hwang HJ. Biogenic amine in jeotkals, Korean salted and fermented fish products. Food Chem. 2002;79:239–43.

Nugrahani A, Anik H, Hariati M. Characterization of bacteriocin *Lactobacillus casei* on histamine-forming bacteria. J Life Sci Biomed. 2016;6:15–21.

Papagianni M, Anastasiadou S. Pediocins: the bacteriocins of Pediococci. Sources, production, properties and applications. Microb Cell Fact. 2009;8:1–16.

Rodtong S, Nawong S, Yongsawatdigul J. Histamine accumulation and histamine-forming bacteria in Indian anchovy (Stolephorus indicus). Food Microbiol. 2005;22:475–82.

Santos MHS. Biogenic amines: their importance in foods. Int J Food Microbiol. 1996;29:213–31.

Schillinger U, Geisen R, Holzapfel WH. Potential of antagonistic microorganisms and bacteriocins for the biological preservation of foods. Trends Food Sci Technol. 1996;7:158–64.

Shalaby AR. Significance of biogenic amines to food safety and human health. Food Res Int. 1996;29:675–90.

Tabanelli G, Montanari C, Bargossi E, Lanciotti R, Gatto V, Felis G, Torriani S, Gardini F. Control of tyramine and histamine accumulation by lactic acid bacteria using bacteriocin forming lactococci. Int J Food Microbiol. 2014;190:14–23.

Taylor SL. Histamine food poisoning: toxicology and clinical aspects. Crit Rev Toxicol. 1986;17:91–117.

Toit MD, Franz CMAP, Dicks LMT, Holzapfel WH. Preliminary characterization of bacteriocins produced by *Enterococcus faecium* and *Enterococcus faecalis* isolated from pig faeces. J Appl Microbiol. 2000;88:482–94.

Visciano P, Schirone M, Tofalo R, Suzzi G. Biogenic amines in raw and processed seafood. Front Microbiol. 2012;3:1–10.

Zacharof MP, Lovitt RW. Bacteriocins produced by lactic acid bacteria. APCBEE Procedia. 2012;2:50–6.

Zaman MZ, Bakar FA, Jinap S, Bakar J. Novel starter cultures to inhibit biogenic amines accumulation during fish sauce fermentation. Int J Food Microbiol. 2011;145:84–91.

Zamfir M, Callewaert R, Cornea PC, Savu L, Vatafu I, De Vuyst L. Purification and characterization of a bacteriocin produced by *Lactobacillus acidophilus* IBB 801. J Appl Microbiol. 1999;87:923–31.

6

Butyrate and taurine exert a mitigating effect on the inflamed distal intestine of European sea bass fed with a high percentage of soybean meal

Simona Rimoldi[1], Giovanna Finzi[2], Chiara Ceccotti[1], Rossana Girardello[1], Annalisa Grimaldi[1,3], Chiara Ascione[1] and Genciana Terova[1,3]* (iD)

Abstract

Background: Due to the paucity of oceanic resources utilized in the preparation of diets for cultured fish, commercial feed producers have been trying to replace fishmeal (FM) using alternative protein sources such as vegetable protein meals (VMs). One of the main drawbacks of using VMs in fish feed is related to the presence of a variety of anti-nutritional factors, which could trigger an inflammation process in the distal intestine. This reduces the capacity of the enterocytes to absorb nutrients leading to reduced fish growth performances.

Methods: We evaluated the mitigating effects of butyrate and taurine used as feed additives on the morphological abnormalities caused by a soybean meal (SBM)-based diet in the distal intestine of sea bass (*Dicentrarchus labrax*). We used three experimental diets, containing the same low percentage of FM and high percentage of SBM; two diets were supplemented with either 0.2% sodium butyrate or taurine. Histological changes in the intestine of fish were determined by light and transmission electron microscopy. Infiltration of CD45+ leucocytes in the lamina propria and in the submucosa was assessed by immunohistochemistry. We also quantified by One-Step Taqman® real-time RT-PCR the messenger RNA (mRNA) abundance of a panel of genes involved in the intestinal mucosa inflammatory response such as *TNFα* (tumor necrosis factor alpha) and interleukins: *IL-8*, *IL-1β*, *IL-10*, and *IL-6*.

Results: Fish that received for 2 months the diet with 30% soy protein (16.7% SBM and 12.8% full-fat soy) developed an inflammation in the distal intestine, as confirmed by histological and immunohistochemistry data. The expression of target genes in the intestine was deeply influenced by the type of fish diet. Fish fed with taurine-supplemented diet displayed the lowest number of mRNA copies of *IL-1β*, *IL-8*, and *IL-10* genes in comparison to fish fed with control or butyrate-supplemented diets. Dietary butyrate caused an upregulation of the *TNFα* gene transcription. Among the quantified interleukins, *IL-6* was the only one to be not influenced by the diet.

Conclusions: Histological and gene expression data suggest that butyrate and taurine could have a role in normalizing the intestinal abnormalities caused by the SBM, but the underling mechanisms of action seem different.

Keywords: Aquaculture, Intestinal inflammation, TEM, Light microscopy, Gene expression, Cytokines, Interleukins, Soybean meal, Butyrate, Taurine

* Correspondence: genciana.terova@uninsubria.it
[1]Department of Biotechnology and Life Sciences (DBSV), University of Insubria, Via Dunant, 3-21100 Varese, Italy
[3]Inter-University Centre for Research in Protein Biotechnologies "The Protein Factory", Polytechnic University of Milan and University of Insubria, Varese, Italy
Full list of author information is available at the end of the article

Background

Due to the paucity of oceanic resources utilized in the preparation of diets for cultured fish, commercial feed producers have been trying to replace fishmeal (FM) by using alternative protein sources such as vegetable protein meals (VMs) (Tacon and Metian 2008). Vegetable products include soy, canola (rape), corn, wheat, cottonseed, lupin, sunflower, flax linseed, and peas. Of these, soybeans are promising because of their higher protein content, higher digestibility, and better amino acid profile in comparison to other grains and oilseeds (Hardy 1999; Gatlin et al., 2007; Zhang et al., 2014).

The main drawbacks of using plant-derived proteins in fish feed are related to the low level of indispensable amino acids (in particular, methionine and lysine) and to the presence of a wide variety of anti-nutritional factors, such as saponins, lectins, phytate, trypsin inhibitors, phenols, and tannins (Francis et al. 2001), which could damage the intestinal tract thus reducing nutrient absorption and fish growth (Van den Ingh et al. 1991; Knudsen et al. 2008; Urán et al. 2008a, b).

Indeed, studies on Atlantic salmon (*Salmo salar*), rainbow trout (*Oncorhychus mykiss*), common carp (*Cyprinus carpio*), and gilthead sea bream (*Sparus aurata*) have indicated that the inclusion of less-refined plant products such as soybean meal (SBM) in the diet triggers an inflammation process in the distal intestine, referred to as SBM-induced enteropathy (Baeverfjord and Krogdahl 1996; Olli and Krogdahl 1994; Van den Ingh et al. 1991; Knudsen et al. 2008; Urán, et al. 2008a, b; Venou et al. 2006). Although the provoking mechanism remains to be established, the SBM-induced enteritis is believed to be caused by a disruption of the intestinal barrier, with subsequent exposure of otherwise shielded layers of the mucosa to luminal ingredients, including food-derived and microbial antigens (Romarheim et al. 2011). The typical signs of such inflammation are a shortening of the primary and secondary intestinal mucosal folds, an increase in the number of Goblet cells, and the infiltration of inflammatory cells, particularly macrophages and eosinophilic granulocytes into the lamina propria. This situation reduces the capacity of the enterocytes lining the epithelium to absorb nutrients (Baeverfjord and Krogdahl 1996; Van den Ingh et al. 1991; Buttle et al., 2001). These effects proved to be dose-dependent in Atlantic salmon; the worst symptoms were observed at the highest inclusion level (30%), but even diets containing as low as 7.6% SBM produced morphological changes at the intestinal level (Krogdahl et al. 2003).

It has been well documented that the responsible for the off-target effects of SBM is not the soy protein but other components present in the SBM, such as the anti-nutritional factor saponin, in combination with at least one or more unidentified components (Van den Ingh et al. 1991; Baeverfjord and Krogdahl 1996; Bakke-Mckellep et al. 2000; Krogdahl et al. 2003; Knudsen et al. 2008).

Once an inflammatory immune response is initiated, proinflammatory cytokines and a panel of chemokines (secreted proteins that play a major role in the inflammatory response) instigate the coordinated expression of downstream genes (Martin et al. 2006; Kortner et al. 2012; Grammes et al. 2013; Sahlmann et al. 2013; Marjara et al. 2012; Krogdahl et al. 2015; De Santis et al. 2015). The expression of pro- and anti-inflammatory cytokine genes was quantified by Urán et al. (2008a) in the isolated intraepithelial lymphocytes of common carp (*Cyprinus carpio* L.), in which an intestinal inflammation was observed. The enteropathy was developed when carp continuously fed on animal protein were transferred to a diet in which 20% of the protein was replaced by SBM. After 3 weeks of feeding, the pro-inflammatory interleukin 1β (*IL-1β*) and tumor necrosis factor α1 (*TNF-α1*) genes were upregulated whereas the anti-inflammatory interleukin 10 (*IL-10*) was downregulated after an initial upregulation at the first week of feeding.

The effects of replacement of FM with VM, often accompanied by reduced fish performance, are not restricted to SBM inclusion solely but have been observed after inclusion of many plant protein sources in several teleost species such as gilthead sea bream, turbot, Atlantic cod, and parrot fish (Gomez-Requeni et al. 2004; Sitja-Bobadilla et al. 2005; Yun et al. 2011; Hansen et al. 2007; Lim and Lee 2009). Baeza-Ariño et al. (2014) described liver and gut alterations of gilthead sea bream, *S. aurata* L., fed on diets in which FM was replaced by a mixture of rice and pea protein concentrates. The results of the histological analysis showed significant changes in the case of the 90% substitution in parameters such as thickness of the gut layers, number of Goblet cells, and villi's length and thickness, whereas the integrity of the gut structure was not significantly affected by a diet with up to 60% of replacement. In some cases, severe vacuolization was encountered, which consequently deformed enterocytes and displaced the nucleus.

The short chain fatty acid of butyrate may promote the healing of inflamed intestine through its major role in enhancing epithelial cell proliferation and differentiation and in improving the intestinal absorptive function (Canani et al. 2012; Gálfi and Neogrády 2002; Wong et al. 2006). Like other short chain or volatile fatty acids (acetic, propionic, valeric, and caproic), butyric acid is produced during the fermentation of dietary fibers by the anaerobic microbiota associated with the epithelium of the animals' digestive tract. In addition to being the main respiratory fuel source of the intestinal bacteria, and preferred to glucose or glutamine, this four-carbon chain organic acid molecule has potential immunomodulatory and anti-inflammatory properties (Vinolo et al. 2009; Toden et al. 2007; Terova et al. 2016), and exert multiple other beneficial effects on host energy metabolism (Hamer et al. 2007; Den Besten et al. 2013; Liu et al. 2014; da Silva et al. 2016). Although the mechanisms underlying these effects are still enigmatic and subject of intense scrutiny, it is believed that they encompass the complex interplay between diet, gut microbiota, and host energy metabolism.

However, much of the research on butyrate has been focused on terrestrial vertebrates, including humans, while very few studies have been conducted in fish. In particular, little is known about the effects of butyrate used as a feed additive on fish intestinal integrity. In terrestrial farmed animals such as pig and chicken, butyrate included in the diet has had a positive influence on body weight gain, feed utilization, and composition of intestinal microflora. It exerted trophic effects on the intestinal epithelium through an increase in the villi length and crypt depth, too (Gálfi and Bokori 1990; Kotunia et al. 2004; Hu and Guo 2007). In fish, Robles et al. (2013) reported an effect of butyrate used as a feed additive in increasing the availability of several essential amino acids and nucleotide derivatives, which have been demonstrated to increase fish growth when they are added individually to the diet.

Another feed additive that has been shown to have a restorative effect on soya saponin- and soya lectin-induced enteritis in fish (Iwashita et al. 2008, 2009) is the organic acid taurine. A number of recent studies reviewed by El-Sayed (2014) and Salze and Davis (2015) have demonstrated the essentiality of dietary taurine for many commercially relevant cultured species, including marine teleosts. According to these studies, taurine is involved in many physiological functions in fish and represents an essential nutrient, which exerts powerful antioxidant and anti-inflammatory properties, and is required as a supplement in the feed when a relevant percentage of vegetable protein sources are utilized. Taurine is a neutral β-amino acid found in high concentrations in animal tissues, whereas plants contain less than 1% of the taurine levels found in animals. In mammals, taurine is mainly synthesized in liver and brain through enzymatic oxidation and direct conversion of cysteine (derived from methionine) to cysteine sulfinic acid and hypotaurine. In this pathway, the enzymes cysteine dioxygenase (CDO) and cysteinsulfinate decarboxylase (CDS) have important roles (Griffith 1987; El-Sayed 2014). Therefore, in animal taxa with limited or no activity of the CDS, such as marine fish species, a continuous supply of taurine should be provided with the diet (El-Sayed 2014; Johnson et al. 2015; López et al. 2015). Indeed, Yokoyama et al. (2001), by comparing the CSD activity

and hypotaurine production in several teleost and mammalian species, found that the only fish species with a high CSD activity were rainbow trout and tilapia. However, even in these species, the enzyme activity was an order of magnitude lower than that found in mammalian species. Other data (Salze and Davis 2015) indicate that both species benefit from dietary taurine supplementation thus suggesting that levels of CSD activity are insufficient to provide the necessary amounts of taurine for maximum growth of these species.

Accordingly, the objective of the present research was to evaluate the mitigating effects of butyrate and taurine used as feed additives on the morphological abnormalities caused by a SBM-based diet in the intestine of European sea bass (*Dicentrarchus labrax*). In order to attain this aim, we utilized both light and transmission electron microscopy (TEM) to determine intestinal changes in fish. We also quantified the intestinal messenger RNA (mRNA) abundance of several genes involved in the mucosal inflammatory response such as *TNFα*, which is a cell-signaling protein (cytokine) that makes up the inflammatory acute-phase reaction, and interleukins such as *IL-8*, *IL-1β*, *IL-10*, and *IL-6*, which are well-known cytokines that regulate immune responses, inflammatory reactions, and hematopoiesis.

Methods

Feeding trial

The feeding experiment was conducted at the water recirculating system facility of the Department of Biotechnology and Life Sciences, University of Insubria, Varese, Italy. After being individually weighted and tagged, 30 European sea bass (*D. labrax*) of an average weight of 514 ± 67.4 g were transferred into six circular 750-L tanks with 5 fish/tank for 1-week acclimation before the start of the feeding trial. During the feeding trial, fish were fed two times a day at apparent satiety with three different diets (Table 1) used in duplicate (two tanks per diet).

Fish were sampled at the end of the feeding experiment, which lasted 60 days. The experimental tanks were connected to a 20-m^3 water recirculation system. The water parameters such as pH, temperature, and dissolved oxygen (DO) were strictly controlled during all the experiment. Temperature was maintained at 20 ± 0.5 °C; salinity at 22 g L^{-1}; pH 7.2; total N–NH$_3 \leq 0.1$ mg L^{-1}; N–NO$_2^- \leq 0.02$ mg L^{-1}; N–NO$_3^- \leq$ 5 mg L^{-1}, DO 8–8.5 mg L^{-1}, and DO saturation, over 97%.

Fish were euthanized by an overdose (320 mg/L at 22 °C) of anesthetic (tricaine-methasulfonate MS-222, Sigma-Aldrich, Italy) and then weighed individually. The specific growth rate (SGR), the feed conversion ratio (FCR), and the condition factor (K) were calculated using the following formula:

Table 1 Formulation (%), and proximate composition of experimental diets

	Diet		
	C	B	T
Ingredient (%)			
Full-fat soy	12.8	12.8	12.8
SPC	13.6	13.6	13.6
Wheat	8.0	8.0	8.0
Wheat gluten meal	8.19	8.19	8.19
DCP	1.72	1.72	1.72
Mixed oil[a]	12	10	12
Lysine (98%)	0.29	0.29	0.29
Vitamins and mineral premix[b]	0.4	0.4	0.4
Corn gluten	16.0	16.0	16.0
Soybean meal (48%)	16.7	16.7	16.7
Fish meal (65%)	10.0	10.0	10.0
Anti molds	0.1	0.1	0.1
Taurine	–	–	0.2
Sodium butyrate	–	0.2	–
Filler	0.2	–	–
Proximate composition (%)			
Crude protein	45.0	45.0	45.0
Fat	16.0	16.0	16.0
Fiber	2.3	2.3	2.3
Ash	6.4	6.4	6.4
Total calcium	1.0	1.0	1.0
Total phosphorus	0.95	0.95	0.95
Methionine	0.9	0.9	0.9
Methionine + cysteine	1.6	1.6	1.6
Lysine	2.3	2.3	2.3

SPC soy protein concentrate, *DCP* dicalcium phosphate
[a]Mixed oil: 40% corn 60% soy
[b]Vitamin and mineral premix (quantities in 1 kg of mix): vitamin A, 4,000,000 IU; vitamin D3, 800,000 IU; vitamin C, 25,000 mg; vitamin E, 15,000 mg; inositol, 15,000 mg; niacin, 12,000 mg; choline chloride, 6000 mg; calcium pantothenate, 3000 mg; vitamin B1, 2000 mg; vitamin B3, 2000 mg; vitamin B6, 1800 mg; biotin, 100 mg; manganese, 9000 mg; zinc, 8000 mg; iron, 7000 mg; copper, 1400 mg; cobalt, 160 mg; iodine 120 mg; anticaking and antioxidant + carrier, making up to 1000 g

$$\text{SGR } (\%/\text{day})$$
$$= 100 \times \frac{[\ln (\text{final body weight}) - \ln (\text{initial body weight})]}{\text{days}};$$

$$\text{FCR} = \frac{\text{dry feed intake}}{\text{wet weight gain}};$$

$$K = 100 \left[\frac{\text{wet weight(g)}}{\text{total length(cm)}^3} \right].$$

The distal intestine of all sampled fish (six fish/diet, three from each tank) was dissected out. A small part of it was fixed in 4% buffered formalin, pH 7.2, for histological, immunohistochemical or electron microscopy examination, whereas the remaining part was conserved at −80 °C for gene expression analyses.

The experimental feed
We used three experimental diets (C, B, and T), which contained the same low percentage of fishmeal and high percentage of soybean meal, as shown in Table 1. Diets B and T were supplemented with 0.2% of sodium butyrate or taurine, respectively, whereas Diet C, which represented the control diet, was not supplemented.

Light microscopy
Sampled intestines were fixed by immersion in 4% buffered formal, dehydrated with a graded ethanol series (20, 30, 50, 70, and 95%), embedded in paraffin, and sectioned at 4 μm. After dewaxing and rehydrating the sections, they were consecutively stained with hematoxylin-eosin. Periodic Acidic Schiff (PAS) and Alcian blue stain at pH 2.5 were used to reveal neutral and acidic mucosubstances, respectively.

The thickness of intestinal wall was measured by analysing six intestines/diet and using the average value obtained from 3 tissue sections/intestine ($n = 6$). The length of muscle layer and mucosa was measured (in micrometers) by using the Image J software package and then the percentages were calculated by referring to the thickness of the whole intestine wall. Statistical differences were calculated by ANOVA followed by Duncan's post hoc test, and $P < 0.05$ was considered as level of significance. Statistical analysis was performed using Statistica 7.0 software (StatSoft Inc., Tulsa, OK, USA).

Immunohistochemistry for leukocyte assessment
Paraffin sections of distal intestine were dewaxed and rehydrated with a standard step of decreasing ethanol series. After washing with PBS, sections were pre-incubated for 30 min in blocking solution (2% bovine serum albumin, BSA, 0.1% Tween in PBS) and then incubated overnight at 4 °C with the polyclonal antibody rabbit anti-human CD45 (GenScript NJ, USA) diluted 1:100 in blocking solution. The washed specimens were incubated for 1 h at room temperature with the secondary antibody goat anti-rabbit (dilution 1:100 in PBS/BSA) conjugated with alkaline phosphatase (Jackson, Immuno Research Laboratories, West Grove, PA, USA). Secondary antibody alkaline phosphatase conjugated was visualized using 5-Bromo-4-Chloro-3-Indolyl Phosphate/Nitro Blue Tetrazolium tablets (BCIP/NBT, Sigma). Mounted slides were examined under an optical microscope (Nikon Eclipse Ni, Nikon, Tokyo, Japan). Images were acquired with a DS-5 M-L1 digital camera system (Nikon). Control sections were incubated in PBS/BSA

without the primary antibody. To avoid false positives due to the chromogen reaction with endogenous alkaline phosphatase, sections were treated with CH_3COOH 20% applied for 15 min at 4 °C.

Electron microscopy

Specimens already fixed in formalin were transferred to a 2% Karnovsky solution at 4 °C for 2 h, post-fixed in osmium tetroxide for 1 h, and then embedded in Epoxy resins. Ultra-thin sections were stained with uranyl acetate and lead citrate and examined with a TEM Morgagni Philips/FEI electron microscope.

One-Step Taqman® real-time RT-PCR quantification of target genes' mRNA copies

Preparation of total RNA and first-strand (cDNA) synthesis

Total RNA was extracted from 125 mg of sea bass distal intestine (6 fish/diet). Tissue lysis and homogenization were performed in special disposable sterile tubes (GentleMACS M tubes™, MiltenyiBiotec), in order to minimize the possibility of cross-contamination between samples, and using the gentleMACSDissociator (MiltenyiBiotec). After an automated purification process by using the Maxwell® 16 Instrument and Maxwell® 16 Tissue LEV total RNA purification Kit (Promega, Milan, Italy), the RNA was isolated. This RNA purification procedure provides an easy method for efficient, automated purification of highly concentrated total RNA from up to 25 mg of tissue. Purified RNA for up to 16 samples is obtained in less than 45 min of hands-free instrument operation. No post-purification treatment with nuclease, cleanup, or concentration is required to achieve superior performance in downstream applications. The low elution volume (30–100 μL) is used to generate purified RNA in a more concentrated format for use in downstream applications such as qRT-PCR, RT-PCR, and complementary DNA (cDNA) synthesis.

The integrity and purity of total RNA were determined by a spectrophotometer NanoDrop ™ (Thermo Scientific), measuring the absorbance at 260 nm and the absorbance ratio 260/280, respectively. The integrity of RNA was verified by electrophoresis on 1% agarose gel stained with ethidium bromide.

Reverse transcription of 1 μg total RNA from each fish intestine tissue was performed with random decamers in a volume of 20 μL by using the High-Capacity cDNA Archive Kit (ThermoFisher Scientific, Milan, Italy) following the manufacturer's instructions.

Generation of in vitro-transcribed mRNAs for target genes

The number of transcript copies of target genes *IL-1β* (GenBank acc. nr. AJ269472), *IL-6* (GenBank AM490062), *IL-8* (AM490063), *IL-10* (AM268529), and *TNFα* (DQ070246) was quantified by One-Step Taqman® real-

time RT-PCR technique using the standard curve method. Standard curves were constructed using the known copy number of synthetic mRNAs of each gene. To obtain synthetic mRNAs, a forward and a reverse primer were designed based on the mRNA sequence of each gene and used to create templates for the in vitro transcription of mRNAs (Table 2).

The forward primer was engineered to contain a T3 phage polymerase promoter gene sequence to its 5′ end and used together with the reverse primer in a conventional RT-PCR of total RNA extracted from the sea bass intestine.

PCR products were then evaluated on a 2.5% agarose gel, cloned using pGEM®-T Easy cloning vector system (Promega, Milan, Italy), and subsequently sequenced. In vitro transcriptions were performed using T3 RNA polymerase and other reagents supplied in the Promega RiboProbe In Vitro Transcription System kit according to the manufacturer's protocol.

Generation of standard curves

The mRNAs produced by in vitro transcription were used as quantitative standards in analyzing experimental (biological) samples (Terova et al. 2009). Defined amounts of mRNAs at tenfold dilutions were subjected to real-time PCR using One-Step Taqman® EZ RT-PCR Core Reagents (ThermoFisher Scientific, Milan, Italy), including 1× Taqman® buffer, 3 mM Mn(OAc)2, 0.3 mM dNTP except dTTP, 0.6 mM dUTP, 0.3 μM forward primer, 0.3 μM reverse primer, 0.2 μM FAM-6 (6-carboxyfluorescein-labeled probe), 5 units rTH DNA polymerase, and 0.5 units AmpErase® UNG enzyme in a 25-μL reaction volume. RT-PCR conditions were 2 min at 50 °C, 30 min at 60 °C, and 5 min at 95 °C,

Table 2 Sequences and melting temperatures (Tm) of the primers used for in vitro synthesis of standard mRNAs

Gene	GenBank acc. nr.	Nucleotide sequence (5′–3′)	Tm (°C)
IL-1β	AJ269472	F: gtaatacgactcactatagggTGC CATGGAGAGACTGAAGG	>70
		R: ACTGGGTGTACGGTCCAAGT	60.7
IL-6	AM490062	F: gtaatacgactcactatagggACTT CCAAAACATGCCCTGA	>70
		R: CCGCTGGTCAGTCTAAGGAG	59.5
IL-8	AM490063	F: gtaatacgactcactatagggTCAGT GAAGGGATGAGTCTGA	>70
		R: CTCGGGGTCCAGGCAAAC	60.3
IL-10	AM268529	F: gtaatacgactcactatagggCAGT GCTGTCGTTTTGTGGA	>70
		R: TCACTCTTGAGCTGGTCGAAG	59.7
TNFα	DQ070246	F: gtaatacgactcactatagggCACTA CACACTGAAGCGCAT	>70
		R: CTGTAGCTGTCCTCCTGAGC	59.5

followed by 40 cycles consisting of 20 s at 94 °C, and 1 min at 62 °C. The Ct values obtained by amplification were used to create standard curves for target genes.

Quantification of IL-1β, IL-6 IL-8, IL-10, and TNFa in biological samples

One hundred nanograms of total RNA extracted from the experimental intestines was subjected to One-step Taqman® quantitative real-time RT-PCR, in parallel to tenfold-diluted defined amounts of standard mRNA, under the same experimental conditions as used to establish the standard curves. Real-time Assays-by-Design SM PCR primers and gene-specific fluorogenic probes were designed by Life Technologies. Primer sequences and TaqMan® probe used for each target gene amplification are shown in Table 3. TaqMan® PCR reactions were performed on a StepOne Real-Time PCR System (Life Technologies). To reduce pipetting errors, master mixes were prepared to set up duplicate reactions (2 × 30 μL) for each sample.

Sample quantification

Data from the TaqMan® PCR runs were collected with StepOne™ Software v 2.0. Cycle threshold (Ct) values corresponded to the number of cycles at which the fluorescence emission monitored in real time exceeded the threshold limit. The Ct values were used to create standard curves to serve as a basis for calculating the absolute amounts of mRNA in total RNA.

Statistical analyses

All statistical analysis were performed by using the software IBM SPSS Statistics 21. Growth performance data were submitted to two-way analysis of variance (two-way ANOVA) considering both tank and diet effects. One-way ANOVA was applied to analyze the morphology and gene expression data. In all statistical analyses, each individual fish was considered as an experimental unit. Duncan's test was used for post hoc analysis. Differences were considered to be statistically significant when $P < 0.05$.

Results

Fish growth

As reported in Table 4, statistical analysis by two-way ANOVA showed significant differences in fish growth performance parameters (SGR, and FCR) related to diet, whereas no tank effect was observed. Growth performances of fish fed for 60 days with diet T were significantly higher ($P < 0.05$) than in fish fed with diets C and B, whereas there were no significant differences between fish fed diet C and diet B. The condition factor (K) of fish fed with diet T was significantly higher ($P < 0.05$) than in fish fed with diet B, whereas there were no significant differences between fish fed diets C and T and diets C and B.

Light microscopy and immunohistochemistry

By light microscopy, we evaluated different aspects of the various anatomical portions of the distal intestine: the thickness of muscle and mucosal layers and the shape and length of the villi of the intestinal mucosa. In addition, we paid attention to the presence and distribution of Goblet

Table 3 Primers and probes used for one-step quantitative real-time RT-PCR

Gene	Nucleotide sequence (5′–3′)
IL-1β	F: TTGTGTTTGAGCGCGGAACA
	R: TGTCGGTCACGCTGCATTG
	Probe: 6-Fam-CTCCAACAGCGCAGTACAGCAAGC-BHQ1
IL-6	F: GCCTGCTCACTTACACAGCTCTTC
	R: TCTTGAAACTGTGGCCCTCTGA
	Probe: 6-Fam-AGAAGGAGTCCCCCAGCTCGATCCG-BHQ1
IL-8	F: CTGTCGCTGCATCCAAACAGA
	R: GCAATGGGAGTTAGCAGGAATCAG
	Probe: 6-Fam-AGCAAACCCATCGGCCGCCAC-BHQ1
IL-10	F: AGCGCTGCTAGACCAGACTGT
	R: CGGCAGAACCGTGCTTAGAT
	Probe: 6-Fam-AGACACTTTAAAGACCCCGTTCGCTTGC-BHQ1
TNFa	F: AAACCGGCCTCTACTTCGTCTA
	R: TCCCGCACTTTCCTCTTCAC
	Probe: 6-Fam-AGCCAGGCGTCGTTCAGAGTCTCC-BHQ1

Table 4 Growth performance indexes of European sea bass fed three different diets during the feeding trial

	Diet C	Diet B	Diet T
Biomass (g) ± SD			
Initial	508.20 ± 60.02	502.60 ± 73.82	528.00 ± 67.76
Final	546.80 ± 61.14	535.20 ± 62.51	641.20 ± 60.95
SGR (%/day)	0.12 ± 0.08[a]	0.11 ± 0.05[a]	0.32 ± 0.12[b]
FCR	2.92 ± 1.54[a]	2.81 ± .0.51[a]	1.30 ± 0.25[b]
K	1.11 ± 0.11[ab]	1.07 ± 0.03[b]	1.23 ± 0.08[a]

Fish were fed for 60 days with three experimental diets (C, B, and T), which contained the same low percentage of fishmeal and high percentage of soybean meal. Diets B and T were supplemented with 0.2% of sodium butyrate or taurine, respectively, whereas Diet C (control), was not supplemented. Values are mean ± SD ($n = 10$). Statistical differences were calculated by two-way ANOVA followed by Duncan's post hoc test. The values in each row with different superscript letters are significantly different ($P < 0.05$)

$$\text{SGR } (\%/\text{day}) = 100 \times \frac{[\ln (\text{final body weight}) - \ln (\text{initial body weight})]}{\text{days}}$$

$$\text{FCR} = \frac{\text{dry feed intake}}{\text{wet weight gain}}$$

$$K = 100 \left[\frac{\text{wet weight(g)}}{\text{total length(cm)}^3} \right]$$

cells and to the presence, size, and content of supranuclear vacuoles of enterocytes. The presence and extent of any inflammatory infiltrates were also examined.

In the distal intestines of sea bass fed with diet C, the thickness of the intestinal wall was formed by 48% of muscle and 52% of villi (Fig. 1a). In enterocytes, we observed abundant vacuoles of various sizes not regularly aligned, eosinophils and PAS-positive, occupying much of the cytoplasm and forcing the nucleus at the base of the cell (Fig. 1b, c). We observed a few Alcian blue positive Goblet cells that were uniformly distributed (Fig. 1c). An important leukocyte infiltrate CD45$^+$ was found in the lamina propria, at the level of the submucosa and also at the level of the muscle layers (Fig. 1d).

In the intestines of sea bass receiving the diet B in which 0.2% of butyrate was added, the thickness of muscular layer in the intestinal wall was significantly reduced in comparison to fish fed on diet C, whereas the villi layer was elongated and irregularly shaped (Fig. 2a, b; Fig. 3). The intestinal villi were long and branched (Fig. 2a, b). Supranuclear vacuoles diverse in form, size, and content were observed (Fig. 2b). The enterocytes vacuoles were PAS-negative while numerous Goblet cells were found well distributed along the villi, as demonstrated specifically by Alcian blue staining (Fig. 2c). In several areas, the typical single layer showed an abnormal pseudostratification (Fig. 2c). An inflammatory infiltrate CD45$^+$ was identified by immunohistochemistry in the lamina propria and in the submucosa (Fig. 2d).

In the sea bass fed with diet T which was supplemented with 0.2% of taurine, the muscle layer and intestinal mucosa had a 1:1 ratio (Figs. 3 and 4a). Supranuclear vacuoles, characterized by a heterogeneous content, were evident (Fig. 4b). Alcian blue positive Goblet cells were abundant, preferentially concentrated in the medium-apical villi, while vacuoles were faintly PAS+ (Fig. 4c). The CD45$^+$ cell infiltrate was detected both in the lamina propria and in the submucosa, as confirmed by immunohistochemistry (Fig. 4d).

Electron microscopy

Further investigation by using transmission electron microscopy allowed us to highlight the cell differences between the three types of samples. Different aspects of the enterocytes were investigated. From the lumen to the lamina propria of the fold, we considered the brush border, the tubulo-vesicular system occupying the apical area of enterocytes, the supranuclear vacuoles, the nuclei, and infiltrating inflammatory cells.

The distal intestines of sea bass receiving diet C showed brush border areas with microvilli that appeared irregular and damaged (Fig. 5a). The supranuclear cytoplasm was occupied by numerous vacuoles of different dimensions, with an electrondense content from homogenous to granular (Fig. 5a, b). Some vacuoles showed an irregular contour, suggesting a rupture of their membrane (Fig. 5a). Images of fusion between vacuoles were observed (Fig. 5b). Some nuclei of enterocytes appeared pyknotic (Fig. 5c). An infiltrate of leukocytes was shown between epithelial cells (Fig. 5a, c).

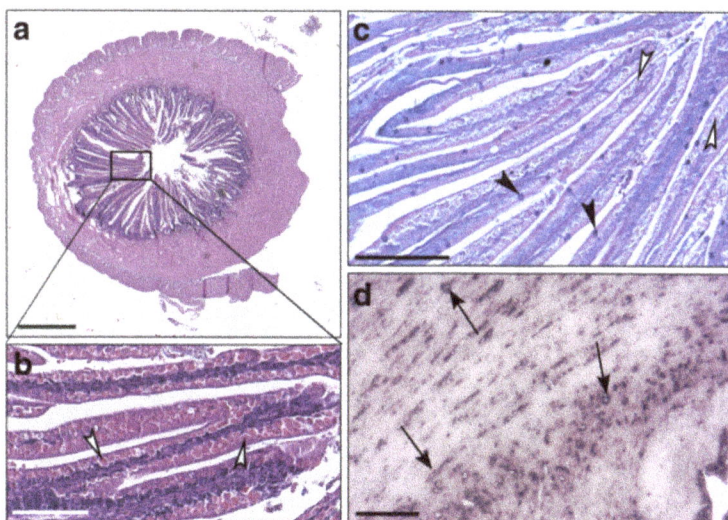

Fig. 1 Light microscope images obtained from the distal intestine of sea bass fed with diet C. **a** Morphology of a complete section of intestine at low magnification, hematoxylin-eosin staining. **b** In the epithelial skinfold, at higher magnification, numerous eosinophils vacuoles are visible (*white arrowheads*). **c** Alcian blue-PAS staining showing the presence of numerous PAS-positive enterocytes vacuoles (*white arrowheads*) and a few Alcian blue positive Goblet cells (*black arrowheads*). **d** Immunohistochemical analysis showing numerous infiltrating CD45$^+$ leukocytes in the lamina propria, submucosa, and muscle layers (*black arrows*). *Bar* in **a** 1 mm; *bar* in **b** 50 μm; *bars* in **c**, **d** 100 μm

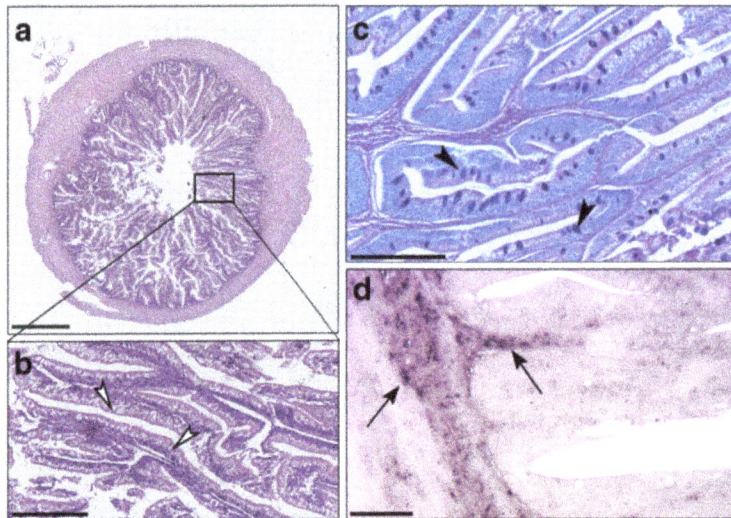

Fig. 2 Light microscope images obtained from the distal intestine of sea bass fed with diet B. **a** Morphology of a section of intestine, EE staining. **b** Higher magnification of intestinal villi showing their irregular structure and enterocytes supernuclear vacuoles (*white arrowheads*). **c** Alcian blue-PAS staining highlighting intraepithelial Goblet cells (*black arrowheads*). **d** Anti-CD45 immunohistochemistry showing numerous positive cells infiltrating in the submucosa and in the lamina propria (*black arrows*). *Bar* in **a** 1 mm; *bar* in **b** 50 μm; *bars* in **c**, **d** 100 μm

In the intestines of fish fed with diet B, the microvilli were well developed (Fig. 6a, c), even if sometimes interrupted for alteration of the apical plasma membrane (Fig. 6a). In the apical cytoplasm, an expanded tubulo-vesicular system was observed (Fig. 6c). In the supranuclear cytoplasm, many characteristic vacuoles that had an irregular shape and heterogeneous content, with clear areas mixed with dense material, were evident (Fig. 6b, c). Between enterocytes some Goblet cells were present (Fig. 6a).

Sections of distal intestines from fish fed with diet T presented microvilli, which were long and well-shaped, sometimes topped by a layer of mucus and a tubulo-vesicular system in the apical cytoplasm (Fig. 7a, c). In the supranuclear cytoplasm, there were many vacuoles with a heterogeneous content showing some lamellar profiles (Fig. 7b). Some Goblet cells were observed between enterocytes (Fig. 7a).

Gene expression analysis

The expression levels of the proinflammatory interleukins (*IL-1β*, *IL-6*, *IL-8*), *TNFα* and the anti-inflammatory interleukin *IL-10* genes were evaluated in the distal portion of sea bass intestine by One-Step Taqman® real-time RT-PCR. The results of this analysis (Table 5) showed that the expression of target genes in the intestine was deeply influenced by the type of fish diet.

Although *IL-1β* was expressed at relatively low levels in all analyzed fish, sea bass fed with diet T (supplemented with taurine), displayed the lowest number of mRNA copies of *IL-1β* gene in comparison to fish fed with the other two diets. Also, the

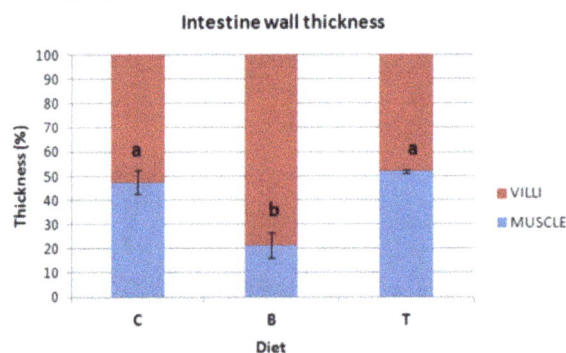

Fig. 3 Histogram showing thickness (in percentage) of different parts of intestinal wall in fish fed on different diets. The thickness of intestinal wall was measured by analysing the average value obtained from 3 tissue sections from each intestine (6 intestines/diet). Values are mean ± SD (*n* = 6). Statistical differences were calculated by ANOVA followed by Duncan's post-hoc test. *Different letters* indicate statistically significant differences (*P* < 0.05)

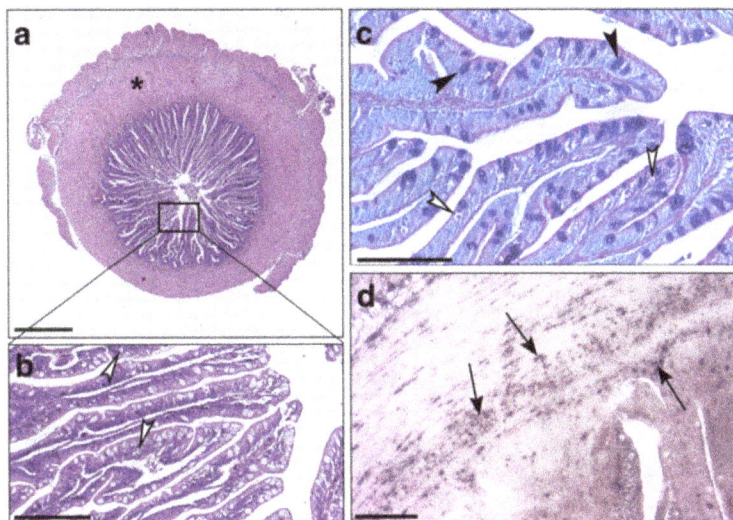

Fig. 4 Light microscope images obtained from the distal intestine of sea bass fed with diet T. **a** Morphology of the intestine observed at low magnification, EE staining. **b** At higher magnification, enterocytes vacuoles are visible (*white arrowheads*). **c** Alcian blue-PAS staining highlights abundant Alcian blue positive Goblet cells (*black arrowheads*) and some faintly PAS$^+$vacuoles (*white arrowheads*). **d** Immunohistochemistry showing a large number of lamina propria and submucosa infiltrating cells, CD45-positive (*black arrows*). Bar in **a** 1 mm; *bar* in **b** 50 μm; *bars* in **c**, **d** 100 μm

expression of *IL-8* and *IL-10* genes in fish fed diet T was significantly lower ($P < 0.05$) with respect to the fish that received control or butyrate-supplemented diets. Dietary butyrate caused an upregulation of *TNFα* gene transcription (Table 5); indeed, fish fed on diet B showed higher levels of TNFα mRNA copies than fish fed with the other diets ($P < 0.05$). Finally, among the interleukins quantified in this study, *IL-6* was the only one, to be not influenced by the diet (Table 5).

Discussion

The feeding experiment conducted in this study lasted 60 days which is a short period for SGR and FCR reliable evaluation. Nevertheless, SGR of fish fed with diet T was 2.7 times higher ($P < 0.05$) than in fish fed with diet C or B. This result is also in line with findings of Brotons Martinez et al. (2004) on sea bass fingerlings fed on a diet containing SBM and supplemented with taurine for 30 days. The dramatic differences in growth

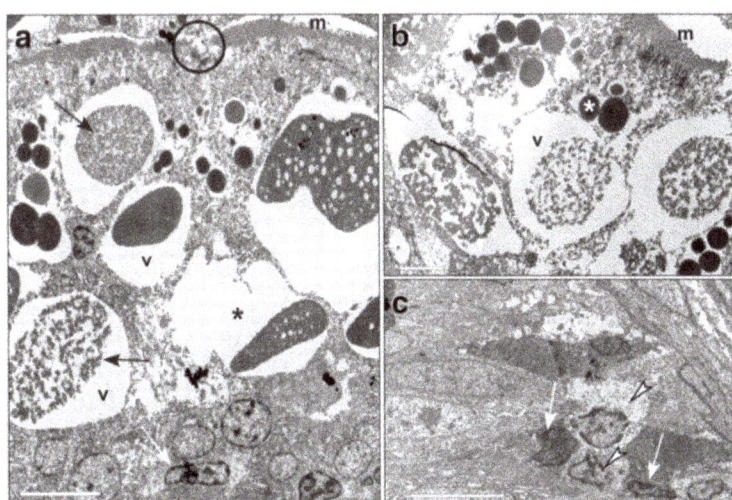

Fig. 5 Electron microscope images obtained from the distal intestine of sea bass fed with diet C. **a** A portion of the epithelium, with an advanced degeneration and enterocytes with focal alterations of the microvilli (*circle*), numerous supranuclear vacuoles with an electrondense content from homogenous to granular (*black arrows*), some of which are partially broken (*black asterisk*), and incipient leukocyte infiltration (*white arrow*). **b** At higher magnification, a small vacuole (*white asterisk*) pouring its contents into a larger one is visible. **c** Pyknotic nuclei of enterocytes (*white arrows*) and leukocyte infiltration between enterocytes (*white arrowhead*). v vacuoles, m microvilli. Bars in **a**, **c** 5 μm; bar in **b** 200 nm

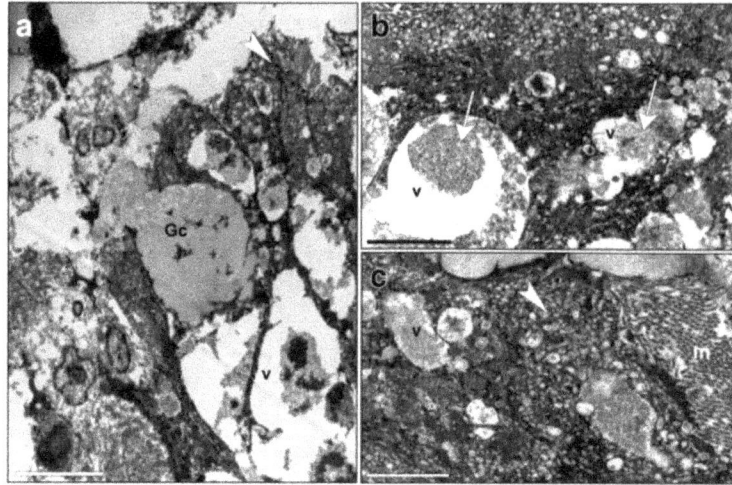

Fig. 6 Electron microscope images obtained from the distal intestine of sea bass fed with diet B. **a** In the apex portion, well developed, partly damaged microvilli are present (*white arrowhead*); in the median portion of the enterocytes, there are large heterogeneous vacuoles. The cytoplasm of a Goblet cell is also visible. **b** Vacuoles at higher magnification showing a heterogeneous content (*white arrows*). **c** Apical portion of a cytoplasm in which there are long microvilli and a well-developed tubulo-vesicular system (*white arrowhead*). Vacuoles with heterogeneous content are also visible. *v* vacuoles, *m* microvilli, *Gc* Goblet cell. *Bar* in **a** 5 μm; *bars* in **b**, **c** 200 nm

observed in our study between the taurine-supplemented group (T) and the experimental groups B and C may be due to taurine deficiency in the latter groups, thus indicating that taurine is an essential dietary nutrient, as reported in several studies reviewed by Salze and Davis (2015).

Histological samples from distal intestine of fish fed with diet B (supplemented with sodium butyrate) showed that the muscle layers were thinner than in fish fed with diets C or T. This suggests that butyrate facilitated the formation of softer fecal pellets in fish of this group. Furthermore, the complex structure of the intestinal villi included a high number of Goblet cells, and it was not pseudostratified which means that it was iperproliferative. This result is in line with previous studies (Gálfi and Bokori 1990; Kotunia et al. 2004; Hu and Guo 2007), which have demonstrated the trophic effects of butyrate on the intestinal epithelium of terrestrial animals through an increase in the villi length and crypt depth. The number of supranuclear vacuoles was less than in fish fed with diet C, and they were often heterogeneous. We observed a few intestinal Goblet cells, in

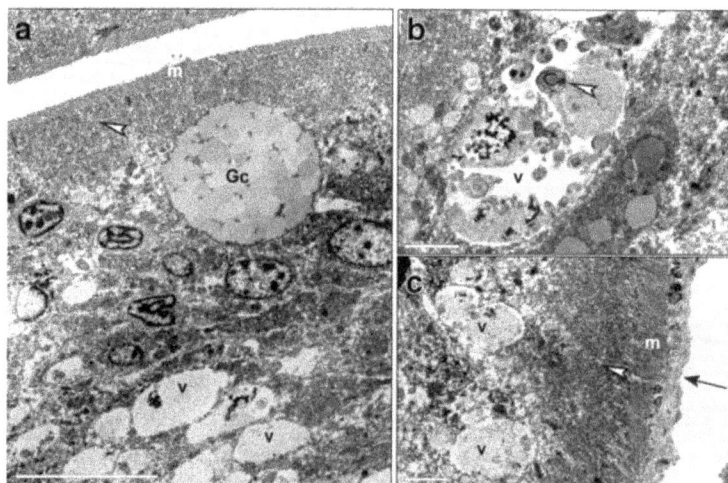

Fig. 7 Electron microscope images obtained from the distal intestine of sea bass fed with diet T. **a** The enterocytes show long microvilli, an apical tubulo-vesicular system (*white arrowhead*), and some vacuoles. Between enterocytes, a Goblet cell is observed. **b** An enterocyte vacuole containing lamellar body (*white arrowhead*) is visible. **c** Epithelium with well-developed microvilli covered by mucus (*black arrow*), a tubulo-vesicular system (*white arrowhead*), and supranuclear vacuoles. *v* vacuoles, *m* microvilli, *Gc* Goblet cell. *Bar* in **a** 10 μm; *bars* in **b**, **c** 200 nm

Table 5 Expression levels of pro and inflammatory interleukins genes (*IL-1β*, *IL-6*, *IL-8*, *IL-10*) and *TNFα* gene in the distal intestine of European sea bass fed different diets

Gene	Diet C	Diet B	Diet T
	Gene expression levels (transcript copies/ng total RNA)		
IL-1β	323.19 ± 33.55^a	321.30 ± 15.96^a	215.09 ± 41.94^b
IL-6	647.14 ± 325.05	873.26 ± 176.68	603.50 ± 163.84
IL-8	548.88 ± 196.73^a	763.04 ± 190.70^a	285.90 ± 63.58^b
IL-10	257.18 ± 66.55^a	326.67 ± 64.73^a	110.80 ± 67.70^b
TNFα	370.07 ± 104.81^b	664.20 ± 109.52^a	377.25 ± 118.41^b

The number of transcript copies of target genes was quantified by One-Step Taqman® real-time RT-PCR technique using the standard curve method. Values are mean ± SD ($n = 6$). The values in each row with different superscript letters are significantly different ($P < 0.05$)

fish fed with diet C with SBM, and this result is in agreement with other studies conducted in Atlantic salmon (*S. salar*) and rainbow trout (*O. mykiss*) (Baeverfjord and Krogdahl 1996; Iwashita et al. 2008). However, we found also high concentrations of absorptive vacuoles containing PAS-positive substances (probably glicoproteins, considering their acidophilic properties) in the supranuclear region of the enterocytes from fish fed with diet C. This could indicate an interruption of the digestive processes occurring within the enterocytes, since these cells are involved not only in the uptake of different nutrients but also in the digestion. For example, once inside the enterocyte, the absorbed di- and tripeptides are digested into amino acids by cytoplasmic peptidases and transported from the cell into blood; only a very small number of these small peptides enter blood intact. In contrast, other authors have reported a reduction of the same supranuclear vesicles in salmonids and in carps. In fish fed with diet C, lamina propria, muscle layer, and epithelial lining were heavily infiltrated with inflammatory cells, including macrophages and polymorphonuclear leukocytes. The important signs of inflammation in our samples were also confirmed by immunohistochemistry staining. This analysis showed a strong positivity for anti-CD45 antibody indicating a relevant diffusion of leukocytes in all the distal intestinal tissue. The fact that an antibody directed against the human protein reacted also with sea bass tissue might confirm a high structural conservation of this protein in fish as well. This hypothesis is also supported by the discovery of a CD45 homolog (Marozzi et al. 2012) in sea bass (*D. labrax*), whose sequence obtained from transcriptome of gills (kindly provided by Francesco Buonocore, University of Tuscia, Italy) shows 52% of identity with human CD45 isoform 1 precursor. Since a common rule of thumb is that two sequences are homologous if they share more than 30% of identity over their entire length (Pearson, 2013), we are confident that 52% is a high enough identity to support the similarity

between human and sea bass CD45 sequences. Of note, a 30% identity threshold for homology underestimates the number of homologs detected by sequence similarity between humans and yeast by 33% (Pearson 2013).

The TEM observations showed partly damaged microvilli and numerous large vacuoles occupying the supranuclear cytoplasm and containing a homogeneous electrondense substance, with partial break of their membranes probably leading to the release of lytic enzymes and recall of leukocytes. Some enterocytes with pyknotic nuclei were also observed. According to Noaillac-Depeyre and Gas (1973), the supranuclear body is of complex origin. Indeed, in the adult carp, the apical cytoplasm of the enterocytes of the medium intestine shows the presence of a dense tubolovesicular network that form a voluminous supranuclear body. However, the supranuclear bodies are not related to the presence of food in the intestinal lumen since they are also present in fasting animals. Therefore, it seems that they do not result only from the running together of the food vacuoles but probably arise from the fusion of the Golgi vesicles. The supranuclear bodies would thus have a dual origin, endogenous for the hydrolytic enzymes that contain, and exogenous for the di- and tripeptides supplied by the food which are then further digested inside them (Noaillac-Depeyre and Gas 1973).

Furthermore, the aforementioned acidophilic and PAS-positive substance that we found in the vacuoles of fish fed with diet C was absent in the distal intestine of fish fed with diet B, suggesting that butyrate supplementation of the diet have had a mitigation effect in the intestine of fish.

Although the inflammatory cells were infiltrated less than in the distal intestine of fish fed with diet C and were more concentrated at the mucosal and submucosa level, we found a significant number of leukocytes as confirmed by the positive staining with the CD45 antibody. TEM analysis of enterocytes showed a focal alteration at the microvilli level. However, other areas appeared much less compromised, microvilli were regularly displayed, and a developed apical tubulo-vesicular system was evident, suggesting an intense phagocitosys activity The cells showed supranuclear vacuoles with dishomogeneous content: partly dense, like that found in the fish of group C, and partly clear; this aspect suggests that in these vacuoles a partial digestion of the content occurs.

In fish fed with diet T, the muscle layers were rather developed. The villi folds were long and regular, with enterocytes showing a limited number of vacuoles and an increased number of Goblet cells confirming a strong inflammation reaction, which were mainly concentrated in the medioapical part of the fold. Leukocytes, immunoreactive to the CD45 antibody, were still present in both, the mucosa and the submucosa, similar to those observed in fish fed with diet B. The TEM images of enterocytes were very different from those in fish fed with diets C and B, with cells showing well-displayed microvilli frequently covered by a protective mucus layer, no alterations in the structure, an apical tubulo-vesicular system, and supranuclear vacuoles filled with a heterogeneous substance, including membranous bodies suggesting that an autophagic process was in progress. To our knowledge, no other example of taurine used as a feed additive causing autophagic processes has been reported in the literature, and thus this finding should be further investigated. However, this process seems to enable enterocytes of fish fed with diet T to cope with the negative effects induced by the SBM diet and contributes to the protection of the intestinal mucosa.

The expression pattern of molecular markers of inflammatory processes such as the inflammatory cytokines supported the light and TEM microscopy observations. The highest transcription level of the pro-inflammatory *IL-1β* gene was found in the distal intestine of sea bass fed with diet C, indicating a local tissue reaction to SBM insult, which was less evident in fish fed on diets B or T. Cytokine *IL-8* was less expressed in fish fed with diet T than in those fed with diet C or B, suggesting a protective action of taurine against oxidative stress through the reduction of the inflammatory cascade induced by oxidative stress. The transcription activity of *IL-10* gene, which is known as anti-inflammatory cytokine, was strongly reduced in fish fed with diet T, remaining ,however, relatively high in fish fed with diets C and B. Among the target cytokines, only *IL-6* gene expression was not affected by dietary butyrate or taurine supplementation. Proinflammatory IL-6 is of particular interest since it is expressed to a small extent in normal colonic epithelium but to a much greater degree in colonic carcinomas (Shirota et al. 1990). In human colonic cells, increased expression of both IL-6 and its receptor IL-6R have mainly an anti-apoptotic effect (Yuan et al. 2004). IL-6, indeed, upregulates the expression of anti-apoptotic protein Bcl-xl (Fujio et al. 1997; Scwarze and Hawly 1995). Yuan and colleagues (2004), clearly demonstrated that butyrate downregulates IL-6 signaling in human colonocytes by inhibiting IL-6 receptor rather than IL-6 expression. Similarly, in our study, butyrate did not influence *IL-6* expression, but we cannot exclude that, even in fish, butyrate could act principally on IL-6 receptor rather than in IL-6 expression. So, this is an issue that requires further investigation.

It is well documented that taurine inhibits the over-production of inflammatory mediators such as TNFα (Kim and Cha 2014; Locksley et al. 2001). TNFα is known to stimulate further production of itself and to induce the expression of other pro-inflammatory mediators, such as IL-6, IL-8. *TNFα* was expressed at higher levels in the intestine of sea bass fed with diet B, which induced an upregulation also of *IL-8* but not of *IL-6*. Moreover, higher expression of *TNFα* in fish receiving B diet supported the TEM results indicating a promotive effect of butyrate in increasing cellular turnover.

Our findings on cytokine activity are mainly in agreement with the literature. Indeed, in the intestine of carps receiving soy as protein source (Urán et al. 2008a), a peak of *IL-1β* was observed after the first week of feeding, and the expression of this gene as well as that of *TNFα* remained high over the control levels for the entire experiment. In the same study, a strong upregulation of *IL-10* was observed after one week of SBM feeding, but at weeks 3 and 5, expression levels were downregulated at values either lower or similar to the control (Urán et al. 2008a). In zebrafish fed on diets containing different amounts of SBM, Hedrera et al. (2013) reported a significant increase in the intestinal transcription of the mRNA coding for proinflammatory cytokines, such as *IL-1β* and *IL-8*. The transcriptional activity of *IL-10*, known to have an anti-inflammatory action, was increased in the group receiving feed with soy protein, too.

Conclusions

European sea bass fed on a diet containing a concentration of a soy protein source close to 30% (16.7% as SBM and 12.8 as full-fat soy) for 2 months developed an inflammatory status in the distal intestine. Both butyrate and taurine, at concentrations as low as 0.2%, had some mitigating effect on the symptoms of inflammation, but the underling mechanisms of action seem different. The mechanisms underlying these effects should be further investigated along with any possible synergistic effect of butyrate and taurine in the feed in normalizing the intestinal abnormalities caused by soybean meal.

Acknowledgements
CA is a student of the Ph.D. course in "Biotechnology, Bioscience and Surgical Technology" at the "Università degli Studi dell'Insubria", Varese, Italy.

Funding
This work was funded under the seventh EU Framework Programme by the ARRAINA project N°288925: Advanced Research Initiatives for Nutrition & Aquaculture. The views expressed in this work are the sole responsibility of the authors and do not necessarily reflect the views of the European Commission. The funding body had no role in the design of the study, collection, analysis, and interpretation of the data and in writing the manuscript.

Authors' contributions

SR carried out the molecular biology and statistical analysis and participated in the feeding trial. CC and CA carried out the feeding trial and participated in the sequence alignment. GF carried out the histological study and was involved in drafting the manuscript. RG and AG carried out the immunoassays and participated in the analysis and interpretation of the data. GT conceived of the study, participated in the experimental design, and wrote the manuscript. All authors read and approved the final manuscript.

Competing interests

The authors declare that they have no competing interests.

Author details

[1]Department of Biotechnology and Life Sciences (DBSV), University of Insubria, Via Dunant, 3-21100 Varese, Italy. [2]Department of Pathology, Ospedale di Circolo, Varese, Italy. [3]Inter-University Centre for Research in Protein Biotechnologies "The Protein Factory", Polytechnic University of Milan and University of Insubria, Varese, Italy.

References

Baeverfjord G, Krogdahl A. Development and regression of soybean meal induced enteritis in Atlantic salmon, Salmo salar, distal intestine: a comparison with the intestines of fasted fish. J Fish Dis. 1996;19(5):375–87.

Baeza-Ariño R, Silvia Martínez-Llorens S, Nogales-Mérida S, Jover-Cerda M, Tomás-Vidal A. Study of liver and gut alterations in sea bream, Sparus aurata, fed a mixture of vegetable protein concentrates. Aquacult Res. 2014;47(2):460-71.

Bakke-Mckellep M, Mcl Press C, Baeverfjord G, Krogdahl A, Landsverk T. Changes in immune and enzyme histochemical phenotypes of cell in the intestinal mucosa of Atlantic salmon, Salmo salar L., with soybean meal-induced enteritis. J Fish Dis. 2000;23:115–27.

Brotons Martinez J, Chatzifotis S, Divanach P, Toshiotakeuchi A. Effect of dietary taurine supplementation on growth performance and feed selection of sea bass, Dicentrarchus labrax, fry fed with demand-feeders. Fish Sci. 2004;70:74–9.

Buttle LG, Burrells AC, Good JE, Williams PD, Southgate PJ, Burrells C. The binding of soybean agglutinin (SBA) to the intestinal epithelium of Atlantic salmon, Salmo salar and rainbow trout, Oncorhynchus mykiss, fed high levels of soybean meal. Vet Immunol Immunopathol. 2001;80:237e44.

Canani RB, Di Costanzo M, Leone L. The epigenetic effects of butyrate: potential therapeutic implications for clinical practice. Clin Epigenetics. 2012;4(1):4.

da Silva BC, Vieira FN, Mourino JLP, Bolivar N, Seiffert WQ. Butyrate and propionate improve the growth performance of Litopenaeus vannamei. Aquac Res. 2016;47:612–23.

De Santis C, Bartie KL, Olsen RE, Taggart JB, Tocher DR. Nutrigenomic profiling of transcriptional processes affected in liver and distal intestine in response to a soybean meal-induced nutritional stress in Atlantic salmon (Salmo salar). Comp Biochem Physiol Part D Genomics Proteomics. 2015;15:1–11.

Den Besten G, Van Eunen K, Groen AK, Venema K, Reijngoud DJ, Bakker BM. The role of short-chain fatty acids in the interplay between diet, gut microbiota, and host energy metabolism. J Lipid Res. 2013;54:2325–40.

El-Sayed A-FM. Is dietary taurine supplementation beneficial for farmed fish and shrimp? a comprehensive review. Rev Aquac. 2014;6(4):241–55.

Francis G, Makker HPS, Becker K. Antinutritional factors present in plant-derived alternate fish feed ingredients and their effects in fish. Aquaculture. 2001;199: 197–227.

Fujio Y, Kunisada K, Hirota H, Yamauchi-Takihara K, Kishimoto T. Signals through gp130 upregulate bcl-x gene expression via STAT1-binding cis-element in cardiac myocytes. J Clin Invest. 1997;99:2898–905.

Gálfi P, Bokori J. Feeding trial in pigs with a diet containing sodium n-butyrate. Acta Vet Hung. 1990;38(1–2):3–17.

Gálfi P, Neogrády S. The pH-dependent inhibitory action of n-butyrate on gastrointestinal epithelial cell division. Food Res Int. 2002;34:581–6.

Gatlin DM, Barrows FB, Brown P, Dabrowski K, Gaylord TG, Hardy RW, Herman E, Hu G, Krogdahl A, Nelson R, Overturf K, Rust M, Sealey W, Skonberg D, Souza EJ, Stone D, Wilson R, Wurtele E. Expanding the utilization of sustainable plant products in aquafeeds: a review. Aquacult Res. 2007;38(6):551–79.

Gomez-Requeni P, Mingarro M, Calduch-Giner JA, Médale F, Martin SAM, Houlihan DF, Kaushik S, Pérez-Sánchez J. Protein growth performance, amino acid utilization and somatotropic axis responsiveness to fish meal replacement by plant protein sources in gilthead sea bream (Sparus aurata). Aquaculture. 2004;232:493–510.

Grammes F, Reveco FE, Romarheim OH, Landsverk T, Mydland LT, Øverland M. Candida utilis and Chlorella vulgaris counteract intestinal inflammation in Atlantic salmon (Salmo salar L.). PLoS One. 2013;8:e83213.

Griffith OW. Mammalian sulphuramino acid metabolism: an overview. In: Jakoby WB, Griffith OW, editors. Methods in Enzymology, vol. 143. New York: Academic; 1987. p. 366–76.

Hamer HM, Jonkers D, Venema K, Vanhoutvin S, Troost FJ, Brummer R-J. Review Article: The role of butyrate on colonic function. Aliment Pharm Ther. 2007; 27(2):104–19.

Hansen AC, Rosenlund G, Karlsen O, Koppe W, Hemre GI. Total replacement of fish meal with plant proteins in diets for Atlantic cod (Gadus morhua L.) I – effects on growth and protein retention. Aquaculture. 2007;272:599–611.

Hardy RW. Collaborative opportunities between fish nutrition and other disciplines in aquaculture: an overview. Aquaculture. 1999;177(1):217–30.

Hedrera MI, Galdames JA, Jimenez-Reyes MF, Reyes AE, Avendaño-Herrera R, Romero J, Feijóo CG. Soybean Meal Induces Intestinal Inflammation in Zebrafish Larvae. PLoS One. 2013;8(7):e69983.

Hu Z, Guo Y. Effects of dietary sodium butyrate supplementation on the intestinal morphological structure, absorptive function and gut flora in chickens. Anim Feed Sci Technol. 2007;132:240–9.

Iwashita Y, Suzuki N, Yamamoto T, Shibata J-i, Isokawa K, Soon A, Ikehata Y, Furuita H, Sugita T, Goto T. Supplemental effect of cholytaurine and soybean lecithin to a soybean meal based fish meal-free diet on hepatic and intestinal morphology of rainbow trout Oncorhynchus mykiss. Fish Sci. 2008; 74(5):1083–95.

Iwashita Y, Suzuki N, Matsunari H, Sugita T, Yamamoto T. Influence of soya saponin, soya lectin, and cholytaurine supplemented to a casein-based semipurified diet on intestinal morphology and biliary bile status in fingerling rainbow trout Oncorhynchus mykiss. Fish Sci. 2009;75(5):1307–15.

Johnson RB, Kim S-K, Watson AM, Barrows FT, Kroeger EL, Nicklason PM, Goetz GW, Place AR. Effects of dietary taurine supplementation on growth, feed efficiency, and nutrient composition of juvenile sablefish (Anoplopoma fimbria) fed plant based feeds. Aquaculture. 2015;445:79–85.

Kim C, Cha Y-N. Taurine chloramine produced from taurine under inflammation provides anti-inflammatory and cytoprotective effects. Amino Acids. 2014;46: 89–100.

Knudsen D, Jutfelt F, Sundh H, Sundell K, Koppe W, Frøkiær H. Dietary soya saponins increase gut permeability and play a key role in the onset of soyabean-induced enteritis in Atlantic salmon (Salmo salar L.). Br J Nutr. 2008; 100(01):120–9.

Kortner TM, Skugor S, Penn MH, Mydland LT, Djordjevic B, Hillestad M, Krasnov A, Krogdahl Å. Dietary soyasaponin supplementation to pea protein concentrate reveals nutrigenomic interactions underlying enteropathy in Atlantic salmon (Salmo salar). BMC Vet Res. 2012;8:101.

Kotunia A, Wolinski J, Laubitz D, Jurkowska M, Rome V, Guilloteau P, Zabielski R. Effect of sodium butyrate on the small intestine development in neonatal piglets fed correction of feed by artificial sow. J Physiol Pharmacol. 2004; 55(2):59–68.

Krogdahl Å, Bakke-McKellep AM, Baeverfjord G. Effects of graded levels of standard soybean meal on intestinal structure, mucosal enzyme activities, and pancreatic response in Atlantic salmon (Salmo salar L.). Aquacult Nutr. 2003;9:361–71.

Krogdahl A, Gajardo K, Kortner TM, Penn M, Gu M, Berge GM, Bakke AM. Soya saponins induce enteritis in Atlantic salmon (Salmo salar L.). J Agric Food Chem. 2015;63:3887–902.

Lim SJ, Lee KJ. Partial replacement of fish meal by cottonseed meal and soybean meal with iron and phytase supplementation for parrot fish Oplegnathus fasciatus. Aquaculture. 2009;290:283–9.

Liu W, Yang Y, Zhang J, Gatlin DM, Ringø E, Zhou Z. Effects of dietary microencapsulated sodium butyrate on growth, intestinal mucosal morphology, immune response and adhesive bacteria in juvenile common carp (Cyprinus carpio) prefed with or without oxidised oil. Br J Nutr. 2014;112: 15–29.

Locksley RM, Killeen N, Lenardo MJ. The TNF and TNF receptor superfamilies: integrating mammalian biology. Cell. 2001;104(4):487–501.

López LM, Flores-Ibarra M, Bañuelos-Vargas I, Galaviz MA, True CD. Effect of fishmeal replacement by soy protein concentrate with taurine supplementation on growth performance, hematological and biochemical status, and liver histology of totoaba juveniles (Totoaba macdonaldi). Fish Physiol Biochem. 2015;41:921–36.

Marjara IS, Chikwati EM, Valen EC, Krogdahl A, Bakke AM. Transcriptional regulation of IL-17A and other inflammatory markers during the development of soybean meal-induced enteropathy in the distal intestine of Atlantic salmon (Salmo salar L.). Cytokine. 2012;60:186–96.

Marozzi C, Bertoni F, Randelli E, Buonocore F, Timperio AM, Scapigliati G. A monoclonal antibody for the CD45 receptor in the teleost fish Dicentrarchus labrax. Dev Comp Immunol. 2012;37:342–53.

Martin SAM, Blaney SC, Houlihan DF, Secombes CJ. Transcriptome response following administration of a live bacterial vaccine in Atlantic salmon (Salmo salar). Mol Immunol. 2006;43:1900–11.

Noaillac-Depeyre J, Gas N. Absorption of protein macromolecules by the enterocytes of the carp (Cyprinus carpio L.). Z Zellforsch. 1973;146:525–41.

Olli JJ, Krogdahl Å. Nutritive value of 4 soybean products as protein sources in diets for rainbow trout (Oncorhynchus mykiss, Walbaum) reared in fresh-water. Acta Agricult Scand A Anim Sci. 1994;44:185–92.

Pearson WR. An introduction to sequence similarity ("homology") searching. Curr Protoc Bioinformatics. 2013;Chapter 3:Unit3.1.

Robles R, Lozano AB, Sevilla A, Márquez L, Nuez-Ortín W, Moyano FJ. Effect of partially protected butyrate used as feed additive on growth and intestinal metabolism in sea bream (Sparus aurata). Fish Physiol Biochem. 2013;39(6):1567–80.

Romarheim OH, Øverland M, Mydland LT, Skrede A, Landsverk T. Bacteria grown on natural gas prevent soybean meal-induced enteritis in Atlantic salmon. J Nutr. 2011;141:124–30.

Sahlmann C, Sutherland BJ, Kortner TM, Koop BF, Krogdahl A, et al. Early response of gene expression in the distal intestine of Atlantic salmon (Salmo salar L.) during the development of soybean meal induced enteritis. Fish Shellfish Immunol. 2013;34:599–609.

Salze G, Davis DA. Taurine: a critical nutrient for future fish feed. Aquaculture. 2015;437:215–29.

Scwarze MMK, Hawly PG. Prevention of myeloma cell apoptosis by ectopic bcl-2 expression or interleukin 6-mediated upregulation of bcl-xl. Cancer Res. 1995;55:2262–6.

Shirota K, LeDuy L, Yuan SY, Jothy S. Interleukin-6 and its receptor are expressed in human intestinal epithelial cells. Virch Archiv B Cell Pathol. 1990;58:303–8.

Sitja-Bobadilla A, Pena-Llopis S, Gomez-Requeni P, Medale F, Kaushik S, Pérez-Sánchez J. Effect of fish meal replacement by plant protein sources on non-specific defense mechanisms and oxidative stress in gilthead sea bream (Sparus aurata). Aquaculture. 2005;249:387–400.

Tacon AGJ, Metian M. Global overview on the use of fish meal and fish oil in industrially compounded aquafeeds: trends and future prospects. Aquaculture. 2008;285(1–4):146–58.

Terova G, Corà S, Verri T, Rimoldi S, Bernardini G, Saroglia M. Impact of feed availability on PepT1 mRNA expression levels in sea bass (Dicentrarchus labrax). Aquaculture. 2009;294:288–99.

Terova G, Rimoldi S, Díaz N, Ceccotti C, Gliozheni E, Piferrer F. Effects of sodium butyrate treatment on histone modifications and the expression of genes related to epigenetic regulatory mechanisms and immune response in European sea bass (Dicentrarchus labrax) fed a plant-based diet. PLoS One. 2016;11(7):e0160332.

Toden S, Bird AR, Topping DL, Conlon MA. Dose-dependent reduction of dietary protein-induced colonocyte DNA damage by resistant starch in rats correlates more highly with caecal butyrate than with other short chain fatty acids. Cancer Biol Ther. 2007;6(2):e1–6.

Urán PA, Goncalves AA, Taverne-Thiele JJ, Schrama JW, Verreth JAJ, Rombout JHWM. Soybean meal induces intestinal inflammation in common carp (Cyprinus carpio). Fish Shellfish Immunol. 2008a;25:751–60.

Urán PA, Schrama JW, Obach A, Jensen L, Koppe W, Verreth JAJ. Soybean meal-induced enteritis in Atlantic salmon (Salmo salar L.) at different temperatures. Aquacult Nutr. 2008b;14:324–30.

Van den Ingh TSGAM, Krogdahl A, Olli JJ, Hendriks HGCJM, Koninkx JGJF. Effects of soybean-containing diets on the proximal and distal intestine in Atlantic salmon (Salmo salar): a morphological study. Aquaculture. 1991;94(4):297–305.

Venou B, Alexis MN, Fountoulaki E, Haralabous J. Effects of extrusion and inclusion level of soybean meal on diet digestibility, performance and nutrient utilization of gilthead sea bream (Sparus aurata). Aquaculture. 2006;261:343–56.

Vinolo MA, Hatanaka E, Lambertucci RH, Curi R. Effects of short chain fatty acids on effector mechanisms of neutrophils. Cell Biochem Funct. 2009;27(1):48–55.

Wong JMWRD, de Souza RRD, Kendall CWC, Emam A, Jenkins DJA. Colonic health: fermentation and short chain fatty acids. J Clin Gastroenterol. 2006;40(3):235–43.

Yokoyama M, Takeuchi T, Park GS, Nakazoe J. Hepatic cysteinesulphinate decarboxylase activity in fish. Aquacult Res. 2001;32:216–20.

Yuan H, Liddle FJ, Mahajan S, Frank DA. IL-6-induced survival of colonrectal carcinoma cells is inhibited by butyrate through down-regulation of the IL-6 receptor. Carcinogenesis. 2004;25(11):2247–55.

Yun BA, Mai KS, Zhang WB, Xu W. Effects of dietary cholesterol on growth performance, feed intake and cholesterol metabolism in juvenile turbot (Scophthalmus maximus L.) fed high plant protein diets. Aquaculture. 2011;319:105–10.

Zhang Z, Xu L, Liu W, Yang Y, Du Z, Zhou Z. Effects of partially replacing dietary soybean meal or cottonseed meal with completely hydrolyzed feather meal (defatted rice bran as the carrier) on production, cytokines, adhesive gut bacteria, and disease resistance in hybrid tilapia (Oreochromis niloticus♀ × Oreochromis aureus ♂). Fish Shellfish Immunol. 2014;41(2):517–25.

7

Toxic effects of ammonia exposure on growth performance, hematological parameters, and plasma components in rockfish, *Sebastes schlegelii*, during thermal stress

Ki Won Shin[2], Shin-Hu Kim[1], Jun-Hwan Kim[1], Seong Don Hwang[2] and Ju-Chan Kang[1*]

Abstract

Rockfish, *Sebastes schlegelii* (mean length 14.53 ± 1.14 cm and mean weight 38.36 ± 3.45 g), were exposed for 4 weeks with the different levels of ammonia in the concentrations of 0, 0.1, 0.5, and 1.0 mg/L at 19 and 24 °C. The indicators of growth performance such as daily length gain, daily weight gain, condition factor, and hematosomatic index were significantly reduced by the ammonia exposure and high temperature. The ammonia exposure induced a significant decrease in hematological parameters, such as red blood cell (RBC) count, white blood cell (WBC) count, hemoglobin (Hb), and hematocrit (Ht), whose trend was more remarkable at 24 °C. Mean corpuscular volume (MCV), mean corpuscular hemoglobin (MCH), and mean corpuscular hemoglobin concentration (MCHC) were also notably decreased by the ammonia exposure. Blood ammonia concentration was considerably increased by the ammonia concentration exposure. In the serum components, the glucose, glutamic oxalate transaminase (GOT), and glutamic pyruvate transaminase (GPT) were substantially increased by the ammonia exposure, whereas total protein was significantly decreased. But, the calcium and magnesium were not considerably changed.

Key words: Ammonia, Hematological parameters, Growth performance, Plasma components, Rockfish

Background

Ammonia is one of the nitrogenous wastes especially in water. It is generated from the catabolism of amino acids, purines, and pyrimidines (Ruyet et al. 1995). In an aquatic environment, ammonia exists as two main forms such as unionized ammonia (NH_3) and ionized ammonium (NH_4^+) (Randall and Tsui 2002). The toxicity of ammonia is significantly affected by the levels of pH; the increase in pH induces the concentration of NH_3 increase (Richardson 1997). The toxic effects of ammonia exposure to aquatic animals strongly occur by the high concentration of unionized ammonium (NH_3) because it can readily diffuse through the gill membranes (Sinha et al. 2012). Excessive ammonia can cause the growth

performance decrease, tissue erosion and degeneration, immune suppression, and high mortality in aquatic animals, which acts as toxicity by increasing ammonia levels in blood and tissues (Lemarie et al. 2004; Li et al. 2014). In addition, the ammonia exposure also induces the neurotoxicity, oxidative stress, and oxygen delivery impairment as well as hyperactivity, convulsions, and coma (Wilkie 1997).

Ammonia toxicity can be affected by various environmental parameters like temperature, pH, salinity, and oxygen (Lemarie et al. 2004). Among the environmental indicators, temperature is one of the most major parameters to influence ammonia toxicity, and Richardson (1997) reported that the temperature increase caused a significant elevation in ammonia toxicity. Generally, the increase in temperature in aquatic animals induces the higher toxic effects under toxicity exposure (Patra et al. 2015), because the high temperature elevates the

* Correspondence: jckang@pknu.ac.kr

[1]Department of Aquatic Life Medicine, Pukyong National University, Busan 608-737, Republic of Korea

Full list of author information is available at the end of the article

diffusion rate, bioavailability, and chemical reactions of the aquatic animals (Delos and Erickson 1999). Barbieri and Bondioli (2015) also reported the lower LC$_{50}$ for the ammonia exposure of Pacu fish, *Piaractus mesopotamicus* by increasing water temperature, which means that the higher temperature causes the higher ammonia toxicity.

The exposure to toxic substances in aquatic environment can induce the negative effects on reproduction and growth performance in fish (Kim and Kang 2015). Among various toxicants, ammonia is one of the most toxic substances to cause the growth inhibition in fish farming, and the ammonia toxicity can be a main reason in fish mortality (El-Shafai et al. 2004). Given that the toxicants generally inhibit the growth performance in aquatic animals, the growth performance can be a good indicator to assess the toxicity in the animals.

Considering the exposure to toxicants induces the alterations of fish blood indicators, the hematological parameters can be a sensitive and reliable indicator to assess the toxicity on the exposed animals (Kim and Kang 2014). The ammonia exposure negatively causes the alterations in blood chemistry in aquatic animals as well as the decrease in reproductive capacity and growth rate (Vosyliene and Kazlauskiene 2004). Ajani (2008) reported a significant decrease in blood parameters such as red blood cell (RBC) counts, hemoglobin, and hematocrit of African catfish, *Clarias gariepinus*, exposed to ammonia, which may be due to anemia and haemodilution of haemolysis for RBC.

Rockfish, *Sebastes schlegelii*, is a commonly cultured fish in the marine net cages of South Korea because of its rapid growth performance and high demand, which is one of the three largest cultured fish in South Korea. But, the study about the ammonia exposure depending on water temperature has insufficiently been conducted, although the ammonia is ubiquitous in the marine net cages and highly toxic to cultured animals. Therefore, the purpose of this study was to evaluate the toxic effects for the ammonia exposure depending on water temperature to the *S. schlegelii* on hematological parameters and plasma components.

Methods
Experimental animals and conditions
Rockfish, *S. schlegelii* (mean length 14.53 ± 1.14 cm and mean weight 38.36 ± 3.45 g), were obtained from a commercial farm (Tongyeong, Korea). Fish were held for 3 weeks in seawater at 19 °C to ensure that all individuals were healthy and feeding, and also to reset the thermal history (19 and 24 °C) of the animals prior to initiating temperature acclimations (temperature; 19.0 ± 0.6 and 23 ± 0.5 °C, pH; 7.9 ± 0.6, salinity; 33.1 ± 0.5‰, dissolved oxygen; 7.4 ± 0.5 mg/L). The fish were fed a

commercial diet twice daily (Woosung Feed, Daejeon City, Korea). The water temperature was adjusted from ambient at a rate of ±1 °C/day until a final temperature of 24 °C was reached. The acclimation period commenced once the final temperature had been sustained for 24 h and animals were feeding while showing no sign of stress. Ammonia exposure took place in 40-L glass tanks containing 13 fish per treatment group. Ammonia chloride (NH$_4$Cl) (Sigma, St. Louis, MO, USA) solution was dissolved in the respective glass tanks. The annual report on marine environment monitoring in Korea 2014 showed the ammonia levels 0.46 mg/L in Gunsan, 0.64 mg/L in Ulsan, and 1.39 mg/L in Busan during the summer season. By our surveying of the fish farm in Tongyeong, it showed the over 0.5 mg/L ammonia level in summer. Therefore, our studies established the experimental concentrations of ammonia 0, 0.1, 0.5, and 1.0 mg/L. The ammonia concentrations in the glass tanks were 0, 0.1, 0.5, and 1.0 mg/L, and the actual ammonia concentration is demonstrated in Table 1. The glass tank water was thoroughly exchanged once per 2 days and made the same concentration in the respective glass tank. At the end of each period (at 2 and 4 weeks), animals were anesthetized in buffered 3-aminobenzoic acid ethyl ester methane sulfonate (Sigma Chemical, St. Louis, MO).

Growth performance
The weight and length of rockfish were measured just before exposure, at 2 and 4 weeks. Daily length gain, daily weight gain, condition factor, and Hepatosomatic Index (HIS) were calculated by the following method.

Daily growth gain = W$_f$–W$_i$/day

(W$_f$ = final or weight, W$_i$ = Initial length or weight)

Condition factor (%) = $\left(W/L^3\right) \times 100$

(W = weight (g), L = length (cm))

HIS = (liver weight/total fish weight) × 100

Hematological parameters
Blood samples were collected within 35–40 s through the caudal vein of the fish in 1-mL disposable heparinized syringes. The blood samples were kept at 4 °C until the blood parameters were completely studied. The total

Table 1 Analyzed waterborne ammonia concentration from each source

Ammonia concentration (mg/L)				
Ammonia levels	0	0.1	0.5	1.0
Actual ammonia levels	0.05	0.18	0.67	1.21

red blood cell (RBC) count, white blood cell (WBC), hemoglobin (Hb) concentration, and hematocrit (Ht) value were determined immediately. Total RBC and WBC counts were counted using an optical microscope with a hemocytometer (Improved Neubauer, Germany) after being diluted with Hendrick's diluting solution. The Hb concentration was determined using the cyanmethemoglobin technique (Asan Pharm. co., Ltd.). The Ht value was determined by the microhematocrit centrifugation technique. Erythrocyte indices like mean corpuscular volume (MCV), mean corpuscular hemoglobin (MCH), and mean corpuscular hemoglobin concentration (MCHC) were also calculated according to standard formulas.

$$MCV \text{ (fl)} = \frac{Ht(\%) \times 10}{RBC(10^6/uL)}$$

$$MCH \text{ (}\mu\mu g) = \frac{Hb(g/dL)X10}{RBC(10^6/uL)}$$

$$MCHC \text{ (\%)} = \frac{Hb(g/dL)X100}{Ht(\%)}$$

Serum components

The blood samples were centrifuged to separate serum from blood samples at 3000 g for 5 min at 4 °C. The serum samples were analyzed for inorganic substances, organic substances and enzyme activity using clinical kit (Asan Pharm. Co., Ltd.). In inorganic substances assay, calcium and magnesium were analyzed by the o-cresolphthalein complexone technique and xylidyl blue technique. In an organic substance assay, glucose and total protein were analyzed by GOD/POD technique and biuret technique. In an enzyme activity assay, glutamic oxalate transaminase (GOT) and glutamic pyruvate transaminase (GPT) were analyzed by Kind-King technique using a clinical kit.

Blood ammonia concentration

Blood samples were collected within 35–40 s through the caudal vein of the fish in 1-mL disposable heparinized syringes. The blood samples were kept at 4 °C until the blood ammonia was completely studied. The blood ammonia was determined by indophenol method using a clinical kit (Asan Pharm. Co.,Ltd.).

Statistical analysis

The experiment was conducted in exposure periods for 4 weeks and performed triplicate. Statistical analyses were performed using the SPSS/PC+ statistical package (SPSS Inc, Chicago, IL, USA). Significant differences between groups were identified using one-way ANOVA and Duncan's test for multiple comparisons or Student's

t test for two groups (Duncan, 1955). The significance level was set at $P < 0.05$.

Results

Growth performance

The indicators of the growth performance of *S. schlegelii* such as daily length gain, daily weight gain, condition factor, and hepatosomatic index are demonstrated in Fig. 1. In daily length gain, a notable reduction was observed over 0.5 mg/L at 19 and 24 °C after 2 weeks. After 4 weeks, the daily length gain was significantly decreased over 0.5 mg/L at 19 °C and over control at 24 °C. In daily weight gain, a considerable decrease was observed over 0.5 mg/L at 19 and 24 °C. After 4 weeks, the daily weight gain was markedly reduced over 0.5 mg/L at 19 °C and over 0.1 mg/L at 24 °C. The condition factor after 2 weeks was substantially decreased in the concentration of 1.0 mg/L at 19 and 24 °C, and a considerable decrease after 4 weeks was observed in the concentration in 1.0 mg/L at 19 °C and over 0.5 mg/L at 24 °C. In the Hepatosomatic Index, a significant reduction was observed in the concentration of 1.0 mg/L at 19 °C and over 0.5 mg/L at 24 °C. After 4 weeks, the Hepatosomatic Index was notably decreased over 0.5 mg/L at 19 and 24 °C. In the growth performance, the growth indicators affected the concentration of ammonia exposure and temperature.

Hematological parameters

The hematological parameters (RBC count, WBC count, Hb, and Ht; MCV, MCH, and MCHC) of *S. schlegelii* is demonstrated in Figs. 2 and 3. In RBC count, a notable decline was shown in the concentration of 1.0 mg/L at 19 °C and over 0.5 mg/L at 24 °C after both 2 and 4 weeks. In WBC count, a substantial decrease was observed over 0.5 mg/L at 19 °C and over 0.1 mg/L at 24 °C after 2 weeks. After 4 weeks, the WBC count was markedly increased in the concentration of 1.0 mg/L at 19 °C and over 0.1 mg/L at 24 °C. In hemoglobin, a significant decrease was observed in the concentration of 1.0 mg/L at 19 °C and over 0.5 mg/L at 24 °C after 2 weeks. After 4 weeks, the Hb concentration was decreased over 0.5 mg/L of ammonia exposure. In hematocrit, a considerable decrease was observed in the concentration of 1.0 mg/L at 19 °C and over 0.5 mg/L at 24 °C after both 2 and 4 weeks. The values of RBC count, WBC count, Hb, and Ht were notably decreased by the ammonia exposure, and the high temperature catalyzed the decline in the values.

In MCV value, a significant decrease was observed in the concentration of 1.0 mg/L at 19 °C and over 0.5 mg/L 24 °C after 2 and 4 weeks. The MCH value was considerably decreased in the concentration of 1.0 mg/L at 19 °C and over 0.5 mg/L 24 °C after 2 and 4 weeks. In

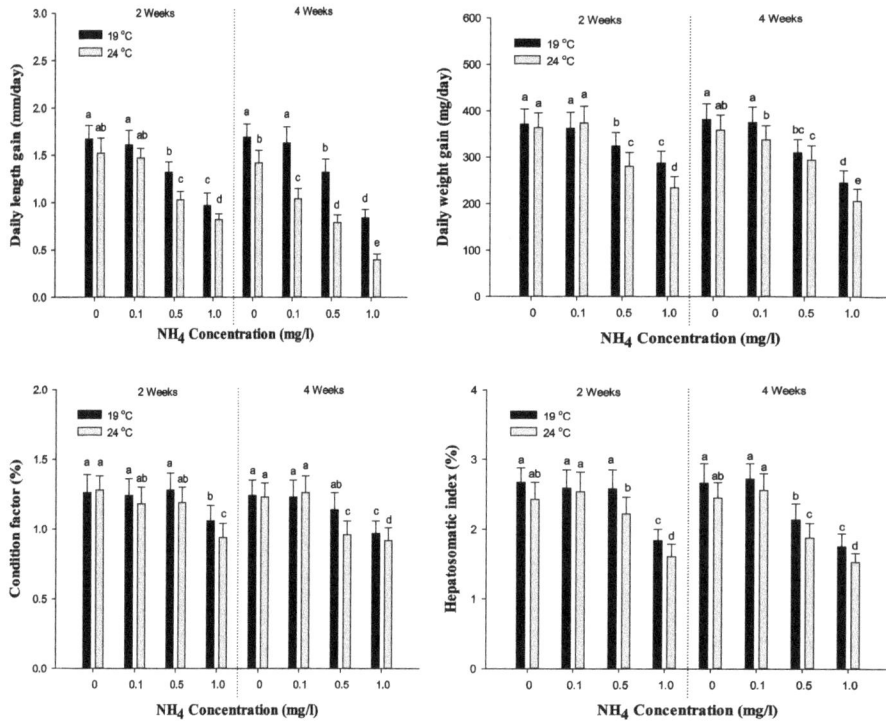

Fig. 1 Change of growth rate in rockfish, *Sebastes schlegelii*, exposed to the different ammonia concentrations and water temperatures. Values with different *superscripts* are significantly different ($P < 0.05$) as determined by Duncan's multiple range test

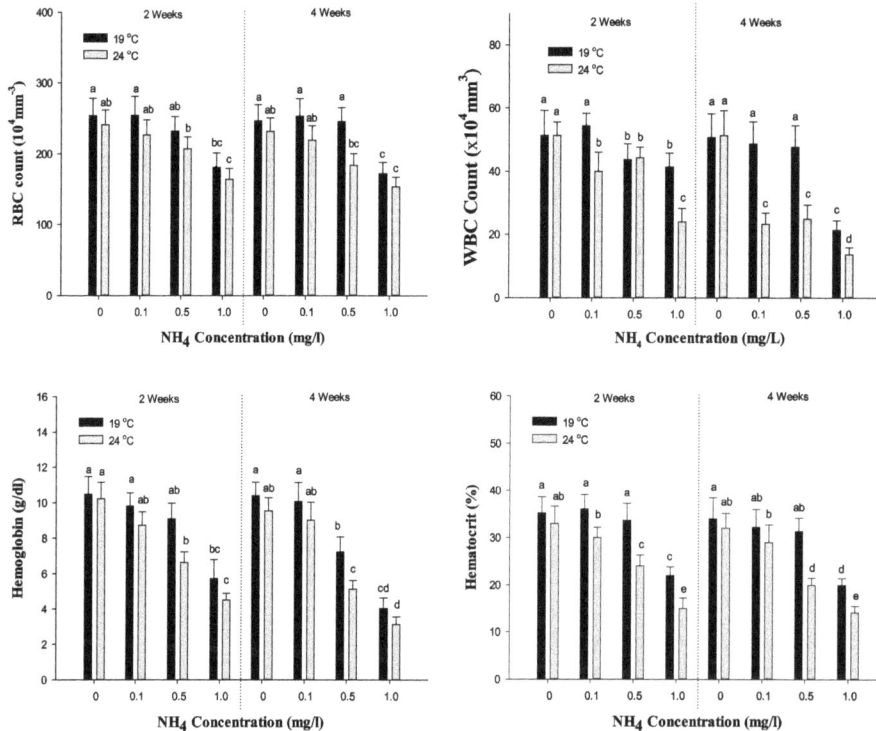

Fig. 2 Change of hematological parameter in rockfish, *Sebastes schlegelii*, exposed to the different ammonia concentrations and water temperatures. Values with different *superscripts* are significantly different ($P < 0.05$) as determined by Duncan's multiple range test

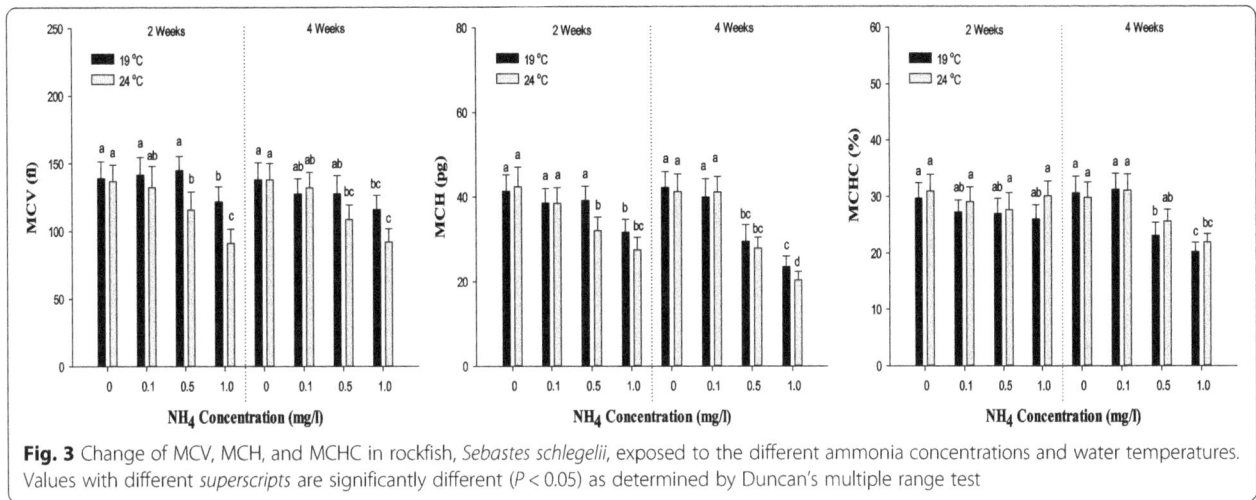

Fig. 3 Change of MCV, MCH, and MCHC in rockfish, *Sebastes schlegelii*, exposed to the different ammonia concentrations and water temperatures. Values with different *superscripts* are significantly different (P < 0.05) as determined by Duncan's multiple range test

MCHC value, there was no notable alteration after 2 weeks. But, a substantial decline was observed over 0.5 mg/L of ammonia exposure after 4 weeks. The values of MCV, MCH and MCHC were markedly reduced by the high ammonia exposure and temperature.

Blood ammonia concentration

The blood ammonia concentration of *S. schlegelii* is shown in Fig. 4. The blood ammonia concentration of *S. schlegelii* was considerably increased over 0.5 mg/L at 19 °C and 0.1 mg/L at 24 °C after 2 and 4 weeks. There was no notable alteration according to the difference of temperature except for the concentration of 0.1 mg/L of ammonia exposure.

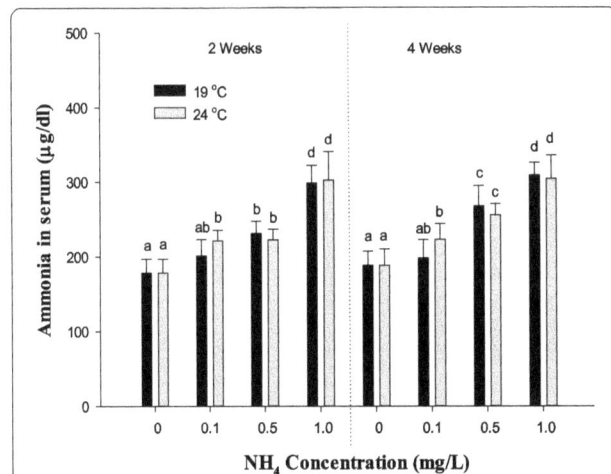

Fig. 4 Change of serum ammonia concentration in rockfish, *Sebastes schlegelii*, exposed to the different ammonia concentrations and water temperatures. Values with different *superscripts* are significantly different (P < 0.05) as determined by Duncan's multiple range test

Serum components

The serum inorganic components such as calcium and magnesium of *S. schlegelii* are shown in Table 2. There was no alteration in calcium and magnesium of *S. schlegelii* by ammonia exposure depending on temperature. The serum organic components such as total protein and glucose of *S. schlegelii* are demonstrated in Table 3. In total protein, a notable decrease was observed in the concentration of 1.0 mg/L at 19 °C and over 0.5 mg/L at 24 °C. The total protein was considerably decreased over 0.5 mg/L at 19 and 24 °C after 4 weeks. In glucose, a substantial increase was observed in the concentration of 1.0 mg/L at 19 °C and over 0.5 mg/L at 24 °C. The glucose was substantially reduced over 0.5 mg/L at 19 and 24 °C after 4 weeks. The serum enzyme components such as GOT and GPT of *S. schlegelii* are shown in Table 4. The GOT was significantly increased over 0.5 mg/L at 19 and 24 °C after 2 and 4 weeks. In GPT after 2 weeks, a notable increase was observed over 0.5 mg/L at 24 °C, whereas there was no change at 19 °C. After 4 weeks, the GPT was considerably increased in the concentration of 1.0 mg/L at 19 °C and over 0.5 mg/L at 24 °C.

Discussion

The toxicant exposure can induce the inhibition of growth performance in aquatic animals. Erickson et al. (2010) reported a significant reduction of growth performance of rainbow trout, *Oncorhynchus mykiss*, exposed to arsenic. In this study, the ammonia exposure caused a notable decrease in growth performance of *S. schlegelii*. And, the reduction of growth performance may result from the demand for energy to detoxicate the ammonia which effects the drop in the energy for growth (Clearwater et al. 2002). Also, the growth performance of *S. schlegelii* was affected by the water

Table 2 Change of serum inorganic substances in rockfish, *Sebastes schlegelii*, exposed to the different ammonia concentrations and water temperatures. Values with different *superscripts* are significantly different (*P* < 0.05) as determined by Duncan's multiple range test

Parameters	Period (weeks)	Ammonia concentration (mg/L)							
		19 °C				24 °C			
		0	0.1	0.5	1.0	0	0.1	0.5	1.0
Calcium (mg/dL)	2	8.44 ± 0.82[a]	8.59 ± 0.67[a]	8.76 ± 0.94[a]	7.73 ± 0.66[a]	8.72 ± 0.79[a]	7.90 ± 0.84[a]	8.39 ± 0.56[a]	8.28 ± 0.73[a]
	4	8.19 ± 0.94[a]	8.53 ± 0.61[a]	7.93 ± 0.59[a]	8.09 ± 0.77[a]	8.19 ± 0.76[a]	8.16 ± 0.46[a]	7.48 ± 0.94[a]	7.51 ± 0.70[a]
Magnesium (mg/dL)	2	3.18 ± 0.34[a]	3.41 ± 0.36[a]	3.56 ± 0.33[a]	3.42 ± 0.35[a]	3.30 ± 0.38[a]	3.42 ± 0.44[a]	3.57 ± 0.39[a]	3.51 ± 0.31[a]
	4	3.26 ± 0.37[a]	3.41 ± 0.36[a]	3.45 ± 0.37[a]	3.47 ± 0.27[a]	3.37 ± 0.37[a]	3.50 ± 0.35[a]	3.71 ± 0.31[a]	3.49 ± 0.32[a]

temperature change. Carvalho and Fernandes (2006) reported that the high temperature causes the severe accumulation in fish by the toxicant exposure, which may need the more energy for detoxification.

The biochemical and physiological alterations in fish blood can be occurred by the toxic substances in an aquatic environment, and the blood parameters can be a sensitive and reliable indicator to evaluate the physiological status of fish (Mazon et al. 2002. Vosyliene and Kazlauskiene (2004) reported a negative change in blood chemistry of rainbow trout, *O. mykiss*, exposed to ammonia. In this study, the RBC count, WBC count, hemoglobin, and hematocrit of *S. schlegelii* were substantially decreased by ammonia exposure. Tilak et al. (2007) also reported a substantial decrease in hemoglobin of common carp, *Cyprinus carpio*, exposed to ammonia, which is caused by the increase in the oxygen intake and elevation in methemoglobin by gill damage. Thangam et al. (2014) reported a notable reduction in RBC and WBC count of common carp, *C. carpio*, exposed to ammonia. RBC count decreased due to anemia leading to inhibition of erythropoietin, and WBC count also decreased by the leucopenia coupled with stress for toxicants in aquatic animals. Knoph and Thorud (1996) reported that Atlantic salmon, *Salmo salar*, when exposed to ammonia showed decreased hematocrit resulting from reduction in RBC count. The ammonia exposure also caused a notable reduction in MCV, MCH, and MCHC of *S. schlegelii*. Saravanan et al. (2011) suggested that the diazinon pesticide exposure to

European catfish, *Cyprinus carpio* resulted in a considerable decrease in MCV, MCH, and MCHC, which is due to the increase of immature red blood cells by the toxicant exposure. In this study, the temperature in addition to the ammonia concentration substantially affected the hematological parameters of *S. schlegelii*. Adeyemo et al. (2003) reported that temperature decreases both the quantity and quality of erythrocytes and hemoglobin, which cause deteriorated oxygen supply. Carvalho and Fernandes (2006) suggested that a high temperature considerably affected the hematological values of *Prochilodus scrofa* exposed to copper result in the increased diffusion rate, chemical reactions, and increased oxygen transport at a high temperature.

The ammonia exposure induced a significant accumulation in the blood of *S. schlegelii*. Lemarie et al. (2004) also reported a considerable ammonia accumulation in the blood of juvenile sea bass, *Dicentrarchus labrax*, exposed to ammonia, which should negatively affect the experimental animal. The ammonia concentration notably affected the ammonia accumulation in the blood of *S. schlegelii*, but there was no significant change in ammonia accumulation in the blood according to the temperature.

The inorganic serum components such as calcium and magnesium have been considered as critical indicators to assess the toxicity of substances, which also act as the ion regulator for homeostasis (Kim and Kang 2015). In inorganic serum components, there was no alteration in calcium and magnesium of *S. schlegelii* exposed

Table 3 Change of serum organic substances in rockfish, *Sebastes schlegelii*, exposed to the different ammonia concentrations and water temperatures. Values with different *superscripts* are significantly different (*P* < 0.05) as determined by Duncan's multiple range test

Parameters	Period (weeks)	Ammonia concentration (mg/L)							
		19 °C				24 °C			
		0	0.1	0.5	1.0	0	0.1	0.5	1.0
Total protein (mg/mL)	2	0.39 ± 0.01[a]	0.37 ± 0.02[a]	0.37 ± 0.02[a]	0.31 ± 0.04[b]	0.39 ± 0.02[a]	0.38 ± 0.03[a]	0.31 ± 0.03[b]	0.30 ± 0.03[b]
	4	0.37 ± 0.02[a]	0.38 ± 0.03[a]	0.30 ± 0.02[b]	0.29 ± 0.02[b]	0.39 ± 0.04[a]	0.39 ± 0.03[a]	0.29 ± 0.02[b]	0.28 ± 0.02[b]
Glucose (mg/mL)	2	30.70 ± 4.10[a]	31.74 ± 4.50[a]	30.59 ± 4.80[a]	49.47 ± 3.20[b]	32.46 ± 4.26[a]	32.84 ± 3.46[a]	47.65 ± 6.80[b]	50.32 ± 5.10[b]
	4	29.57 ± 5.70[a]	31.66 ± 5.90[a]	48.12 ± 6.20[b]	49.04 ± 4.30[b]	29.91 ± 4.60[a]	30.58 ± 3.60[a]	50.72 ± 3.80[b]	50.79 ± 5.10[b]

Table 4 Change of serum enzyme activity in rockfish, *Sebastes schlegelii*, exposed to the different ammonia concentrations and water temperatures. Values with different superscript are significantly different ($P < 0.05$) as determined by Duncan's multiple range test

Parameters	Period (weeks)	Ammonia concentration (mg/L)							
		19 °C				24 °C			
		0	0.1	0.5	1.0	0	0.1	0.5	1.0
GOT (Karmen/mL)	2	4.25 ± 0.64^a	4.65 ± 0.45^a	6.32 ± 0.33^b	6.29 ± 0.65^b	4.21 ± 0.64^a	4.06 ± 0.43^a	5.95 ± 0.48^b	8.44 ± 1.06^c
	4	4.51 ± 0.68^a	4.61 ± 0.63^a	5.80 ± 0.67^b	8.13 ± 0.73^c	4.36 ± 0.53^a	4.65 ± 0.42^a	6.23 ± 0.54^b	8.63 ± 0.78^c
GPT (Karmen/mL)	2	5.41 ± 0.67^a	5.30 ± 0.56^a	5.34 ± 0.62^a	5.12 ± 0.44^a	5.37 ± 0.43^a	4.97 ± 0.35^a	5.15 ± 0.57^b	9.47 ± 0.78^c
	4	5.76 ± 0.55^a	5.45 ± 0.58^a	5.79 ± 0.38^a	10.94 ± 1.30^b	5.65 ± 0.75^a	5.83 ± 0.61^a	13.02 ± 1.26^c	11.04 ± 1.32^b

ammonia. Knoph and Thorud (1996) also reported no significant change in calcium and magnesium of Atlantic salmon, *S. salar*, exposed to ammonia for 2 weeks, whereas a notable increase in calcium and magnesium was observed after 2–3 days of exposure time. Thus, our studies showed that exposure at 2 and 4 weeks was not influenced in plasma, calcium, and magnesium concentrations. The organic components such as total protein and glucose can be a reliable biomarker to detect animal health (Oner et al. 2007). In organic serum components, the total protein of *S. schlegelii* was significantly decreased by ammonia exposure, whereas the glucose of *S. schlegelii* was increased. Gopal et al. (1997) suggested that total plasma has been notably changed under stress situations. The increase in glucose may be a consequence of glycogenolytic activity of catecholamines and gluconeogenetic effect of glucocorticoids by the stress response under toxic substance exposure (Dobsikova et al. 2011). The GOT and GPT in serum components can be generally used to assess the tissue damage of the liver and kidney (Agrahari et al. 2007). In enzyme serum components, the GOT and GPT of *S. schlegelii* was significantly increased by the ammonia exposure. Vedel et al. (1998) also reported a considerable increase in the GOT and GPT of rainbow trout, *O. mykiss*, exposed to ammonia, indicated some degree of tissue necrosis. The temperature as well as the concentration of ammonia exposure notably affected the alterations of serum components of *S. schlegelii*, showing that temperature also can be a critical factor to affect the experimental animals.

Conclusions

The ammonia exposure to *S. schlegelii* depending on water temperature induced notable decreases in growth performance (daily length gain, daily weight gain, condition factor, and hepatosomatic index) hematological parameters (RBC count, WBC count, hemoglobin, and hematocrit) and significant alterations in serum components (total protein, glucose, GOT, and GPT). Considering the results of this study, the ammonia depending on water temperature should negatively influence the experimental fish, *S. schlegelii*.

Abbreviations
GOT: Glutamic oxalate transaminase; GPT: Glutamic pyruvate transaminase; Hb: Hemoglobin; HIS: Hepatosomatic Index; Ht: Hematocrit; MCH: Mean corpuscular hemoglobin; MCHC: Mean corpuscular hemoglobin concentration; MCV: Mean corpuscular volume; RBC: Red blood cell; WBC: White blood cell

Acknowledgements
This work was supported by a grant from the National Institute of Fisheries Science (R2016069).

Authors' contributions
SH and KW carried out the environmental toxicity studies and manuscript writing. JH participated in the design of the study and data analysis. SD participated in the collection and assembly of the data; JC participated in its design and coordination and helped to draft the manuscript. All authors read and approved the final manuscript.

Competing interests
The authors declare that they have no competing interests.

Disclosure
The dataset(s) supporting the conclusions of this article is not included in the article.

Author details
[1]Department of Aquatic Life Medicine, Pukyong National University, Busan 608-737, Republic of Korea. [2]Aquatic life disease control division, National Fisheries Research and Development Institute, Busan 619-902, Republic of Korea.

References
Adeyemo OK, Agbede SA, Olaniyan AO, Shoaga OA. The haematological response of *Clarias gariepinus* to change in acclimation temperature. Afr J Biomed Res. 2003;6:105–8.
Agrahari S, Pandey KC, Gopal K. Biochemical alteration induced by monocrotophos in the blood plasma of fish, *Channa punctatus* (Bloch). Pestic Biochem Physiol. 2007;88:268–72.
Ajani F. Hormonal and haematological responses of *Clarias gariepinus* (Burchell 1822) to nitrite toxicity. Afr J Biotechnol. 2008;7(19):3466–71.
Barbieri E, Bondioli ACV. Acute toxicity of ammonia in Pacu fish (*Piaractus mesopotamicus*, Holmberg, 1887) at different temperatures levels. Aquac Res. 2015;8:565–71.

Carvalho CS, Fernandes MN. Effect of temperature on copper toxicity and hematological responses in the neotropical fish *Prochilodus scrofa* at low and high pH. Aquaculture. 2006;251:109–17.

Clearwater SJ, Farag AM, Meyer JS. Bioavailability and toxicity of dietborne copper and zinc to fish. Comparative Biochemistry and Physiology - C Toxicology and Pharmacology. 2002;132(3):269–313.

Delos C, Erickson R. Update of ambient water quality criteria for ammonia. EPA/822/R-99/014. Final/technical report. Washington, DC: U.S. Environmental ProtectionAgency; 1999.

Dobsikova R, Blahova J, Modra H, Skoric M, Svobodova Z. The effect of acute exposure to herbicide Gardoprim Plus Gold 500 SC on haematological and biochemical indicators and histopathological changes in common carp (*Cyprinus carpio* L.). Journal of the University of Veterinary and Pharmaceutical Sciences in Brno. 2011;80:359–63.

Duncan DB. Multiple range and multiple f tests. Biometrics. 1955;11:1–42.

El-Shafai SA, El-Gohary FA, Nasr FA, Van Der Steen NP, Gijzen HJ. Chronic ammonia toxicity to duckweed-fed tilapia (*Oreochromis niloticus*). Aquaculture. 2004;232(1-4):117–27.

Erickson RJ, Mount DR, Highland TL, Hockett JR, Leonard EN, Mattson VR, Dawson TD, Lott KG. Effects of copper, cadmium, lead, and arsenic in a live diet on juvenile fish growth. Can J Fish Aquat Sci. 2010;67(11):1816–26.

Gopal V, Parvathy S, Balasubramanian PR. Effect of heavy metals on the blood protein biochemistry of the fish *Cyprinus carpio* and its use as abio-indicator of pollution stress. Environ Monit Assess. 1997;48:117–24.

Kim JH, Kang JC. The selenium accumulation and its effect on growth, and Haematological parameters in red seabream, *Pagrus major*, exposed to water borne selenium. Ecotoxicol Environ Saf. 2014;104:96–102.

Kim JH, Kang JC. The lead accumulation and hematological findings in juvenile rock fish *Sebastes schlegelii* exposed to the dietary lead(II) concentrations. Ecotoxicol Environ Saf. 2015;115:33–9.

Knoph MB, Thorud K. Toxicity of ammonia to Atlantic salmon (*Salmo salar* L.) in seawater—effects on plasma osmolality, ion, ammonia, urea and glucose levels and hematologic parameters. Comp Biochem Physiol A Physiol. 1996; 113(4):375–81.

Lemarie G, Dosdat A, Coves D, Dutto G, Gasset E, Person-Le Ruyet J. Effect of chronic ammonia exposure on growth of European seabass (*Dicentrarchus labrax*) juveniles. Aquaculture. 2004;229(1-4):479–91.

Li M, Yu N, Qin JG, Li E, Du Z, Chen L. Effects of ammonia stress, dietary linseed oil and *Edwardsiella ictaluri* challenge on juvenile darkbarbel catfish *Pelteobagrus vachelli*. Fish and Shellfish Immunology. 2014;38(1):158–65.

Mazon AF, Monteiro EAS, Pinheiro GHD, Fernandes MN. Hematological and physiological changes induced by short-term exposure to copper in the freshwater fish, *Prochilodus scrofa*. Brazilian Journal of Biology = Revista Brasleira de Biologia. 2002;62(4A):621–31.

Oner M, Alti G, Canli M. Changes in serum in biochemical parameters of freshwater fish Oreochromis niloticus following prolonged metal (Ag, Cd, Cr, Cu, Zn) exposures. Environ Toxicol Chem. 2007;27(2):360–6.

Patra RW, Chapman JC, Lim RP, Gehrke PC, Sunderam RM. Interactions between water temperature and contaminant toxicity to freshwater fish. Environ Toxicol Chem. 2015;34(8):1809–17.

Randall DJ, Tsui TKN. Ammonia toxicity in fish. Mar Pollut Bull. 2002;45(1-12):17–23.

Richardson J. Acute ammonia toxicity for eight New Zealand indigenous freshwater species. N Z J Mar Freshw Res. 1997;31(2):185–90.

Ruyet PLJ, Chartois H, Quemener L. Comparative acute ammonia toxicity in marine fish and plasma ammonia response. Aquaculture. 1995;136(1-2): 181–94.

Saravanan M, Prabhu Kumar K, Ramesh M. Haematological and biochemical responses of freshwater teleost fish *Cyprinus carpio* (Actinopterygii: Cypriniformes) during acute and chronic sublethal exposure to lindane. Pestic Biochem Physiol. 2011;100(3):206–11.

Sinha AK, Liew HJ, Diricx M, Blust R, De BG. The interactive effects of ammonia exposure, nutritional status and exercise on metabolic and physiological responses in gold fish (*Carassius auratus* L.). Aquat Toxicol. 2012;109:33–46.

Thangam Y, Perumayee M, Jayaprakash S, Umavathi S, Basheer SK. Studies of ammonia toxicity on haematological parameters to freshwater fish *Cyprinus carpio* (common carp). International jounal of current mocrobiology and applied sciences. 2014;3(12):535–42.

Tilak KS, Veeraiah K, Milton Prema Raju J. Effects of ammonia, nitrite and nitrate on hemoglobin content and oxygen consumption of freshwater fish, *Cyprinus carpio* (Linnaeus). J Environ Biol. 2007;28(1):45–7.

Vedel NE, Korsgaard B, Jensen FB. Isolated and combined exposure to ammonia and nitrite in rainbow trout (*Oncorhynchus mykiss*): effects on electrolyte status, blood respiratory properties and brain glutamine:glutamate concentrations. Aquat Toxicol. 1998;41:325–42.

Vosyliene MZ, Kazlauskiene N. Comparative studies of sublethal effects of ammonia on rainbow trout (*Oncorhynchus mykiss*) at different stages of its development. Acta Zoologica Lituanica. 2004;14(1):13–8.

Wilkie MP. Mechanisms of ammonia excretion across fish gills. Comp Biochem Physiol A Physiol. 1997;Comp. Biochem. Physiol. A: Physiol. 118(1):39–50.

8

A novel method to depurate β-lactam antibiotic residues by administration of a broad-spectrum β-lactamase enzyme in fish tissues

Young-Sik Choe, Ji-Hoon Lee, Soo-Geun Jo and Kwan Ha Park[*]

Abstract

As a novel strategy to remove β-lactam antibiotic residues from fish tissues, utilization of β-lactamase, enzyme that normally degrades β-lactam structure-containing drugs, was explored. The enzyme (TEM-52) selectively degraded β-lactam antibiotics but was completely inactive against tetracycline-, quinolone-, macrolide-, or aminoglycoside-structured antibacterials. After simultaneous administration of the enzyme with cefazolin (a β-lactam antibiotic) to the carp, significantly lowered tissue cefazolin levels were observed. It was confirmed that the enzyme successfully reached the general circulation after intraperitoneal administration, as the carp serum obtained after enzyme injection could also degrade cefazolin ex vivo. These results suggest that antibiotics-degrading enzymes can be good candidates for antibiotic residue depuration.

Keywords: β-Lactamase, Carp, Antibiotics, Depuration

Background

Concerns have risen over the presence of antimicrobial residues in food fishes because the contaminants can cause human pathogenic bacteria to develop resistance to therapeutically valuable antimicrobials (Cabello, 2006). Development of antibacterial resistance will then require higher doses of the same agent, or more potent drugs for efficacious treatment of bacterial pathogens. Of several known mechanisms for resistance development, degradation of antimicrobial agents is regarded as the most frequent one (De Pascale and Wright 2010).

Soon after the introduction of β-lactam antibiotics such as penicillins and cephalaosporins into clinical use, bacterial resistance to these drugs were reported (Abraham and Chain, 1988). Those bacteria have been found to harbor plasmid-borne genes producing enzymes which can catalyze a few β-lactam antibiotics nullifying the efficacy of such drugs (Matagne et al. 1999). However, the efficacy has become increasingly limited by the advent of β-lactamase that catalyzes a broader range of β-lactam antibiotics compared with previous enzymes (Livermore 1995). Those enzymes are called extended-spectrum β-lactamase (ESBL) and they are usually represented by TEM-52.

If β-lactamase can catalyze β-lactam antibiotics in fish tissues, it could be useful to remove residual antibiotics remaining after use to food fish species. To test this possibility, we assessed the catalytic activity in vitro of purified TEM-52 enzymes obtained by recombinant DNA techniques, followed by an in vivo activity test in carp after administration of cefazolin, a β-lactam antibiotic.

Methods

β-Lactamase enzyme preparations

The gene for TEM-52 was obtained from a Korean clinical isolate after necessary cloning processes (Pai et al., 1999). Its sequence was confirmed to match 100% by query coverage tests with those of extended-spectrum enzymes reported from *Salmonella enterica* subspp. enterica serovar Typhimurium (gb|AY883411.1|) and *Klebsiella pneumoniae* (emb|Y13612.1). This gene was used to build a recombinant plasmid and inserted into *E. coli*

---
* Correspondence: khpark@kunsan.ac.kr
Department of Aquatic Life Medicine, College of Ocean Sciences, Kunsan National University, San-68 Miryong-Dong, Gunsan City, Jeonbuk, South Korea

and then cultured at 30 °C for about 48 h in 2-L complex media. When the culture reached the density of ca. 0.7 optical density at 650 nm, cells were harvested by centrifugation (×13,000g, 10 min, 3 °C). Cells were disrupted by sonication and the lysate was processed for purification purposes with Q-Sepharose FF column chromatography after refolding steps. The refolding process was composed of decreasingly stepwise dilutions at 8, 2, 1, and 0.4 M in denaturant urea. The TEM-52 enzyme was eluted in 20 mM Tris buffer (pH 7.0) and adjusted to contain 500 mU/ml β-lactamase activity when assayed for 5 min using nitrocefin (50 µg/ml, Oxoid, Basingstoke, UK) as the substrate (O'Callaghan et al., 1972). The final purified TEM-52 enzyme is known to have a molecular mass of 28 KDa (Perilli et al., 2002).

In vitro degradation activities against various antimicrobials

Degradation activity of TEM-52 against various classes of antimicrobials was assessed incubating the β-lactamase enzyme solution (25 mU/ml) with various antimicrobials. Ten different antimicrobials (cefotaxime-Na, cefazolin-Na, penicillin-G-K salt, amoxicillin, pivampicillin, oxytetracycline-HCl, ciprofloxacin, erythromycin, josamycin, and streptomycin) were incubated with TEM-52 in 50 mM sodium phosphate buffer (pH 7.0) for 30 min at 25 °C, and then followed by bioassay for semi-quantitative analysis of remaining drug levels. These antimicrobials were obtained from Sigma (St. Louis, MO, USA) except pivampicillin (Dong-Hwa Pharmaceuticals, Seoul, Korea) and josamycin (Wako Chemicals, Japan). Paper disks (8 mm diameter, Advantec Toyo, Dublin, CA, USA) were soaked with 50 µl of incubated enzyme-antimicrobial mixtures and placed on top of BHI agar in Petri dishes streaked with *Staphylococcus epidermidis* ATCC10145 (precultured in BHI to 1.8×10^7 CFU/ml). The dishes were incubated at 36 °C for 12 h, and inhibitory zones beyond the disk areas were measured.

In vivo and ex vivo degradation activities against cefazolin

Of several β-lactam antibiotics, cefazolin was selected as the model antibiotic because of the simplicity in chemical analysis for both in vivo and ex vivo studies. To assess degradation activity of TEM-52 more quantitatively than the above in vitro study, cefazolin was incubated with TEM-52 (25 mU/ml in 50 mM phosphate buffer, pH 7.0) for 30 min and remaining concentration of cefazolin was chemically analyzed with a high-performance liquid chromatography (HPLC) method.

The HPLC-UV method for cefazolin analysis (Nadai et al., 1993) was composed of a Waters 2690 separation module and Waters 2487 dual wavelength UV-visible detector at 274 nm (Waters, Milford, MA, USA). Samples were injected into a C_{18} reverse-phase HPLC column

(250 × 4.6 mm, 5 µm particle size, Shiseido, Japan) at 10–50 µl range and eluted with a mobile phase composed of 30 mM sodium phosphate buffer (pH 5.0) and methanol (88:12 ratio). Flow rate of the mobile phase was 1.2 ml/min.

For in vivo cefazolin degradation experiment, common carp *Cyprinus carpio* weighing 40–100 g maintained at 23 °C were used. Cefazolin (sodium salt, Sigma) was dissolved in sterile saline and administered intramuscularly (im) around the lateral line to render doses of 10 and 30 mg/kg of drug base. TEM-52 enzyme was immediately injected into the peritoneal cavity at 75 mU/100 g body weight. This intraperitoneal (ip) injection volume was 150 µl/100 g and an equivalent volume of saline was also administered to control fish. One hour after TEM-52 injections, fish were anesthetized with MS-222 (Sigma) and blood was sampled for cefazolin analysis through the caudal vessels. Serum was subsequently obtained following centrifugation at ×3600g for 20 min under refrigeration (3 °C). Liver and muscle were also isolated for cefazolin analyses. Serum and fish tissues were mixed with HPLC mobile phase in tenfold volumes and vigorously vortex-shaken before filtering through membrane filters, and finally injected into the HPLC column. This method led to an almost complete cefazolin recovery (>90%, $n = 3$). With cefazolin-spiked liver and muscle tissues (10, 20, 50, 100, and 200 ng/g), linearity of the standard curve was confirmed ($r^2 = 0.967$). The limits of detection (LOD) and quantification (LOQ) were 10 and 30 ng/g, respectively.

In some fish, TEM-52 was administered into the peritoneal cavity (ip) at 75 mU/100 g to naive carp, and serum was obtained 1 h later to assess ex vivo degradation activity against cefazolin. All experiments using carp were performed in accordance with the guidelines approved by the Institutional Animal Care and Use Committee, Kunsan National University, Korea.

Statistics
Data were expressed as mean ± S.D. Statistical significance was examined with unpaired t tests at the significant level of $p < 0.05$.

Results and discussion
Table 1 shows in vitro enzymatic activity of TEM-52 to degrade different classes of antibacterial agents. TEM-52 is a wide spectrum β-lactamase obtained from *E. coli* through recombinant DNA technique. The enzyme preparation reduced antibacterial activity of β-lactam antibiotics without effects on drugs of other classes, indicating highly selective degradation activity toward β-lactams. Other classes of antibacterials, i.e., tetracycline, quinolone, macrolide, and aminoglycoside, were not influenced at all. These data reflect that the enzyme preparation contains

Table 1 Bioassay of degradation activity of TEM-52 enzyme against various classes of antimicrobials

Class	Antimicrobials	Inhibition zone (mm, mean ± SD)	
		Control drug disk	TEM-52-treated drug disk
β-Lactam	Cefotaxime	9.3 ± 0.6	0.0 ± 0.0
	Cefazolin	6.0 ± 1.7	2.3 ± 1.2
	Penicillin G	8.0 ± 0.6	0.0 ± 0.0
	Amoxicillin	8.7 ± 0.6	0.0 ± 0.0
	Pivampicillin	7.3 ± 0.6	0.0 ± 0.0
Tetracycline	Oxytetracycline	9.8 ± 0.9	9.7 ± 0.6
Quinolone	Ciprofloxacin	8.2 ± 0.0	8.0 ± 0.4
Macrolide	Erythromycin	10.0 ± 1.0	10.0 ± 2.0
	Josamycin	8.6 ± 0.6	8.7 ± 0.5
Aminoglycoside	Streptomycin	8.5 ± 0.4	8.7 ± 0.6

Concentrations of antimicrobials in incubation were adjusted to produce 6–11 mm clear zones beyond the disk areas (0.5–500 µg/disk)
Triplicate determinations

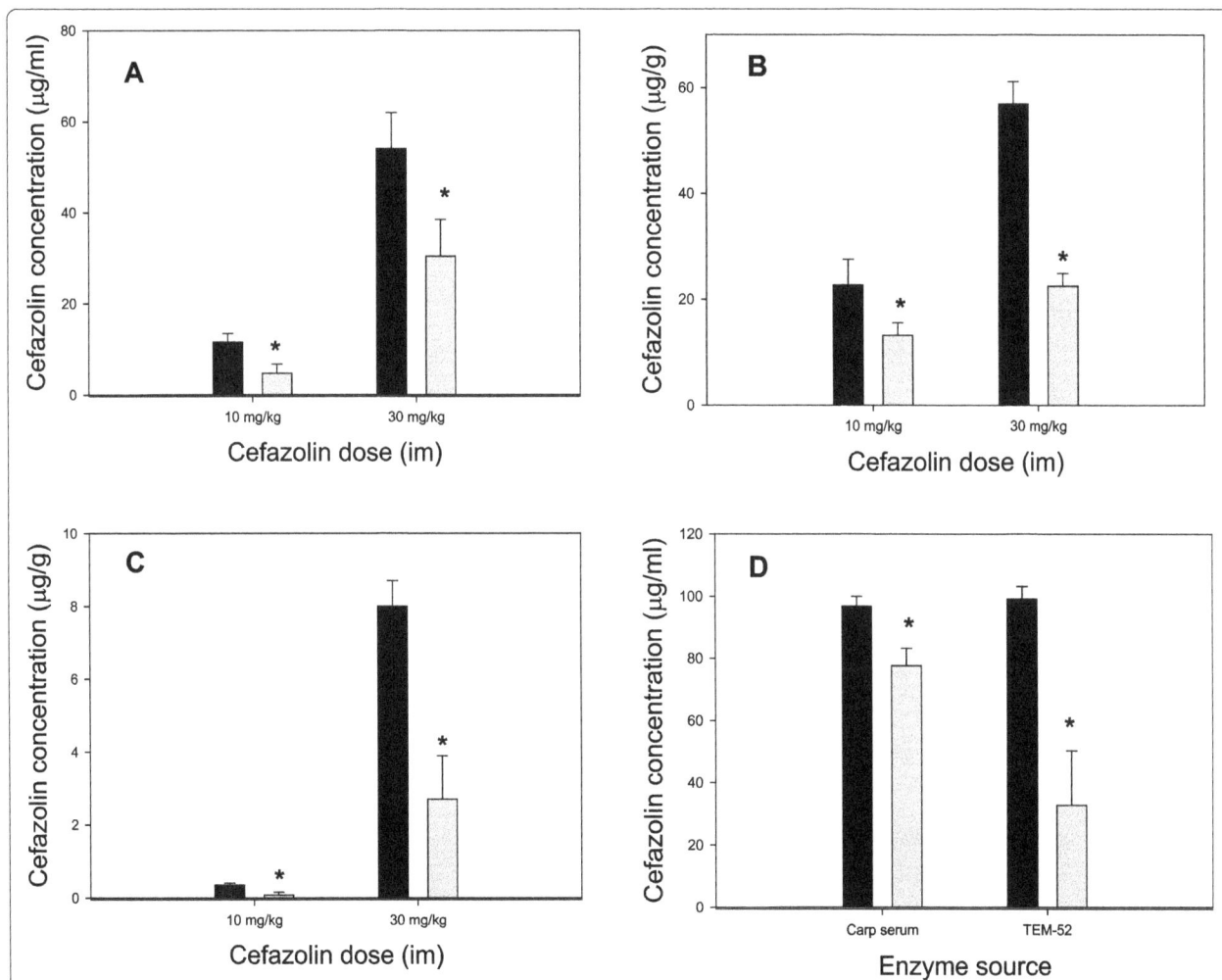

Fig. 1 In vivo and ex vivo cefazolin degradation activity of TEM-52 in carp. **a–c** In vivo cefazolin changes in serum (**a**), liver, (**b**) and muscle (**c**) following intraperitoneal injection of TEM-52, N = 5–10; filled bar, control carp; shaded bar, TEM-52-injected carp (**d**) ex vivo cefazolin changes when TEM-52-injected carp serum were incubated with cefazolin, N = 5; filled bar, without enzyme; shaded bar, incubated with TEM-52 itself or TEM-52-injected carp serum. *p < 0.05 with unpaired t tests when compared to control groups

exclusively β-lactamase efficacy and hydrolyzes β-lactams, cephalosporins, and penicillins (Bush et al., 1995; Pai et al., 1999; Poyart et al., 1998; Shahada et al., 2010), without any activity on other antibacterials.

Of the five β-lactam antibiotics examined with the in vitro bioassay, cefazolin was chosen for in vivo studies mostly because the procedures for HPLC analysis are quite simple. The results obtained after simultaneous administration of TEM-52 and cefazolin to carp are demonstrated in Fig. 1. In all three tissue samples examined, the levels of cefazolin were significantly lower in TEM-52 co-administered fish compared with the cefazolin-alone group. With cefazolin doses 10 and 30 mg/kg, and regardless of tissues, the degree of reduction was ~50% (Fig. 1a–c). In this in vivo experiment, TEM-52 was administered into the peritoneal cavity whereas cefazolin into the muscle. These results therefore indicate that β-lactamase activity included in TEM-52 and the β-lactam antibiotic cefazolin could come into direct contact in the fish body leading to cefazolin degradation. These results thus clearly demonstrate that TEM-52 is active not only in vitro system but also in the fish body, i.e., in vivo following injections.

It was additionally checked whether TEM-52 was actually present in carp serum after intraperitoneal injection to the fish. For this, TEM-52 was injected and serum was sampled 1 h later for ex vivo degradation tests. After incubation of TEM-52-pretreated carp serum with cefazolin ex vivo, a significant reduction ($p < 0.05$) in cefazolin residue was observed (Fig. 1d, Carp serum columns). In a parallel test, it was observed that if a higher reduction of cefazolin concentration occurred, cefazolin was incubated directly with TEM-52 in the presence of naive serum (Fig. 1d, TEN-52 columns). Although it is difficult to compare directly because these two test conditions were not identical, a significant portion of intact TEM-52 enzymes seemed to reach in serum when blood was taken (1 h post-injection).

Because TEM-52 was active for cefazolin degradation in carp, it can be deduced that the administered enzyme must have been sufficiently absorbed from the peritoneal cavity. It is not well described in fishes whether there is an inhibiting barrier to proteins injected into the peritoneal cavity. However, the fact that peptides are systemically active following intraperitoneal injection in fishes (Hong and Secombe, 2009; Murashita et al., 2010) suggests a loose barrier within the peritoneum in fishes. In rats too, absorption of proteins from the peritoneum is comparable to that of isotonic solution (Flessner, 2005).

Conclusions

In summary, this study is the first attempt to seek a utilization of sorts of antibiotics degradation enzymes to depurate drug residues in animals. We observed that

TEM-52, an enzyme active against a broad range of β-lactam antibiotics, can be a novel tool to remove residues from drug-treated fish bodies. It may be needed to expand tests against other β-lactam antibiotics and also in different fish species. There has not been any attempt in idea similar to this up to now. Various aspects of studies need to be carried out in order to establish practical usefulness. Safety of the test enzyme to humans is not known at all, for example.

Conclusively, however, a similar strategy will also be applicable to various classes of antimicrobials to which antibiotic-degrading enzymes have been reported.

Acknowledgements

This work was supported by the fund of Fisheries Research Institute of Kunsan National University in the year 2016. The authors are grateful for the production and supply of enzyme preparations to Dr. CS Shin at Advanced Protein Technologies Corp. (Gyeonggi Bio-center, Suwon, Korea) and Dr. SK Moon at MTC Korea (Gyeonggi Techno-Park, Ansan City, Gyeonggi-Do, Korea).

Authors' contribution

YSC carried out the in vitro and in vivo tests with the antimicrobials. JHL performed the HPLC analysis for the cefazolin residues. SGJ and KHP designed the overall experiments and prepared the manuscript. All authors read and approved the final manuscript.

Competing interests

The authors declare that they have no competing interests.

References

Abraham EP, Chain E. An enzyme from bacteria is able to destroy penicillin. 1940. Rev Infect Dis. 1988;10:677–8.

Bush K, Jacoby GA, Medeiros AA. A functional classification scheme for β-lactamases and its correlation with molecular structure. Antimicrob Agents Chemother. 1995;39:1211–33.

Cabello FC. Heavy use of prophylactic antibiotics in aquaculture: a growing problem for human and animal health and for the environment. Environ Microbiol. 2006;8:1137–44.

De Pascale G, Wright GD. Antibiotic resistance by enzyme inactivation: from mechanisms to solutions. Chem Biol Chem. 2010;11:1326–34.

Flessner MF. The transport barrier in intraperitoneal therapy. Am J Ren Physiol. 2005;288:F433–42.

Hong S, Secombe CJ. Two peptides derived from trout IL-1β have different stimulatory effects on immune gene expression after intraperitoneal administration. Comp Biochem Physiol Part B. 2009;153:275–80.

Livermore DM. β-lactamases in laboratory and clinical resistance. Clin Microbiol Rev. 1995;8:557–84.

Matagne A, Dubus A, Galleni M, Frere JM. The beta-lactamase cycle: a tale of selective pressure and bacterial ingenuity. Nat Prod Rep. 1999;16:1–19.

Murashita K, Jordal AO, Nilsen TO, Stefansson SO, Kurokawa T, Björnsson BT, Moen AG, Rønnestad I. Leptin reduces Atlantic salmon growth through the entral pro-opiomelanocortin pathway. Comp Biochem Physiol Part A. 2010;158:79–86.

Nadai M, Hasegawa T, Kato K, Wang L, Nabeshima T, Kato N. Antimicrob Agents Chemother. 1993;37:1781–5.

O'Callaghan CH, Morris A, Kirby SM, Shingler AH. Novel method for detection of beta-lactamases by using a chromogenic cephalosporin substrate. Antimicrob Agents Chemother. 1972;4:283–8.

Pai H, Lyu S, Lee JH, Kim J, Kwon Y, Kim JW, Choe KW. Survey of extended-spectrum β-lactamases in clinical isolates of Escheria coli and Klebsiella pneumoniae: prevalence of TEM-52 in Korea. J Clin Microbiol. 1999;37:1758–63.

Perilli M, Segatore B, De Massis MR, Pagani L, Luzzaro F, Rossolini GM,

Amicosante G. Biochemical characterization of TEM-92 extended-spectrum b-lactamase, a protein differing from TEM-52 in the signal peptide. Antimicrob Agent Chemother. 2002;46:3981–3.

Poyart C, Mugnier P, Quesne G, Berche P, Trieu-Cuot P. A novel extended-spectrum TEM-type β-lactamse (TEM-52) associated with decreased susceptibility to moxalactam in *Klebsiella pneumoniae*. Antimicrob Agents Chemother. 1998;42:108–13.

Shahada F, Chuma T, Dahshan H, Akiba M, Sueyoshi M, Okamoto K. Detection and characterization of extend-spectrum β-latamase (TEM-52)-producing *Salmonella* serotype infantis from broilers in Japan. Foodborne Pathogens Dis. 2010;7:515–21.

9

A survey of epiphytic organisms in cultured kelp *Saccharina japonica* in Korea

Jong-Oh Kim[1], Wi-Sik Kim[1], Ha-Na Jeong[1], Sung-Je Choi[2], Jung-Soo Seo[3], Myoung-Ae Park[3] and Myung-Joo Oh[1*]

Abstract

A survey was conducted to investigate the presence of epiphytic organisms in four kelp *Saccharina japonica* farms in the coastal area of Korea from 2014 to 2015. Of 740 kelp samples that were taken, 208 exhibited six kinds of epiphytic organisms, including hydroid (detection rate: 11.6%), bryozoan (6.4%), polychaete (3.4%), algae (3.2%), caprellid (3%), and oyster (0.5%). The infestation rate for hydroid, bryozoan, and polychaete was significantly higher in the Wando farm, Busan farm, and Pohang farm, respectively. Epiphytic organisms were generally observed during May to September and not January to April, indicating that their infestation was significantly higher when the water had a higher temperature. The histopathogical examination revealed that hydroid and bryozoan organisms were attached on the cuticula of the thallus while some algae were attached on the cuticula of the thallus or had penetrated the epidermis. These results indicate that hydroid and bryozoan were the most predominant epiphytic organisms in Korean kelp farms, even though the infested thallus had not been broken.

Keywords: Epiphytic organisms, Bryozoan, Hydroid, Kelp, *Saccharina japonica*

Background

Within the last 30 years, seaweed cultivation has attained large-scale production levels and has become a major industry in China, Korea, and Japan. Korea had a harvest of 1,022,326 tons in 2012, which makes it the fourth largest seaweed-producing country in the world (FAO, 2014). The economically important seaweed species harvested in Korea are laver *Porphyra* spp., wakame *Undaria* spp., kelp *Laminaria* spp., fusiforme, *Hizikia fusiformis*, and green laver *Monostroma* spp. (KOSIS, 2015).

Kelp *Saccharina japonica* cultivation was industrialized in Korea in the 1970s, and from 2005, it has been extensively developed together with the abalone industry. During the 1970s to 1980s, kelp production reached less than 12,000 tons, and production steadily increased since then with about 370,000 tons harvested in 2013 (KOSIS, 2015). Kelp production now accounts for about 25% of all Korean seaweed cultivation (KOSIS, 2015). However, the increase in production has also resulted in kelp farms becoming severely infected by several

epiphytic organisms (Gong et al., 2010; Park and Hwang, 2012). An infestation is characterized by the appearance of numerous colonies on the thallus of the kelp, and therefore, the affected kelp appears unsightly and cannot be sold. Although systematic research on fish disease has been conducted in Korea since the 1990s, seaweed disease has not yet been systematically studied. Therefore, little information is available on kelp disease. The present study consists of a survey that was conducted to investigate the presence of epiphytic organisms in Korean kelp farms. In addition, a histopathological examination was conducted to understand the effect by epiphytic organisms on the thallus.

Methods

Kelp samples and epiphytic organism examination

Kelp samples were obtained from four private farms in Wando (Southern Sea), Busan (Southern Sea), and Pohang (Eastern Sea) during 2014 and 2015 (Table 1, Fig. 1). Wando is the most popular area for kelp cultivation, producing about 97% of the kelp crop of Korea. The kelps were randomly selected by a farmer, transported on ice, and immediately subjected an examination of the epiphytic organisms. The total length (from apex to holdfast) of the thallus was measured, and the

* Correspondence: ohmj@Jnu.ac.kr
[1]Department of Aqualife Medicine, Jeonnam National University, Yeosu 59626, South Korea
Full list of author information is available at the end of the article

Table 1 Detection of epiphytic organisms in four kelp farms in Korea

Place	Farm	Date	Water temperature (°C)	Mean thallus length (cm)	Detection rate of epiphytic organisms % (detection no./total no.)					
					Hydroid	Bryozoan	Polychaete	Algae	Caprellid	Oyster
Wando	WDA	Apr 2014	11.5	340 (200–380)	0 (0/10)	0 (0/10)	0 (0/10)	0 (0/10)	0 (0/10)	0 (0/10)
		May 2014	15.5	260 (260–270)	0 (0/2)	0 (0/2)	0 (0/2)	0 (0/2)	0 (0/2)	0 (0/2)
		Jun 2014	17.4	NT	20 (1/5)	0 (0/5)	0 (0/5)	0 (0/5)	20 (1/5)	0 (0/5)
		Jul 2014	21.7	190 (100–240)	100 (6/6)	50 (3/6)	0 (0/6)	100 (6/6)	0 (0/6)	0 (0/6)
		Aug 2014	23	200 (100–240)	71.4 (5/7)	57.1 (4/7)	0 (0/7)	71.4 (5/7)	71.4 (5/7)	0 (0/7)
		Sep 2014	22.1	120 (90–160)	57.1 (4/7)	71.4 (5/7)	0 (0/7)	14.3 (1/7)	0 (0/7)	57.1 (4/7)
		Jan 2015	9	41 (20–71)	0 (0/71)	0 (0/71)	0 (0/71)	0 (0/71)	0 (0/71)	0 (0/71)
		Jan 2015	8.3	82 (39–122)	0 (0/75)	0 (0/75)	0 (0/75)	0 (0/75)	0 (0/75)	0 (0/75)
		Apr 2015	11	236 (200–280)	0 (0/30)	0 (0/30)	0 (0/30)	0 (0/30)	0 (0/30)	0 (0/30)
	Total detection rate in WDA				7.5 (16/213)	5.6 (12/213)	0 (0/213)	5.6 (12/213)	2.8 (6/213)	0.9 (4/213)
	WDB	May 2014	15.4	220 (100–330)	0 (0/14)	0 (0/14)	0 (0/14)	0(0/14)	0 (0/14)	0 (0/14)
		Jun 2014	21	260 (180–320)	57.1 (4/7)	0 (0/7)	0 (0/7)	28.6 (2/7)	0 (0/7)	0 (0/7)
		Jul 2014	21.5	160 (130–200)	100 (14/14)	0 (0/14)	35.7 (5/14)	0 (0/14)	7.1 (1/14)	0 (0/14)
		Sep 2014	21.6	130 (60–230)	100 (15/15)	33.3 (5/15)	0 (0/15)	13.3 (2/15)	0 (0/15)	0 (0/15)
		Feb 2015	9	57 (30–100)	0 (0/100)	0 (0/100)	0 (0/100)	0 (0/100)	0 (0/100)	0 (0/100)
		Mar 2015	11	97 (50–180)	0 (0/73)	0 (0/73)	0 (0/73)	0 (0/73)	0 (0/73)	0 (0/73)
		Apr 2015	12.5	127 (90–190)	0 (0/30)	0 (0/30)	0 (0/30)	0 (0/30)	0 (0/30)	0 (0/30)
	Total detection rate in WDB				13 (33/253)	2 (5/253)	2 (5/253)	1.6 (4/253)	0.4 (1/253)	0 (0/253)
Busan	BS	Apr 2014	11	170 (50–270)	0 (0/10)	10 (1/10)	0 (0/10)	30 (3/10)	0 (0/10)	0 (0/10)
		May 2014	14.5	230 (130–280)	100 (9/9)	88.9 (8/9)	0 (0/9)	0 (0/9)	0 (0/9)	0 (0/9)
		Jun 2014	21	220 (180–300)	100 (7/7)	71.4 (5/7)	0 (0/7)	71.4 (5/7)	0 (0/7)	0 (0/7)
		Jul 2014	20.4	160 (120–180)	100 (10/10)	90 (9/10)	0 (0/10)	0 (0/10)	100 (10/10)	0 (0/10)
		Feb 2015	11.5	76 (26–175)	0 (0/58)	0 (0/58)	0 (0/58)	0 (0/58)	0 (0/58)	0 (0/58)
		Feb 2015	12	143 (80–219)	0 (0/37)	0 (0/37)	0 (0/37)	0 (0/37)	0 (0/37)	0 (0/37)
		Mar 2015	12.5	182 (110–260)	0 (0/70)	0 (0/70)	0 (0/70)	0 (0/70)	0 (0/70)	0 (0/70)
		Apr 2015	13	203 (120–270)	0 (0/35)	0 (0/35)	0 (0/35)	0 (0/35)	0 (0/35)	0 (0/35)
	Total detection rate in BS				11 (26/236)	9.7 (23/236)	0 (0/236)	3.4 (8/236)	4.2 (10/236)	0 (0/236)
Pohang	PH	Jul 2014	20.5	140 (50–220)	28.6 (4/14)	7.1 (1/14)	28.6 (4/14)	0 (0/14)	0 (0/14)	0 (0/14)
		Aug 2014	23.4	110 (80–150)	50 (5/10)	60 (6/10)	20 (2/10)	0 (0/10)	0 (0/10)	0 (0/10)
		Sep 2014	23.1	70 (30–110)	14.3 (2/14)	0 (0/14)	100 (14/14)	0 (0/14)	0 (0/14)	0 (0/14)
	Total detection rate in PH				28.9 (11/38)	18.4 (7/38)	52.6 (20/38)	0 (0/38)	0 (0/38)	0 (0/38)
	Total detection rate in 4 farms				11.6 (86/740)	6.4 (47/740)	3.4 (25/740)	3.2 (24/740)	3 (22/740)	0.5 (4/740)

NT not tested

surfaces of the specimens were examined macroscopically and microscopically for the presence of epiphytic organisms. The detection rate of the epiphytic organisms was measured by calculating the percentages of the kelp infested by epiphytic organisms, and their infestation area was calculated as the percentage of the thallus that had been colonized. The number of caprellid was measured in a 2 × 2 cm kelp area infected with hydroid because their infestation area cannot be calculated. The epiphytic organisms that were obtained were photographed and identified according to their morphological characteristics (Gong et al., 2010; Kim and Lee, 1975; Lee, 1994; Park, 2010; 2011; Saunders and Metaxas, 2009; Schwaninger, 1999; The Korean Society of Systematic Zoology, 2014).

Histology
The epiphytic organisms attached to the thallus were removed and immediately fixed in 10% neutral buffered formalin. After fixation, standard histological procedures

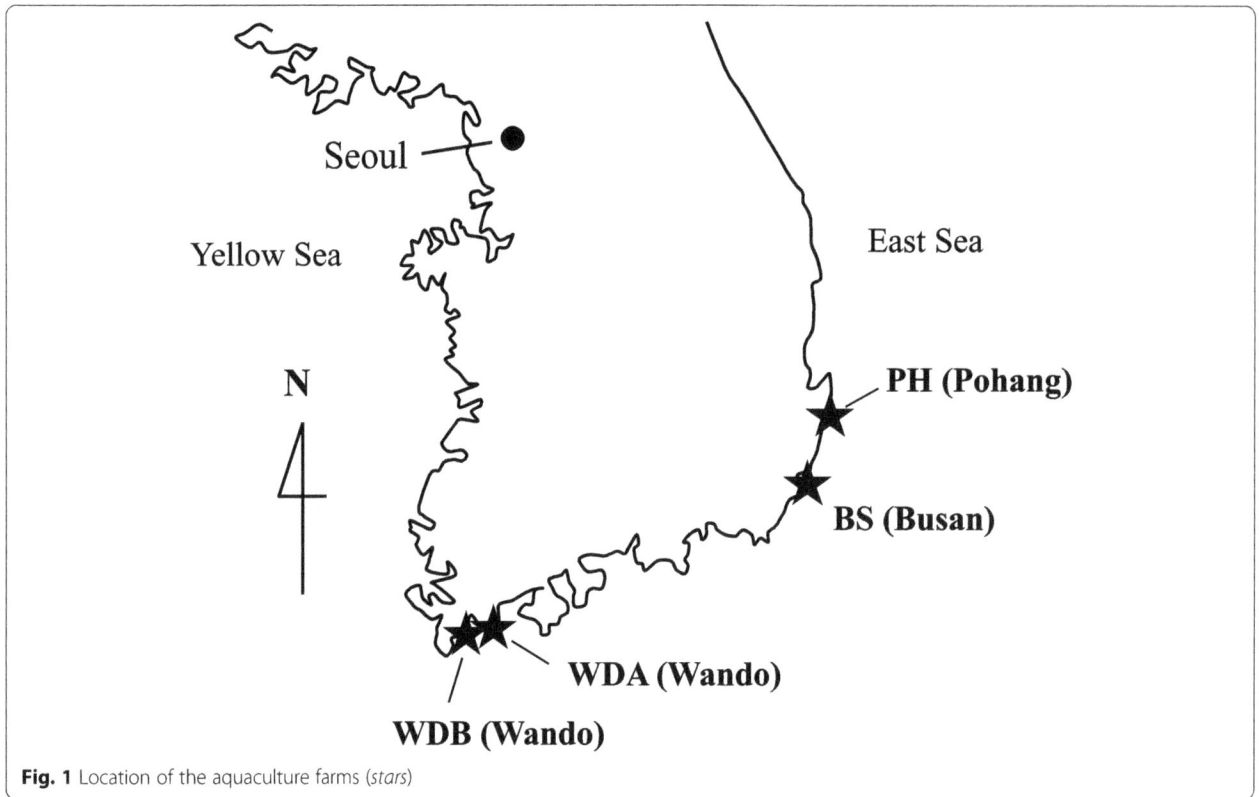

Fig. 1 Location of the aquaculture farms (*stars*)

were used for tissue dehydration and paraffin embedding. The tissue sections were then stained with hematoxylin and eosin.

Results

Detection of epiphytic organisms

Table 1 shows the results of the detection of epiphytic organisms. Of the 740 kelp samples, 208 had six kinds of epiphytic organisms attached, including hydroid, bryozoan, polychaete, algae, caprellid, and oyster (Fig. 2). Of these, hydroid, bryozoan, polychaete, algae, and caprellid were frequently observed and were detected in the kelp samples with rates of 11.6% (86/740), 6.4% (47/740), 3.4% (25/740), 3.2% (24/740), and 3% (22/740), respectively.

Two hundred thirteen samples were examined in the Wando A farm (Table 1). Hydroid, bryozoan, algae, caprellid, and oyster were detected in 7.5% (16/213), 5.6% (12/213), 5.6% (12/213), 2.8% (6/213), and 0.9% (4/213) kelp, respectively, and these organisms were only observed during the period from June to September in 2014. Twenty to 100% were found to be hydroid-positive, 50 to 71.4% were bryozoan-positive, and 14.3 to 100% were algae-positive. Twenty and 71.4% of the caprellid were observed on hydroids, but not on kelp. The oyster (57.1%) was observed in the September 2014 sample. Epiphytic organisms were not observed among the kelp during April to May in 2014 and January to April in 2015.

The infestation areas for hydroid, bryozoan, and algae were 10 to 85%, 0.3 to 17%, and 20 to 30%, respectively (Table 2). Two and eight caprellids were observed in the June sample and in the August sample of 2014, respectively.

Two hundred fifty-three samples were examined in the Wando B farm (Table 1). Hydroid, bryozoan, polychaete, algae, and caprellid were detected in 13% (33/253), 2% (5/253), 2% (5/253), 1.6% (4/253), and 0.4% (1/253) kelp, respectively, and these organisms were observed from June to September in 2014. 57.1 to 100% were found to be hydroid-positive, and 33.3, 35.7, and 13.3 to 28.6% were bryozoan-positive, polychaete-positive, and algae-positive, respectively. The caprellid (7.1%) was observed on the hydroids in the July 2014 sample. No epiphytic organisms were observed in the kelp in May 2014 and from February to April 2015. The infestation areas for hydroid gradually increased from 10 to 55% as the water temperature increased, but these were lower than those of the Wando A farm (Table 2). The infestation areas for bryozoan, polychaete, and algae were observed in 0.13% (0.06–0.5%) in the September sample, 3% (1–5%) in the July sample, and 5% (3–6%) in the June sample, 2014, respectively. Two caprellids were observed in the July 2014 sample.

Two hundred thirty-six samples were examined in the Busan farm (Table 1). Hydroid, bryozoan, algae, and

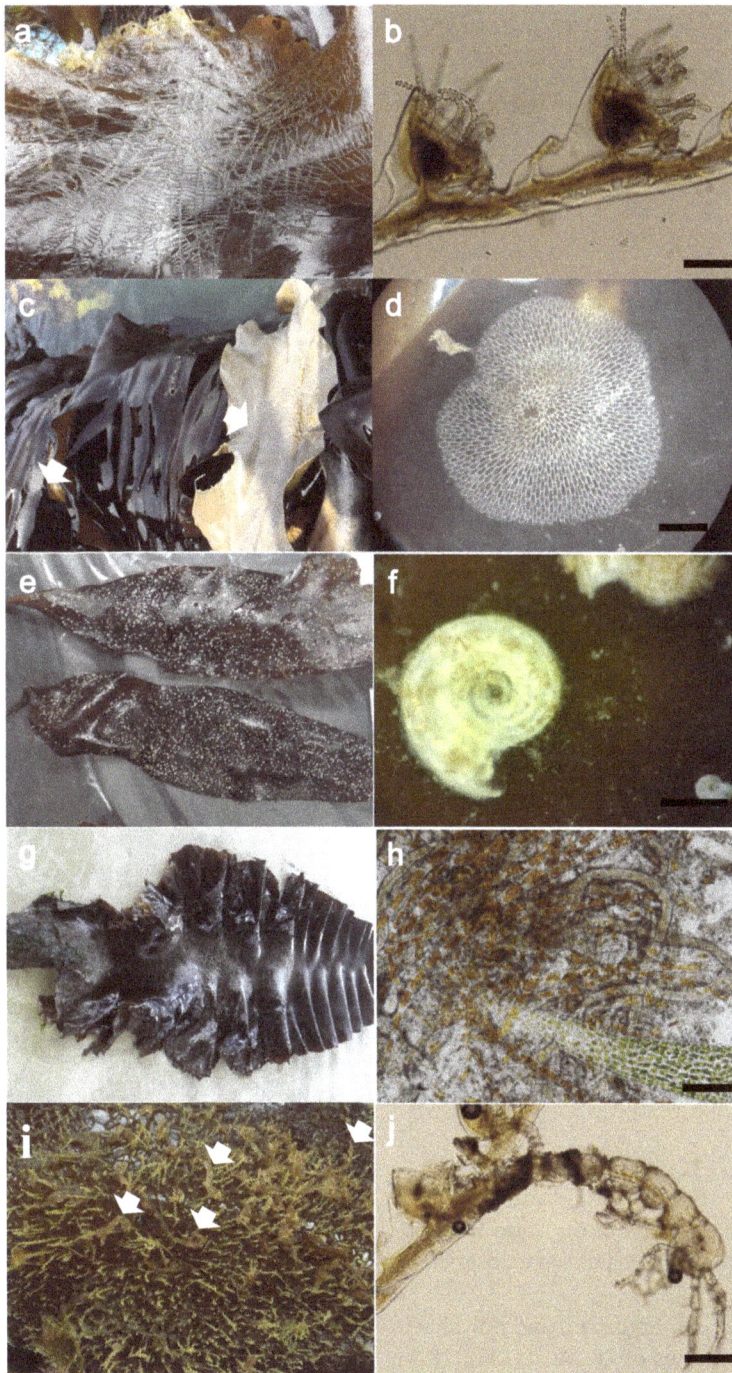

Fig. 2 Photos of epiphytic organisms on cultured kelp. Hydroid (**a**), bryozoan (**c**), polychaete (**e**), algae (**g**), and caprellid (**i**) encrusting blades of kelp. Magnification of hydroid (**b**), bryozoan (**d**), polychaete (**f**), algae (**h**), and caprellid (**j**). *Bar* = 100 μm (**b**, **h**, and **j**) and 1000 μm (**d**, **f**)

caprellid were detected in kelp with an incidence rate of 11% (26/236), 9.7% (23/236), 3.4% (8/236), and 4.2% (10/236), respectively, and these organisms were observed from April to July 2014. One hundred percent of the samples were found to be hydroid-positive from May to July 2014. Ten to 90% and 30 to 71.4% were bryozoan-positive and algae-positive, respectively. No epiphytic organisms were observed among the kelp from February to April 2015. The infestation areas of the hydroid were 14.4 to 37%, which are lower than those of the two farms in Wando (Table 2). In contrast, the infestation areas for bryozoan were of 0.04 to 37.9%, which were

Table 2 Infestation areas for epiphytic organisms in kelp

Place	Farm	Date	Water temperature (°C)	Mean thalli length (cm)	Mean infection area of epiphytic organisms % (epiphytic area/total area)					
					Hydroid	Bryozoan	Polychaete	Algae	Caprellid[a]	Oyster
Wando	WDA	Apr 2014	11.5	340 (200–380)	0	0	0	0	0	0
		May 2014	15.5	260 (260–270)	0	0	0	0	0	0
		Jun 2014	17.4	NT	10 (5–20)	0	0	0	2 (1–5)	0
		Jul 2014	21.7	190 (100–240)	85 (70–100)	4 (0.8–7.9)	0	30 (20–50)	0	0
		Aug 2014	23	200 (100–240)	81 (60–95)	0.3 (0.2–0.6)	0	20 (10–35)	8 (3–17)	0
		Sep 2014	22.1	120 (90–160)	NT	17 (0.2–48.7)	0	0	0	2.35 (0.4–4.3)
		Jan 2015	9	41 (20–71)	0	0	0	0	0	0
		Jan 2015	8.3	82 (39–122)	0	0	0	0	0	0
		Apr 2015	11	236 (200–280)	0	0	0	0	0	0
	WDB	May 2014	15.4	220 (100–330)	0	0	0	0	0	0
		Jun 2014	21	260 (180–320)	10 (5–25)	0	0	5 (3–6)	0	0
		Jul 2014	21.5	160 (130–200)	41.8 (10–80)	0	3 (1–5)	0	2 (1–5)	0
		Sep 2014	21.6	130 (60–230)	55 (20–90)	0.13 (0.06–0.5)	0	0	0	0
		Feb 2015	9	57 (30–100)	0	0	0	0	0	0
		Mar 2015	11	97 (50–180)	0	0	0	0	0	0
		Apr 2015	12.5	127 (90–190)	0	0	0	0	0	0
Busan	BS	Apr 2014	11	170 (50–270)	0	0.04 (0.04)	0	0	0	0
		May 2014	14.5	230 (130–280)	14.4 (5–30)	7.3 (1.8–14.1)	0	0	0	0
		Jun 2014	21	220 (180–300)	34 (10–60)	2.8 (0.03–11.6)	0	34.1 (5–70)	0	0
		Jul 2014	20.4	160 (120–180)	37 (20–60)	37.9 (1.1–86)	0	0	3 (1–5)	0
		Feb 2015	11.5	76 (26–175)	0	0	0	0	0	0
		Feb 2015	12	143 (80–219)	0	0	0	0	0	0
		Mar 2015	12.5	182 (110–260)	0	0	0	0	0	0
		Apr 2015	13	203 (120–270)	0	0	0	0	0	0
Pohang	PH	Jul 2014	20.5	140 (50–220)	3.3 (3–5)	0.5 (0.5)	9.5 (3–20)	0	0	0
		Aug 2014	23.4	110 (80–150)	5.3 (1–10)	1.2 (0.2–3.6)	50 (40–60)	0	0	0
		Sep 2014	23.1	70 (30–110)	3 (2–4)	0	90 (80–95)	0	0	0

NT not tested
[a]Mean number of caprellid

the highest in the four farms. 34.1% of the algae were observed in June 2014, and three caprellids were observed in July 2014.

In the Pohang farm, hydroid, bryozoan, and polychaete were detected in kelp with rates of 28.9% (11/38), 18.4% (7/38), and 52.6% (20/38), respectively, from July to September 2014 (Table 1). The infestation areas for hydroid and bryozoan were less than 5.5%, which were the lowest in the four farms (Table 2). In contrast, the infestation areas for polychaete were 9.5 to 90%, which were the highest of the four farms.

Histopathology
The hydroid, bryozoan, and algae that were attached to the thallus were examined (Fig. 3). The hydroid and

bryozoan were attached to the cuticula of the thallus but did not penetrate into the thallus (Fig. 3a–d). In contrast, some of the algae penetrated the epidermis and attached to the cuticula of thallus (Fig. 3e, f).

Discussion
Korean farms have been cultivating kelp since the 1970s, but to date, no systematic research on their disease has been conducted. In this study, we investigated the presence of epiphytic organisms in four kelp farms in the coastal area of Korea. Of the 740 kelp samples, 208 samples had epiphytic organisms attached, including hydroid, bryozoan, polychaete, algae, caprellid, and oyster. The predominant epiphytic organisms were hydroid (detection rate: 11.6%) followed by bryozoan (6.4%),

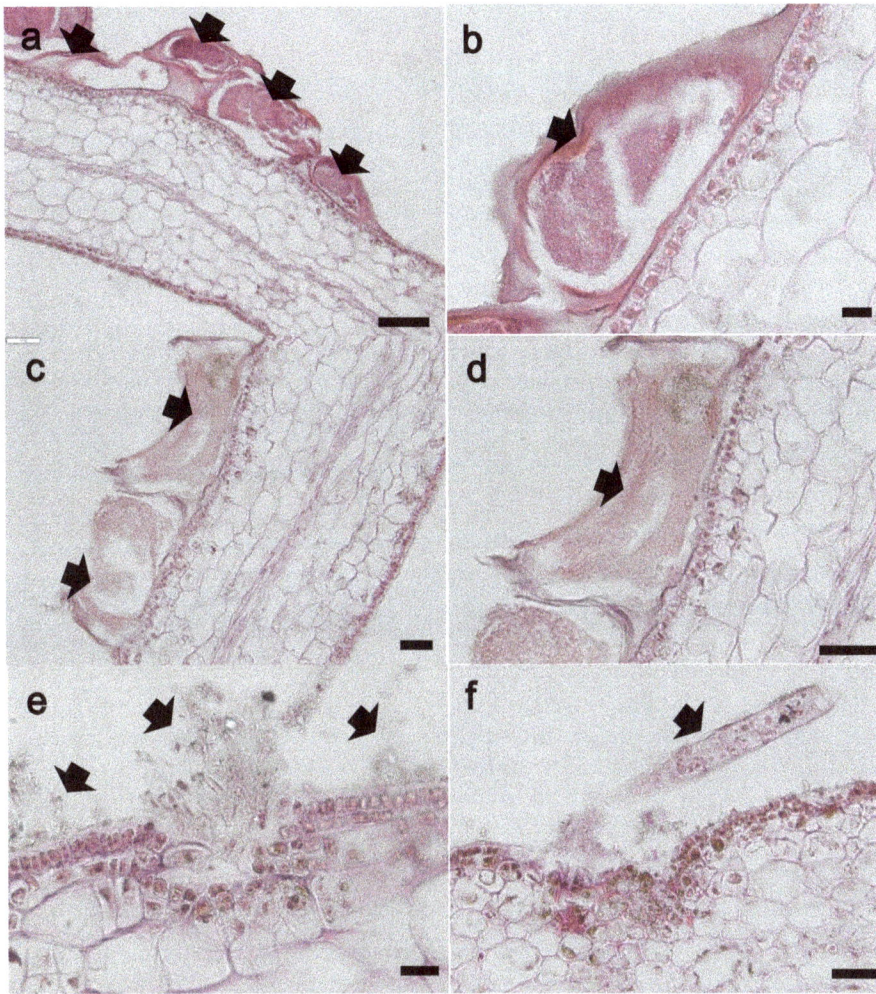

Fig. 3 Histological section of kelp infected with hydroid (**a**, **b**), bryozoan (**c**, **d**), and algae (**e**, **f**). The *arrows* indicate the epiphytic organisms. *Bar* = 20 μm (**b**, **e**), 50 μm (**c**, **d**, and **f**), 100 μm (**a**)

polychaete (3.4%), algae (3.2%), and caprellid (3%), as observed from May to September. No epiphytic organisms were observed among the kelp from January to April, except for one sample from Busan. These results indicate that at least six kinds of epiphytic organisms were observed in kelp farms in Korea, and their infestation was significantly higher in water with a higher temperature. Moreover, encrusting hydroid and bryozoan were the most predominant form of infestation in kelp farms, even though their infestation rates were different among the kelp farms. Encrusting by hydroid has been reported to be abundant on farmed kelp in Korea at higher water temperatures (Park and Hwang, 2012). The infestation rate from May to July was of about 97%, and it was below 26% from February to April. These results are similar to the results of the present study. Encrusting by bryozoan (*Membranipora*) has been reported to be abundant on kelp blades in Atlantic Nova Scotia (Canada) and

the Gulf of Maine (USA) (Berman et al., 1992; Saunders and Metaxas, 2008; Scheibling and Gagnon, 2009). The calcified *Membranipora* zooids present a firm mechanical barrier on epidermal tissue. This barrier can affect exchange processes such as mineral nutrient uptake between kelp epidermis and surrounding seawater (Hurd et al., 1994; 2000), or it can interfere with the photophysiology of the host alga (Oswald et al., 1984; Cancino et al., 1987; Muñoz et al., 1991). In this study, even though the effect of the bryozoan on kelp is unknown, Korean kelp farm is suffered from infestation by bryozoan.

It is unclear how the epiphytic organisms attach to the thallus tissue. A histopathogical examination revealed that hydroid and bryozoan were attached on the cuticula of thallus while some algae attached to the cuticula of the thallus or penetrated the epidermis. These results indicate that kelp tissue can be broken by some algae, but not by hydroid and bryozoan.

The cultivation period for kelp in Korea had traditionally spanned from December to July, with a harvest of thalli occurring between June and July for use with edible marine vegetable food. Recently, the harvest season has been shortened from March to May due to the presence of these epiphytic organisms. Moreover, kelp that is severely infected by hydroid and bryozoan is used as abalone feed during the summer, even though their effect on abalone remains. In addition, there are many environmental factors that can affect the prevalence of the epiphytes on kelps. Therefore, further studies are necessary to elucidate which environmental factors are related to the epiphytic organisms' infestation, to understand how to prevent the epiphytic organisms from infesting kelp farms and to assess their effect when used as abalone feed.

Conclusion

In conclusion, we investigated the presence of epiphytic organisms in four kelp Saccharina japonica farms in the coastal area of Korea from 2014 to 2015. The infestation rate for hydroid, bryozoan, and polychaete was significantly higher in the Wando farm, Busan farm, and Pohang farm, respectively. Epiphytic organisms were generally observed during May to September. The histopathogical examination revealed that hydroid and bryozoan organisms were attached on the cuticula of the thallus. These results indicate that hydroid and bryozoan were the most predominant epiphytic organisms in Korean kelp farms.

Acknowledgements
This study was supported by the National Fisheries Research and Development Institute (R2016069).

Funding
This study was supported by the National Fisheries Research and Development Institute (R2016069).

Authors' contributions
JOK, WSK, HNJ, and SJC carried out the experiments. JOK and WSK participated to write the manuscript. MJO conceived of the study and participated in its design. JSS and MAP analyzed the data. All authors read and approved the final manuscript.

Competing interests
The authors declare that they have no competing interests.

Author details
[1]Department of Aqualife Medicine, Jeonnam National University, Yeosu 59626, South Korea. [2]Algae Research Institute, JeollaNamdo, Wando 59146, South Korea. [3]Aquatic Life Disease Control Division, Fundamental Research Department, National Fisheries Research and Development Institute, Busan 46083, South Korea.

References
Berman J, Harris L, Lambert W, Buttrick M, Dufresne M. Recent invasions of the Gulf of Maine: three contrasting ecological histories. Conserv Biol. 1992;6:435–41.
Cancino JM, Muñoz J, Muñoz M, Orellana MC. Effects of the bryozoan Membranipora tuberculata (Bosc.) on the photosynthesis and growth of Gelidium rex Santelices et Abbott. J Exp Mar Biol Ecol. 1987;113:105–12.
FAO (Food and Agriculture Organization of the United Nations). The state of world fisheries and aquaculture. Rome: Food and Agriculture Department. Food and Agriculture Organization of the United Nations; 2014.
Gong YG, Hwang IG, Ha DS, Hwang MS, Hwang EK, Lee SY, Park EJ. Study on the Saccharina culture technique for abalone feed, Report of National Fisheries Research & Development Institute. 2010. p. 42–9.
Hurd CL, Durante KM, Chia FS, Harrison PJ. Effect of bryozoan colonization on inorganic nitrogen acquisition by the kelps Agarum fimbriatum and Macrocystis integrifolia. Mar Biol. 1994;121:167–73.
Hurd CL, Durante KM, Harrison PJ. Influence of bryozoan colonization on the physiology of the kelp Macrocystis integrifolia (Laminariales, Phaeophyta) from nitrogen-rich and -poor sites in Barkley Sound, British Columbia, Canada. Phycologia. 2000;39:435–40.
Kim HS, Lee KS. Faunal studies on the genus Caprella (Crustacea: Amphipoda, Caprellidae) in Korea. Kor J Zool. 1975;18:115–26.
KOSIS (Korean statistical information service). 2015. Fishery production survey: statistics by type of fishery and species. Retrieved from http://kosis.kr/eng/statisticsList/statisticsList_01List.jsp?vwcd=MT_ETITLE&parmTabId=M_01_01#SubCont. On March 2016.
Lee CM. A systematic study on Korean Caprellids (Crustacea, Amphipoda) in the east sea. Yongin: Thesis for Master's degree. Dan Kook University; 1994
Muñoz J, Cancino JM, Molina MX. Effect of encrusting bryozoans on the physiology of their algal substratum. J Mar Biol Assoc UK. 1991;71:877–82.
Oswald RC, Telford N, Seed R, Happey-Wood CM. The effect of encrusting bryozoans on the photosynthetic activity of Fucus serratus L. Estuar Coast Shelf Sci. 1984;19:697–702.
Park JH. Invertebrate fauna of Korea. Athecates. National institute of biological resources ministry of environment; 2011
Park JH. Invertebrate fauna of Korea. Thecates. National institute of biological resources ministry of environment; 2010
Park CS, Hwang EK. Seasonality of epiphytic development of the hydroid Obelia geniculata on cultivated Saccharina japonica (Laminariaceae, Phaeophyta) in Korea. J Appl Phycol. 2012;24:433–9.
Saunders M, Metaxas A. High recruitment of the introduced bryozoan Membranipora membranacea is associated with kelp bed defoliation in Nova Scotia, Canada. Mar Ecol Prog Ser. 2008;369:139–51.
Saunders M, Metaxas A. Effects of temperature, size, and food on the growth of Membranipora membranacea in laboratoy and field studies. Mar Biol. 2009;156:2267–76.
Scheibling RE, Gagnon P. Temperature-mediated outbreak dynamics of the invasive bryozoan Membranipora membranacea in Nova Scotian kelp beds. Mar Ecol Prog Ser. 2009;390:1–13.
Schwaninger H. Population structure of the widely dispersing marine bryozoan Membranipora membranacea (Cheilostomata): implications for population history, biogeography, and taxonomy. Mar Biol. 1999;135:411–23.
The Korean Society of Systematic Zoology. Seoul: Zoological taxonomy. Jiphyupsa; 2014

10

The first record of a frogfish, *Fowlerichthys scriptissimus* (Antennariidae, Lophiiformes), from Korea

Song-Hun Han[1], Joon Sang Kim[2] and Choon Bok Song[3*]

Abstract

This is the first report of *Fowlerichthys scriptissimus* (Lophiiformes, Antennariidae) from Korea. A single specimen (291.0 mm SL) was collected off the coast of Jejudo Island by gill net on 28 March 2012 and identified with morphological and molecular approaches. The specimen is characterized by having all five pelvic fin rays bifurcate and possessing 20 vertebrae, 13 pectoral-fin rays, and a basidorsal ocellus on the side of the body. This species is distinguishable from other Korean taxa by the number of pectoral fin rays, the bifurcate form of the pelvic rays, and the vertebral count. We add this species to the Korean fish fauna and suggest new Korean names, "Byeol-ssin-beng-i-sok" and "Byeol-ssin-beng-i" for the genus and species, respectively.

Keywords: *Fowlerichthys scriptissimus*, Antennariidae, New record, Jejudo Island, Korea

Background

The frogfishes (Antennariidae), which belong to order Lophiiformes, occur in all tropical and subtropical seas except the Mediterranean Sea (Nelson 2006). Worldwide, the family includes 46 species in 13 genera (Arnold and Pietsch 2012) and four species in two genera in Korea (Kim et al. 2005; Kim et al. 2011). All members of Antenariidae have the first dorsal fin spine modified into a fishing pole (illicium) and gill openings below or behind the base of the pectoral fin (Nelson 2006; Arnold and Pietsch 2012). The genus *Fowlerichthys* was originally suggested by Barbour (1941) when he described the new species, *Fowlerichthys floridanus*. However, as some of his morphological descriptions were not clear, most ichthyologists regarded the genus *Fowlerichthys* as a synonym of *Antennarius* (Pietsch 1984; Senou 2002; Manilo and Bogorodsky 2003). Later, Arnold and Pietsch (2012) reconstructed evolutionary relationships within Antennariidae with molecular phylogenetics and recognized that *Fowlerichthys* as a valid genus. They demonstrated that it can be separated from *Antennarius* by having all five pelvic fin rays bifurcate instead of one in the genus *Antennarius*.

A single specimen of *Fowlerichthys scriptissimus*, which is otherwise unknown from Korea, was collected in the coastal waters of Jejudo Island by gill net on 28 March 2012. Here, we describe the morphological characters of *F. scriptissimus* and report the results of a molecular barcode determination of the specimen's identification using the COI gene.

Methods

Counts and measurements followed the method of Hubbs and Lagler (1964). When conducting this study, we adhered to the ethical guideline of the International Council for Laboratory Animal Science (ICLAS) for researchers. The specimen was fixed in 10% buffered formalin and then transferred to 70% ethanol. Vertebrae were counted from radiographs (REX-525R, listem). The examined specimen was deposited at the Jeju National University (JNU), Korea, and is available from the corresponding author by reasonable request.

Total DNA was extracted from 25 mg of the muscle tissue with an AccuPrep Genomic DNA Extraction Kit (Bioneer Inc.) according to manufacturer's protocol. The mitochondrial cytochrome *c* oxidase subunit I (COI) gene was PCR-amplified with the primers, Asn-F1 (AAA HWC

* Correspondence: cbsong@jejunu.ac.kr
[3]College of Ocean Sciences, Jeju National University, Jeju 63243, Korea
Full list of author information is available at the end of the article

TTA GTT AAC AGC TAA) and Ser-R1 (GGG GTT CDA YTC CYC CCT TTC T). The polymerase chain reaction (PCR) was performed on a thermal cycler (TP600, Takara Bio Inc.) with a final volume of 40 µL in a 0.2 mL PCR tube containing 10 mM Tris-HCl (pH 8.3), 50 mM KCl, 2 mM MgCl2, 0.2 mM dNTP mix, 0.5 mM of each primer, 50 ng template DNA solution, and 1 U Ex *Taq* DNA polymerase (Takara Bio Inc.) The PCR cycles consisted of an initial denaturing step of 94 °C for 2 min, followed by 30 cycles of 30 s at 94 °C for denaturation, 1 min at 54 °C for primer annealing, and 1 min and 50 s at 72 °C for extension, and an additional 7 min interval at 72 °C for a final extension. The PCR products were sequenced using the BigDye Terminator v3.1 Cycle Sequencing Kit (Applied Biosystems Inc.) and ABI PrismTM 3730XL DNA Analyzer. Sequencing reactions were conducted with the amplification primers in two directions. The DNA sequence obtained was deposited in the National Center for Biotechnology Information (NCBI) as GenBank accession number, KY195977. For molecular identification, we compared the specimen's COI sequence with the GenBank DNA sequences of the seven anglerfish species: *Antennarius hispidus* (FJ582855), *Antennarius pictus* (FJ582858), *Antennarius striatus* (AB282828), *Histrio histrio* (AB282829), *Fowlerichthys avalonis* (DQ0279840), *F. scriptissimus* (GU188480), and *Lophius litulon* (KJ020931). The DNA sequences were aligned and edited using Clustal W (Thompson et al. 1994) and BioEdit version 7 (Hall 1999). Genetic distances were calculated using MEGA 6 (Tamura et al. 2013) based on the Kimura two-parameter (K2P) model (Kimura 1980). A neighbor-joining (NJ) tree was constructed using the K2P model and 10,000 bootstrap replications in MEGA 6.

Results

Genus *Fowlerichthys* Barbour 1941
(New Korean genus name, Byeol-ssin-beng-i-sok)
Fowlerichthys Barbour 1941; 12 (type species, *Fowlerichthys floridanus* Barbour 1941). Arnold and Pietsch 2012,

128 (Rarotanga, Cook Islands); Stewart 2015, 886 (New Zealand).

The genus *Fowlerichthys* includes five species diagnosed by the following combination of characters: one or three darkly pigmented ocelli on side of body; illicium about as long as second dorsal spine; esca in form of tuft of slender filaments, a simple, oval-shaped appendage, escal pigment spots absent; dorsal rays 12–14 (usually 13); pectoral rays 11–14 (usually 12 or 13); all five pelvic rays bifurcate; anal rays 7–10 (usually 8), all bifurcate; vertebrae 20 (Pietsch and Grobecker 1987; Arnold and Pietsch 2012).

F. scriptissimus (Jordan 1902)
(New Korean name, Byeol-ssin-beng-i) (Table 1; Fig. 1)
Antennarius scriptissimus Jordan, 1902, 373 (type locality, Bōsō Peninsula, Chiba Prefecture, Japan); Pietsch 2000, 597 (South China Sea); Senou 2002, 456 (Japan); Manilo and Bogorodsky 2003, S99 (Oman).

Antennarius sarasa Pietsch 1984, 36 (Japan); Araga 1984, 103 (Japan); Pietsch and Grobecker 1987, 123 (Réunion Island, New Zealand, and Philippines).

F. scriptissimus Arnold and Pietsch 2012, 128 (Rarotanga, Cook Islands)

Material examined
JNU-637, one specimen, 291.0 mm in standard length (SL), gill net, Hanlim-eup, Jeju-si, Jejudo Island, Korea, 28 March 2012.

Description
Meristic counts appear in Table 1. Measurements as a percentage of SL are as follows: body depth 67.6; body width 21.3; snout length 24.0; eye diameter 5.1; interorbital length 14.7; illicium length 6.9; first predorsal fin length 18.3; second predorsal fin length 20.1; third predorsal fin length 35.5; prepectoral fin length 36.7; preanal fin length 67.3; prepelvic fin length 9.5; length of second dorsal spine 8.4; length of third dorsal spine 10.8; length of pectoral fin ray 13.0; length of anal fin ray

Table 1 Morphological characters compared between the present specimen and previous studies on *F. scriptissimus*

Morphological characters	Present study	Jordon (1902)	Araga (1984)	Pietsch and Grobecker (1987)	Senou (2002)
Total length (mm)	350.0 ($n = 1$)	–	350.0 ($n = 1$)	- ($n = 5$)	–
Standard length (mm)	291.0	–	–	99.0–280.0	280.0
Count					
Dorsal fin rays (bifurcate rays)	13 (7)	III-12	III-13	13 (7–9)	13
Pectoral fin rays	13	–	13	13	13
Pelvic fins rays	I, 5	–	5	5	I, 5
Anal fin rays	8	8	8	8	8
Caudal fin rays	9	–	–	9	9

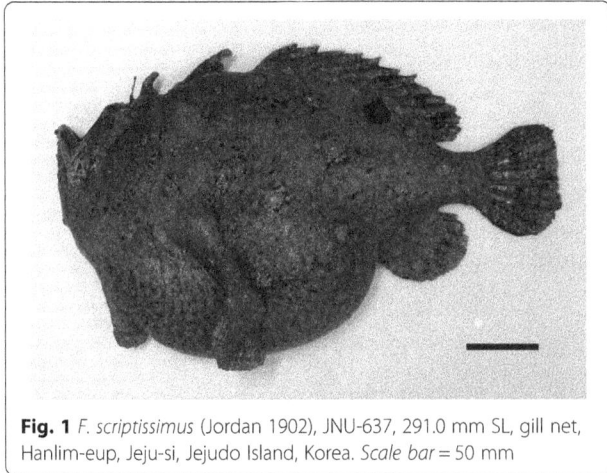

Fig. 1 *F. scriptissimus* (Jordan 1902), JNU-637, 291.0 mm SL, gill net, Hanlim-eup, Jeju-si, Jejudo Island, Korea. *Scale bar* = 50 mm

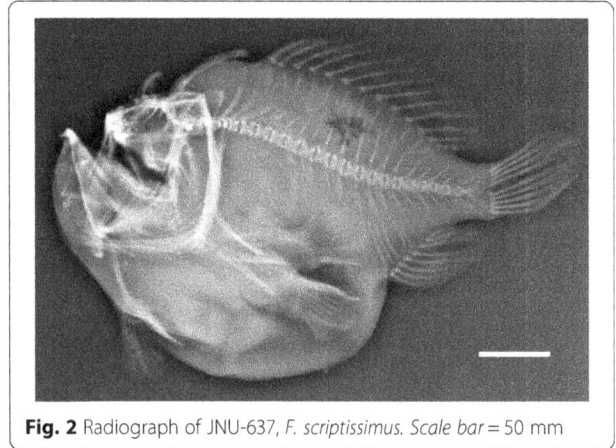

Fig. 2 Radiograph of JNU-637, *F. scriptissimus*. *Scale bar* = 50 mm

12.5; caudal peduncle depth 12.1; and caudal peduncle length 9.2.

Body round and compressed; skin of body very rough and covered with close-set dermal spinules; head and eyes small; eyes lateral; mouth large and extremely oblique; first dorsal spine (illicium) shorter than second spine; esca simple and oval-shaped with a tuft; second and third dorsal spine rough and curved posteriorly; gill opening located below base of pectoral fin; depression between second and third dorsal spines; membrane behind second dorsal spine extending posteriorly, dividing area between second and third dorsal spines and nearly reaching to base of third (Fig. 3).

Live coloration
Whole body uniformly greenish brown; entire head, body, and fins with mottled dark brown reticulations except for inner surface of paired fins; a single, darkly pigmented, basidorsal ocellus on each side of body; scattered beige spots on body; esca with dark pigment at base.

Color in preservative
Body uniformly pale greenish brown; mottled dark brown spots and ocellus; beige spots no longer apparent.

Distribution
Widely known from Indo-West Pacific, including Réunion Island, New Zealand, the Philippines (Pietsch and Grobecker 1987), the Cook Islands (Arnold and Pietsch 2012), the South China Sea (Pietsch 2000), Japan (Jordan 1902; Pietsch 1984; Araga 1984), and Korea (Jejudo Island, present study).

Molecular identification
To confirm and verify the accuracy of the morphological species identification, we also analyzed 634 base pairs of the mitochondrial COI gene. The DNA sequence of the specimen was almost identical to a previously published sequence (GU188480) of *F. scriptissimus* (genetic distance, d = 0.003). The NJ tree also clustered the specimen with the GenBank sample of *F. scriptissimus*, with a strong 100% bootstrap value (Fig. 4).

Discussion
The specimen in question has all the previously reported diagnostic morphological characters of *F. scriptissimus*, including five bifurcate pelvic fin rays, 20 vertebrae, 13 pectoral fin rays, and a basidorsal ocellus on the side of the body (Table 1 and Fig. 2). The form of the esca as a simple, oval-shaped appendage with numerous, more or less parallel, vertically aligned folds also matches the morphology of *F. scriptissimus*. Additionally, the form of the membrane behind the second dorsal spine matches *F. scriptissimus*, by lacking division into naked dorsal

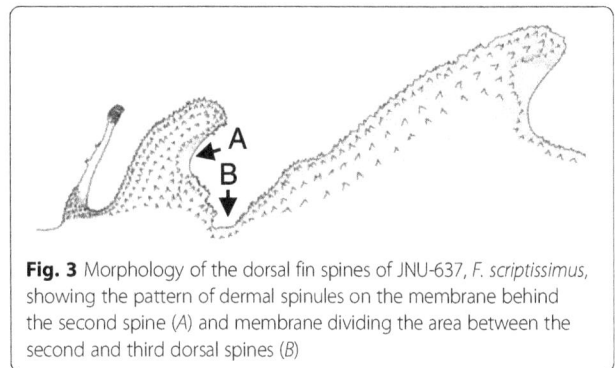

Fig. 3 Morphology of the dorsal fin spines of JNU-637, *F. scriptissimus*, showing the pattern of dermal spinules on the membrane behind the second spine (*A*) and membrane dividing the area between the second and third dorsal spines (*B*)

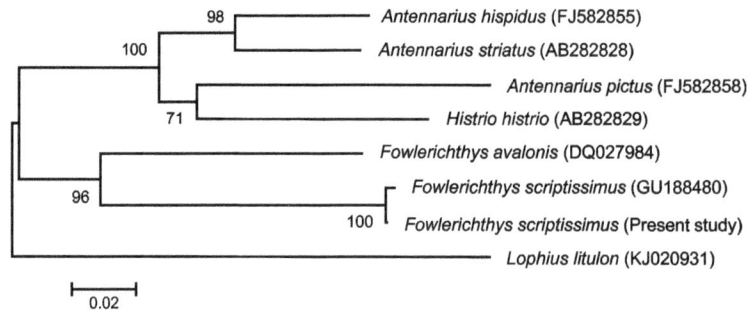

Fig. 4 Neighbor-joining tree showing the relationship of the specimen with the seven species of anglerfishes. *Numbers above nodes* indicate bootstrap probabilities based on 10,000 replications. *Bar* indicates a K2P genetic distance of 0.02

and ventral portions, extending posteriorly to the area between the second and third dorsal spines, and nearly reaching to the base of the third spine (Fig. 3). Thus, the morphological characteristics of the specimen clearly fit the species descriptions given by previous studies (Table 1). In the molecular analysis of the mitochondrial COI gene, the small genetic distance ($d = 0.003$) between the new specimen and a previously sequenced specimen of *F. scriptissimus*, and the close clustering of those two samples in a neighbor-joining tree (Fig. 4) including the seven anglerfish species confirmed the morphological identification. Thus, both morphological and molecular approaches indicated the specimen in the present study to be *F. scriptissimus*.

The genus *Fowlerichthys* was firstly suggested when Barbour (1941) described *Fowlerichthys floridanus*, but his description and diagnosis of the genus were unclear. He mentioned that the strong and sharp dorsal spines of the species were unlike the dorsal fin rays in the species of *Antennarius*, but whether these spines were bare in life is difficult now to determine certainly. Thus, most ichthyologists have regarded the genus *Fowlerichthys* as a synonym of *Antennarius*. Recently, Arnold and Pietsch (2012) demonstrated that *Fowlerichthys* is a valid genus that can be separated from *Antennarius* by having all five pelvic fin rays bifurcate (all other antennariids have

four simple and one bifurcate rays). It can be further separated from all members of *Antennarius* except *Antennarius commerson* by possessing 20 vertebrae (Table 2). Currently, the genus *Fowlerichthys* is recognized as valid, with five known species worldwide *Fowlerichthys radiosus*, *F. avalonis*, *Fowlerichthys senegalensis*, *Fowlerichthys ocellatus*, and *F. scriptissimus* (Froese and Pauly 2016).

F. scriptissimus is easily distinguished from the other Korean frogfishes (*A. pictus*, *A. commerson*, *A. maculatus*, and *H. histrio*) by having 13 pectoral fin rays (vs. 10 in *A. commerson*, 11 in *A. maculatus*, 10 in *A. pictus*, and 10 in *H. histrio*), five bifurcate pelvic rays (vs. one in the genus *Antennarius* and none in *H. histrio*), and 20 vertebrae (vs. 19 in *H. histrio and* all members of *Antennarius* except *A. commerson*) (Table 2).

We propose new Korean names, "Byeol-ssin-beng-i-sok" and "Byeol-ssin-beng-i" for the genus and species, respectively. The Korean name "Byeol-ssin-beng-i" was given for *F. scriptissimus* because of its scattered beige spots on the body that look alike stars. The Korean words "byeol" and "ssin-beng-i" mean star and frogfish, respectively.

Conclusions
Not applicable.

Table 2 Comparison in the number of pectoral fin rays, bifurcate pelvic fin rays, and vertebrae among five frogfish species inhabiting Korea

Species	Pectoral fin rays	Bifurcate pelvic rays	Vertebrae	References
F. scriptissimus	13	5	20	Pietsch and Grobecker (1987)
A. commerson	10–11 (usually 11)	1	19–20 (usually 19)	
A. maculatus	10–11 (usually 10)	1	19	
A. pictus	9–11 (usually 10)	1	19	
H. histrio	9–11 (usually 10)	0	18–19 (usually 19)	

Abbreviations
COI: Cytochrome c oxidase subunit 1; d: Genetic distance

Acknowledgements
This work was supported by a grant from the National Institute of Fisheries Science (R2016034). We thank Prof. B. Sidlauskas, Oregon State University, for the valuable discussion.

Funding
This study was funded by a grant from the National Institute of Fisheries Science (R2016034).

Authors' contributions
SHH performed experiment and wrote the manuscript. JSK collected the sample. CBS conceived of the study and helped to write the manuscript. All authors read and approved the final manuscript.

Competing interests
The authors declare that they have no competing interests.

Author details
[1]Jeju Fisheries Research Institute, National Institute of Fisheries Science, Jeju 63068, Korea. [2]Korea Fisheries Resources Agency, Jeju Branch, Jeju 63005, Korea. [3]College of Ocean Sciences, Jeju National University, Jeju 63243, Korea.

References
Araga C. Family Antennarioidae. In: Masuda H, Amaoka K, Araga C, Uyeno U, Yoshino T, editors. The fishes of the Japanese Archipelago. Tokyo: Tokai Univ Press; 1984. p. 102–3.

Arnold RJ, Pietsch TW. Evolutionary history of frogfishes (Teleostei: Lophiiformes: Antennariidae): a molecular approach. Mol Phylogenet Evol. 2012. doi:10.1016/j.ympev.2011.09.012.

Barbour T. Notes on pediculate fishes. Proc New England Zool Club. 1941;19:7–14.

Froese R, Pauly D, editors. FishBase. World Wide Web electronic publication. www.fishbase.org, version (10/2016); 2016.

Hall TA. BioEdit: a user-friendly biological sequence alignment editor and analysis program for Windows 95/98/NT. Nucleic Acids Symp Ser. 1999;41:95–8.

Hubbs CL, Lagler KF. Fishes of the Great Lakes region. Bull Granbrook Inst Sci. 1964;26:19–27.

Jordan DS. A review of the pediculate fishes or anglers of Japan. Proc US Nat Mus. 1902. doi:10.5479/si.00963801.24-1261.361.

Kim IS, Choi Y, Lee CL, Lee YJ, Kim BJ, Kim JH. Illustrated book of Korean fishes. Seoul: Kyo-Hak Publishing; 2005. p. 615.

Kim BY, Kim MJ, Song CB. First record of the frogfishes Antennarius pictus (Antennariidae, Lophiiformes). Korean J Ichthyol. 2011;23:168–71.

Kimura M. A simple method for estimating evolutionary rates of base substitutions through comparative studies of nucleotide sequences. J Mol Evol. 1980;16:111–20.

Manilo LG, Bogorodsky SV. Taxonomic composition, diversity and distribution of coastal fishes of the Arabian Sea. J Ichthyol. 2003;43:S75–S149.

Nelson JS. Fishes of the world. 4th ed. New Jersey: Wiley; 2006. p. 601.

Pietsch TW. The genera of frogfishes (family Antennariidae). Copeia. 1984. doi:10.2307/1445032.

Pietsch TW. Antennariidae. In: Randall JE, Lim KKP, editors. A checklist of the fishes of the South China Sea, Raffles Bull Zool Suppl, vol. 28. 2000. p. 569–667.

Pietsch TW, Grobecker DB. Frogfishes of the world: systematics, zoogeography, and behavioral ecology. California: Stanford Univ Press; 1987. p. 420.

Senou H. Antennariidae. In: Nakabo T, editor. Fishes of Japan with pictorial keys to the species. Tokyo: Tokai Univ Press; 2002. p. 454–8.

Stewart AL. Family Antennariidae. In: Roberts CD, Stewart AL, Struthers CD, editors. The Fishes of New Zealand, vol. 3. 2015. p. 577–1152.

Tamura K, Stecher G, Peterson D, Filipski A, Kumar S. MEGA6: molecular evolutionary genetics analysis version 6.0. Mol Bio Evol. 2013;30:2725–9.

Thompson JD, Higgins DG, Gibson TJ. CLUSTAL W: improving the sensitivity of progressive multiple sequence alignment through sequence weighting, position specific gap penalties and weight matrix choice. Nucl Acids Res. 1994;22:4673–80.

11

Anti-fatigue activity of a mixture of seahorse (*Hippocampus abdominalis*) hydrolysate and red ginseng

Nalae Kang[1], Seo-Young Kim[1], Sum Rho[2], Ju-Young Ko[1*†] and You-Jin Jeon[1*†]

Abstract

Seahorse, a syngnathidae fish, is one of the important organisms used in Chinese traditional medicine. *Hippocampus abdominalis*, a seahorse species successfully cultured in Korea, was validated for use in food by the Ministry of Food and Drug Safety in February 2016; however. the validation was restricted to 50% of the entire composition. Therefore, to use *H. abdominalis* as a food ingredient, *H. abdominalis* has to be prepared as a mixture by adding other materials. In this study, the effect of *H. abdominalis* on muscles was investigated to scientifically verify its potential bioactivity. In addition, the anti-fatigue activity of a mixture comprising *H. abdominalis* and red ginseng (RG) was evaluated to commercially utilize *H. abdominalis* in food industry. *H. abdominalis* was hydrolyzed using Alcalase, a protease, and the effect of *H. abdominalis* hydrolysate (HH) on the muscles was assessed in C2C12 myoblasts by measuring cell proliferation and glycogen content. In addition, the mixtures comprising HH and RG were prepared at different percentages of RG to HH (20, 30, 40, 50, 60, 70, and 80% RG), and the anti-fatigue activity of these mixtures against oxidative stress was assessed in C2C12 myoblasts. In C2C12 myoblasts, H_2O_2-induced oxidative stress caused a decrease in viability and physical fatigue-related biomarkers such as glycogen and ATP contents. However, treatment with RG and HH mixtures increased cell viability and the content of fatigue-related biomarkers. In particular, the 80% RG mixture showed an optimum effect on cell viability and ATP synthesis activity. In this study, all results indicated that HH had anti-fatigue activity at concentrations approved for use in food by the law in Korea. Especially, an 80% RG to HH mixture can be used in food for ameliorating fatigue.

Keywords: *Hippocampus abdominalis*, Anti-fatigue activity, C2C12 myoblast, Mixture of *Hippocampus abdominalis* hydrolysate and red ginseng

Background

Seahorse is a well-known ingredient in traditional Chinese medicine and is used as an invigorator for the treatment of erectile dysfunction, impotence, wheezing, and nocturnal enuresis. Modern scientific research has proven the pharmaceutical effects of seahorse. *Hippocampus kuda* has various bioactivities such as anti-tumor, anti-aging, and anti-fatigue as well as Ca^{2+} channel blocking properties (Kumaravel et al. 2010). A peptide derived from *H. kuda* has been shown to be effective in chondrocytes and inflammatory arthritis (Kumaravel et al. 2012). In addition,

seahorses have a putative free radical scavenging effect in controlling aging process (Kumaravel et al. 2012). However, the natural source of seahorse has dramatically reduced owing to overfishing, unsustainable trade, and habitat destruction (Qian et al. 2012). Therefore, seahorses became the first commercially valuable marine genus to be protected and included in Appendix II of the Convention on International Trade in Endangered Species (CITES) in 2004 (Segade et al. 2015).

Hippocampus abdominalis is one of the largest seahorse species growing up to 35 cm in length (Perera et al. 2016). It was validated for use as a food ingredient by the Ministry of Food and Drug Safety in February 2016. However, the validation was restricted to 50% of the entire composition. For use in food, we should try to prepare a mixture of *H. abdominalis* by

---

* Correspondence: herolegend@hanmail.net; youjin2014@gmail.com
[†]Equal contributors
[1]Department of Marine Life Sciences, Jeju National University, Jeju 63243, Korea
Full list of author information is available at the end of the article

adding other materials. In addition, biological activities of *H. abdominalis* have rarely been reported so far.

Fatigue is a common distressing condition accompanied by a feeling of extreme physical or mental tiredness that often results in diverse disorders such as anemia, thyroid dysfunction, premature aging, and depression. It could also have adverse effects on work efficiency, physical activities, life quality, and social relationships (Huang et al. 2011). Fatigue is caused by sleep deprivation, inadequate rest, low mood, stress, nutritional imbalance, insufficient exercise, as well as side effects of medications. Chronic fatigue is a persistent unexplainable fatigue lasting for more than 6 months, and it is considered a complex symptom of various neurological, psychiatric, and systemic diseases (Huang et al. 2014). Recently, many researchers have presented the results on the anti-fatigue activity of natural products (Yu et al. 2008; Zhang et al. 2006). Especially, red ginseng has been mainly focused on its anti-fatigue activity with the ability mitigating exercise-related muscle damage, maintaining homeostasis of the body and enhancing vital energy (Kim et al. 2013; Kim et al. 2016). On the other hand, anti-fatigue activity of seahorse has not been scientifically proven although seahorse is a well-known traditional Chinese medicine.

Oxidative stress is caused by an imbalance between reactive oxygen species (ROS) and antioxidant molecules. Excess accumulation of ROS causes oxidative damage by reacting with biomolecules including DNA, membrane lipids, cellular proteins, and diverse pathological states (Kang et al. 2013). Oxidative stress and ROS are the most important causes of exercise-induced disturbances (Fan et al. 2016). In particular, an oxidative imbalance in the skeletal muscle results in increased muscle fatigability. Thus, antioxidants can be used to alleviate fatigue by counteracting the oxidative stress (Nam et al. 2016).

In this study, the effect of *H. abdominalis* on muscles was investigated to scientifically verify its potential bioactivity. Also, the anti-fatigue activity of a mixture comprising *H. abdominalis* and red ginseng was investigated to evaluate the synergy effect and to utilize *H. abdominalis* in the food market. The anti-fatigue activity of *H. abdominalis* and a mixture was evaluated by measuring the levels of physical fatigue-related biomarkers such as serum glycogen and ATP contents.

Methods
Materials
H. abdominalis was kindly donated by Corea Center of Ornamental Reef & Aquariums CCORA (Jeju, Korea) and lyophilized at –70 °C using a freeze dryer. The lyophilized *H. abdominalis* powder was stored at –80 °C until use. Red ginseng extract containing 30% saponin was purchased from ILHWA Co., LTD. (Gyeonggi, Korea) and

lyophilized at –70 °C using a freeze dryer. The lyophilized red ginseng powder was stored at –80 °C until use. Alcalase, a commercial food-grade protease, was purchased from Novozyme Co. (Novozyme Nordisk, Bagsvaerd, Denmark). The other chemicals and reagents used were of analytical grade.

Preparation of *H. abdominalis* hydrolysate (HH)
The enzymatic hydrolysis of *H. abdominalis* was performed using Alcalase under optimal conditions (50 °C and pH 8). The dried *H. abdominalis* powder was homogenized in distilled water and hydrolyzed using the enzyme at an enzyme/substrate (E/S) ratio of 1:100 for enzymatic reactions. The optimal pH of the homogenates was adjusted before enzymatic hydrolysis. The mixture was incubated for 24 h at the optimal temperature for each homogenate, with stirring, and then boiled for 10 min at 100 °C to inactivate the enzyme. After filtration, all hydrolysates were stored at –70 °C for further experiments.

Preparation of the *H. abdominalis* mixture
The *H. abdominalis* mixtures were prepared by adding different concentrations of red ginseng (RG). The lyophilized *H. abdominalis* powder and RG powder were mixed as indicated in Table 1, and these seven mixtures were labeled as % of RG.

Cell culture
The C2C12 myoblasts obtained from American Type Culture Collection (ATCC, Manassas, VA, USA) were cultured in Dulbecco's Modified Eagle Medium (DMEM) supplemented with 10% heat-inactivated fetal bovine serum (FBS), streptomycin (100 mg/mL), and penicillin (100 u/mL) at 37 °C in a 5% CO_2 humidified incubator. To induce differentiation, 80% confluent cultures were switched to DMEM containing 2% horse serum (HS) for 6 days with medium changes every other day.

Cell viability
The cytotoxicity of the samples on C2C12 myoblasts was determined by colorimetric MTT assays. The cells

Table 1 Percentage of HH and RG to prepare the *H. abdominalis* mixtures

HH (%)	RG (%)	The mixture (%)
80	20	100
70	30	100
60	40	100
50	50	100
40	60	100
30	70	100
20	80	100

were seeded at 5×10^4 cells per well into 48-well plates. After the induction of differentiation, the cells were treated with various concentrations of the sample and incubated for an additional 24 h at 37 °C. MTT stock solution (100 µL; 2 mg/mL in PBS) was then added to each well. After incubating for 4 h, the plate was centrifuged at 500 g for 10 min, and the supernatant was aspirated. The formazan crystals in each well were dissolved in dimethyl sulfoxide (DMSO). The amount of purple formazan was determined by measuring the absorbance at 540 nm.

Cell proliferation assay
The cell proliferation effect of the samples on C2C12 myoblasts was determined by using 5-bromo-2′-deoxyuridine (BrdU) assay (Millipore, Billerica, MA, USA). The cells were seeded at 1×10^4 cells per well into 48-well plates. After the induction of differentiation by switching media, the cells were treated with various concentrations of the sample and incubated for an additional 72 h at 37 °C. Then, the cell proliferation was determined by BrdU reagent following manufacture protocol. In briefly, 10 µL of BrdU reagent was added to each well and the cells were incubated for 2 h. After incubation, the cells were fixed using 100 µL fixing solution. Then, the cells were washed using wash buffer and 50 µL of anti-BrdU monoclonal was added to each well and the cells were incubated for 1 h at RT. The cells were washed using wash buffer and 50 µL of goat anti-mouse IgG was added to each well, and the cells were incubated for 30 min at RT. Also, 50 µL of TMB substrate was added to each well and then, 50 µL of stop solution was added to each well. Finally, the cell proliferation was calculated by comparison with the absorbance at 450 nm of standard solutions of BrdU in the non-treated cells.

Anti-fatigue activity in oxidative stress-induced C2C12 myoblasts
The anti-fatigue activity was determined by measuring cell proliferation as well as the glycogen, ATP contents in H_2O_2-treated C2C12 myoblasts. The cells were seeded into 48-well plates. Then, they were treated with various concentrations of the sample during the differentiation period. After differentiation, fatigue was induced by adding H_2O_2 to each well at a concentration of 100 µM; then, the cells were incubated for an additional 24 h at 37 °C.

Measurement of fatigue-related biochemical parameters
To investigate the effect of the samples on muscle growth, we determined several factors such as glycogen and ATP contents in C2C12 myoblasts. For analysis of the effects of the sample on glycogen accumulation, the glycogen content in the cells was measured via glycogen assay (Abcam, Cambridge, MA, USA). Glucoamylase hydrolyzes glycogen to glucose, which was then specifically oxidized to form an intermediate product that reacts with OxiRed probe to generate color. The color was detected by measuring the absorbance at 450 nm. To determine ATP contents, the cell lysates were deproteinized with 4 M perchloric acid (PCA) and 2 M KOH, and the supernatant was assessed using ATP assay kits (Abcam, Cambridge, MA, USA).

Statistical analysis
All measurements were made in triplicate, and all values were represented as means ± SE. The results were subjected to analysis of variance using Tukey's test to analyze the differences. $p < 0.05$ and $p < 0.01$ were considered significant.

Results and discussion
Cytotoxicity of HH
Cell viability was estimated using the MTT assay, which is a test of metabolic competence predicated upon the assessment of mitochondrial performance. It is a colorimetric assay, which is dependent on the conversion of yellow tetrazolium bromide to its purple formazan derivative by mitochondrial succinate dehydrogenase in the viable cells (Kang et al. 2012). The viabilities of C2C12 myoblasts treated with different concentrations of HH (50, 100, 150, and 200 µg/mL) were expressed to represent 100% viability (the viability of control cells; Fig. 1). In a preliminary experiment, HH concentrations up to 200 µg/mL showed no significant cytotoxicity for 24 h.

Effect of HH on cell proliferation
HH significantly enhanced cell proliferation in C2C12 myoblasts compared with the control cells (Fig. 2). In particular, HH treatment induced cell proliferation in a concentration-dependent manner in the range of 100–200 µg/mL. The cell numbers increased approximately 1.8-fold by HH treatment at 200 µg/mL concentration (Fig. 2).

Effect of HH on glycogen contents
The skeletal muscles are the major site of glycogen storage in the body (Deshmukh et al. 2015). The glycogen content in C2C12 myoblasts was increased by HH treatment at concentrations of 50 and 100 µg/mL (Fig. 3). In C2C12 myoblasts, HH (100 µg/mL) increased glycogen content by 1.5-fold compared with that in the control cells. However, HH treatment at higher concentrations (150 and 200 µg/mL) decreased glycogen content. Thus, it can be suggested that high concentrations of HH suppressed glycogen content.

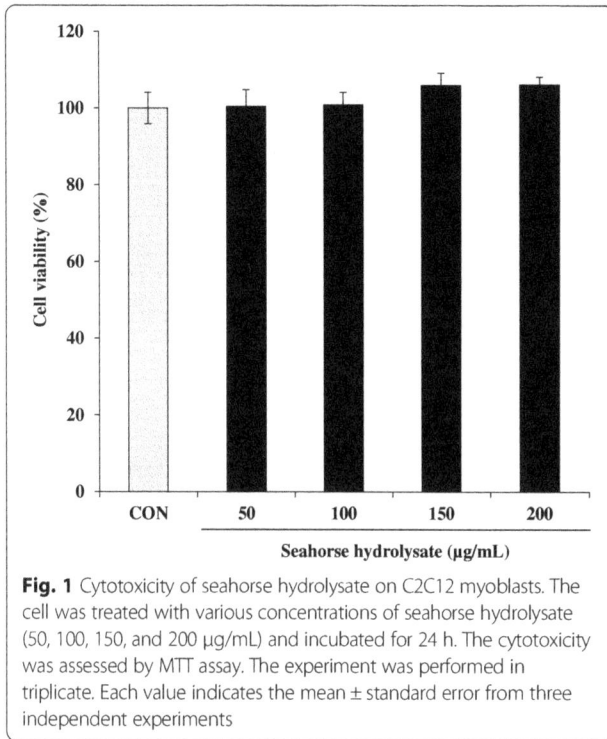

Fig. 1 Cytotoxicity of seahorse hydrolysate on C2C12 myoblasts. The cell was treated with various concentrations of seahorse hydrolysate (50, 100, 150, and 200 μg/mL) and incubated for 24 h. The cytotoxicity was assessed by MTT assay. The experiment was performed in triplicate. Each value indicates the mean ± standard error from three independent experiments

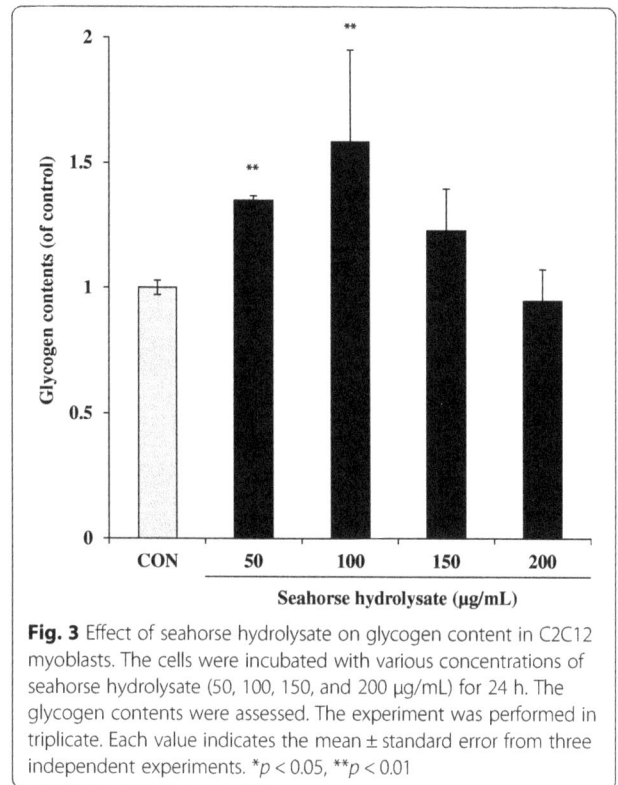

Fig. 3 Effect of seahorse hydrolysate on glycogen content in C2C12 myoblasts. The cells were incubated with various concentrations of seahorse hydrolysate (50, 100, 150, and 200 μg/mL) for 24 h. The glycogen contents were assessed. The experiment was performed in triplicate. Each value indicates the mean ± standard error from three independent experiments. $*p < 0.05$, $**p < 0.01$

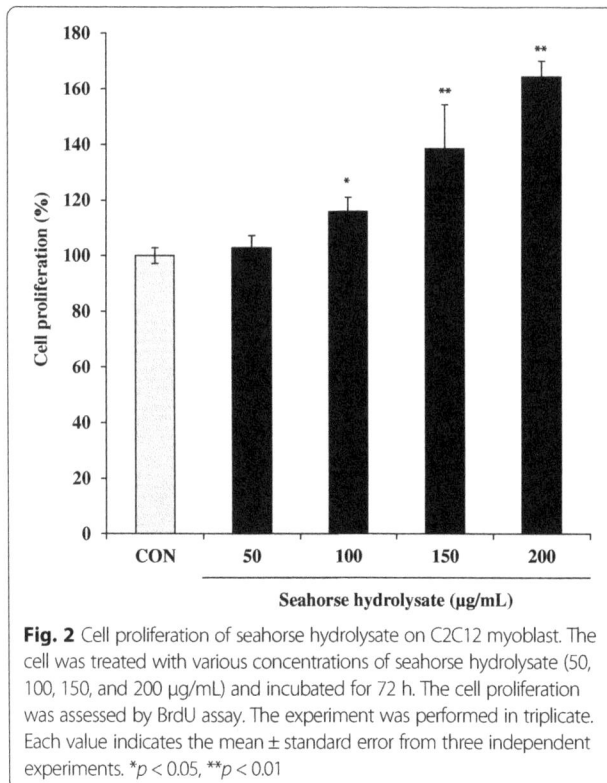

Fig. 2 Cell proliferation of seahorse hydrolysate on C2C12 myoblast. The cell was treated with various concentrations of seahorse hydrolysate (50, 100, 150, and 200 μg/mL) and incubated for 72 h. The cell proliferation was assessed by BrdU assay. The experiment was performed in triplicate. Each value indicates the mean ± standard error from three independent experiments. $*p < 0.05$, $**p < 0.01$

Cytotoxicity of the mixture of HH and RG

Although *H. abdominalis* was validated for use in food by the Ministry of Food and Drug Safety in February 2016, the validation was restricted to 20% of the entire composition. To use *H. abdominalis* as a food ingredient, we should prepare an *H. abdominalis* mixture by adding other materials.

RG has been frequently used in traditional Asian medicine to treat many disorders, such as debility, aging, stress, diabetes, and insomnia (Tang et al. 2008). Especially, RG has been mainly focused on its anti-fatigue activity with the ability mitigating exercise-related muscle damage, maintaining homeostasis of the body, and enhancing vital energy (Kim et al. 2013; Kim et al. 2016). Thus, *H. abdominalis* mixtures were prepared by adding different concentrations of RG (20, 30, 40, 50, 60, 70, and 80% of RG) to investigate the synergy effect between *H. abdominalis* and RG on anti-fatigue activity.

Effect of these mixtures on cell viability was estimated using the MTT assay. The viabilities of C2C12 myoblasts treated with the mixtures at different concentrations (50, 100, 200, 250, and 500 μg/mL) were expressed to represent over 90% viability, which was similar to that of the control cells (Fig. 4). Thus, mixtures up to 500 μg/mL concentration did not show any significant cytotoxicity for 24 h.

Effect of the mixture of HH and RG on cell proliferation

To assess the effect of HH (0% of RG) and RG mixture on muscle growth, cell proliferation was measured as

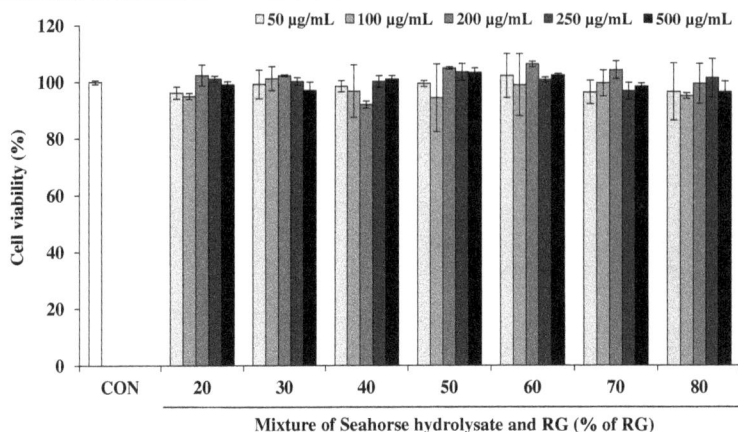

Fig. 4 Cytotoxicity of the six mixtures of seahorse hydrolysate and RG on C2C12 myoblasts. The cell was treated with various concentrations of the six mixtures (50, 100, 150, 200, 250, and 500 µg/mL) and incubated for 24 h. The cytotoxicity was assessed by MTT assay. The experiment was performed in triplicate. Each value indicates the mean ± standard error from three independent experiments

shown in Fig. 5. HH and RG mixtures showed the significant effect on cell proliferation. Especially, at the low concentrations (200 and 250 µg/mL), HH significantly enhanced proliferation of C2C12 myoblasts compared with that of mixtures as well as that of the control cells. However, at the high concentrations (400 and 500 µg/mL), HH and RG mixtures did not show any significant effects on cell proliferation to each other.

Anti-fatigue activity of the mixtures of HH and RG
Several biomarkers such as lactate dehydrogenase (LDH), glycogen, aspartate transaminase (AST), and alanine transaminase (ALT) have been used to investigate muscle injury during exhaustive exercise (Huang et al. 2015). Also, fatigue is related to mitochondrial dysfunction and diminished ATP levels (Singh and Singh 2014).

The anti-fatigue activity of the mixtures of HH and RG was assessed in H_2O_2-treated C2C12 myoblasts by measuring cell proliferation as well as the glycogen and ATP contents. Severe and continuous exercise may elevate the formation of ROS, thereby increasing oxidative stress. A sustained elevated oxidative stress can hamper mitochondrial function resulting in low ATP synthesis and increased lactic acid in the muscles, consequently decreasing the physical efficiency. These observations suggest that improving the antioxidant status may enhance the overall physical performance by maintaining the pro-oxidant/antioxidant balance (Swamy et al. 2011). To induce oxidative stress in C2C12 myoblasts, the cells were incubated with H_2O_2 at a concentration of 100 µM. After H_2O_2 treatment, the viability of C2C12 myoblasts decreased to less than 60% compared to that

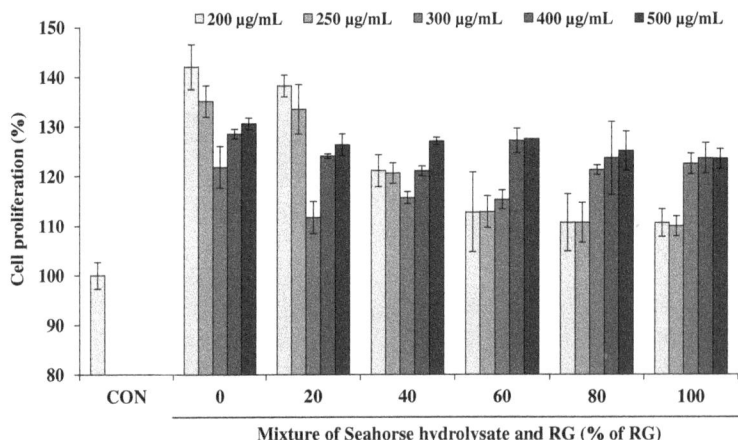

Fig. 5 Cell proliferation of the six mixtures of seahorse hydrolysate and RG on C2C12 myoblast. The cell was treated with various concentrations of the six mixtures (50, 100, 150, and 200 µg/mL) and incubated for 72 h. The cell proliferation was assessed by BrdU assay. Effect of different mixtures of seahorse hydrolysate and RG on C2C12 myoblast proliferation. The experiment was performed in triplicate. Each value indicates the mean ± standard error from three independent experiments

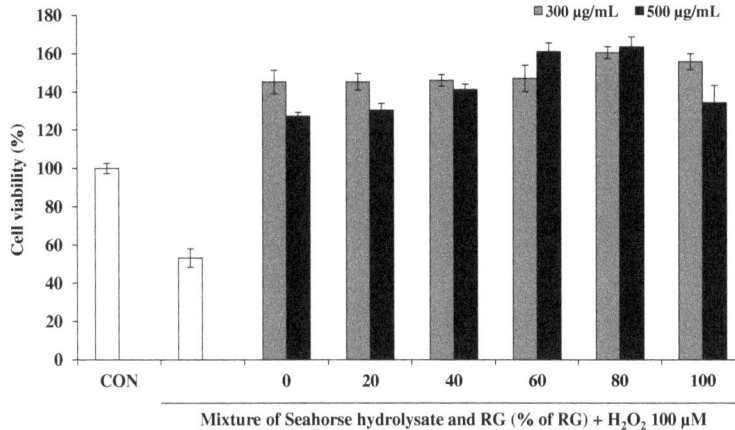

Fig. 6 Protective effect of the six mixtures of seahorse hydrolysate and RG against H_2O_2-treated C2C12 myoblasts. The treatment of H_2O_2 induced a decrease in cell viability. All mixtures showed protective effect on H_2O_2-induced oxidative stress in C2C12 myoblasts. The experiment was performed in triplicate. Each value indicates the mean ± standard error from three independent experiments

in the control cells (Fig. 6). However, C2C12 myoblasts treated with HH and RG mixtures showed increased viability compared with that reported for the control cells. Notably, at a sample concentration of 500 μg/mL, cell viability increased steadily with increasing percentage of RG except at 100% RG, where only RG was present in the mixture.

Glycogen contents

Energy expenditure during exercise leads to physical fatigue, which is mainly caused by energy consumption and deficiency. Catabolized fat and carbohydrates are considered the main sources of energy in the skeletal muscles during exercise, and glycogen is the predominant source of glycolysis for energy production. Therefore, glycogen storage directly affects exercise ability (Wu et al. 2013). The glycogen content of the H_2O_2-

treated cells was lower than that of the control cells. However, in C2C12 myoblasts, the treatment with HH and RG mixtures increased the glycogen content to more than double as compared to the values reported for the control cells. In particular, RG60, RG80, and RG100 showed increased glycogen content at a sample concentration of 300 and 500 μg/mL (Fig. 7).

ATP contents

Muscular exercise causes rapid ATP consumption, and energy deficiency is a critical reason for physical fatigue. Therefore, compounds that promote ATP production could be candidates for alleviating physical fatigue. The skeletal muscle mainly catabolizes fat and carbohydrates as sources of energy during exercise (Nozawa et al. 2009). ATP content in the H_2O_2-treated cells was lower than that in the control cells (Fig. 8). Although HH and

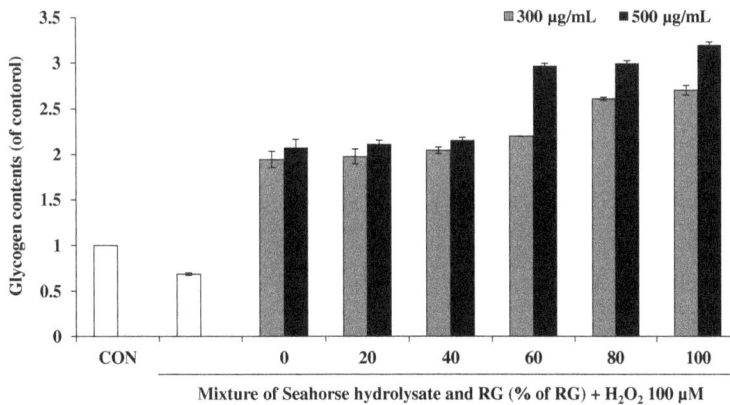

Fig. 7 Effect of the six mixtures of seahorse hydrolysate and RG on glycogen content in H_2O_2-treated C2C12 myoblasts. H_2O_2 treatment induced a decrease in glycogen contents. All mixtures showed protective effect on H_2O_2-induced oxidative stress in C2C12 myoblasts. The experiment was performed in triplicate. Each value indicates the mean ± standard error from three independent experiments

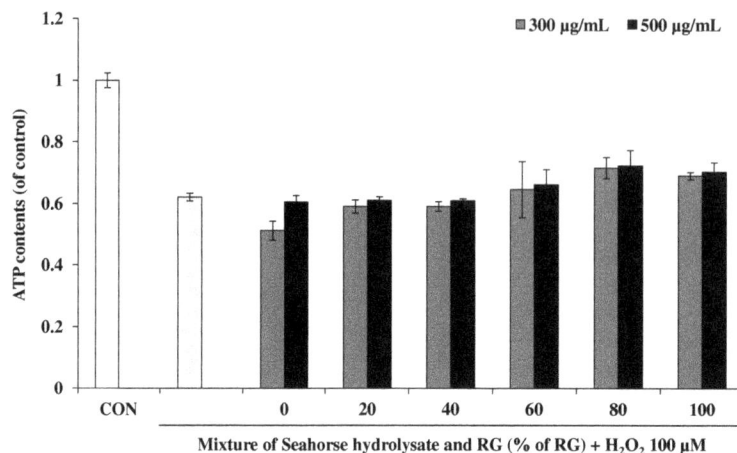

Fig. 8 Effect of the six mixtures of seahorse hydrolysate and RG on ATP synthesis in H_2O_2-treated C2C12 myoblasts. The treatment of H_2O_2 induced a decrease in ATP contents. All mixtures showed a protective effect against H_2O_2-induced oxidative stress in C2C12 myoblasts. The experiment was performed in triplicate. Each value indicates the mean ± standard error from three independent experiments

RG mixtures did not increase the ATP content, RG80 relatively increased the ATP content at 300 and 500 μg/mL concentrations of the mixture.

Exercise-induced oxidative stress can cause the increased muscle fatigability. Thus, antioxidants can decrease the oxidative stress and improve the physiological condition (You et al. 2011). Some reports showed that a loach peptide has not only antioxidant activities but also an anti-fatigue effect in mice (You et al. 2011). Actually, the peptide showing in vitro antioxidant activity possesses the in vivo anti-fatigue activity. The peptide acts as the scavenger for DPPH and hydroxyl radicals. Also, the anthocyanins of mulberry fruit have been assessed in vitro antioxidant activity and in vivo anti-fatigue activity (Jiang et al. 2013). These studies showed values of in vitro study to evaluate the potential anti-fatigue activity through in vivo study. In the present study, the mixtures of HH and RG acted as the antioxidant for hydrogen peroxide and showed the anti-fatigue activity on C2C12 myoblast. Furthermore, the mixtures have valuable needs to be investigated through in vivo animal study.

Conclusions

In this study, the effect of *H. abdominalis* on the muscles was investigated to scientifically verify its potential bioactivity. Also, the anti-fatigue activity of a mixture comprising HH and RG was evaluated to commercially utilize *H. abdominalis* in food industry. The treatment of HH to C2C12 myoblast induced the cell proliferation and glycogen contents. These results indicated that *H. abdominalis* had anti-fatigue activity on C2C12 myoblast. Moreover, the treatment of the mixture comprising HH and RG increased cell viability and the content of fatigue-related biomarkers such as glycogen and ATP

contents. In particular, the 80% RG mixture showed an optimum effect on cell viability and ATP synthesis activity. These results indicated that HH had anti-fatigue activity at concentrations approved for use in food by the law in Korea. Especially, an 80% RG to HH mixture has the potential to ameliorate fatigue condition induced by oxidative stress by increasing the fatigue-related biochemical parameters such as glycogen and ATP contents in C2C12 myoblasts. Therefore, 80% RG to HH mixture can be used in food for ameliorating fatigue in Korea.

Abbreviations
HH: *Hippocampus abdominalis*; RG: Red ginseng

Acknowledgements
This research was financially supported by the Ministry of Trade, Industry, and Energy (MOTIE), Korea, under the "Regional Specialized Industry Development Program" supervised by the Korea Institute for Advancement of Technology (KIAT).

Funding
This study was funded from the Ministry of Trade, Industry, and Energy (MOTIE), Korea.

Authors' contributions
NK contributed to conduct the research and prepare the draft manuscript. SYK contributed to conduct the cell experiments and analyze the biomarker levels. SR managed aquaculture of the seahorse and supported it for this study. JYK contributed to design the study and conduct the experiments. YJJ contributed to monitor the experiments and finalize the manuscript. All authors read and approved the final manuscript.

Competing interests
The authors declare that they have no competing interests.

Author details
[1]Department of Marine Life Sciences, Jeju National University, Jeju 63243, Korea. [2]Center of Ornamental Reefs and Aquariums, Jeju 63354, Korea.

References

Deshmukh AS, Murgia M, Nagaraj N, Treebak JT, Cox J, Mann M. Deep proteomics of mouse skeletal muscle enables quantitation of protein isoforms, metabolic pathways, and transcription factors. Mol Cell Proteomics. 2015;14:841–53.

Fan H, Tan Z, Hua Y, Huang X, Gao Y, Wu Y, Liu B, Zhou Y. Deep sea water improves exercise and inhibits oxidative stress in a physical fatigue mouse model. Biomed Rep. 2016;4:751–7.

Huang L-Z, Huang B-K, Ye Q, Qin L-P. Bioactivity-guided fractionation for anti-fatigue property of Acanthopanax senticosus. J Ethnopharmacol. 2011;133: 213–9.

Huang W-C, Lin C-I, Chiu C-C, Lin Y-T, Huang W-K, Huang H-Y, Huang C-C. Chicken essence improves exercise performance and ameliorates physical fatigue. Nutrients. 2014;6:2681–96.

Huang W-C, Chiu W-C, Chuang H-L, Tang D-W, Lee Z-M, Wei L, Chen F-A, Huang C-C. Effect of curcumin supplementation on physiological fatigue and physical performance in mice. Nutrients. 2015;7:905–21.

Jiang D-Q, Guo Y, Xu D-H, Huang Y-S, Yuan K, Lv Z-Q. Antioxidant and anti-fatigue effects on anthocyanins of mulberry juice purification (MJP) and Mulberry marc purification (MMP) from different varieties mulberry fruit in China. Food Chem Toxicol. 2013;59:1–7.

Kang S-M, Heo S-J, Kim K-N, Lee S-H, Jeon Y-J. Isolation and identification of new compound, 2, 7″-phloroglucinol-6, 6′-bieckol from brown algae, Ecklonia cava and its antioxidant effect. J Funct Foods. 2012;4:158–66.

Kang N, Ko S-C, Samarakoon K, Kim E-A, Kang M-C, Lee S-C, Kim J, Kim Y-T, Kim J-S, Kim H. Purification of antioxidative peptide from peptic hydrolysates of Mideodeok (Styela clava) flesh tissue. Food Sci Biotechnol. 2013;22:541–7.

Kim H-G, Cho J-H, Yoo S-R, Lee J-S, Han J-M, Lee N-H, Ahn Y-C, Son C-G. Antifatigue effects of Panax ginseng CA Meyer: a randomised, double-blind, placebo-controlled trial. PLoS One. 2013;8:e61271.

Kim S, Kim J, Lee Y, Seo MK, Sung DJ. Anti-fatigue effects of acute red ginseng intake in recovery from repetitive anaerobic exercise. Iran J Public Health. 2016;45:387–9.

Kumaravel K, Ravichandran S, Balasubramanian T, Siva Subramanian K, Bilal A. Antimicrobial effect of five seahorse species from Indian coast. Br J Pharmacol Toxicol. 2010;1:62–6.

Kumaravel K, Ravichandran S, Balasubramanian T, Sonneschein L. Seahorses—a source of traditional medicine. Nat Prod Res. 2012;26:2330–4.

Nam S-Y, Kim H-M, Jeong H-J. Anti-fatigue effect by active dipeptides of fermented porcine placenta through inhibiting the inflammatory and oxidative reactions. Biomed Pharmacother. 2016;84:51–9.

Nozawa Y, Yamada K, Okabe Y, Ishizaki T, Kuroda M. The anti-fatigue effects of the low-molecular-weight fraction of bonito extract in mice. Biol Pharm Bull. 2009;32:468–74.

Perera N, Godahewa G, Lee J. Copper-zinc-superoxide dismutase (CuZnSOD), an antioxidant gene from seahorse (Hippocampus abdominalis); molecular cloning, sequence characterization, antioxidant activity and potential peroxidation function of its recombinant protein. Fish Shellfish Immunol. 2016;57:386–99.

Qian Z-J, Kang K-H, Kim S-K. Isolation and antioxidant activity evaluation of two new phthalate derivatives from seahorse, Hippocampus kuda Bleeler. Biotechnol Bioprocess Eng. 2012;17:1031–40.

Segade Á, Robaina L, Otero-Ferrer F, García Romero J, Molina Domínguez L. Effects of the diet on seahorse (Hippocampus hippocampus) growth, body colour and biochemical composition. Aquacult Nutr. 2015;21:807–13.

Singh T, Singh K. Mitochondrial dysfunction and chronic fatigue syndromes: issues in clinical care (modified version). IOSR J Dental Med Sci. 2014;13:30–3.

Swamy M, Naveen S, Singsit D, Naika M, Khanum F. Anti-fatigue effects of polyphenols extracted from pomegranate peel. Int J Integr Biol. 2011;11:69–72.

Tang W, Zhang Y, Gao J, Ding X, Gao S. The anti-fatigue effect of 20 (R)-ginsenoside Rg3 in mice by intranasally administration. Biol Pharm Bull. 2008; 31:2024–7.

Wu R-E, Huang W-C, Liao C-C, Chang Y-K, Kan N-W, Huang C-C. Resveratrol protects against physical fatigue and improves exercise performance in mice. Molecules. 2013;18:4689–702.

You L, Zhao M, Regenstein J-M, Ren J. In vitro antioxidant activity and in vivo anti-fatigue effect of loach (Misgurnus anguillicaudatus) peptides prepared by papain digestion. Food Chem. 2011;124:188–94.

Yu B, Lu Z-X, Bie X-M, Lu F-X, Huang X-Q. Scavenging and anti-fatigue activity of fermented defatted soybean peptides. Eur Food Res Technol. 2008;226:415–21.

Zhang Y, Yao X, Bao B, Zhang Y. Anti-fatigue activity of a triterpenoid-rich extract from Chinese bamboo shavings (Caulis bamfusae in taeniam). Phytother Res. 2006;20:872–6.

12

Alterations of growth performance, hematological parameters, and plasma constituents in the sablefish, *Anoplopoma fimbria* depending on ammonia concentrations

Jun-Hwan Kim[1], Hee-Ju Park[2], In-Ki Hwang[2], Jae-Min Han[2], Do-Hyung Kim[2], Chul Woong Oh[3], Jung Sick Lee[4] and Ju-Chan Kang[2*]

Abstract

Juvenile *Anoplopoma fimbria* (mean length 16.8 ± 2.2 cm, and mean weight 72.8 ± 5.4 g) were exposed for 2 months with different levels of ammonia (0, 0.25, 0.50, 0.75, 1.00, and 1.25 mg/L). Growth performances such as daily length gain, daily weight gain, condition factor, and hepatosomatic index were significantly decreased by ammonia exposure. Hematological parameters such as red blood cell (RBC) count, hematocrit, and hemoglobin were also significantly decreased. In plasma inorganic components, calcium and magnesium were significantly decreased by ammonia exposure. In plasma organic components, there was no alteration in cholesterol and total protein. In enzyme plasma components, glutamic oxalate transaminase (GOT) and glutamic pyruvate transaminase (GPT) were significantly increased. The results of this study indicated that ammonia exposure can induce significant growth reduction and blood biochemistry alterations of *A. fimbria*.

Keywords: Sablefish, Ammonia, Growth performance, Hematological parameters, Plasma components

Background

In Korea, aquaculture is a major industry in food security dimension because it can supply high-quality protein to the public by stably breeding aquatic organisms. However, ammonia hypergenesis by high density breeding in aquaculture environment is a critical environmental toxic factor to induce death. Exposure to excessive ammonia in aquatic animals induces depolarization in neuron because increased NH_4^+ displaces K^+, which results in cell death in central nervous system. Therefore, it induces convulsions, coma, and death by the cell death (Thangam et al. 2014). In addition, acute ammonia exposure induces gill ventilation increase, equilibrium loss, convulsions, ionic balance failure, and hyper-excitability in aquatic animals (Kim et al. 2015).

Growth factor in aquaculture is one of the most basic and critical parameters to assess toxic effects by harmful substances in aquaculture environment. In fish exposed to toxic substances, growth performance is generally decreased by energy transition from use for growth and development to use for tissue damage recovery (Wendelaar Bonga, 1997). In aquatic environment, excessive ammonia concentrations can be accumulated in body fluids in fish, which results in growth inhibition, tissue erosion and degeneration, immune suppression, and high mortality (Liang et al., 2015).

Hematological and biochemical parameters in fish can be a critical indicator to assess alterations in circulatory system by toxic substances in external environment (Vinodhini and Narayanan, 2009). Ammonia especially affects hematological parameters in fish by blocking oxygen transfer from gill to blood (Thangam et al. 2014).

Sablefish, *Anoplopoma fimbria* used in this study is recognized as a high value fish species around the globe.

* Correspondence: jckang@pknu.ac.kr
[2]Department of Aquatic Life Medicine, Pukyong National University, Busan, Korea
Full list of author information is available at the end of the article

In liberalization trend of the world market, aquatic products are also involved in the trend. Therefore, development in aquaculture of a high value fish species is urgent. The purpose of this study was to assess toxic effects of *A. fimbria* exposed to ammonia a part of complete sablefish culture, and to build standard breeding guidelines of sablefish aquaculture.

Methods

Experimental fish and conditions

Juvenile sablefish were obtained from Troutlodge Inc. in USA. During the acclimation period, the fish were fed diet twice daily and maintained on a 24-h dark cycle and constant condition at all times (Table 1). After acclimatization, 72 fish (body length, 16.8 ± 2.2 cm; body weight, 72.8 ± 5.4 g) were randomly selected for the study. The acclimation period commenced once the final temperature had been sustained for 24 h and animals were feeding, while showing no sign of stress. Ammonia exposure took place in tanks containing six fish per treatment group. Ammonia chloride (NH_4Cl) (Sigma, St. Louis, MO, USA) solution was dissolved in the respective tanks. The ammonia concentrations in the tanks were 0, 0.25, 0.50, 0.75, 1.00, and 1.25 mg/L, and actual ammonia concentrations are demonstrated in Table 2. Diluted 100 mg/L ammonia chloride (NH_4Cl) in 20 L glass tank to make respective concentrations (50, 0.25 mg/L, 100 ml; 0.50 mg/L, 150 ml; 0.75 mg/L, 200 ml; 1.00 mg/L, 250 ml; 1.25 mg/L). After the exposure experiment, feed was given at a rate of 2% body weight daily (as two 1% meals per day). There was no water flow rate. The tank water was thoroughly exchanged once per 2 days and made the same concentration in the respective 500 L circular tank. At the end of each period (at 1 and 2 months), animals were anesthetized in buffered 3-aminobenzoic acid ethyl ester methanesulfonate (Sigma Chemical, St. Louis, MO).

Growth

The weight and length of sablefish was measured just before exposure, at 1 and 2 months. Daily length gain, daily weight gain, condition factor, and hepatosomatic index (HSI) were calculated by the following method.

Table 1 The chemical components of seawater and experimental condition used in the experiments

Item	Value
Temperature (°C)	13.0 ± 1.0
pH	8.2 ± 0.5
Salinity (‰)	33.5 ± 0.6
Dissolved oxygen (mg/L)	7.8 ± 0.5
Chemical oxygen demand (mg/L)	1.21 ± 0.14

Table 2 Analyzed waterborne ammonia concentration from each source

Waterborne ammonia concentration (mg/L)						
Waterborne ammonia concentrations	0	0.25	0.50	0.75	1.00	1.25
Actual ammonia concentrations	0.02 ± 0.01	0.28 ± 0.07	0.57 ± 0.14	0.81 ± 0.26	1.04 ± 0.32	1.32 ± 0.27

Daily growth gain $= W_f - W_i / \text{day}$

(W_f = Final length or weight, W_i = Initial length or weight)

Condition factor $(\%) = (W / L^3) \times 100$

($W = \text{weight(g)}, L = \text{length(cm)}$)

HSI $= (\text{liver weight} / \text{total fish weight}) \times 100$

Blood samples and hematological assay

Blood samples were collected within 35–40 s through the caudal vein of the fish in 1-ml disposable heparinized syringes at the end of 1 and 2 months. The blood samples were kept at 4 °C until the blood parameters were completely studied. The total red blood cell (RBC) count, hemoglobin (Hb), concentration, and hematocrit (Ht) value were determined immediately. Total RBC counts were counted using optical microscope with hemo-cytometer (Improved Neubauer, Germany) after diluted by Hendrick's diluting solution. The Hb concentration was determined using Cyan-methemoglobin technique (Asan Pharm. Co., Ltd.). The Ht value was determined by the microhematocrit centrifugation technique. The blood samples were centrifuged to separate plasma from blood samples at $3000\,g$ for 5 min at 4 °C. The plasma samples were analyzed for inorganic substances, organic substances, and enzyme activity using clinical kit (Asan Pharm. Co.,Ltd.). In inorganic substances assay, calcium and magnesium were analyzed by the o-cresolphthalein-complexon technique and xylidyl blue technique. In organic substances assay, cholesterol and total protein were analyzed by enzyme method and by biuret technique. In enzyme activity assay, glutamic oxalate transaminase (GOT) and glutamic pyruvate transaminase (GPT) were analyzed by Kind-king technique.

Statistical analysis

The experiment was conducted in exposure periods for 2 months and performed triplicate. Statistical analyses were performed using the SPSS/PC+ statistical package (SPSS Inc, Chicago, IL, USA). Significant differences between groups were identified using one-way ANOVA and Tukey's test for multiple comparisons. The significance level was set at $P < 0.05$.

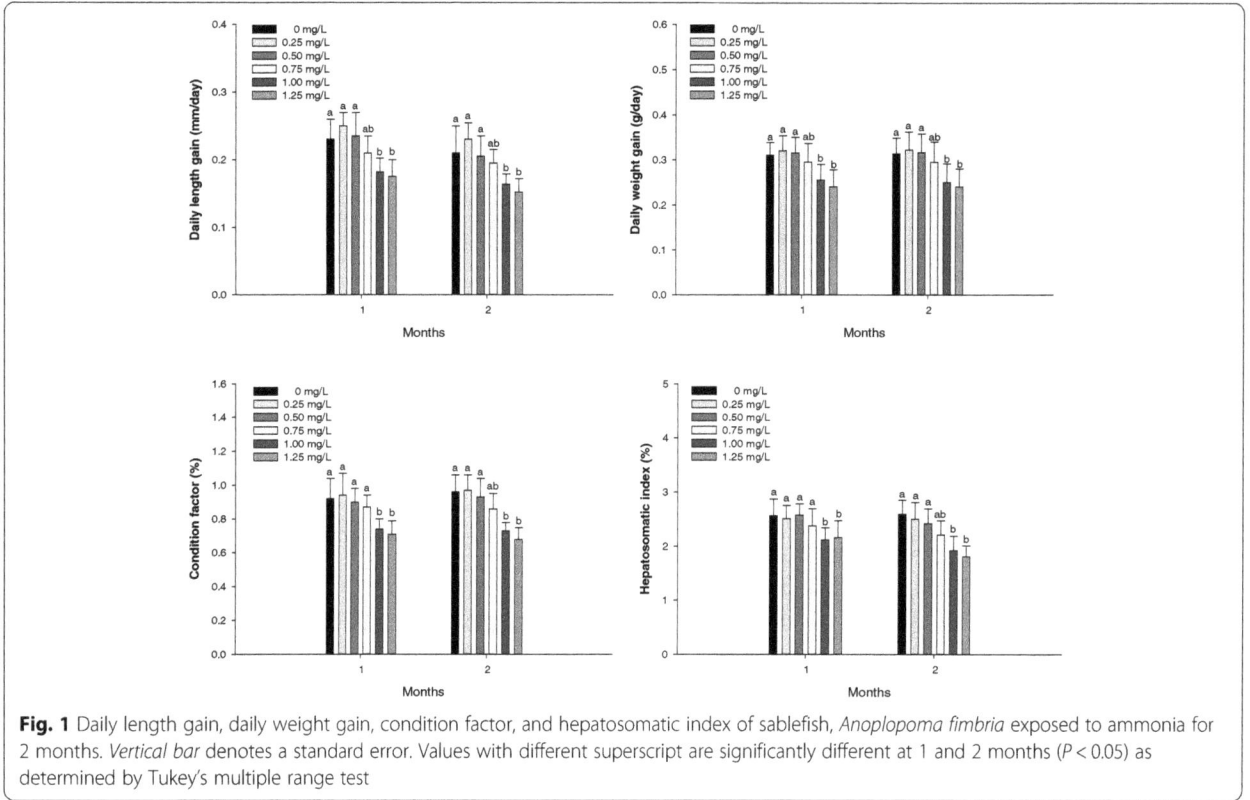

Fig. 1 Daily length gain, daily weight gain, condition factor, and hepatosomatic index of sablefish, *Anoplopoma fimbria* exposed to ammonia for 2 months. *Vertical bar* denotes a standard error. Values with different superscript are significantly different at 1 and 2 months ($P < 0.05$) as determined by Tukey's multiple range test

Results

Growth

No mortality was observed for the exposure periods. The growth performance, condition factor, and hepatosomatic index of *A. fimbria* is demonstrated in Fig. 1. Significant decreases in daily length gain and daily weight gain were observed at ammonia exposure greater than 1.00 mg/L both in 1 and 2 months. Condition factor was significantly decreased at ammonia exposure greater than 1.00 mg/L both in 1 and 2 months. Hepatosomatic index was also significantly decreased at ammonia exposure greater than 1.00 mg/L both in 1 and 2 months. However, there was no change in daily length, daily weight gain, and condition factor and

hepatosomatic index from 0 to 0.75 mg/L ammonia exposure after 1 and 2 months.

Hematological parameters

RBC count, hematocrit value, and hemoglobin concentration of *A. fimbria* exposed to different concentrations of waterborne ammonia are demonstrated in Fig. 2. RBC count was significantly decreased at ammonia exposure greater than 1.00 mg/L in 1 month and greater than 0.75 mg/L in 2 months. Hematocrit value was significantly decreased at ammonia exposure greater than 1.00 mg/L in 1 month and greater than 0.75 mg/L in 2 months. Hemoglobin concentration was significantly

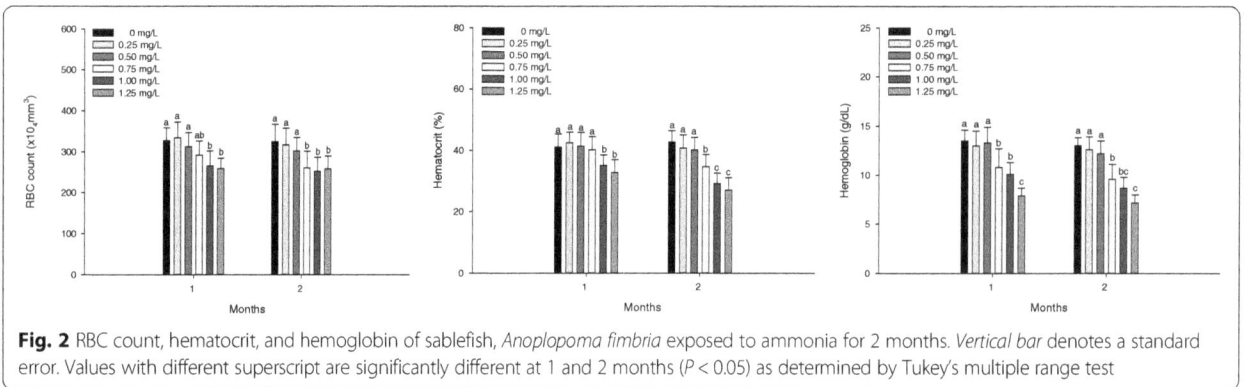

Fig. 2 RBC count, hematocrit, and hemoglobin of sablefish, *Anoplopoma fimbria* exposed to ammonia for 2 months. *Vertical bar* denotes a standard error. Values with different superscript are significantly different at 1 and 2 months ($P < 0.05$) as determined by Tukey's multiple range test

Table 3 Changes of inorganic plasma components in sablefish, *Anoplopoma fimbria* exposed to ammonia for 2 months

Parameters	Period (month)	Ammonia (mg/L)					
		0	0.25	0.50	0.75	1.00	1.25
Calcium (mg/dL)	1	1.84 ± 0.23[a]	1.81 ± 0.26[a]	1.75 ± 0.31[a]	1.71 ± 0.30[ab]	1.59 ± 0.25[b]	1.55 ± 0.21[b]
	2	1.81 ± 0.27[a]	1.78 ± 0.23[a]	1.62 ± 0.25[ab]	1.64 ± 0.21[ab]	1.49 ± 0.27[b]	1.42 ± 0.24[b]
Magnesium (mg/dL)	1	3.46 ± 0.42[a]	3.51 ± 0.32[a]	3.41 ± 0.27[a]	3.29 ± 0.35[ab]	2.89 ± 0.22[b]	2.95 ± 0.32[b]
	2	3.41 ± 0.36[a]	3.46 ± 0.38[a]	3.28 ± 0.41[ab]	3.06 ± 0.25[b]	2.74 ± 0.30[b]	2.71 ± 0.24[b]

Values are mean ± SE. Values with different superscript are significantly different at 1 and 2 months ($P < 0.05$) as determined by Tukey's multiple range test

decreased at ammonia exposure greater than 0.75 mg/L in 1 and 2 months.

Plasma components

Plasma inorganic components of *A. fimbria* are demonstrated in Table 3. Calcium was significantly decreased at ammonia exposure greater than 1.00 mg/L in 1 and 2 months. Magnesium was also significantly decreased at ammonia exposure greater than 1.00 mg/L in 1 month and greater than 0.75 in 2 months. Plasma organic components are demonstrated in Table 4. No alterations in cholesterol and total protein were observed by waterborne ammonia exposure. Plasma enzyme components are demonstrated in Table 5. GOT was significantly increased at ammonia exposure greater than 1.00 mg/L in 1 month and greater than 0.75 in 2 months. GPT was also significantly increased at ammonia exposure greater than 1.00 mg/L in 1 month and greater than 0.75 in 2 months.

Discussion

Ammonia exposure to fish is a critical environmental limited factor to inhibit growth performance by decreasing feed intake and feed utilization (Foss et al., 2003). Many authors reported that high concentrations of ammonia exposure induced growth inhibition of spotted wolfish, *Anarhichas minor* Olafsen (Foss et al., 2003), turbot, *Scophthalmus maximus* (Foss et al., 2009), Atlantic halibut, *Hippoglossus hippoglossus* (Paust et al., 2011). In this study, high concentrations of ammonia induced a significant decrease in growth of sablefish, *A. fimbria*, which may be due to energy transition from growth and development to detoxification. Hepatosomatic index (HSI) is considered as a critical indicator to evaluate

health status by toxic substance exposure (Datta et al., 2007), and HSI of sablefish, *A. fimbria* was significantly decreased by ammonia exposure.

Blood cells in fish are generated from hematopoietic tissues of kidney and spleen, and changes in hematological parameters indicate physiological effects by stress responses (Das et al., 2004). Jeney et al. (1992) suggest that high levels of ammonia exposure induce oxygen-free condition by increasing affinity of hemoglobin to combine with ammonia molecules, thereby elevating ammonia concentration in blood. Knoph and Thorud (1996) reported a significant decrease in RBC count and hematocrit of Atlantic salmon, *Salmo salar* exposed to ammonia. Das et al. (2004) also reported a significant decrease in hemoglobin of Mrigal carp, *Cirrhinus cirrhosus* exposed to ammonia. In this study, ammonia exposure caused a significant decrease in RBC count, hematocrit, and hemoglobin of sablefish, *A. fimbria*, which may be due to hematopoietic cell damage according to hypoxic status by ammonia exposure.

Calcium and magnesium in plasma inorganic components are critical indicators of osmotic pressure alterations, and these can be increased or decreased by environmental changes (Hur et al., 2001). Person-Le Ruyet et al., (2003) reported that ammonia exposure to turbot, *Scophthalmus maximus* induced changes in osmotic pressure by altering Na^+, Cl^-, K^+, Ca^{2+} concentrations in plasma. In this study, calcium and magnesium in sablefish, *A. fimbria* were significantly decreased by ammonia exposure, which indicate that ammonia exposure affected the osmotic ion regulation of sablefish. Cholesterol and total protein of plasma organic components in fish have been considered as a major component to assess fish health. However, there

Table 4 Changes of organic plasma components in sablefish, *Anoplopoma fimbria* exposed to ammonia for 2 months

Parameters	Period (month)	Ammonia (mg/L)					
		0	0.25	0.50	0.75	1.00	1.25
Cholesterol (mg/dL)	1	131.5 ± 20.5[a]	139.1 ± 18.2[a]	140.5 ± 21.3[a]	137.2 ± 16.8[a]	142.2 ± 21.1[a]	146.2 ± 18.3[a]
	2	135.1 ± 17.1[a]	131.5 ± 17.1[a]	138.2 ± 17.1[a]	143.2 ± 21.3[a]	148.3 ± 18.3[a]	142.6 ± 20.2[a]
Total protein (g/dL)	1	4.42 ± 0.53[a]	4.35 ± 0.39[a]	4.26 ± 0.45[a]	4.51 ± 0.48[a]	4.33 ± 0.37[a]	4.59 ± 0.41[a]
	2	4.36 ± 0.47[a]	4.56 ± 0.35[a]	4.37 ± 0.31[a]	4.62 ± 0.52[a]	4.24 ± 0.45[a]	4.39 ± 0.51[a]

Values are mean ± SE. Values with different superscript are significantly different at 1 and 2 months ($P < 0.05$) as determined by Tukey's multiple range test

Table 5 Changes of enzymatic plasma components in sablefish, *Anoplopoma fimbria* exposed to ammonia for 2 months

Parameters	Period (month)	Ammonia (mg/L)					
		0	0.25	0.50	0.75	1.00	1.25
GOT (karmen unit)	1	2.67 ± 0.33^a	2.72 ± 0.29^a	2.75 ± 0.35^a	2.81 ± 0.28^a	3.36 ± 0.41^b	3.45 ± 0.37^b
	2	2.72 ± 0.25^a	2.61 ± 0.32^a	2.91 ± 0.21^{ab}	3.24 ± 0.32^b	3.46 ± 0.40^{bc}	3.62 ± 0.41^c
GPT (karmen unit)	1	1.72 ± 0.24^a	1.79 ± 0.19^a	1.84 ± 0.18^a	1.93 ± 0.26^{ab}	2.19 ± 0.31^b	2.28 ± 0.27^b
	2	1.75 ± 0.28^a	1.82 ± 0.23^a	1.99 ± 0.28^a	2.23 ± 0.31^b	2.41 ± 0.35^c	2.40 ± 0.35^c

Values are mean ± SE. Values with different superscript are significantly different at 1 and 2 months ($P < 0.05$) as determined by Tukey's multiple range test

was no significant alteration in sablefish, *A. fimbria*. GOT and GPT in enzymatic plasma components can be easily increased by hepatic tissue damage, and these are used to evaluate hepatic tissue damage (Agrahari et al., 2007). Le Ruyet et al. (1998) reported that a significant increase in GOT and GPT of turbot, *Scophthalmus maximus* and seabream, *Sparus aurata* exposed to ammonia. In this study, GOP and GPT in sablefish, *A. fimbria* were significantly increased by ammonia exposure, which may be due to hepatic tissue damage by ammonia.

Conclusion

The results of this study indicate that ammonia exposure at the higher than proper concentrations affected growth performance and hematological parameters of sablefish, *A. fimbria*, and these changes should influence the health of sablefish, *A. fimbria*. In conclusion, ammonia concentrations at the higher than 0.75 mg/L can affect various physiological effects of sablefish, *A. fimbria*, and the high concentrations of ammonia exposure require special attention in sablefish aquaculture. In addition to this environmental study, various environmental standards should be established for stable sablefish aquaculture.

Abbreviations
GOT: Glutamic oxalate transaminase; GPT: Glutamic pyruvate transaminase; HIS: Hepatosomatic index; RBC: Red blood cell

Acknowledgements
This research was a part of the project titled 'Development of practical techniques for the artificial seeding production of sablefish', funded by the MOF, Korea.

Authors' contributions
HJ, IK, and JM carried out the environmental toxicity studies and manuscript writing. JH, DH, CW, and JS participated in the design of the study and data analysis. JC participated in its design and coordination and helped to draft the manuscript. All authors read and approved the final manuscript.

Competing interests
The authors declare that they have no competing interests.

Disclosure
The dataset(s) supporting the conclusions of this article is not included in the article.

Author details
[1]West Sea Fisheries Research Institute, National Institute of Fisheries Science, Incheon, Korea. [2]Department of Aquatic Life Medicine, Pukyong National University, Busan, Korea. [3]Department of Marine Biology, Pukyong National University, Busan, Korea. [4]Department of Aqualife Medicine, Chonnam National University, Yeosu, Korea.

References
Agrahari S, Pandey KC, Gopal K. Biochemical alteration induced by monocrotophos in the blood plasma of fish, *Channa punctatus* (Bloch). Pestic Biochem Physiol. 2007;88:268–72.

Das PC, Ayyappan S, Jena JK, Das BK. Acute toxicity of ammonia and its sub-lethal effects on selected haematological and enzymatic parameters of mrigal, *Cirrhinus mrigala* (Hamilton). Aquac Res. 2004;35:134–43.

Datta S, Saha DR, Ghosh D, Majumdar T, Bhattacharya S, Mazumder S. Sub-lethal concentration of arsenic interferes with the proliferation of hepatocytes and induces *in vivo* apoptosis in *Clarias batrachus* L. Comp Biochem Physiol C Toxicol Pharmacol. 2007;145:339–49.

Foss A, Evensen TH, Vollen T, Øiestad V. Effects of chronic ammonia exposure on growth and food conversion efficiency in juvenile spotted wolffish. Aquaculture. 2003;228:215–24.

Foss A, Imsland AK, Roth B, Schram E, Stefansson SO. Effects of chronic and periodic exposure to ammonia on growth and blood physiology in juvenile turbot (*Scophthalmus maximus*). Aquaculture. 2009;296:45–50.

Hur JW, Chang YJ, Lim HK, Lee BK. Stress responses of cultured fishes elicited by water level reduction in rearing tank and fish transference during selection process. J Korean Fish Soc. 2001;34:465–72.

Jeney G, Nemcsok J, Zs J, Olah J. Acute effect of sublethal ammonia concentrations on common carp (*Cyprinus carpio* L.). II Effect of ammonia on blood plasma transminases (GOT, GPT), G1DH enzyme activity, and ATP value. Aquaculture. 1992;104:149–56.

Kim SH, Kim JH, Park MA, Hwang SD, Kang JC. The toxic effects of ammonia exposure on antioxidant and immune responses in Rockfish, *Sebastes schlegelii* during thermal stress. Environ Toxicol Pharmacol. 2015;40:954–9.

Knoph MB, Thorud K. Toxicity of ammonia to Atlantic salmon (*Salmo salar* L.) in seawater—Effects on plasma osmolality, ion, ammonia, urea and glucose levels and hematologic parameters. Comp Biochem Physiol A Physiol. 1996;113:375–81.

Le Ruyet JP, Boeuf G, Infante JZ, Helgason S, Le Roux A. Short-term physiological changes in turbot and seabream juveniles exposed to exogenous ammonia. Comp Biochem Physiol A Mol Integr Physiol. 1998;119:511–8.

Liang Z, Liu R, Zhao D, Wang L, Sun M, Wang M, Song L. Ammonia exposure induces oxidative stress, endoplasmic reticulum stress and apoptosis in epatopancreas of pacific white shrimp (*Litopenaeus vannamei*). Fish Shellfish Immunol. 2015;54:523–8.

Alterations of growth performance, hematological parameters, and plasma constituents in the sablefish...

93

Paust LO, Foss A, Imsland AK. Effects of chronic and periodic exposure to ammonia on growth, food conversion efficiency and blood physiology in juvenile Atlantic halibut (*Hippoglossus hippoglossus* L.). Aquaculture. 2011;315:400–6.

Person-Le Ruyet J, Lamers A, Roux AL, Severe A, Boeuf G, Mayer-Gostan N. Long-term ammonia exposure of turbot: effects on plasma parameters. J Fish Biol. 2003;62:879–94.

Thangam Y, Perumayee M, Jayaprakash S, Umavathi S, Basheer SK. Studies of ammonia toxicity on haematological parameters to freshwater fish *Cyprinus carpio* (common carp). Int J Curr Microbiol App Sci. 2014;3:535–42.

Vinodhini R, Narayanan M. The impact of toxic heavy metal on the hematological parameters in common carp (*Cyprinus carpio* L.). Iran J Environ Health Sci Eng. 2009;6:23–8.

Wendelaar Bonga SE. The stress response in fish. Physiol Rev. 1997;77:591–625.

13

Cloning and characterization of ADP-ribosylation factor 1b from the olive flounder *Paralichthys olivaceus*

So-Hee Son[1], Jin-Hyeon Jang[1], Hyeon-Kyeong Jo[1], Joon-Ki Chung[2] and Hyung-Ho Lee[1*]

Abstract

Small GTPases are well known as one of the signal transduction factors of immune systems. The ADP-ribosylation factors (ARFs) can be classified into three groups based on the peptide sequence, protein molecular weight, gene structure, and phylogenetic analysis. ARF1 recruits coat proteins to the Golgi membranes when it is bound to GTP. The class I duplicated ARF gene was cloned and characterized from the olive flounder (*Paralichthys olivaceus*) for this study. PoARF1b contains the GTP-binding motif and the switch 1 and 2 regions. PoARF1b and PoARF1b mutants were transfected into a Hirame natural embryo cell to determine the distribution of its GDP/GTP-bound state; consequently, it was confirmed that PoARF1b associates with the Golgi body when it is in a GTP-binding form. The results of the qPCR-described PoARF1b were expressed for all of the *P. olivaceus* tissues. The authors plan to study the gene expression patterns of PoARF1b in terms of immunity challenges.

Keywords: ADP-ribosylation factor, Olive flounder, GTPase, Duplication, Immune system, Golgi complex

Background

The aquatic culture of the olive flounder (*Paralichthys olivaceus*) has been widespread in Korea. The farming of juvenile olive flounder, however, has caused a lot of problems due to the occurrence of various diseases (Ototake and Matsusato 1986; Park 2009). The juvenile flounder is difficult to manage and is weak against diseases, and the mortality rates have been economically damaging (Jee et al. 2001).

Small GTPases are well known as one of the signal transduction factors of immune systems (Narumiya 1996; Scheele et al. 2007). Some documents indicated that small GTPases are related to virus infection in the shrimp (Wu et al. 2007; Liu et al. 2009; Zhang et al. 2010). Also, the small GTPases of zebra fish have provided a firm basis of an innate immune system in vertebrates (Salas-Vidal et al. 2005). The authors therefore studied on ADP-ribosylation factor, which is a member of the GTP-binding proteins, from the olive flounder to

investigate the relation between cytoskeleton remodeling and the olive flounder immune system.

The ADP-ribosylation factor (ARF) proteins are small GTP-binding proteins, and they are involved in membrane dynamics and the regulation of actin cytoskeleton organization (D'Souza-Schorey and Chavrier 2006; Myers and Casanova 2008). The ARF can be classified into three groups based on the peptide sequence, protein molecular weight, gene structure, and phylogenetic analysis, as follows: class I including ARF1, ARF2, and ARF3; class II including ARF4 and ARF5; and class III including only ARF6 (Myers and Casanova 2008; Tsuchiya et al. 1991). The class I and class II ARFs are mainly associated with the Golgi complex, although they also function in endosomal compartments (Myers and Casanova 2008). In addition, ARF proteins were identified as activators of phospholipase D (PLD) (Luo et al. 1998). ARF1 was shown to recruit coat proteins to the Golgi membranes when it is bound to GTP (Balch et al. 1992). The hydrolysis and binding of GTP by ARF1 were originally linked to the assembly and disassembly of vesicle coats (Nie and Randazzo 2006).

The radiation of teleosts has been attributed to a genome-DNA event during the evolution of the teleosts

* Correspondence: hyunghl@pknu.ac.kr
[1]Department of Biotechnology, Pukyong National University, Busan 608-737, South Korea
Full list of author information is available at the end of the article

(Venkatesh 2003). Although a number of ARFs, from those of micro-organisms to those of mammals, have been studied, a lack of studies on the duplicated ARF genes in the olive flounder persists. The authors therefore isolated and characterized one of the class I duplicated ARF genes.

Methods
cDNA cloning and phylogenetic analysis of *Paralichthys olivaceus* ARF1b
Total RNA was extracted by using GeneAll® Hybrid-R™ Total RNA (GeneAll Biotechnology Co., Ltd., Korea) following the manufacturer's instructions from 12 tissues, including the brain, eye, gullet, heart, liver, stomach, muscle, kidney, spleen, pyloric ceca, intestine, and gill tissues, of healthy *Paralichthys olivaceus*. And then, we carried out 5′- and 3′-rapid amplification of cDNA ends (RACE) by using SMART™ RACE cDNA Amplification Kit (Clontech laboratories, Inc.) according to the manufacturer's instructions. For obtaining full-length cDNA sequence, new gene-specific sense and antisense primers were designed (Table 1). The primers were used to PCR for obtaining full-length cDNA sequence. Nucleotide sequences and deduced amino acid sequence aligned with their respective homologues using Genetyx 7.0 software (GENETYX Corporation, Tokyo, Japan) and sequence alignment editor (BioEdit) (Hall 2011).

The phylogenetic tree was constructed using the Neighbor-Joining Method by MEGA6 (Tamura et al. 2013). Different DNA and protein sequences in Ensembl sequence database were used to carry out phylogenetic tree generation, sequence alignments, and database searching (Additional file 1) (Flicek et al. 2011).

Tissue distribution of PoARF1b by qPCR analysis
The tissue distribution of PoARF1b in different tissues was measured by RT-qPCR using a LightCycler 480 Real-Time PCR System (Roche, Mannheim, Germany) with LightCycler 480 SYBR green master I (Roche). The total RNA was extracted from the brain, gullet, eye, heart, stomach, liver, kidney, spleen, pyloric ceca, muscle, intestine, and gill from healthy *P. olivaceus* specimens. cDNA was synthesized with random hexamer primers and oligo(dT)18 using the PrimeScript™ 1st strand cDNA Synthesis Kit (TaKaRa), according to the manufacturer's instructions. The specific primer for internal control was used 18s rRNA (Table 1) (Ahn et al. 2008). The quantitative real-time PCR followed program: pre-incubation at 95 °C for 5 min, 45 cycles at 95 °C for 10 s, 60 °C for 10 s, and 72 °C for 10 s. The qPCR reaction mixture consists of the following elements: 10 μl of 2× SYBR (Roche), 7.5 μl of SYBR water (Roche), 1 μl each of sense and antisense primers, and 0.5 μl diluted first-strand cDNA (diluted at 1:20). The ΔΔCt method

Table 1 Oligonucleotide primers used in PCR amplification of *ARF1b* of *P. olivaceus*; F, Forward; R, reverse

Primer name	5'-3' sequence	Information
DgARF1b-F1	ATGGGDRMYWTBKCYWSC	Primers for cDNA library screening
DgARF1b-F2	GACVACCATCYTGTACAARCTCAAAC	
DgARF1b-R1	CAYTTBVYDTTYYTBAGCTKG	
DgARF1b-R2	CRTCYTCWGMCARCATTCGCATC	
3′GSP-PoARF1b-F1	GTTGAGACAGTAGAGTACAGGAACATCAG	
3′GSP-PoARF1b-F2	GGAGCGAATCGGTGAGGCGAGAGAGGAGC	
5′GSP-PoARF1b R1	CAGTTCCGCCGGGTGAGGTCGTGCAGGC	
5′GSP-PoARF1b R2	GCTCCTCTCTCGCCTCACCGATTCGCTCC	
PoARF1b-RT-F	GCAGCAAAGTACTTCAAAGCCC	Primers for qPCR
PoARF1b-RT-R	CTCAGCCAGCATTCGCATC	
Po18S rRNA-RT-For2	ATGGCCGTTCTTAGTTGGTG	GenBank accession no. EF126037.1
Po18S rRNA-RT-Rev2	CACACGCTGATCCAGTCAGT	
PoARF1b-ORF F	ATGGGACAGTTCTTTAGCCTGTTTAAAG	Primers for the construction of pEGFP Cl
PoARF1b-ORF R	TCATTTTATGTTCTTGAGCTGGTTTGAAAG	
XhoI-PoARF1b-F	CCGCTCGAGCTATGGGACAGTTCTTTA	
XpnI-PoARF1b-R	CGGGGTACCTCATTTTATGTTCTTGAGCTG	
PoARF1bT30N-F	CTGCATGCTGGAAAGAACACCATCCTGTACAAA	Primers for the site-directed mutation
PoARF1bT30N-R	TTTGTACAGGATGGTGTTCTTTCCAGCATGCAG	
PoARF1bQ70L-F	GTCGGTGGTCTGGACAAGATCAGGCCACTC	
PoARF1bQ70L-R	GAGTGGCCTGATCTTGTCCAGACCACCGAC	

was applied to calculate the data, and $2^{-\Delta\Delta Ct}$ method was applied to calculate the relative quantitative value (Giulietti et al. 2001).

Statistics

All qPCR data were statistically analyzed using SPSS 21 program (SPSS, Chicago, IL, USA). One-way ANOVA was used to study the PoARF1b expression, followed by Duncan's Multiple Range test. A p value with $p < 0.05$ was considered to be significant (Sokal and Rohlf 1969).

Cell culture and transfection

Hirame natural embryo cell line (HINAE) was grown in Leibovitz's L-15 medium (Gibco BRL, Grand Island, NY) containing 10% fetal bovine serum (Gibco) and 1% antibiotics (Gibco) at 20 °C (Kasai and Yoshimizu 2001). The transfection was performed using PolyPlus (JetPrime, New York, NY, USA) kit to transient transfection of PoARF1b and its mutants according to the manufacturer's instructions in six-well test plates. The PoARF1b and mutants were observed by EGFP fluorescence signal under confocal microscopy after 48 h post-transfection.

Site mutation of PoARF1b

PoARF1b(T30N) and PoARF1b(Q70L) were performed using QuikChange II Site-Directed Mutagenesis Kit (Agilent Technologies), according to the manufacturer's instructions (Wang and Malcolm 1999). For PoARF1b mutants, we used specific primers (Table 1). The pEGFP-C1 (Clontech) was used to construct the green fluorescent protein-fused PoARF1b and PoARF1b mutants.

The Golgi body in HINAE was stained using GOLGI ID® Green assay kit containing a Golgi apparatus-selective dye.

Results and Discussion

Cloning and sequence analysis of PoARF1b

To identify the initial sequence of PoARF1b, we obtained databases of other ARF1b by using Ensembl sequence data. These sequences were used to design the forward and reverse primers (Table 1). The initial sequence was obtained from PCR amplification of olive flounder cDNA including the brain, eye, gullet, heart, liver, muscle, stomach, kidney, spleen, pyloric ceca, intestine, and gill. The partial sequence was used for isolation of full-length flounder ARF1b by using 3′ and 5′ GeneRace with flounder ARF1b specific primers (Table 1). As the result, the full nucleotide sequence of PoARF1b is 1677 bp (GenBank accession no. KX668134).

The sequence comprised a 108 bp 5′-untranslated region (5′-UTR), a 544 bp coding region, and a 1025 bp 3′-untranslated region (3′-UTR). Also, the PoARF1b has

180 amino acid residues and the molecular weight is approximately 20,561 Da (Fig. 1a). PoARF1b contains GTP-binding motif, switch 1 and 2 regions (Fig. 1a) (Pasqualato et al. 2002). The GTP-binding motif is shaded with gray box and conserved sequence in other ARFs. The switch 1 and 2 regions were indicated with blue and red letters. The switch regions were significantly assumed to conformational change classical structural GDP/GTP switch that bind tightly to GTP but poorly or not at all to the GDP nucleotide (Pasqualato et al. 2002).

The aligned amino acid sequence appeared in PoARF1b had well conserved domains such as GTP-binding motif, switch 1 and 2 regions, and shared the high homology with ARFs from other species (Fig. 1b). It showed 90% homology with ARF1b from *Takifugu rubripes*.

Phylogenetic tree of PoARF1b

To determine the evolutionary relationship of PoARF1b with other ARFs, phylogenetic tree was performed using the Ensembl sequence data using the Neighbor-Joining Method by MEGA (version 6) with bootstrapping 2000 times (Flicek et al. 2011; Tamura et al. 2013). The result of phylogenetic tree was contained fish which was grouped with the tetrapod and human. The ARF tree consists of three major groups: (i) class I, (ii) class II, and (iii) class III. This result indicated that PoARF1b is closely related to ARF1b of class I (Fig. 2).

Tissue distribution of PoARF1b by qPCR analysis

Real-time PCR showed tissue distribution of PoARF1b. The results of qPCR-described PoARF1b was expressed for all mRNA transcripts in different organs which include the brain, gullet, eye, heart, stomach, liver, kidney, spleen, pyloric ceca, muscle, intestine, and gill (Fig. 3). The expression of the PoARF1b gene was the highest level in the gill and the lowest level in the muscle.

Site mutation analysis of PoARF1b

To determine the distribution of PoARF1b, PoARF1b, and PoARF1b mutants, those were constructed into pEGFP-C1 (Clontech) and transfected into HINAE cell. The punctuate morphology of PoARF1b-EGFP resembled the Golgi complex distribution in HINAE cells (Fig. 4b). This result examines PoARF1b that may act in the Golgi body like as human ARF1 complexed with GDP (Amor et al. 1994). Also, PoARF1b mutants examine the distribution which depends on each GDP or GTP-binding form. The mutants designed PoARF1b(T30N) and PoARF1b(Q70L) (Fig. 4a), according to other reports (Chavrier and Goud 1999; Teal et al. 1994). PoARF1b(T30N) was designed by exchanging the Thr amino acid at position 30 with Asn amino acid. It

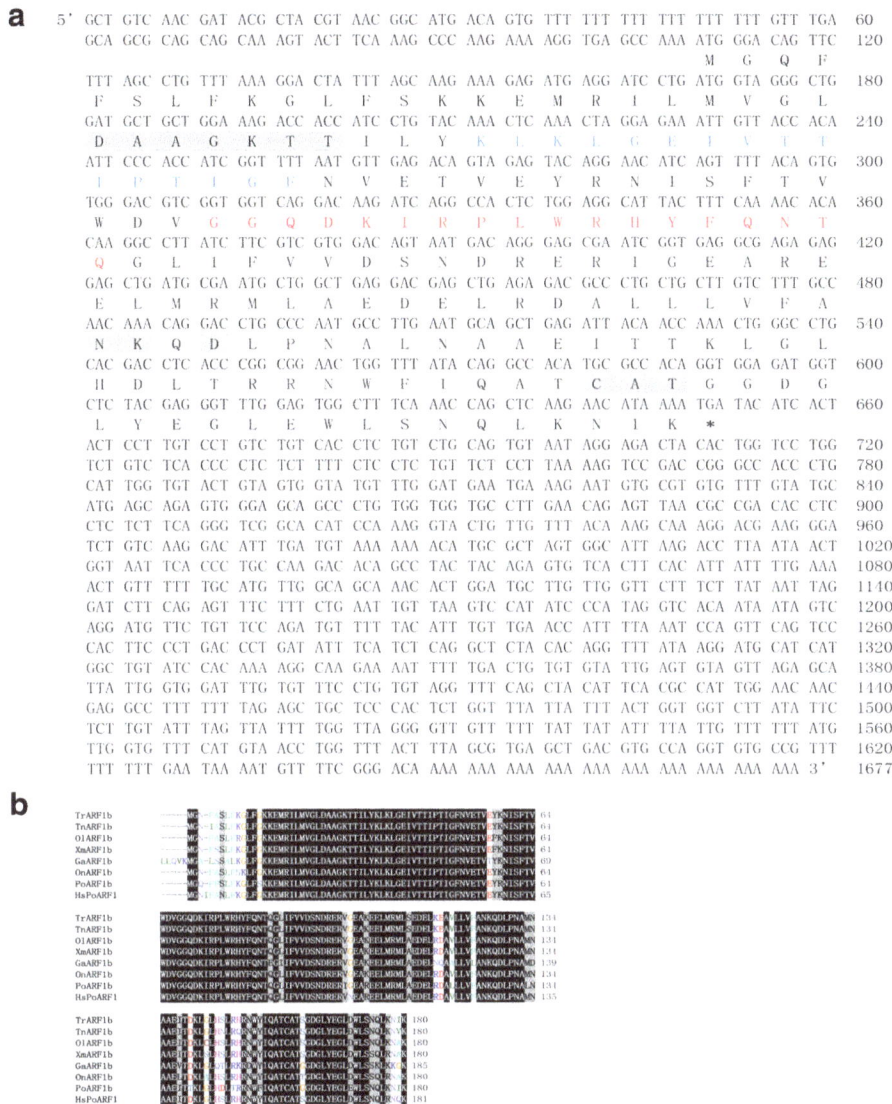

Fig. 1 a Cloning analysis of *PoARF1b*. GTP-binding site is shaded with *gray box*; switch 1 region is indicated with *blue letter*; switch 2 region is indicated with *red letter*. **b** Amino acid sequence analysis of ARFs. The identical conserved amino acid residues are *shaded in black*. The *letters* in front of ARF are species names for acronym. *Tr Takifugu rubripes*, *Tn Tetraodon nigroviridis*, *Ol Oryzias latipes*, *Xm Xiphophorus maculatus*, *Ga Gasterosteus aculeatus*, *On Oreochromis niloticus*, *Po Paralichthys olivaceus*, and *Hs Homo sapiens*

was expected to function in a dominant-negative manner and retain the GDP-binding form. PoARF1b(Q70L) was designed by replacing the Gln at position 70 with Leu. It was expected to function in a dominant-positive manner and keep the GTP-binding form. The result of PoARF1b(T30N) showed a clearly disassembled punctuate morphology (Fig. 4a). When PoARF1b is in a GDP-binding form, it examines to disassociate from the Golgi complex. On the other hand, the result of PoARF1b(Q70L) examined more expanded morphology than that of normal PoARF1b and PoARF1b(T30N) (Fig. 4a). When PoARF1b is in a GTP-binding form, it shows to associate from the Golgi complex.

Conclusion

The small GTPases regulate multiple signaling processes including cell growth, survival, and differentiation (Johnson and Chen 2012). The ARF1 function of the Golgi complex may be important and plays a significant role in the secretory pathway (Radhakrishna and Donaldson 1997). In this paper, *Paralichthys olivaceus* ARF (PoARF) was cloned. The deduced amino acid sequence of PoARF contains the GTP-binding motif, and the switch 1 and 2 regions are present as mammalian ARF. The PoARF is highly conserved in the other amino acid sequences from teleosts and humans. The PoARF is indicated from approximately 76 to 85% of the overall

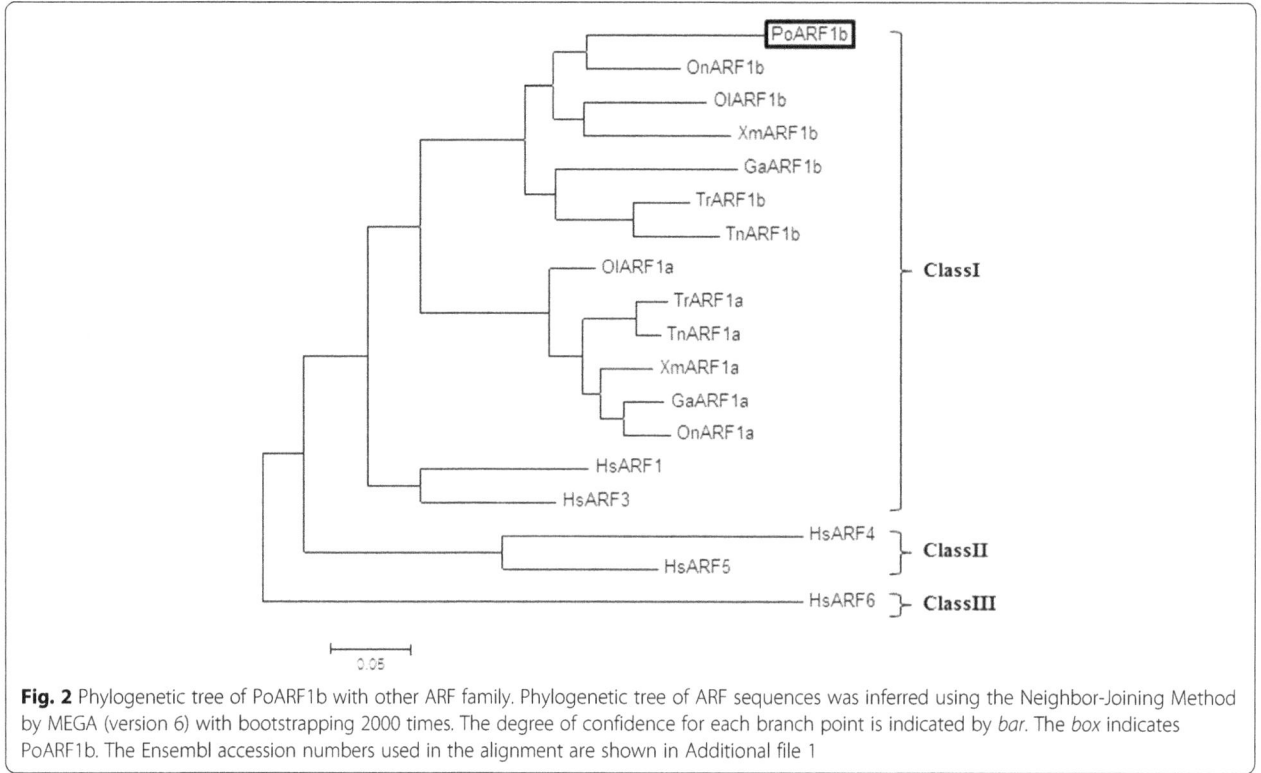

Fig. 2 Phylogenetic tree of PoARF1b with other ARF family. Phylogenetic tree of ARF sequences was inferred using the Neighbor-Joining Method by MEGA (version 6) with bootstrapping 2000 times. The degree of confidence for each branch point is indicated by *bar*. The *box* indicates PoARF1b. The Ensembl accession numbers used in the alignment are shown in Additional file 1

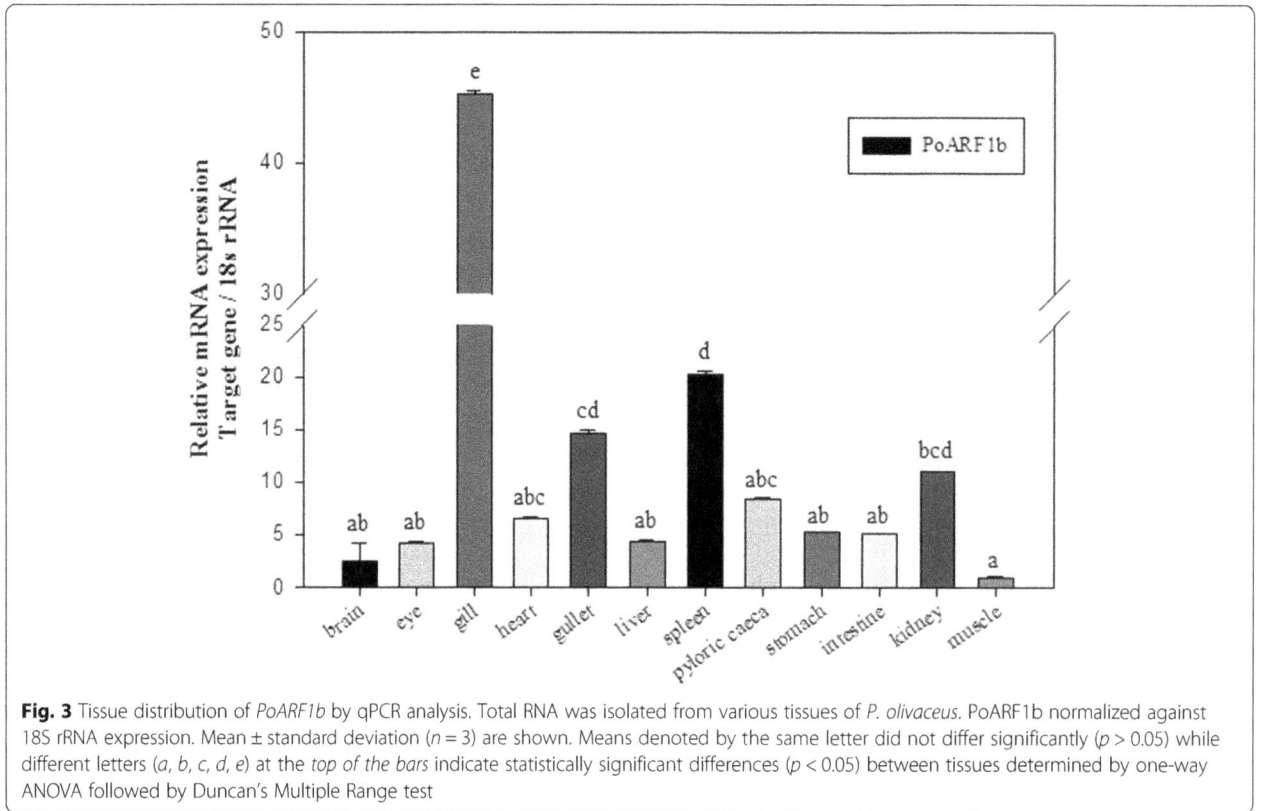

Fig. 3 Tissue distribution of *PoARF1b* by qPCR analysis. Total RNA was isolated from various tissues of *P. olivaceus*. PoARF1b normalized against 18S rRNA expression. Mean ± standard deviation ($n = 3$) are shown. Means denoted by the same letter did not differ significantly ($p > 0.05$) while different letters (*a, b, c, d, e*) at the *top of the bars* indicate statistically significant differences ($p < 0.05$) between tissues determined by one-way ANOVA followed by Duncan's Multiple Range test

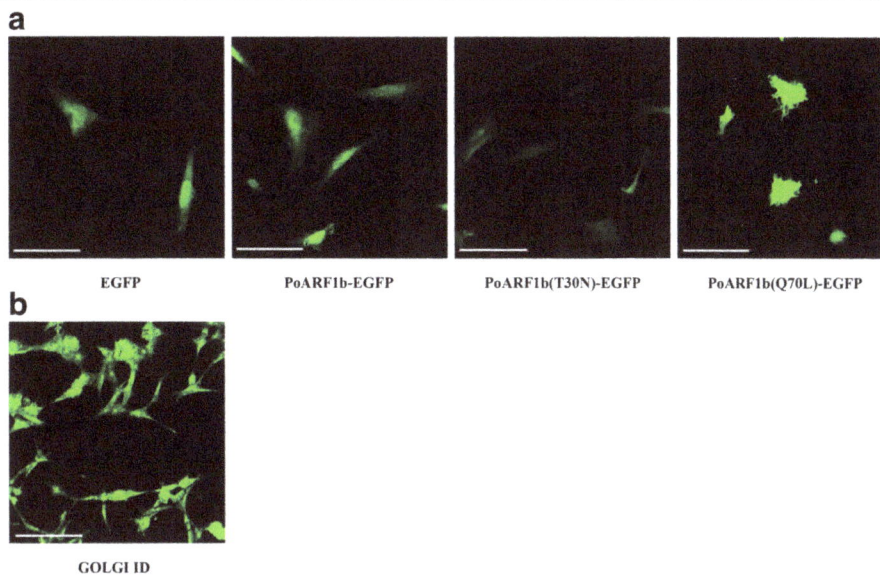

Fig. 4 a Punctuate morphology of PoARF1b and its mutants. The amounts of plasmids (2 μg) were transfected into HINAE cells in 6 wells. EGFP, control; PoARF1b-EGFP, wild-type; PoARF1b(T30N)-EGFP, negative mutant; and PoARF1b(Q70L)-EGFP, positive mutant. **b** Intracellular distribution of the Golgi complex in HINAE cells using GOLGI ID. *Bar* means 50 μm

identities from the other ARF isozymes (data not shown). PoARF shares approximately 85% with the identity of *Oreochromis niloticus* ARF1b (OnARF1b) and approximately 79% with the identity of *Gasterosteus aculeatus* ARF1b (GaARF1b). Also, PoARF shares 76% with the identity of *Homo sapiens* ARF1 (HsARF1). In addition, the phylogenetic tree showed that PoARF is more closely related to ARF1b than ARF1a. These results indicate that PoARF is PoARF1b. OnARF1b, which shares a high percentage with the identity of PoARF1b, shares 76% with the identity of HsARF1.

As is known, the PoARF1b is expressed in all of the tissues of the olive flounder. The PoARF1b mRNA has a high expression level in the gill and a low expression level in the muscle. This finding resembles the ARF1 expression from the shrimp (*Marsupenaeus japonicus*), which shows the lowest expression level in the muscle (Ma et al. 2010). It will be necessary to further study why such an outcome.

The Golgi-binding distribution of PoARF1b depends on the GTP- or GDP-bound state. The PoARF1b-EGFP showed a punctuate morphology that resembles the morphology of the Golgi body in HINAE cells (Fig. 4). The GOLGI ID of the Golgi bodies can be detected using a microscopy.

PoARF1b(T30N) showed a clearly disassembled punctuate morphology, and PoARF1b(Q70L) showed a more expanded morphology; these results resemble those of the mammalian ARF1. The results of this study indicate that PoARF1b functions within the Golgi complex.

Further studies are needed to be carried out to explain the highest expression of PoARF1b in gill.

Abbreviations
ARF: ADP-ribosylation factor; HINAE: Hirame natural embryo cell line; RACE: Rapid amplification of cDNA ends; UTR: Untranslated region

Funding
This work was supported by a Research Grant of Pukyong National University (year 2015).

Authors' contributions
S-HS carried out the whole process, participated in the whole experiments and drafted the manuscript. J-HJ carried out the cDNA cloning and phylogenetic analysis. J-KC participated in its statistical analysis. HHL participated in the design of the study and coordination. All authors read and approved the final manuscript.

Competing interests
The authors declare that they have no competing interests.

Author details
[1]Department of Biotechnology, Pukyong National University, Busan 608-737, South Korea. [2]Department of Aquatic Life Medicine, Pukyong National University, Busan 608-737, South Korea.

References

Ahn SJ, Kim NY, Jeon SJ, Sung JH, Je JE, Seo JS, … and Lee HH. Molecular cloning, tissue distribution and enzymatic characterization of cathepsin X from olive flounder (*Paralichthys olivaceus*). Comp Biochem Physiol Part B Biochem Mol Biol. 2008;151(2):203–212.

Amor JC, Harrison DH, Kahn RA, Ringe D. Structure of the human ADP-ribosylation factor 1 complexed with GDP. Nature. 1994;372(6507):704–8.

Balch WE, Kahn RA, Schwaninger R. ADP-ribosylation factor is required for vesicular trafficking between the endoplasmic reticulum and the cis-Golgi compartment. J Biol Chem. 1992;267(18):13053–61.

Chavrier P, Goud B. The role of ARF and Rab GTPases in membrane transport. Curr Opin Cell Biol. 1999;11(4):466–75.

D'Souza-Schorey C, Chavrier P. ARF proteins: roles in membrane traffic and beyond. Nat Rev Mol Cell Biol. 2006;7(5):347–58.

Flicek P, Amode MR, Barrell D, Beal K, Brent S, Carvalho-Silva D, … and Gil L. Ensembl 2012. Nucleic Acids Res. 2011;gkr991.

Giulietti A, Overbergh L, Valckx D, Decallonne B, Bouillon R, Mathieu C. An overview of real-time quantitative PCR: applications to quantify cytokine gene expression. Methods. 2001;25(4):386–401.

Hall T. BioEdit: an important software for molecular biology. GERF Bull Biosci. 2011;2(1):6.

Jee BY, Kim YC, Park MS. Morphology and biology of parasite responsible for scuticociliatosis of cultured olive flounder *Paralichthys olivaceus*. Dis Aquat Org. 2001;47(1):49–55.

Johnson DS, Chen YH. Ras family of small GTPases in immunity and inflammation. Curr Opin Pharmacol. 2012;12(4):458–63.

Kasai H. and Yoshimizu M. Establishment of two Japanese flounder [*Paralichthys olivaceus*] embryo cell lines. Bulletin of Fisheries Sciences, Hokkaido University (Japan). 2001.

Liu W, Han F, Zhang X. Ran GTPase regulates hemocytic phagocytosis of shrimp by interaction with myosin. J Proteome Res. 2009;8(3):1198–206.

Luo JQ, Liu X, Frankel P, Rotunda T, Ramos M, Flom J, … and Foster DA. Functional association between Arf and RalA in active phospholipase D complex. Proc Natl Acad Sci. 1998;95(7):3632–3637.

Ma J, Zhang M, Ruan L, Shi H, Xu X. Characterization of two novel ADP ribosylation factors from the shrimp *Marsupenaeus japonicus*. Fish Shellfish Immunol. 2010;29(6):956–62.

Myers KR, Casanova JE. Regulation of actin cytoskeleton dynamics by Arf-family GTPases. Trends Cell Biol. 2008;18(4):184–92.

Narumiya S. The small GTPase Rho: cellular functions and signal transduction. J Biochem. 1996;120(2):215–28.

Nie Z, Randazzo PA. Arf GAPs and membrane traffic. J Cell Sci. 2006;119(7):1203–11.

Ototake M, Matsusato T. Notes on Scuticociliata infection of cultured juvenile flounder *Paralichtys olivaceus*. Bull. Natl. Res. Inst. Aquaculture. 1986;9:65–68 (in Japanese).

Park SI. Disease control in Korean aquaculture. Fish Pathol. 2009;44(1):19–23.

Pasqualato S, Renault L, Cherfils J. Arf, Arl, Arp and Sar proteins: a family of GTP-binding proteins with a structural device for 'front–back' communication. EMBO Rep. 2002;3(11):1035–41.

Radhakrishna H, Donaldson JG. ADP-ribosylation factor 6 regulates a novel plasma membrane recycling pathway. J Cell Biol. 1997;139(1):49–61.

Salas-Vidal E, Meijer AH, Cheng X, Spaink HP. Genomic annotation and expression analysis of the zebrafish Rho small GTPase family during development and bacterial infection. Genomics. 2005;86(1):25–37.

Scheele JS, Marks RE, Boss GR. Signaling by small GTPases in the immune system. Immunol Rev. 2007;218(1):92–101.

Sokal RR, Rohlf FJ. The principles and practice of statistics in biological research. San Francisco: WH Freeman and company; 1969. p. 399–400.

Tamura K, Stecher G, Peterson D, Filipski A, Kumar S. MEGA6: molecular evolutionary genetics analysis version 6.0. Mol Biol Evol. 2013;30(12):2725–9.

Teal SB, Hsu VW, Peters PJ, Klausner RD, Donaldson JG. An activating mutation in ARF1 stabilizes coatomer binding to Golgi membranes. J Biol Chem. 1994;269(5):3135–8.

Tsuchiya M, Price SR, Tsai SC, Moss J, Vaughan M. Molecular identification of ADP-ribosylation factor mRNAs and their expression in mammalian cells. J Biol Chem. 1991;266(5):2772–7.

Venkatesh B. Evolution and diversity of fish genomes. Curr Opin Genet Dev. 2003;13(6):588–92.

Wang W, Malcolm BA. Two-stage PCR protocol allowing introduction of multiple mutations, deletions and insertions using QuikChange Site-Directed Mutagenesis. Biotechniques. 1999;26(4):680–2.

Wu W, Zong R, Xu J, Zhang X. Antiviral phagocytosis is regulated by a novel Rab-dependent complex in shrimp *Penaeus japonicus*. J Proteome Res. 2007;7(01):424–31.

Zhang M, Ma J, Lei K, Xu X. Molecular cloning and characterization of a class II ADP ribosylation factor from the shrimp *Marsupenaeus japonicus*. Fish Shellfish Immunol. 2010;28(1):128–33.

14

Effects of astaxanthin on antioxidant capacity of golden pompano (*Trachinotus ovatus*) in vivo and in vitro

Jia-jun Xie[1,2†], Xu Chen[1†], Jin Niu[3*], Jun Wang[1], Yun Wang[1] and Qiang-qiang Liu[1]

Abstract

The objective of this research was to study the effect of astaxanthin (AST) on growth performance and antioxidant capacity in golden pompano (*Trachinotus ovatus*) both in vivo and in vitro. In the in vivo study, two diets were formulated with or without astaxanthin supplementation (D1 and D2; 0 and 200 mg/kg) to feed fish for 6 weeks. In the in vitro study, cells from hepatopancreas of golden pompano were isolated and four treatments with or without astaxanthin and H_2O_2 supplementation were applied (control group: without both astaxanthin and H_2O_2 treated; H_2O_2 group: just with H_2O_2 treated; H_2O_2 + AST group: with both astaxanthin and H_2O_2 treated; AST group: just with AST treated). Results of the in vivo study showed that weight gain (WG) and special growth rate (SGR) significantly increased with astaxanthin supplemented ($P < 0.05$). Feed conversion ratio (FCR) of fish fed D2 diet was significantly lower than that of fish fed D1 diet ($P < 0.05$). Hepatic total antioxidant capacity (T-AOC) and the reduced glutathione (GSH) of golden pompano fed D2 diet were significant higher than those of fish fed D1 diet ($P < 0.05$). Superoxide dismutase (SOD) was significantly declined as astaxanthin was supplemented ($P < 0.05$). Results of the in vitro study showed that the cell viability of H_2O_2 group was 52.37% compared to the control group, and it was significantly elevated to 84.18% by astaxanthin supplementation (H_2O_2 + AST group) ($P < 0.05$). The total antioxidant capacity (T-AOC) and the reduced glutathione (GSH) of cell were significant decreased by oxidative stress from H_2O_2 ($P < 0.05$), but it could be raised by astaxanthin supplementation (H_2O_2 vs H_2O_2 + AST), and the malondialdehyde (MDA) was significant higher in H_2O_2 group ($P < 0.05$) and astaxanthin supplementation could alleviate the cells from lipid peroxidation injury. In conclusion, dietary astaxanthin supplementation can improve the growth performance of golden pompano. Moreover, astaxanthin can improve the golden pompano hepatic antioxidant capacity both in vivo and in vitro study by eliminating the reactive oxygen species.

Keywords: Golden pompano, Astaxanthin, Growth performance, Antioxidant capacity

Background

Reactive oxygen species (ROS) are oxidative products, produced continuously in the course of normal aerobic cellular metabolism and respiratory burst (Chew 1995), which participate in a variety of biological processes, including normal cell growth, induction and maintenance of the transformed state, programmed cell death, and cellular senescence (Finkel 2003). However, ROS can, in turn, damage healthy cells if they are not eliminated (Chew 1995). Under normal physiological conditions, the excessive ROS can be removed by internal antioxidants and anti-oxidative systems (Chen et al. 2015), including counter balance such as enzymes (like superoxide dismutase, catalase, and glutathione peroxidase), functionalized large molecules (albumin, ferritin, and ceruloplasmin) and small molecules (ascorbic acid, α-tocopherol, β-carotene, and uric acid) (Martinez-Alvarez et al. 2005). The dietary antioxidants most widely used include vitamin E, vitamin C, carotenoids, flavanoids, zinc, and selenium (Chew and Park 2004). Among those, carotenoids reach the highest plasma and tissue concentrations, despite their lower intake (Olmedilla et al. 2007).

Carotenoids, more than 600 known types, can be classified into two categories, xanthophyll and carotenes.

* Correspondence: gzniujin2003@163.com; 1657864417@qq.com
†Equal contributors
3Institute of Aquatic Economic Animals, School of Life Science, Sun Yat-sen University, NO.135 at Xingang Xi Road, Haizhu District, Guangzhou 510275, Guangdong Province, China
Full list of author information is available at the end of the article

Astaxanthin (3,3′-dihydroxy-β,β-carotene-4,4′dione, AST) is a xanthophyll carotenoid which is found in many microorganisms and marine animals, such as shrimp, crayfish, crustaceans, salmon, trout, krill, microalgae as well as yeast. Its molecule consists of 40 carbon atoms, divided into a central portion containing 22 carbon atoms linked with 13 conjugated double bonds and two terminal benzene rings containing hydroxyl and ketone groups, giving rise to the higher polar structure of AST compared with other carotenoids (Britton 1995).The antioxidant activity of astaxanthin was found to be approximately 10 times stronger than β-carotene (Shimidzu et al. 1996).

Except for its antioxidant capacity, AST is also recognized to have growth performance and survival rate promoting in Atlantic salmon (*Salmo salar*) (Christiansen and Torrissen 1996) and red porgy (*Pagrus pagrus*) (Kalinowski et al. 2011), skins coloration enhancing in large yellow croaker (*Larimichthys croceus*) (Yi et al. 2014) and Atlantic salmon(Baker et al. 2002), anti-lipid peroxidation (Leite et al. 2010), and immune response reinforcing in *Astronotus ocellatus* (Alishahi et al. 2015) properties.

Golden pompano (*Trachinotus ovatus*) belongs to family carangidae, genus *Trachinotus*. It is a warm-water species (25–32 °C) and a carnivorous fish that preys mainly on zooplankton, small crustaceans, shellfish, and small fish (Liu and Chen 2009). *T. ovatus*is is widely distributed in China, Japan, Australia, and other countries (Huo-sheng 2006). Pompano is considered one of the most desirable food fishes, and it commands a significantly higher price than many other marine and freshwater species (Tutman et al. 2004). Recently, pompano is widely farmed owing to its high price in the market and resilience to salinity and temperature ranges (Tutman et al. 2004), and annual output was over 100,000 tons. The suitable dietary protein and lipid levels for golden pompano are 46.0 and 6.5% (Wang et al. 2013). The optimum carbohydrate level for juvenile golden pompano could be 11.2–16.8% of the diet (Zhou et al. 2015). The optimal requirements of methionine (Niu et al. 2013), arginine (Lin et al. 2015), and n-6 long-chain polyunsaturated fatty acid arachidonic acid (ARA) (Qi et al. 2016) for golden pompano have been determined as well.

However, fewer studies were conducted to investigate the effects of antioxidants on golden pompano. To date, the effects of AST on various kinds of fishes are mainly reported in vivo and rarely found in vitro. We used in vivo and in vitro models to study and compare the effect of astaxanthin on antioxidant ability of golden pompano, respectively.

Methods
Diet preparation and dietary treatments
In this study, two isonitrogenous and isoenergetic semipurified diets were formulated supplementing with or without astaxanthin (D1: 0%; D2: 0.2%; CAROPHYLL

Pink, 10% astaxanthin, DSM Nutritional Products France SAS) (Table 1). The method of diet preparation was the same as described by Niu et al. (2015). Briefly, all dry ingredients were finely ground, weighed, mixed manually for 5 min, and then transferred to a Hobart mixer (A-200 T Mixer Bench Model unit; Resell Food Equipment Ltd., Ottawa, ON, Canada) for another 15 min mixing. During the mixing, 6 N NaOH was added to establish a pH level of 7–7.5. The pH of the diet was obtained by homogenizing a 5-g portion of the diet with 50 mL of distilled water with a glass-electrode pH meter on the supernatant (Robinson et al. 1981). Soya lecithin was added to pre-weighed fish oil and mixed until homogenous. The oil mix was then added to the Hobart mixer slowly while mixing was still continuing. All ingredients were mixed for another 10 min. Then, distilled water (about 30–35%, v/w) was added to the mixture to form dough. Dough of even consistency was passed through a pelletizer with a 2.5-mm-diameter

Table 1 Ingredients and proximate composition of the two experimental diets (%)

Ingredients	Diet 1	Diet 2
Fish meal	32.00	32.00
Soybean meal	30.00	30.00
Wheat flour	20.00	19.80
Krill meal	2.00	2.00
Fish oil	8.00	8.00
Soya lecithin	2.00	2.00
Monocalcium phosphate	2.00	2.00
Pre-vitamin[a]	1.00	1.00
Pre-mineral[b]	1.00	1.00
Choline	0.50	0.50
Vc	0.50	0.50
DL-Met	0.40	0.40
Lys-HCL (78%)	0.60	0.60
AST	0	0.20
Sum	100	100
Nutrient levels		
Moisture	9.00	7.54
Crude protein	40.64	40.55
Crude fat	10.71	10.90
Ash	15.21	15.24

[a]Pre-vitamin (mg or g kg^{-1} diet): thiamin, 25 mg; riboflavin,45 mg; pyridoxine HCl, 20 mg; vitamin B12, 0.1 mg; vitamin K3,10 mg; inositol, 800 mg; pantothenic acid, 60 mg; niacin acid, 200 mg; folic acid, 20 mg; biotin, 1.20 mg; retinal acetate, 32 mg; cholecalciferol, 5 mg; α-tocopherol, 120 mg; ascorbic acid, 2000 mg; choline chloride, 2500 mg; ethoxyquin 150 mg; and wheat middling, 14.012 g (Niu et al. 2013)
[b]Pre-mineral (mg or g kg^{-1} diet): NaF, 2 mg; KI, 0.8 mg; CoCi$_2$6H$_2$O (1%), 50 mg; CuSO$_4$5H$_2$O, 10 mg; FeSO$_4$H$_2$O, 80 mg; ZnSO$_4$H$_2$O, 50 mg; MnSO$_4$H$_2$O, 60 mg; MgSO$_4$7H$_2$O, 1200 mg; Ca(H$_2$PO$_4$)$_2$H$_2$O, 3000 mg; NaCl, 100 mg; and zoelite, 15.447 g (Niu et al. 2013)

die (Institute of Chemical Engineering, South China University of Technology, Guangzhou, China). The diets were dried until the moisture was reduced to <10%. The dry pellets were placed in plastic bags and stored 20 °C until fed.

Animal rearing and experimental procedures

The feeding trial was conducted at an experimental station of South China Sea Fisheries Research Institute of Chinese Academy of Fishery Sciences (Sanya, Hainan). Prior to the start of the trial, juvenile *T. ovatus* were acclimated to a commercial diet for 2 weeks and were fed twice daily to apparent satiation. At the beginning of the feeding trial, the fish were starved for 24 h, weighed, and then fish with similar size (initial body weight 23.65 ± 0.10 g) were randomly allotted to 8 sea cages (1.0 m × 1.0 m × 1.5 m; four cages per diet treatment); each cage was stocked with 30 fish. Each experimental diet was randomly assigned to four cages. The feeding frequency was once daily at 8:00 and lasted for 6 weeks. To prevent the waste of pellets, fish were slowly hand-fed to satiation based on visual observation of their feeding behavior. Feed consumption was recorded for each cage every day. Water quality parameters were monitored daily.

Sample collection

At the end of the feeding trial, fish were starved for 24 h and then weighed and counted the total number. Ten fish from each cage were randomly collected for sampling: four for analysis of whole-body composition and six were anesthetized to obtain weights of individual whole body, viscera, and liver. The livers were rapidly removed and frozen in the liquid nitrogen separately for analysis of lipid peroxidation and antioxidant status.

Biochemical composition analysis

Chemical composition of diets and fish were determined by standard methods (Latimer 2012). Moisture was determined by oven drying at 105 °C until a constant weight was obtained. Crude protein content ($N \times 6.25$) was determined according to the Kjeldahl method after acid digestion using an Auto Kjeldahl System (1030-Autoanalyzer; Tecator, Höganäs, Sweden). Crude lipid was determined by the ether extraction method using a Soxtec extraction System HT (Soxtec System HT6, Tecator). Ash content was determined after samples were placed in a muffle furnace at 550 °C for 4 h.

Isolation of liver cells

Golden pompano was purchased from a market in Guangzhou, China. Hepatocytes were isolated according to the methods of Wan et al. (2004) with some modifications. In the procedure, a fish was kept in 0.01% potassium permanganate solution for half an hour, after that, its skin

was sterilized by alcohol, and its abdomen was dissected with sterile instruments from the anus toward the head. Liver tissue was excised and rinsed three times with phosphate buffer solution. The liver tissue was then minced into pieces of 1 mm^3 and transferred to a 15 mL tube to which a solution of 0.25% trypsin (1:20 *w/v*; Sigma) was added. The mixture was trypsinized on a thermostatic water bath to obtain the cell suspension, which was shaken every 5 min. Then, the mixture was filtered through a 100-mesh sieve. The cell suspensions were pooled and centrifuged at 1000 rpm for 10 min, and the cell pellet was washed and resuspended in a culture medium. The cell number was counted using a haemocytometer, and cell viability was estimated immediately following isolation using the trypan blue exclusion assay.

Cell culture and treatments

A final cell density of hepatocytes was adjusted to 2×10^6 cells mL^{-1} in L-15 culture medium (Jinuo Co, Hangzhou, China) supplemented with 2 mM L-glutamine (Sigma) and 10% foetal bovine serum (Gibco). Cells were seeded into 12-well culture plates with 500 μL cell suspension per well. Cells were cultured in a humidified atmosphere at 28 °C. Once seeded, cells were allowed to attach to culture plates for 24 h. At 24 h, 50% of the culture medium (250 μL) was removed and replaced with fresh medium. Then PBS, 100 mM H_2O_2, 1000 ng/mL astaxanthin dissolved in dimethyl sulfoxide (DMSO) (final concentration 0.01%), and H_2O_2 plus astaxanthin were added in the wells. Every treatment was replicated in three wells. Cell viability was evaluated by trypan blue exclusion test, and cells were harvested for antioxidant capacity analysis after the treatments. As an additional measure of cell viability, lactate dehydrogenase (LDH) activity in the extracellular medium (an indicator of membrane leakage (Misra and Niyogi 2009) was measured.

Antioxidant capacity analysis

Hepatic and cell samples were homogenized in ice-cold phosphate buffer (1:10 dilution) (phosphate buffer; 0.064 M, pH 6.4). The homogenate was then centrifuged for 20 min (4 °C, 3000 g), and aliquots of the supernatant were used to quantify hepatic T-AOC, GSH, SOD, and MDA.

The levels of enzyme activity and lipid peroxidation were measured with commercial ELISA kits (Randox Laboratories Ltd.) in accordance with the instructions of the manufacturer. The assays are briefly described as follows: The T-AOC is the representative of enzyme and nonenzyme original antioxidant in the body; these antioxidants can reduce the ferric ion (Fe^{3+}) to ferrous ion (Fe^{2+}). The latter combines with phenanthroline and produces a stable chelate, which can be measured by spectrophotography at 520 nm (Xiao et al. 2004). The T-AOC was determined in units per milligram of tissue protein.

Total superoxide dismutase (SOD) activity was measured by using a xanthine oxides (Marklund and Marklund 1974). The ratio of autooxidation rates of the samples with or without hepatic homogenate was determined at 550 nm. One unit of SOD activity was calculated using the amount of superoxide dismutase required to inhibit the reduction of nitrobluete trazolium by 50%.

The formation of 5-thio-2-nitrobenzoate (TNB) was followed spectrophotometrically at 412 nm (Vardi et al. 2008). The amount of GSH in the extract was determined as μmol/mg protein utilizing a commercial GSH as the standard. The results are expressed as μmol/mg protein.

Lipid-peroxidation levels were determined based on the malondialdehyde (MDA) level generated by oxidizing fatty acids. In the presence of thiobarbituric acid, malondialdehyde started producing colored thiobarbituric-acid-reacting substances (TBARS) that were measured at 532 nm (Buege and Aust 1978).

Lactate dehydrogenase (LDH) can catalyze lactate into pyruvate, which react with 2,4-dinitrophenylhydrazine and produce a stable compound, which was measured by spectrophotography at 450 nm.

Calculations and statistical analysis

The parameters were calculated as follows:

Weight gain rate (WG, %) = 100
× (final body weight - initial body weight)/initial body weight

Specific growth rate (SGR, % day^{-1}) = 100
× (Ln final mean weight - Ln initial mean weight)/ number of days

Feed conversion ratio (FCR) = dry diet fed/wet weight gain

Survival rate (%) = 100
× (final number of fish)/(initial number of fish)

Viscerosomatic index (VSI, %) = 100
× (viscera weight, g)/(whole body weight, g)

Hepatosomatic index (HSI, %) = 100
× (liverweight, g) /(whole body weight, g)

Condition factor (CF, g/cm3) = 100
× (bodyweight, g)/(body length, cm^3)

Data from each treatment were subjected to one-way analysis of variance (ANOVA). Homogeneity of variance was verified using Bartlett and Levene's test. When overall differences were significant, Tukey's multiple range tests was used to compare the mean values among individual treatments. The level of significant difference was set at $P < 0.05$. Statistical analysis was performed using the SPSS19.0 (SPSS Inc., Michigan Avenue, Chicago, IL, USA) for Windows, and the results are presented as means ± SEM (standard error of the mean).

Results

Growth performance in vivo

Growth, feed utilization, and biometric parameters of juvenile pompano fed different dietary astaxanthin levels are shown in Table 2. Results showed that final body wet weight (FBW), weight gain (WG), and special growth rate (SGR) significantly increased with astaxanthin supplemented ($P < 0.05$). Feed conversion ratio (FCR) of golden pompano fed the diets supplemented with astaxanthin was significantly lower than that of fish fed the control diet ($P < 0.05$), while no significant differences were found in survival rate between the two diet treatments ($P > 0.05$). Hepatosomatic somatic indices (HSI), visceral somatic indices (VSI), and condition factor (CF) were significantly decreased in astaxanthin-supplemented diet treatment.

Growth performance in vitro

In the in vitro study, the cell viability of H_2O_2 group was 52.37% compared to the control group (PBS group), and it could be significantly elevated to 84.18% with astaxanthin supplementation (H_2O_2 + AST group) ($P < 0.05$) (Fig. 1). The highest lactate dehydrogenase (LDH) activity was found in H_2O_2 group, and it was 159.02% compared to the control group, it could be significantly lessened to 122.96% with astaxanthin supplementation (H_2O_2 + AST group) ($P < 0.05$) (Fig. 2).

Whole-body composition

Whole-body composition of golden pompano fed different dietary astaxanthin levels are shown in Table 3. There were no significant differences in whole-body composition of fish between the two diet treatments ($P > 0.05$).

Table 2 Growth performance and survival of golden pompano fed diets with and without supplementation of astaxanthin

Diets (AST mg kg^{-1})	Diet 1 0	Diet 2 200
IBW (g)	32.72 ± 0.21	32.58 ± 0.06
FBW (g)	63.71 ± 0.66a	67.23 ± 0.22b
WG (%)	95.30 ± 0.02a	106.36 ± 0.01b
SGR (% day^{-1})	1.59 ± 0.02a	1.72 ± 0.01b
FCR	1.78 ± 0.02a	1.53 ± 0.02b
SR (%)	97.78 ± 0.01	96.67 ± 0.00
HSI (%)	1.13 ± 0.06a	0.94 ± 0.03b
VSI (%)	5.97 ± 0.11a	5.42 ± 0.15b
CF (g/cm3)	3.4 ± 0.06a	3.04 ± 0.05b

Values are mean ± SEM of four replicates, and values in the same row with different letters are significant different ($P < 0.05$). Diet 1 meant golden pompano groups fed diets without supplementation of astaxanthin (AST). Diet 2 meant groups with supplementation of astaxanthin

IBW initial body weight, *FBW* final body weight, *WG* weight gain, *SGR* special growth rate, *FCR* feed conversion ratio, *SR* survival rate, *HSI* hepatosomatic index, *VSI* viscerosomatic index, *CF* condition factor

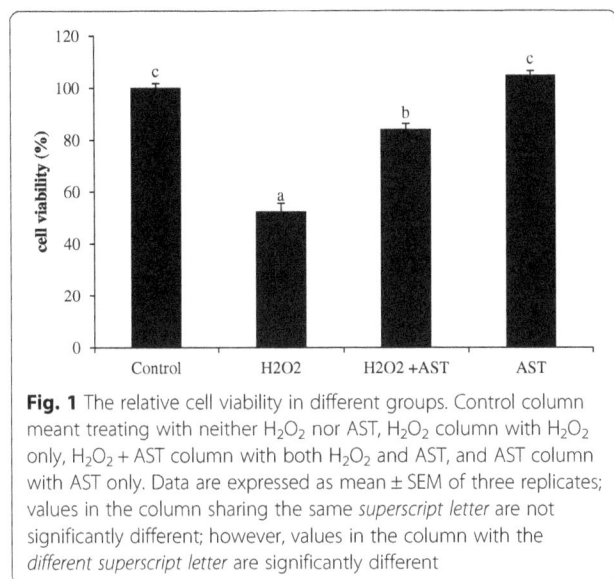

Fig. 1 The relative cell viability in different groups. Control column meant treating with neither H_2O_2 nor AST, H_2O_2 column with H_2O_2 only, H_2O_2 + AST column with both H_2O_2 and AST, and AST column with AST only. Data are expressed as mean ± SEM of three replicates; values in the column sharing the same *superscript letter* are not significantly different; however, values in the column with the *different superscript letter* are significantly different

Antioxidant capacity analysis in vivo

The antioxidant status of juvenile pompano in vivo study are presented in Table 4. The hepatic total antioxidant capacity (T-AOC) and the reduced glutathione (GSH) in fish fed diet supplemented with astaxanthin were significantly higher than that of fish fed the control diet ($P < 0.05$). On the contrary, superoxide dismutase (SOD) declined with astaxanthin supplementation significantly ($P < 0.05$), while hepatic malondialdehyde (MDA) content was not affected by astaxanthin supplementation ($P > 0.05$).

Antioxidant capacity analysis in vitro

The antioxidant status of hepatocytes in the vitro study are shown in Table 5. The H_2O_2 as an oxidizing agent can totally damage the healthy cells, and astaxanthin, an

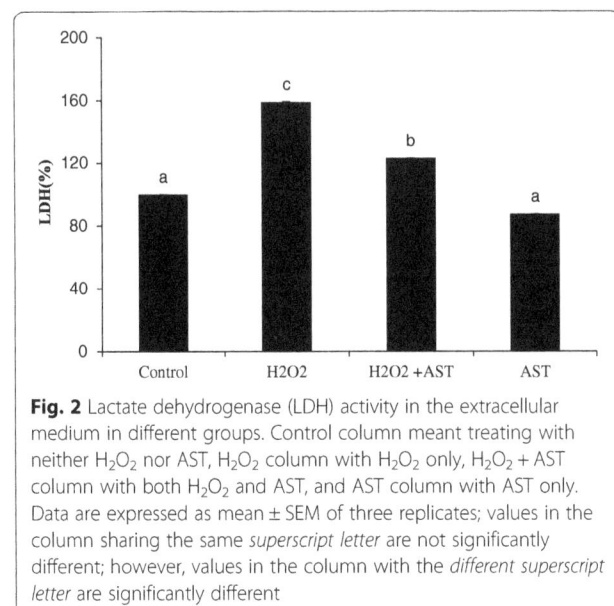

Fig. 2 Lactate dehydrogenase (LDH) activity in the extracellular medium in different groups. Control column meant treating with neither H_2O_2 nor AST, H_2O_2 column with H_2O_2 only, H_2O_2 + AST column with both H_2O_2 and AST, and AST column with AST only. Data are expressed as mean ± SEM of three replicates; values in the column sharing the same *superscript letter* are not significantly different; however, values in the column with the *different superscript letter* are significantly different

antioxidant, can repair the cells from the oxidative stress. The total antioxidant capacity (T-AOC) and the reduced glutathione (GSH) of oxidative stress group (H_2O_2) were the lowest and significantly lower than those of the control group ($P < 0.05$), but it could be significantly improved by astaxanthin supplementation (H_2O_2 + AST group) ($P < 0.05$). The astaxanthin supplemented groups were significant higher than the other ones ($P < 0.05$). The SOD and MDA showed the highest value in H_2O_2 group, which were significantly higher than those in the control group ($P < 0.05$), but it also could be significantly modified by astaxanthin supplementation (H_2O_2 + AST group) ($P < 0.05$). The astaxanthin supplemented groups showed the significantly higher antioxidant capacity than the other groups ($P < 0.05$).

Discussion

Growth performance and proximate composition

Carotenoids are reported to improve growth performance of fish with the reason that carotenoids may exert a positive influence on intermediary metabolism in aquatic animals (Segner et al. 1989), which enhance nutrient utilization, ultimately resulting in improving growth (Amar et al. 2001). The other possible mechanism may be to adjust the intestinal flora breaking down indigestible feed components to extract more nutrients and to stimulate the production of enzymes transporting fats for growth instead of storage (James et al. 2006). Kalinowski et al. (2011) believed that astaxanthin could enhance lipid utilization in whole fish and liver, providing more energy and consequently enhancing growth performance. In the present experiment, the growth performance (FBW, WG, and SGR) and feed utilization of fish fed diet with supplemental astaxanthin were significantly higher than that of fish fed the control diet. This result was in agreement with those in previous studies on Atlantic salmon (Christiansen and Torrissen 1996), red porgy (Kalinowski et al. 2011), *Astronotus ocellatus* (Alishahi et al. 2015), and large yellow croaker (Li et al. 2014). However, effect of carotenoids on fish growth is controversial. Many earlier studies have reported that dietary astaxanthin has no significant influence on growth and flesh composition of fish (Tejera et al. 2007;

Table 3 Whole-body compositions (% dry weight) of golden pompano fed diets with and without supplementation of astaxanthin

	Protein	Lipid	Ash	Moist
Diet 1	61.77 ± 1.17	26.68 ± 0.17	17.71 ± 0.23	70.19 ± 1.17
Diet 2	61.68 ± 1.35	26.33 ± 0.46	17.41 ± 0.94	71.68 ± 1.15

Values are mean ± SEM of four replicates, and values in the same column with different letters are significant different ($P < 0.05$). Diet 1 meant golden pompano groups fed diets without supplementation of astaxanthin (AST). Diet 2 meant groups with supplementation of astaxanthin

Table 4 Hepatic antioxidant statuses of golden pompano fed diets with and without supplementation of astaxanthin

	T-AOC (U/mg protein)	SOD (U/mg protein)	GSH (μmol/g protein)	MDA (nmol/mg protein)
Diet 1	0.11 ± 0.01a	240.87 ± 5.76a	82.44 ± 4.87a	0.41 ± 0.02
Diet 2	0.15 ± 0.01b	214.24 ± 5.71b	118.52 ± 8.93b	0.41 ± 0.06

Values are mean ± SEM of four replicates, and values in the same column with different letters are significant different ($P < 0.05$). Diet 1 meant golden pompano groups fed diets without supplementation of astaxanthin (AST). Diet 2 meant groups with supplementation of astaxanthin
T-AOC total antioxidant capacity, *GSH* reduced glutathione, *SOD* superoxide dismutase, *MDA* malondialdehyde

Zhang et al. 2012; Pham et al. 2014; Yi et al. 2014). Kop and Durmaz (2008) believed that the effectiveness of carotenoids in terms of deposition and physiological function is species-specific in fish and not all fish species possess the same pathways for the metabolism of carotenoids. The mechanisms related to these findings have not yet been clearly elucidated. Our latest research results showed that the dietary astaxanthin can increase the apparent digestibility coefficient of the diet and further promote the expression of insulin-like growth factors (IGFs); moreover, as members of the family of transforming growth factors β, myostatin is affected by dietary astaxanthin (unpublished data).

Antioxidant capacity analysis

H_2O_2 is a strong oxidizer, produced in cell metabolism, but the excessive dose may be cytotoxic. As is shown, cell viability was sharply decreased with H_2O_2 supplemented and the increased LDH leakage into the extracellular media by H_2O_2 indicated the occurrence of oxidative stress membrane damage in our present in vitro study. Cellular antioxidative defense mechanisms can intercept the ROS both enzymatically and nonenzymatically. Total antioxidant capacity (T-AOC) is an overall indicator of the antioxidant status of an individual, representing the level of enzyme and nonenzyme original antioxidantin of the body (Xiao et al. 2004). As the value increases, the antioxidant defense against free radical reaction and reactive oxygen intermediates increases (Chien et al. 2003). In both of the in vivo and in vitro study, the T-AOC in the liver of the fish and in the hepatocytes supplemented with astaxanthin were higher, meaning that astaxanthin can improve the antioxidant status whether in vivo or in vitro. Although

H_2O_2 may decrease the total antioxidant capacity, the supplementation of astaxanthin can repair it to the same level with the control group.

The stress response might increase free radical contents, which may result in the increase of the lipid peroxidation content and lipid peroxidation injury (Liu et al. 2010). Malondialdehyde (MDA) is a product of lipid peroxidation, through crosslinking with the nucleophilic groups of proteins, nucleic acids, and amino phospholipids, accumulation of MDA leads to cell toxicity, accelerating the damage of cells and tissues (Buege and Aust 1978). The antioxidants and antioxidant enzyme system can play a significant role in resisting lipid oxide damage (Liu et al. 2010). Carotenoids may serve as an antioxidant in systems containing unsaturated fatty acids to quench free radicals (Mansour et al. 2006). The results showed that the MDA were not significant different when no stress appeared in the present in vivo study. However, once the cells suffered from oxidative stress in the present in vitro study, the MDA was increased and the cell viability was decreased, but supplemented astaxanthin could totally decrease the MDA value and save cells from the stress. Increased T-AOC and decreased MDA in the in vitro study reflected that supplemented astaxanthin in media can be totally conducive to eliminate the reactive oxygen species and protect the hepatocytes of golden pompano from free radicals. The MDA in $(H_2O_2 + AST)$ group was lower than that in H_2O_2 group, which indicated that AST can alleviate the lipid oxide damage.

Superoxide dismutase (SOD), a cytosolic enzyme that is specific for scavenging superoxide radicals, is the first enzymes to respond against oxygen radicals and important endogenous antioxidants for protection against

Table 5 The antioxidant statuses of hepatocytes treated with or without astaxanthin and H_2O_2 supplementation

	T-AOC (U/mg protein)	SOD (U/mg protein)	GSH (μmol/g protein)	MDA (nmol/mg protein)
Control	0.35 ± 0.01b	2682.76 ± 127.04b	17.81 ± 0.83b	0.13 ± 0.01a
H_2O_2	0.22 ± 0.02a	3264.92 ± 76.26c	5.92 ± 0.91a	0.40 ± 0.01c
H_2O_2 + AST	0.37 ± 0.01bc	2726.34 ± 74.17b	28.24 ± 1.11c	0.23 ± 0.01bc
AST	0.41 ± 0.01c	2312.19 ± 69.94a	136.51 ± 4.11d	0.12 ± 0.01a

Values are mean ± SEM of three replicates, and values in the same column with different letters are significant different ($P < 0.05$). Control meant golden pompano groups fed diets without supplementation of astaxanthin (AST). Control groups meant treating with neither H_2O_2 nor AST, H_2O_2 groups meant treating with H_2O_2 only, H_2O_2 + AST groups meant treating with both H_2O_2 and AST, and AST groups meant treating with AST only
T-AOC total antioxidant capacity, *GSH* reduced glutathione, *SOD* superoxide dismutase, *MDA* malondialdehyde

oxidative stress (Winston and Di Giulio 1991). Lygren et al. showed that high levels of dietary fat-soluble antioxidants, such as astaxanthin and vitamin E, there was a reduced need for endogenous antioxidant enzymes, such as total SOD (Lygren et al. 1999). The higher the SOD value, the more superoxide radicals need to be reacted (Qingming et al. 2010). It was found that the activities of liver SOD were significantly decreased by dietary astaxanthin supplementation in olive flounder (*Paralichthys olivaceus*) (Pham et al. 2014); large yellow croaker (*Pseudosciaena crocea*) (Li et al. 2014) and rainbow trout (*Oncorhynchus mykiss*) (Zhang et al. 2012). In this present study, SOD was significant lower in vivo and vitro study both supplemented with astaxanthin, implying that astaxanthin can eliminate reactive oxygen species to avoid the cells and tissues to produce more SOD. Once suffering from oxidative stress, the cells may produce much more endogenous SOD, as is shown in the study, to protect the body or cells from being hurt.

Glutathione (GSH), ubiquitous non-enzymatic antioxidants in the cells, is known to play an important role in the scavenging of free radicals and thus protect the important cellular macromolecules and organelles from oxidative damage (Misra and Niyogi 2009). Its role in the detoxification of ROS is important (Mallikarjuna et al. 2009). When suffered from oxidative stress, GSH was significantly lower in the present in vitro study. One mechanism for oxidative stress induced GSH depletion may involve enhanced utilization of GSH for the detoxification of free radicals and other oxidants produced as a result of H_2O_2 exposure (Shaw 1989). Vogt suggested that the increase of lipid peroxidation was not apparent until after GSH levels had been depleted (Vogt and Richie 2007). Astaxanthin can improve the GSH content in both our in vivo and in vitro study.

Conclusions

In conclusion, dietary astaxanthin supplementation can improve the growth performance of golden pompano. Moreover, astaxanthin can improve their hepatic antioxidant capacity in both vivo and vitro study by eliminating the reactive oxygen species.

Abbreviations

AST: Astaxanthin; CF: Condition factor; FBW: Final body weight; FCR: Feed conversion ratio; GSH: Reduced glutathione; HSI: Hepatosomatic index; IBW: Initial body weight; LDH: Lactate dehydrogenase; MDA: Malondialdehyde; ROS: Reactive oxygen species; SGR: Special growth rate; SOD: Superoxide dismutase; TAOC: Total antioxidant capacity; TBARS: Thiobarbituric-acid-reacting substances; TNB: 5-Thio-2-nitrobenzoate; WG: Weight gain; VSI: Viscerosomatic index

Acknowledgements
All authors thank the funding of the Project of Science and Technology of Guangdong Province (2013B090500110, 2013B090600045), the Central Institutes of Public Welfare Projects (2014A08YQ02), and the Special Project of Marine Fishery Science and Technology of Guangdong Province (A201601C11).

Funding
The design of the study and collection, analysis, interpretation of data, and writing the manuscript were supported by the Project of Science and Technology of Guangdong Province (2013B090500110, 2013B090600045), the Central Institutes of Public Welfare Projects (2014A08YQ02), and the Special Project of Marine Fishery Science and Technology of Guangdong Province (A201601C11).

Authors' contributions
JN, JW, and YW designed the study. JJX wrote the article. JJX, QQL, and XC performed the experiment. JN and JJX analyzed and interpreted the data. All authors have read, commented upon, and approved the final article.

Competing interests
The authors declare that they have no competing interests.

Author details
[1]South China Sea Fisheries Research Institute, Chinese Academy of Fishery Sciences, Guangzhou 510300, China. [2]School of Life Science and Technology, Shanghai Ocean University, Shanghai 201306, China. [3]Institute of Aquatic Economic Animals, School of Life Science, Sun Yat-sen University, NO.135 at Xingang Xi Road, Haizhu District, Guangzhou 510275, Guangdong Province, China.

Reference
Alishahi M, Karamifar M, Mesbah M. Effects of astaxanthin and Dunaliella salina on skin carotenoids, growth performance and immune response of Astronotus ocellatus. Aquac Int. 2015;23(5):1239–48.

Amar E, Kiron V, Satoh S, Watanabe T. Influence of various dietary synthetic carotenoids on bio-defence mechanisms in rainbow trout, Oncorhynchus mykiss (Walbaum). Aquacult Res. 2001;32(s1):162–73.

Baker R, Pfeiffer A-M, Schöner F-J, Smith-Lemmon L. Pigmenting efficacy of astaxanthin and canthaxanthin in fresh-water reared Atlantic salmon, Salmo salar. Anim Feed Sci Technol. 2002;99(1):97–106.

Britton G. Structure and properties of carotenoids in relation to function. FASEB J. 1995;9(15):1551–8.

Buege JA, Aust SD. Microsomal lipid peroxidation. Methods Enzymol. 1978;52: 302–10.

Chen YY, Lee PC, Wu YL, Liu LY. In vivo effects of free form astaxanthin powder on anti-oxidation and lipid metabolism with high-cholesterol diet. PLoS One. 2015;10(8):e0134733.

Chew BP. Antioxidant vitamins affect food animal immunity and health. J Nutr. 1995;125(6 Suppl):1804S–8S.

Chew BP, Park JS. Carotenoid action on the immune response. J Nutr. 2004; 134(1):257S–61S.

Chien Y-H, Pan C-H, Hunter B. The resistance to physical stresses by Penaeus monodon juveniles fed diets supplemented with astaxanthin. Aquaculture. 2003;216(1):177–91.

Christiansen R, Torrissen OJ. Growth and survival of Atlantic salmon, Salmo salar L. fed different dietary levels of astaxanthin. Juveniles. Aquacult Nutr. 1996; 2(1):55–62.

Finkel T. Oxidant signals and oxidative stress. Curr Opin Cell Biol. 2003;15(2):247–54.

Huo-sheng Y. Studies on the culture of pompano, Trachinotus ovatus [J]. J Fujian Fish. 2006;1:009.

James R., Sampath K., Thangarathinam R. and Vasudevan I. Effect of dietary spirulina level on growth, fertility, coloration and leucocyte count in red swordtail, Xiphophorus helleri. 2006;58(2):97–104.

Kalinowski CT, Robaina LE, Izquierdo MS. Effect of dietary astaxanthin on the growth performance, lipid composition and post-mortem skin colouration of red porgy Pagrus pagrus. Aquac Int. 2011;19(5):811–23.

Kop A, Durmaz Y. The effect of synthetic and natural pigments on the colour of the cichlids (Cichlasoma severum sp., Heckel 1840). Aquac Int. 2008;16(2):117–22.

Latimer GW. Official methods of analysis of AOAC International. Maryland: AOAC International; 2012.

Leite M, De Lima A, Massuyama M, Otton R. In vivo astaxanthin treatment partially prevents antioxidant alterations in dental pulp from alloxan-induced diabetic rats. Int Endod J. 2010;43(11):959–67.

Li M, Wu W, Zhou P, Xie F, Zhou Q, Mai K. Comparison effect of dietary astaxanthin and Haematococcus pluvialis on growth performance, antioxidant status and immune response of large yellow croaker Pseudosciaena crocea. Aquaculture. 2014;434:227–32.

Lin HZ, Tan XH, Zhou CP, Niu J, Xia DM, Huang Z, Wang J, Wang Y. Effect of dietary arginine levels on the growth performance, feed utilization, non-specific immune response and disease resistance of juvenile golden pompano Trachinotus ovatus. Aquaculture. 2015;437:382–9.

Liu C, Chen C. The biology and cultured technology of Pompano (Trachinotus ovatus). Shandong Fisheries. 2009;26:32–3.

Liu B, Xie J, Ge X, Xu P, Wang A, He Y, Zhou Q, Pan L, Chen R. Effects of anthraquinone extract from Rheum officinale Bail on the growth performance and physiological responses of Macrobrachium rosenbergii under high temperature stress. Fish Shellfish Immunol. 2010;29(1):49–57.

Lygren B, Hamre K, Waagbø R. Effects of dietary pro-and antioxidants on some protective mechanisms and health parameters in Atlantic salmon. J Aquat Anim Health. 1999;11(3):211–21.

Mallikarjuna K, Nishanth K, Hou CW, Kuo CH, Sathyavelu RK. Effect of exercise training on ethanol-induced oxidative damage in aged rats. Alcohol. 2009;43(1):59–64.

Mansour N, McNiven MA, Richardson GF. The effect of dietary supplementation with blueberry, α-tocopherol or astaxanthin on oxidative stability of Arctic char (Salvelinus alpinus) semen. Theriogenology. 2006;66(2):373–82.

Marklund S, Marklund G. Involvement of the superoxide anion radical in the autoxidation of pyrogallol and a convenient assay for superoxide dismutase. Eur J Biochem. 1974;47(3):469–74.

Martinez-Alvarez RM, Morales AE, Sanz A. Antioxidant defenses in fish: biotic and abiotic factors. Rev Fish Biol Fish. 2005;15(1–2):75–88.

Misra S, Niyogi S. Selenite causes cytotoxicity in rainbow trout (Oncorhynchus mykiss) hepatocytes by inducing oxidative stress. Toxicol In Vitro. 2009;23(7):1249–58.

Niu J, Du Q, Lin HZ, Cheng YQ, Huang Z, Wang Y, Wang J, Chen YF. Quantitative dietary methionine requirement of juvenile golden pompano Trachinotus ovatus at a constant dietary cystine level. Aquacult Nutr. 2013;19(5):677–86.

Niu J, Figueiredo-Silva C, Dong Y, Yue Y, Lin H, Wang J, Wang Y, Huang Z, Xia D, Lu X. Effect of replacing fish meal with soybean meal and of DL-methionine or lysine supplementation in pelleted diets on growth and nutrient utilization of juvenile golden pompano (Trachinotus ovatus). Aquacult Nutr. 2016;22:606–14.

Olmedilla B, Granado F, Southon S, Wright AJA, Blanco I, Gil-Martinez E, Berg H v d, Corridan B, Roussel A-M, Chopra M, Thurnham DI. Serum concentrations of carotenoids and vitamins A, E, and C in control subjects from five European countries. Br J Nutr. 2007;85(02):227.

Pham MA, Byun HG, Kim KD, Lee SM. Effects of dietary carotenoid source and level on growth, skin pigmentation, antioxidant activity and chemical composition of juvenile olive flounder Paralichthys olivaceus. Aquaculture. 2014;431:65–72.

Qi C-L, Lin H-Z, Huang Z, Zhou C-P, Wang Y, Wang J, Niu J, Zhao S-Y. Effects of dietary arachidonic acid levels on growth performance, whole-body proximate composition, digestive enzyme activities and gut morphology of juvenile golden pompano trachinotus. Isr J Aquacult Bamidgeh. 2016;68: 1275–84.

Qingming Y, Xianhui P, Weibao K, Hong Y, Yidan S, Li Z, Yanan Z, Yuling Y, Lan D, Guoan L. Antioxidant activities of malt extract from barley (Hordeum vulgare L.) toward various oxidative stress in vitro and in vivo. Food Chem. 2010;118(1):84–9.

Robinson EH, Wilson RP, Poe WE. Arginine requirement and apparent absence of a lysine-arginine antagonist in fingerling channel catfish. J Nutr. 1981;111(1):46–52.

Segner H, Arend P, Von Poeppinghausen K, Schmidt H. The effect of feeding astaxanthin to Oreochromis niloticus and Colisa labiosa on the histology of the liver. Aquaculture. 1989;79(1–4):381–90.

Shaw S. Lipid peroxidation, iron mobilization and radical generation induced by alcohol. Free Radic Biol Med. 1989;7(5):541–7.

Shimidzu N, Goto M, Miki W. Carotenoids as singlet oxygen quenchers in marine organisms. Fish Sci. 1996;62(1):134–7.

Tejera N, Cejas JR, Rodríguez C, Bjerkeng B, Jerez S, Bolaños A, Lorenzo A. Pigmentation, carotenoids, lipid peroxides and lipid composition of skin of red porgy (Pagrus pagrus) fed diets supplemented with different astaxanthin sources. Aquaculture. 2007;270(1):218–30.

Tutman P, Glavić N, Kožul V, Skaramuca B, Glamuzina B. Preliminary information on feeding and growth of pompano, Trachinotus ovatus (Linnaeus, 1758)(Pisces; Carangidae) in captivity. Aquac Int. 2004;12(4–5):387–93.

Vardi N, Parlakpinar H, Ozturk F, Ates B, Gul M, Cetin A, Erdogan A, Otlu A. Potent protective effect of apricot and beta-carotene on methotrexate-induced intestinal oxidative damage in rats. Food Chem Toxicol. 2008;46(9):3015–22.

Vogt BL, Richie Jr JP. Glutathione depletion and recovery after acute ethanol administration in the aging mouse. Biochem Pharmacol. 2007;73(10):1613–21.

Wan X, Ma T, Wu W, Wang Z. EROD activities in a primary cell culture of grass carp (Ctenopharyngodon idellus) hepatocytes exposed to polychlorinated aromatic hydrocarbonas. Ecotoxicol Environ Saf. 2004;58(1):84–9.

Wang F, Han H, Wang Y, Ma X. Growth, feed utilization and body composition of juvenile golden pompano Trachinotus ovatus fed at different dietary protein and lipid levels. Aquacult Nutr. 2013;19(3):360–7.

Winston GW, Di Giulio RT. Prooxidant and antioxidant mechanisms in aquatic organisms. Aquat Toxicol. 1991;19(2):137–61.

Xiao N, Wang XC, Diao YF, Liu R, Tian KL. Effect of initial fluid resuscitation on subsequent treatment in uncontrolled hemorrhagic shock in rats. Shock. 2004;21(3):276–80.

Yi XW, Xu W, Zhou HH, Zhang YJ, Luo YW, Zhang WB, Mai KS. Effects of dietary astaxanthin and xanthophylls on the growth and skin pigmentation of large yellow croaker Larimichthys croceus. Aquaculture. 2014;433:377–83.

Zhang J, Li X, Leng X, Zhang C, Han Z, Zhang F. Effects of dietary astaxanthins on pigmentation of flesh and tissue antioxidation of rainbow trout (Oncorhynchus mykiss). Aquac Int. 2012;21(3):579–89.

Zhou CP, Ge XP, Niu J, Lin HZ, Huang Z, Tan XH. Effect of dietary carbohydrate levels on growth performance, body composition, intestinal and hepatic enzyme activities, and growth hormone gene expression of juvenile golden pompano. Trachinotus ovatus Aquaculture. 2015;437:390–7.

15

Molecular cloning of metal-responsive transcription factor-1 (MTF-1) and transcriptional responses to metal and heat stresses in Pacific abalone, *Haliotis discus hannai*

Sang Yoon Lee and Yoon Kwon Nam[*]

Abstract

Background: Metal-responsive transcription factor-1 (MTF-1) is a key transcriptional regulator playing crucial roles in metal homeostasis and cellular adaptation to diverse oxidative stresses. In order to understand cellular pathways associated with metal regulation and stress responses in Pacific abalone (*Haliotis discus hannai*), this study was aimed to isolate the genetic determinant of abalone MTF-1 and to examine its expression characteristics under basal and experimentally stimulated conditions.

Results: The abalone MTF-1 shared conserved features in zinc-finger DNA binding domain with its orthologs; however, it represented a non-conservative shape in presumed transactivation domain region with the lack of typical motifs for nuclear export signal (NES) and Cys-cluster. Abalone MTF-1 promoter exhibited various transcription factor binding motifs that would be potentially related with metal regulation, stress responses, and development. The highest messenger RNA (mRNA) expression level of MTF-1 was observed in the testes, and MTF-1 transcripts were detected during the entire period of embryonic and early ontogenic developments. Abalone MTF-1 was found to be Cd inducible and highly modulated by heat shock treatment.

Conclusion: Abalone MTF-1 possesses a non-consensus structure of activation domains and represents distinct features for its activation mechanism in response to metal overload and heat stress. The activation mechanism of abalone MTF-1 might include both indirect zinc sensing and direct de novo synthesis of transcripts. Taken together, results from this study could be a useful basis for future researches on stress physiology of this abalone species, particularly with regard to heavy metal detoxification and thermal adaptation.

Keywords: Abalone, *Haliotis discus hannai*, MTF-1, Heavy metal, Heat shock

Background

Metal-responsive transcription factor-1 (MTF-1; also termed metal-regulatory transcription factor-1 or metal-responsive-element-binding transcription factor-1) is a key transcriptional regulator playing pivotal roles in metal homeostasis and detoxification (Laity and Andrews 2007; Günther et al. 2012a). In addition to its fundamental role for homeostatic metal regulation, this multitasking transcription factor has also been known to be closely involved in cellular adaptation and protection against oxidative stresses through regulating the transcription of diverse genes related with host defense-related pathways (Günther et al. 2012a; Lichtlen and Schaffner 2001). They include metal reservation/detoxification (e.g., metallothionein, MT; the main target of MTF-1), metal ion transport (e.g., Zn or Cu transporters), iron homeostasis/anti-microbial responses (e.g., hepcidin), cellular redox homeostasis (e.g., selenoproteins and thioredoxin reductase), and glutathione biosynthesis

* Correspondence: yoonknam@pknu.ac.kr
Department of Marine Bio-Materials & Aquaculture, Pukyong National University, Busan 48513, South Korea

(e.g., glutamate cysteine ligase) (Günther et al. 2012a; Lichtlen et al. 2001; Stoytcheva et al. 2010).

In a structural viewpoint, MTF-1 has been considered as a conserved transcription factor to possess six C_2H_2-type zinc fingers as the DNA-binding domain to recognize metal responsive elements (MREs) (Giedroc et al. 2001). As a cellular metal and stress sensor, the activity regulation of MTF-1 is generally characterized by the three successive steps, i.e., nuclear-cytoplasmic shuttling upon stress exposure, DNA-binding, and the interaction(s) with other coactivators to modulate the target gene transcription (Li et al. 2008). To execute the transcriptional regulation, MTF-1 binds to the specific site, called MRE (core sequence = TGCRCNC), in the promoter region of target gene (Günther et al. 2012a). Accordingly, the transcriptional expression of MTF-1 gene itself has been reported to be constitutive and not to be affected by heavy metal and other stressor treatments because its regulatory functions should be controlled mainly at post-translational levels (Auf der Maur et al. 2000; Bi et al. 2006). Structural scheme and functional context of MTF-1 above-described are believed to be widespread in the vertebrate lineage, although the majority of empirical information has come from mammalian MTF-1s.

However, in contrast to richness of knowledge on mammalian orthologs, molluscan MTFs have been narrowly explored barring only couples of previous reports (Qiu et al. 2013; Meng et al. 2015). Nevertheless, it is noteworthy that currently available molluscan MTF-1 sequences from public databases have suggested that mollusc species may show non-canonical features in their MTF-1 structures, which may differ from vertebrate orthologs. For instances, unlike vertebrate orthologs, molluscan MTF-1s often lack several typical motifs in presumed transactivation domains. Furthermore, a recent study has claimed that the nuclear-cytoplasmic shuttling, a key prerequisite step for vertebrate MTF-1s, might not always be an absolute precondition in certain mollusc species (Meng et al. 2015).

Pacific abalone, *Haliotis discus hannai*, is a highly valued seafood mollusc not only in Korea but also in other East-Asian countries. Intensive aquacultural operation for abalone farming using the marine net-cage system has been established in Korean aquaculture domain. During the last decade, the remarkable growth of abalone production in a quantitative term has been achieved (Park and Kim 2013). However, more recently, the sustainable progress of abalone culture has been considerably hurdled by the depressed productivity mainly in relation to frequent outbreaks of high mortality and physiological deformity in many abalone farms (Park and Kim 2013; Kang et al. 2015). Considering that abalone farming in Korea mainly relies on the net-cage facility installed in coastal areas, the heavy metals or other

related pollutants contaminated in both water and sediments could be significant factors to provoke cellular toxicity and oxidative stress in farmed abalones (Kim et al. 2007).

However, despite its importance, adaptive or defensive functions to such environmental perturbations have been limitedly investigated in this abalone species, and almost no information has been available with respect to the coordinated regulations of genes involved in cellular pathways associated with metal regulation and oxidative stress responses (Kim et al. 2007; Lee and Nam 2016a). For this reason, understanding of MTF-1 from abalone species would be much useful to better comprehend orchestrated and coordinated regulations of host defense genes in this abalone species. Based on this need, this study, as a startpoint research, was aimed to characterize the genetic determinant of MTF-1, the superordinate regulator for diverse host defense genes, from the Pacific abalone (*H. discus hannai*). For this, we isolated and characterized the full-length complementary DNA (cDNA) encoding the abalone MTF-1, bioinformatically dissected its 5′-upstream regulatory region, and also scrutinized expression patterns of MTF-1 under both non-stimulated and stress-challenged (i.e., heavy metal exposure and heat shock) conditions.

Methods
Abalone samples and molecular cloning of MTF-1
Abalones (*H. discus hannai*) used in this study were experimental stocks maintained at Experimental Fish Culture Station, Pukyong National University (PKNU), Busan, South Korea. Abalones were maintained with semi-water recirculation system equipped with 3-ton capacity of rectangular culture tanks, in which the tanks were connected with protein skimmers, custom-designed mechanical filters, and 1-μm-mesh filter. Throughout the experiment, water temperature and dissolved oxygen were kept to be ranged within 20 ± 1 °C and 8 ± 1 ppm, respectively. Abalones were fed with frozen or dried seaweeds until 2 days before stress exposure treatments. Daily water exchange rate was about 20%, and in-tank wastes including feces and debris on the bottom were removed twice every day.

Based on the NGS-transcriptome analysis of the juvenile abalone tissues (unpublished data), partial NGS clones representing the significant homology to known animal MTF-1s were selected and assembled into a contig. In order to get full-length cDNA version, rapid amplifications of cDNA ends (RACE) at both 5′- and 3′-directions were carried out using total RNA isolated from a whole body sample and SMARTer® RACE 5′/3′ Kit (Clontech Laboratories Inc., Mountain View, CA, USA) according to the manufacturer's instructions. Oligonucleotide primers used in this study are listed in Additional file 1: Table S1. The amplified fragments were sequenced and again subjected to contig assembly. Based on the assembled sequence in a contig, full-length abalone MTF-1 cDNA was re-isolated by RT-PCR amplification using the same total RNA aforementioned. Amplified RT-PCR products were directly sequenced at both forward and reverse directions to obtain a representative cDNA sequence for abalone MTF-1.

From the cDNA sequence, the 5′-upstream region of abalone MTF-1 gene was cloned by genome walking method. Using the genomic DNA prepared from an individual muscle, genome walking to 5′-upstream region was conducted with designated pairs of gene-specific primers and Universal Genome Walker® Kit (Clontech Laboratories Inc., USA) according to the manufacturer's instructions. Amplified fragments were TA cloned into pGEM-T® easy vector (Promega, Madison, WI, USA), sequenced and assembled into a single contig. Afterward, the continuous, genomic fragment containing the 5′-flanking region was again PCR isolated from the genomic DNA template abovementioned and directly subjected to the sequencing to confirm the representative sequence of the abalone MTF-1 proximal promoter region.

Bioinformatic sequence characterization
With the ORF Finder program (https://www.ncbi.nlm.nih.gov/orffinder/), the open reading frame (ORF) of abalone MTF-1 was predicted and deduced amino acid sequence was obtained. Based on the homology search using NCBI BLASTx (http://blast.ncbi.nlm.nih.gov/Blast.cgi), sequence homology of abalone MTF-1 with its orthologs was examined. Parameter scores for the primary structure of MTF-1 were estimated using ExPASy ProtParam tool (http://web.expasy.org/protparam/). Multiple sequence alignment was done using CLUSTALW program (http://www.genome.jp/tools-bin/clustalw). Identification of putative zinc finger domains was carried out with Simple Modular Architecture Research Tool (SMART; http://smart.embl.de/). Predictions of potential nuclear localization signal (NLS) and nuclear export signal (NES) were conducted with cNLS Mapper (http://nls-mapper.iab.keio.ac.jp/cgi-bin/NLS_Mapper_form.cgi) and NetNES 1.1 Server (http://www.cbs.dtu.dk/services/NetNES/), respectively. The putative zinc finger-DNA binding regions from selected MTF-1 orthologs were subjected to molecular phylogenetic analysis using Molecular Evolutionary Genetics Analysis tool (ver. 7.0.21; http://www.megasoftware.net/). Putative transcription factor binding motifs in the abalone MTF-1 promoter were predicted with TRANSFAC® software (http://genexplain.com/transfac; GeneXplain GmbH, Wolfenbüttel, Germany).

Tissue collection from different developmental stages
Tissue distribution assay was conducted with two age classes of abalones. First, from 1-year-old immature

juveniles (21.5 ± 4.1 g for total weight; $n = 12$), six kinds of tissues including the gill, gut, heart, hemolymph, hepatopancreas, and muscle (foot muscle) were surgically removed individually. For hemolymph, centrifugation (2500 rpm for 10 min at 4 °C) was carried out in order to collect hemocyte pellet. Second, from 3-year-old sexually mature adults showing a clear sign of ovarian and testicular maturation (96.4 ± 13.1 g for total weight; eight each for female and male), same tissue types abovementioned were obtained and additional ovary and testis were obtained. Upon surgically removed, biological samples were immediately frozen on dry ice and stored at −80 °C until used for RNA isolation.

In order to obtain developing embryos and early larvae, artificial insemination of sperm (from three males) to eggs (from eight females) was conducted by using the conventional induced spawning method including an air exposure and ultraviolet-irradiated seawater treatment. Insemination was made with wet-method at 20 °C and incubated at the same temperature until the end of sampling. An aliquot of developmental samples each consisting of approximately 20,000~30,000 embryos or larvae was sampled at six time points: just before insemination (unfertilized eggs), early cleavages (i.e., 2~8 cells stage; at 2 h post insemination; 2 hpi), morula (5 hpi), trochophore (12 hpi), early veliger (18 hpi), and late veliger (42 hpi), based on the microscopic examination. Upon sampling, embryos and larvae were also frozen on dry ice and stored at −80 °C until used. Two replicate samplings were carried out for each time point.

Experimental in vivo stimulatory challenges
Two independent stress exposure experiments were carried out: one was with heavy metal exposure and the other with heat shock treatment. For heavy metal exposure, eight juvenile individuals (24.5 ± 3.6 g; approximately 1 year old) were assigned into one of five experimental tanks (70-L capacity containing 50 L of 1-μm filtered seawater at 20 °C) and acclimated to the tank conditions for 24 h before heavy metal treatment. The heavy metals used for exposure were of analytical grade reagents (Sigma-Aldrich, St. Louis, MO, USA). After 24 h, two tanks were treated with 0.02 mg/L (i.e., 20 ppb) and 0.1 mg/L (i.e., 100 ppb) cadmium (Cd), while two tanks with 20 and 100-ppb zinc (Zn) (Lee and Nam 2016a). Nominal concentration of the metal for each metal-exposed group was adjusted by using $CdCl_2$ or $ZnCl_2$ stock solution. Remaining one tank was treated with only 1-mL distilled water that had been used for reconstitution of the metals (i.e., for non-exposed control). For each group, two replicate tanks were prepared identically. Treatment duration was 24 h. At the end of exposure, gill, hemocyte, hepatopancreas, and foot muscle were sampled individually from six randomly chosen individuals as described above.

On the other hand, for heat shock treatment, 22 individuals (21.1 ± 3.1 g; same-aged as above) were assigned into one of four 100-L tanks (two for heat-stressed groups and two for non-stressed groups) at 20 °C. Each tank was equipped with a custom-designed apparatus for mechanical filtration. After 24 h of acclimation period, water temperature of the two tanks (heat-stressed groups) was elevated using the adjustable thermostat-assisted aquarium heaters (400 W) with an increment rate of 1 °C/h. When the temperature reached 30 °C, the temperature was kept to be constant at 30 °C for additional 24 h. Samplings were made at 20 °C (just before thermal elevation), 25 °C (5 h after elevation started), 30 °C (10 h), 30 °C+12 h (12 h after reaching 30 °C; 22 h after elevation started), and 30 °C+24 h (24 h after reaching 30 °C; 34 h after elevation started). Four individuals were randomly selected from each tank to constitute eight individuals per temperature group at each sampling point. Tissues sampled were gill, hemocyte, hepatopancreas, and foot muscle. Meanwhile, non-stressed control groups were also identically prepared with heat-shock groups, but the temperature (20 °C) was kept to be constant until the end of experiment. At the same sampling point, the identical number of abalones ($n = 4$ per tank) was also obtained from non-stressed control groups. Temperature of each tank was confirmed to be ranged within ± 0.5 °C. Dissolved oxygen levels were adjusted to be ranged from 7.5 to 8.5 ppm for all the experimental tanks. Abalones were not fed during stimulatory challenge experiments.

RT-qPCR assay and statistics

Total RNA was extracted using TriPure® Reagent (Roche Applied Science, Mannheim, Germany) and then purified using RNeasy Mini Plus Kit (Qiagen, Hilden, Germany) including DNase I treatment step. An aliquot (2 µg) of total RNA prepared was reverse transcribed into cDNA by using the Omniscript® Reverse Transcription Kit (Qiagen, Germany) including oligo-dT primer according to the manufacturer's instruction. Synthesized cDNA was fourfold diluted with sterile distilled water, and an aliquot of 2 µL was included in a qPCR reaction as a template. The qPCR reaction was conducted with a LightCycler® 480 Real-Time PCR System and LightCycler® 480 SYBR Green I Master (Roche Applied Science, Germany), according to the manufacturer's instructions. Thermal cycling condition for each gene (i.e., MTF-1 and normalization control genes) can be referred to Additional file 1: Table S1. Based on our preliminary study to evaluate candidate housekeeping genes for the normalization of RT-qPCR amplification (unpublished data; see also (Lee and Nam 2016a; Lee and Nam 2016b)), abalone ribosomal proteins L5 (RPL5; ABO26701) and L7 (RPL7; KP698945) genes were used as reference genes to normalize expression levels of MTF-1 transcripts in tissue samples (i.e., for basal tissue expression

assays and stress exposure treatments), while RPL7 and RPL8 (KP698947) were used to normalize MTF-1 expression across developmental samples (embryos and larvae). Additionally, for heavy-metal exposure groups, the messenger RNA (mRNA) expression levels of MTF-1 were compared with those of MT (the known target gene of MTF-1) in order to examine whether or not there might be any positive or proportional relationship in the metal-mediated modulation patterns between MTF-1 and MT genes. PCR efficiency of primer pair for each gene was validated to be at least higher than 95% based on the standard curve prepared using a fivefold serial dilution of cDNA mix. For each cDNA sample, triplicate assays were carried out in an independent fashion.

Quantitative PCR-based MTF-1 mRNA expression levels across tissue types and developmental stages under non-stressed conditions were presented as ΔCt (Ct of the MTF-1 gene subtracted from the Ct of each internal control gene). On the other hand, differential expression levels of metal-exposed or heat shock-treated group relative to their corresponding non-stressed control groups were presented as the fold difference to the non-stressed controls by using the formula $2^{-\Delta\Delta Ct}$ (Schmittgen and Livak 2008). Expression levels between or among groups were tested using Student's t test or one-way ANOVA (followed by Duncan's multiple ranged tests). Difference was considered to be significant when $P < 0.05$.

Results and discussion
Characteristics of abalone MTF-1 cDNA and deduced amino acid sequences

The full-length cDNA of abalone MTF-1 was comprised of 48-bp 5′-untranslated region (UTR), a 1509-bp single open reading frame (ORF) encoding a polypeptide of 503 amino acids, and 582-bp 3′-UTR including a stop codon and 19-bp poly(A+) tail. A putative polyadenylation signal (AATAAA) was found at 21 bp prior to the poly(A+) tail (GenBank accession number; KT895224) (Additional file 1: Figure S1). The MTF-1 protein based on the deduced amino acid sequence was estimated to have 54.86 kDa of calculated molecular mass and 5.51 of theoretical pI value, respectively. Abalone MTF-1 represented quite a low sequence homology to its vertebrate and invertebrate orthologs where the maximum sequence identity at amino acid level was found to be only 27% with *Biomphalaria glabrata* (air-breathing freshwater snail; Gastropoda; Mollusca). Non-conservative feature without any appreciable sequence similarity to other MTF-1s is also found in the putative activation. Abalone MTF-1 was likely to possess only a shortened fragment (44-aa; pI = 4.06) presumed for the acidic domain and to lack almost entire region corresponding to the proline-rich domain and serine/threonine-rich domain of vertebrate MTF-1 orthologs (Additional file 1: Figure S2).

In contrast, the abalone MTF-1 was proven to share a high structural homology with other MTF-1s in the DNA binding domain (Fig. 1). All the MTF-1 proteins including the abalone MTF-1 (but except for the shortest *Octopus bimaculoides* MTF-1 having only five zinc fingers) were found to show highly conserved six C_2H_2-zinc fingers in their DNA binding domains. In their DNA binding domains, 12 cysteine residues and histidine residues were clearly conserved, in which the zinc finger domain of abalone MTF-1 showed the highest sequence identity (77%) with that of orthologue isoforms from *Crassostrea gigas* (Bivalve, Mollusca). However, the molecular phylogenetic analysis using the DNA binding domains has indicated

that MTF-1s have been largely divergent in the molluscan phylum, which is apparently different form the mono-phyletic clustering of orthologs from Chordata phylum (Additional file 1: Figure S3). Functional partition of the zinc finger domains in abalone MTF-1 has been remained to be further characterized; however, from the mammalian studies, the region from the second to fourth zinc fingers have been proposed to constitute the core DNA binding domain while the first finger has been reported to serve as a metal-sensing domain (Bittel et al. 2000). Within each zinc finger, the His-X-Arg/Lys-X-His [H-X-(R/K)-X-H where X is any amino acid] motif has been known as a key site for zinc binding (Günther et al. 2012a), and it is preserved in the 1st to 5th zinc fingers for all the species examined in the present study. However, in the last finger

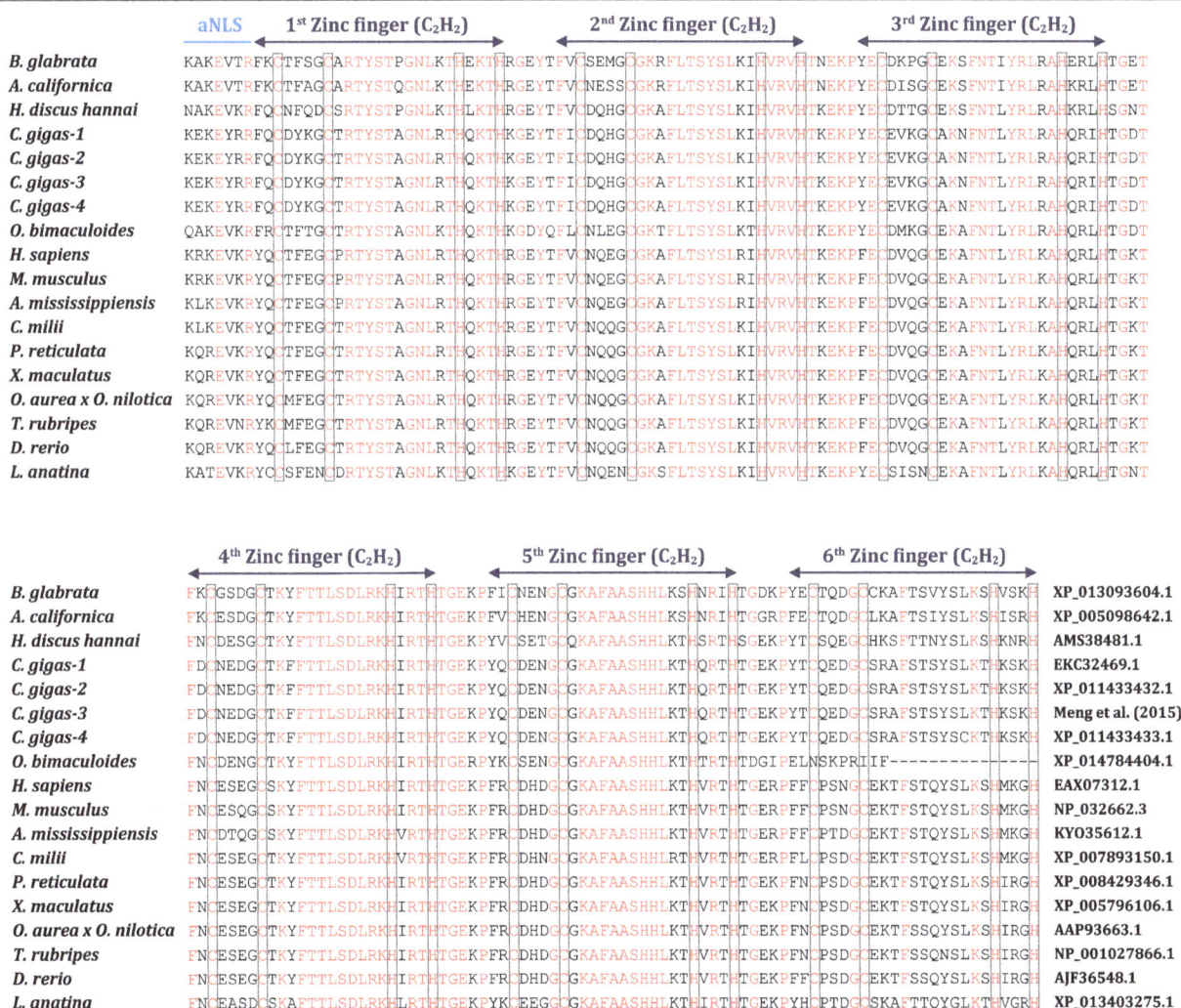

```
                    aNLS       1st Zinc finger (C2H2)              2nd Zinc finger (C2H2)             3rd Zinc finger (C2H2)
B. glabrata            KAKEVTRFKCTFSGCARTYSTPGNLKTHEKTHRGEYTFVCSEMGCGKRFLTSYSLKIHVRVHTNEKPYECDKPGCEKSFNTIYRLRAHERLHTGET
A. californica         KAKEVTRFKCTFAGCARTYSTQGNLKTHEKTHRGEYTFVCNESSCGKRFLTSYSLKIHVRVHTNEKPYECDISGCEKSFNTIYRLRAHKRLHTGET
H. discus hannai       NAKEVKRFQCNFQDCSRTYSTPGNLKTHLKTHRGEYTFVCDQHGCGKAFLTSYSLKIHVRVHTKEKPYECDTTGCEKSFNTLYRLRAHKRLHSGNT
C. gigas-1             KEKEYRRFQCDYKGCTRTYSTAGNLRTHQKTHKGEYTFICDQHGCGKAFLTSYSLKIHVRVHTKEKPYECEVKGCAKNFNTLYRLRAHQRIHTGDT
C. gigas-2             KEKEYRRFQCDYKGCTRTYSTAGNLRTHQKTHKGEYTFICDQHGCGKAFLTSYSLKIHVRVHTKEKPYECEVKGCAKNFNTLYRLRAHQRIHTGDT
C. gigas-3             KEKEYRRFQCDYKGCTRTYSTAGNLRTHQKTHKGEYTFICDQHGCGKAFLTSYSLKIHVRVHTKEKPYECEVKGCAKNFNTLYRLRAHQRIHTGDT
C. gigas-4             KEKEYRRFQCDYKGCTRTYSTAGNLRTHQKTHKGEYTFICDQHGCGKAFLTSYSLKIHVRVHTKEKPYECEVKGCAKNFNTLYRLRAHQRIHTGDT
O. bimaculoides        QAKEVKRFRCTFTGCTRTYSTAGNLKTHQKTHKGDYQFLCNLEGCGKTFLTSYSLKTHVRVHTKEKPYECDMKGCEKAFNTLYRLRAHQRLHTGDT
H. sapiens             KRKEVKRYQCTFEGCPRTYSTAGNLRTHQKTHKGEYTFVCNQEGCGKAFLTSYSLRIHVRVHTKEKPFECDVQGCEKAFNTLYRLKAHQRLHTGKT
M. musculus            KRKEVKRYQCTFEGCPRTYSTAGNLRTHQKTHKGEYTFVCNQEGCGKAFLTSYSLRIHVRVHTKEKPFECDVQGCEKAFNTLYRLKAHQRLHTGKT
A. mississippiensis    KLKEVKRYQCTFEGCPRTYSTAGNLRTHQKTHRGEYTFVCNQEGCGKAFLTSYSLRIHVRVHTKEKPFECDVQGCEKAFNTLYRLKAHQRLHTGKT
C. milii               KLKEVKRYQCTFEGCTRTYSTAGNLRTHQKTHRGEYTFVCNQQGCGKAFLTSYSLKIHVRVHTKEKPFECDVQGCEKAFNTLYRLKAHQRLHTGKT
P. reticulata          KQREVKRYQCTFEGCTRTYSTAGNLRTHQKTHRGEYTFVCNQQGCGKAFLTSYSLKIHVRVHTKEKPFECDVQGCEKAFNTLYRLKAHQRLHTGKT
X. maculatus           KQREVKRYQCTFEGCTRTYSTAGNLRTHQKTHRGEYTFVCNQQGCGKAFLTSYSLKIHVRVHTKEKPFECDVQGCEKAFNTLYRLKAHQRLHTGKT
O. aurea x O. nilotica KQREVKRYQCMFEGCTRTYSTAGNLRTHQKTHRGEYTFVCNQQGCGKAFLTSYSLKIHVRVHTKEKPFECDVQGCEKAFNTLYRLKAHQRLHTGKT
T. rubripes            KQREVNRYKCMFEGCTRTYSTAGNLRTHQKTHRGEYTFVCNQQGCGKAFLTSYSLKIHVRVHTKEKPFECDVQGCEKAFNTLYRLKAHQRLHTGKT
D. rerio               KQREVKRYQCLFEGCTRTYSTAGNLRTHQKTHRGEYTFVCNQQGCGKAFLTSYSLKIHVRVHTKEKPFECDVQGCEKAFNTLYRLKAHQRLHTGKT
L. anatina             KATEVKRYQCSFENCDRTYSTAGNLKTHQKTHKGEYTFVCNQENCGKSFLTSYSLKIHVRVHTKEKPYECSISNCEKAFNTLYRLKAHQRLHTGNT


                          4th Zinc finger (C2H2)          5th Zinc finger (C2H2)          6th Zinc finger (C2H2)
B. glabrata            FKCGSDGCTKYFTTLSDLRKHIRTHTGEKPFICNENGCGKAFAASHHLKSHNRIHTGDKPYECTQDGCCKAFTSVYSLKSHVSKH   XP_013093604.1
A. californica         FKCESDGCTKYFTTLSDLRKHIRTHTGEKPFVCHENGCGKAFAASHHLKSHNRIHTGGRPFECTQDGCLKAFTSIYSLKSHISRH   XP_005098642.1
H. discus hannai       FNCDESGCTKYFTTLSDLRKHIRTHTGEKPYVCSETGCGKAFAASHHLKTHSRTHSGEKPYTCSQEGCHKSFTTNYSLKSHKNRH   AMS38481.1
C. gigas-1             FDCNEDGCTKFFTTLSDLRKHIRTHTGEKPYQCDENGCGKAFAASHHLKTHQRTHTGEKPYTCQEDGCSRAFSTSYSLKTHKSKH   EKC32469.1
C. gigas-2             FDCNEDGCTKFFTTLSDLRKHIRTHTGEKPYQCDENGCGKAFAASHHLKTHQRTHTGEKPYTCQEDGCSRAFSTSYSLKTHKSKH   XP_011433432.1
C. gigas-3             FDCNEDGCTKFFTTLSDLRKHIRTHTGEKPYQCDENGCGKAFAASHHLKTHQRTHTGEKPYTCQEDGCSRAFSTSYSLKTHKSKH   Meng et al. (2015)
C. gigas-4             FDCNEDGCTKFFTTLSDLRKHIRTHTGEKPYQCDENGCGKAFAASHHLKTHQRTHTGEKPYTCQEDGCSRAFSTSYSCKTHKSKH   XP_011433433.1
O. bimaculoides        FNCDENGCTKYFTTLSDLRKHIRTHTGERPYKCSENGCGKAFAASHHLKTHTRTHTDGIPELNSKPRIIF-----------------   XP_014784404.1
H. sapiens             FNCESEGCSKYFTTLSDLRKHIRTHTGEKPFRCDHDGCGKAFAASHHLKTHVRTHTGERPFFCPSNGCEKTFSTQYSLKSHMKGH   EAX07312.1
M. musculus            FNCESQGCSKYFTTLSDLRKHIRTHTGEKPFRCDHDGCGKAFAASHHLKTHVRTHTGERPFFCPSNGCEKTFSTQYSLKSHMKGH   NP_032662.3
A. mississippiensis    FNCDTQGCSKYFTTLSDLRKHVRTHTGEKPFRCDHDGCGKAFAASHHLKTHVRTHTGERPFFCPTDGCEKTFSTQYSLKSHMKGH   KYO35612.1
C. milii               FNCESEGCTKYFTTLSDLRKHVRTHTGEKPFRCDHNGCGKAFAASHHLRTHVRTHTGERPFLCPSDGCEKTFSTQYSLKSHMKGH   XP_007893150.1
P. reticulata          FNCESEGCTKYFTTLSDLRKHIRTHTGEKPFRCDHDGCGKAFAASHHLKTHVRTHTGEKPFNCPSDGCEKTFSTQYSLKSHIRGH   XP_008429346.1
X. maculatus           FNCESEGCTKYFTTLSDLRKHIRTHTGEKPFNCDHDGCGKAFAASHHLKTHVRTHTGEKPFNCPSDGCEKTFSTQYSLKSHIRGH   XP_005796106.1
O. aurea x O. nilotica FNCESEGCTKYFTTLSDLRKHIRTHTGEKPFRCDHDGCGKAFAASHHLKTHVRTHTGEKPFNCPSDGCEKTFSSQYSLKSHIRGH   AAP93663.1
T. rubripes            FNCESEGCTKYFTTLSDLRKHIRTHTGEKPFRCDHDGCGKAFAASHHLKTHVRTHTGEKPFNCPSDGCEKTFSSQNSLKSHIRGH   NP_001027866.1
D. rerio               FNCESEGCTKYFTTLSDLRKHIRTHTGEKPFRCDHDGCGKAFAASHHLKTHVRTHTGEKPFFCPSDGCEKTFSSQYSLKSHIRGH   AJF36548.1
L. anatina             FNCEASDCSKAFTTLSDLRKHLRTHTGEKPYKCEEGGCGKAFAASHHLKTHIRTHTGEKPYHCPTDGCSKAFTTQYGLKTHVGRH   XP_013403275.1
```

Fig. 1 Multiple amino acid sequence alignment of conserved zinc-finger DNA-binding domain from the abalone MTF-1 along with those from representative orthologs. In the alignment, conserved amino acid residues are indicated by *red*-colored letters, and two cysteines and two histidines typical of the C_2H_2 structure for each zinc finger are boxed. The putative auxiliary nuclear localization signal (aNLS) conserved in the front of the first zinc finger is indicated by upper line. The accession code for each sequence is noted at the end of alignment. The alignment with full-length polypeptide sequences and the full name of each species can be referred to Additional file 1: Figure S2 (i.e., the 6th finger), the Arg/Lys residue inside the 5-amino-acid stretch was found to be conserved only in vertebrate MTF-1s. Meanwhile, all the molluscan MTF-1s showed a phenylalanine (Phe) as the first amino acid of the first zinc finger, whereas vertebrate and brachiopod orthologs possessed a tyrosine (Tyr) (Cheung et al. 2010).

Abalone MTF-1 showed both conserved and unique features in the peptide linkers connecting zinc fingers. Conserved peptide linkers were found between 1st and 2nd fingers (Arg-Gly-Glu-Tyr-Thr), between 2nd and 3rd fingers (Thr-Lys-Glu-Lys-Pro), and between 4th and 5th fingers (Thr-Gly-Glu-Lys-Pro). On the other hand, unique linkers were found between 3rd and 4th fingers [Ser-Gly-Asn-Thr in abalone vs. Thr-Gly-Lys-Thr in vertebrates vs. Thr-Gly-(Glu/Asn/Asp)-Thr in other molluscs and a brachiopod species] and between 5th and 6th fingers [Ser-Gly-Glu-Lys-Pro in abalone vs. Thr-Gly-Glu-(Lys/Arg)-Pro in vertebrates vs. Thr-Gly-(Asp/Gly/Glu)-(Lys/Arg)-Pro in others]. Peptide linkers in multi-zinc finger domains have been reported to play roles in not only structural stabilization but also interfinger interactions for DNA-binding affinity of the zinc finger domains (Li et al. 2006). Particularly in the MTF-1, the linker between 1st and 2nd fingers has been known to be crucial in the zinc-sensing ability of the MTF-1 with regard to the formation of the ternary (MTF-1-zinc-DNA) complex for activating MT gene transcription (Li et al. 2006). Taking this into account, the abalone MTF-1 is thought to preserve a fundamental property of zinc-sensing function similarly with mammalian orthologs since it conserves a completely identical linker (i.e., the Arg-Gly-Glu-Tyr-Thr linker between 1st and 2nd fingers) with mammalian MTF-1s. However, the present abalone MTF-1 showed apparent dissimilarity with vertebrate orthologs in the linkers between 3rd and 4th fingers and between 5th and 6th fingers. Particularly because the linker between 3rd and 4th fingers has been reported to be important for the in vivo and in vitro sensitivity of zinc-dependent activation of MTF-1 (Li et al. 2006), this linker could be a good target for future studies to examine the potential difference in the linker-mediated zinc-finger function between abalone MTF-1 and mammalian/vertebrate orthologs.

In the abalone MTF-1, a putative auxiliary nuclear localization signal (aNLS) was identifiable in the front of the first zinc finger, as similarly with all other MTF-1 orthologs. Of seven amino acid residues to comprise the auxiliary NLS, the positions of two amino acids ([3]Glu and [7]Arg) are conserved in all the MTF-1s examined (Fig. 1). However, due to the potential deletion in the putative acidic domain, no canonical nuclear export signal (NES) was observable in the abalone MTF-1 unlike vertebrate orthologs. Within a concept of MTF-1 activation (i.e., a nuclear-cytoplasmic shuttling function

established in mammalian MTF-1), NLS and NES are responsible for balanced subcellular distributions of MTF-1 proteins (i.e., import and export, respectively) under both stressed and non-stressed conditions (Günther et al. 2012a). The NES motif in vertebrate MTF-1 is usually embedded in the acidic activation domain (Günther et al. 2012a; Cheung et al. 2010). However, the abalone MTF-1 does not show any typical NES motif, although a putative NLS motif is predicted in the front of the first zinc-finger domain that are also conserved all MTF-1s examined (i.e., conserved auxiliary NLS). The absence of acidic domain-embedded NES is not limited to abalone MTF-1, i.e., all molluscan MTF-1s are also likely to lack the NES at the corresponding region. Hence, our finding may suggest that molluscan MTF-1s could be different from mammalian MTF-1s in their subcellular localization control under both basal and stimulated conditions. A recent study with an oyster species (C. gigas) has claimed that the MTF-1 would primarily localize in the nucleus even under unstressed conditions and nuclear translocation might be uncritical for the activation of the oyster MTF-1 (Meng et al. 2015).

In addition, the typical motif of cysteine cluster (consensus sequence = Cys-Gln-Cys-Gln-Cys-Ala-Cys) that could be commonly found in the region immediately following the serine/threonine-rich domain of vertebrate MTF-1s was not detected in the abalone MTF-1 (Additional file 1: Figure S2). The cysteine cluster has been reported to be essentially necessary for metal-induced transcriptional activity and homodimerization of mammalian MTF-1 (Günther et al. 2012a; Günther et al. 2012b). Like with NES abovementioned, none of molluscan MTF-1 represents a canonical C-terminus cysteine cluster, suggesting that molluscan MTF-1s might have different mechanism(s) for metal-induced transcription (i.e., recruitments of transcriptional cofactor partners in the promoter/enhancer of target genes).

Prediction of transcription factor binding sites

From the bioinformatic prediction, the 1691-bp 5′-upstream region from the ATG translation start site of the abalone MTF-1 gene represented various transcription factor binding sites (Fig. 2) (Additional file 1: Figure S4). The abalone MTF-1 promoter revealed a canonical TATA box (TATAAA) at −474 bp (from the ATG). Importantly, it represented a copy of MRE (TGCRCNC; −959 bp), suggesting the possible modulation of MTF-1 itself by heavy metal-driven cellular stressors, which is clearly inconsistent with the lack of MRE in many mammalian MTF-1 promoters (Auf der Maur et al. 2000; Bi et al. 2006). In addition, a xenobiotic response element (XRE; TNGCGTG; −736 bp) was predicted in the abalone MTF-1 promoter. XRE is an aryl hydrocarbon receptor (AhR)-targeted motif involved in the ligand

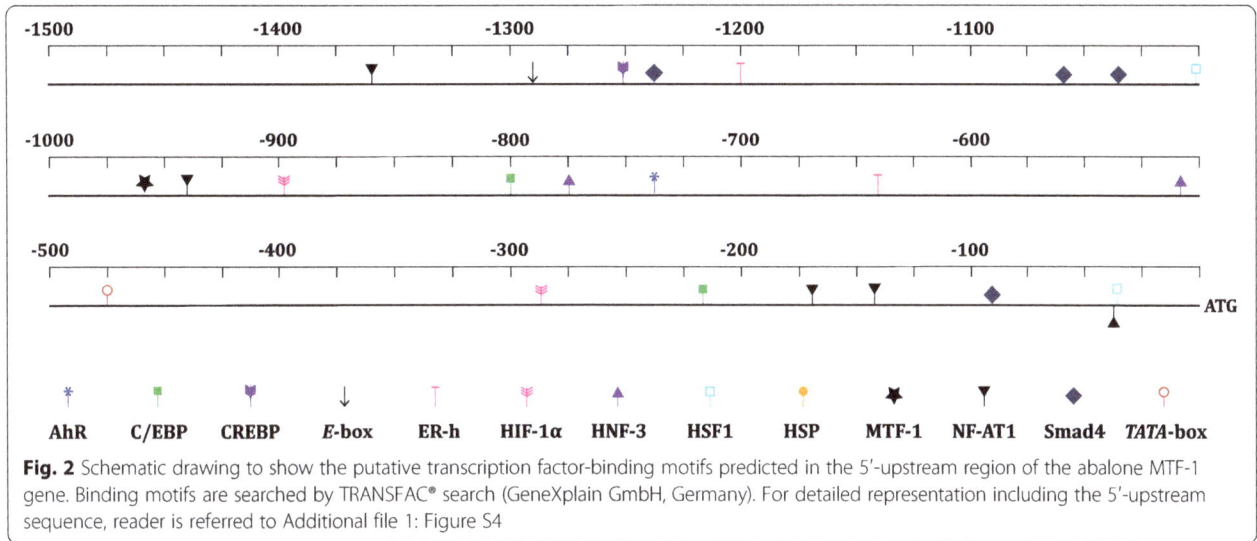

Fig. 2 Schematic drawing to show the putative transcription factor-binding motifs predicted in the 5'-upstream region of the abalone MTF-1 gene. Binding motifs are searched by TRANSFAC® search (GeneXplain GmbH, Germany). For detailed representation including the 5'-upstream sequence, reader is referred to Additional file 1: Figure S4

(i.e., 2,3,7,8-tetrachlorodibenzodioxin (TCDD))-activated pathway to detoxify the effects of TCDD-related compounds. However, because invertebrate AhR homologues have been reported to lack the ability to bind TCDD directly (Hahn et al. 2006), the molecular mechanism on the potential interconnection between MRE/MTF-1 and XRE/AhR paths should be further explored. Nevertheless, recent mammalian studies have also highlighted multitasking roles of AhR in various signaling pathways associated with cell cycle control and antioxidant protection against oxidative stresses (Jackson et al. 2015). Besides, several transcription factor binding sites such as heat shock element (HSE; GAANRTTC; −1003 and −32 bp; targeted by heat shock factor (HSF)), hypoxia response element (HRE; RCGTG; −895 and −285 bp; by hypoxia-inducible factor-1 alpha (HIF-1α)), and other sites recognized by cyclic AMP response element binding protein (CREBP; TGACGY; −1253 bp) and nuclear factor for activated T-cells (NF-AT1; WGGAAA; −941, −169, and −33 bp) were predicted. All of these factors have been known to be related with stress responses of animals (Saydam et al. 2003; Dubé et al. 2011). Abalone MTF-1 promoter also revealed motifs that might be targeted by transcription factors generally known to be involved in development, signal transduction, cell proliferation and/or organ development. They included Smad4, hepatocyte nuclear factor (HNF), CCAAT-enhancer binding protein (C/EBP), E-box binding protein, and estrogen receptor (ER).

Tissue distribution and basal expression levels
Based on the RT-qPCR analysis with immature juvenile abalones, MTF-1 mRNAs were detected in all the tissue types examined; however, basal expression levels were quite variable among tissues (Fig. 3). The ubiquitous detection of MTF-1 transcripts across all the tissues is not surprising when taken into account its housekeeping

and fundamental roles in most cell types (Auf der Maur et al. 2000; O'Shields et al. 2014). The MTF-1 mRNAs were robustly expressed in hemocytes ($P < 0.05$), and this highest expression level was followed by muscles and gills, whereas the least mRNA expression was found in heart ($P < 0.05$). However, this expression pattern was not fully reproducible when measured with the sexually mature adults, although the broad pattern was consistent with findings from juveniles. Unlike in juveniles, the expression level of MTF-1 in the gut was found to be as high as that in the hemocytes in both female and male adults. More noticeably, matured adult abalones displayed a strikingly apparent difference in the MTF-1 expression in gonads where the extraordinarily high expression level was observed in testis while only minute expression in ovary (more than 150-fold difference; $P < 0.05$). Testis-predominant expression pattern of abalone MTF-1 in this study is similar with previous findings made in mouse (Auf der Maur et al. 2000) and hybrid tilapia (*Oreochromis aurea* × *Oreochromis nilotica*) (Cheung et al. 2010), collectively suggesting the possible involvement of MTF-1 in the male reproduction. Yet, the mechanism underlying the robust expression of MTF-1 in the abalone testes is currently unknown and open to hypothesize. In mouse, the reason for the high MTF-1 expression in testis has been explained by that sexually mature mice need to accumulate a large quantity of MTs in their testes (Auf der Maur et al. 2000; De et al. 1991). However, this hypothesis is unlikely to be adopted to this abalone species since the virtual increase of MT expression in sexually mature abalones has been observed in ovary rather than in testis (Lee and Nam 2016a), suggesting the molecular mechanism for the boosted expression of MTF-1 in abalone testis might be distinct from the ones in mammals. Hence, further study to monitor the MTF-1 expression in line with the testis development and

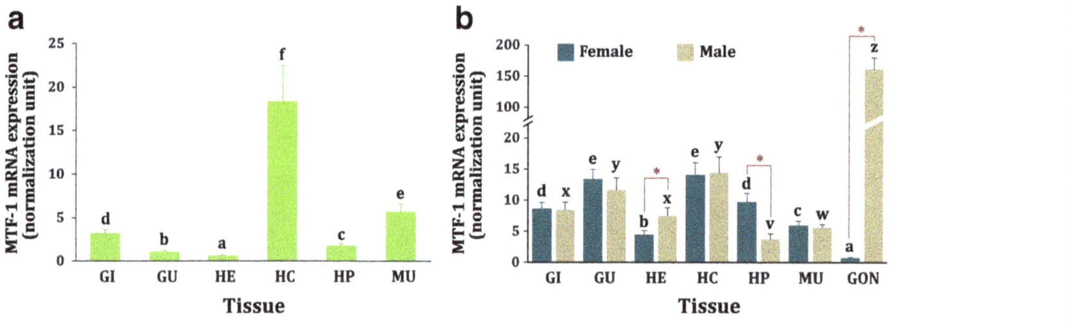

Fig. 3 Tissue distribution patterns and basal expression levels of abalone MTF-1 mRNAs in **a** immature juvenile and **b** sexually matured adult tissues, based on RT-qPCR assays using the normalization against *RPL5* and *PPL7* references. Tissue abbreviations are gill (*GI*), gut (*GU*), heart (*HE*), hemocyte (*HC*), hepatopancreas (*HP*), muscle (*MU*), and gonad (*GON*; *i.e.*, ovary or testis). In **a**, histograms with different letters indicate the significantly different means based on ANOVA followed by Duncan's multiple ranged tests at $P < 0.05$. In **b**, the significantly different means within a sex are indicated by *a–e* (for females) and *v–z* (for males) base on ANOVA followed by Duncan's multiple ranged tests at $P < 0.05$. The significantly different means between sexes observed in heat, hepatopancreas, and gonad tissues are noted by asterisks based on Student's *t* test analysis at $P < 0.05$

maturation cycle would be valuable to get a deeper insight into this phenomenon. Besides the gonadal expression, matured abalones showed a sex-specific difference in the basal expression of MTF-1 in heart and hepatopancreas. Male abalones displayed higher expression in heart than females did whereas female abalones showed higher expression in hepatopancreas than males did. This finding is similar with a previous observation with zebrafish where males should have a higher MTF-1 mRNA expression in the heart than females (O'Shields et al. 2014). Although it has been still inconclusive for such a sex-related dimorphism, the response of zebrafish MTF-1 to Cd exposure has been reported to be gender dependent in some tissues (O'Shields et al. 2014).

Developmental expression

The MTF-1 mRNAs were found to be already present in unfertilized eggs based on RT-qPCR assay, which could be considered as a typical indicative sign of the maternal contribution of MTF-1 to offspring. Fine regulation of metal homeostasis should be one of the prerequisite requirements for developmental success of marine mollusc embryos that undergo external development in metal-residing sea water (Roesijadi et al. 1996; Jenny et al. 2006; Mao et al. 2012). The initial expression level was decreased down in early cleavage stages and rebounded to the initial level at morula stage. Although there was a trend toward increase of MTF-1 mRNAs with the developmental progress up to early veliger stage, the degree of upregulation was only modest. When the development progressed to late veliger stage, the mRNA expression level of MTF-1 was significantly elevated ($P < 0.05$) (Fig. 4). Developmental expression of MTF-1 in marine molluscan animals has not been yet characterized previously. However, the expression pattern observed in this study was generally in congruent with the anticipated roles of MTF-1 in embryonic and early ontogenic

developments, as inferred from mammalian and teleostean cases (Günes et al. 1998; Chen et al. 2002; Chen et al. 2007). The expression pattern of MTF-1 during the development was also in agreement with the modulation pattern of its primary target (e.g., MT) in the same abalone species (Lee and Nam 2016a). Previously, the functional involvements of MTF-1 in the development and organogenesis have been highlighted by the lethality of "MTF-1-knockout" mice (Günes et al. 1998; Wang et al. 2004) and by induced inhibition of MTF-1 signaling followed by transcriptomic profiling in zebrafish embryos (O'Shields et al. 2014).

Expression during Zn and Cd exposure

There was an apparent difference between the two metal ions in the modulation of MTF-1 gene expression. In overall, Cd induced potently the mRNA expression of

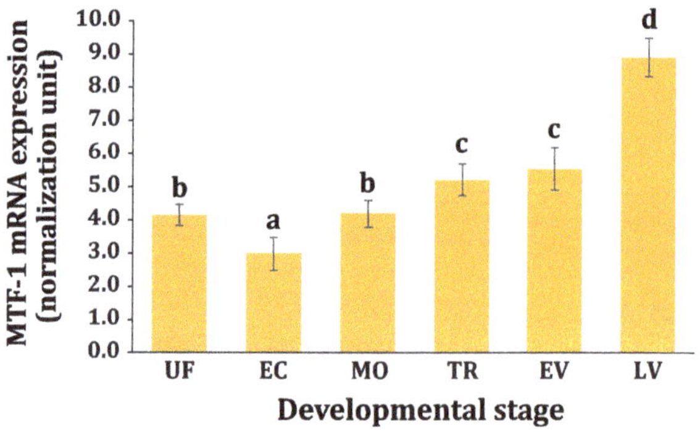

Fig. 4 Expression of abalone MTF-1 in developing embryos and early larvae assessed by RT-qPCR analysis using *RPL7* and *PPL8* reference genes as normalization controls. Abbreviations for developmental samples are unfertilized eggs (*UF*), eggs at early cleavages (*EC*; 2~8 cells stage), eggs at morula (*MO*), trochophore (*TR*), early veliger (*EV*), and late veliger (*LV*). Histograms with different letters indicate the significantly different means based on ANOVA followed by Duncan's multiple ranged tests at $P < 0.05$

MTF-1 while on the contrary, Zn repressed the MTF-1 expression (Fig. 5). Differential expression patterns of MTF-1 in response to Zn and Cd (20 and 100 ppb for both metals) were also dependent upon tissue types examined (gill, hepatopancreas, muscle, and hemocyte). In the gill, the exposure with 20-ppb Cd strongly induced the MTF-1 ($P < 0.05$) but higher exposure dose (100 ppb) did not give rise to the significant modulation of MTF-1 ($P > 0.05$). However, exposure with both doses of Zn significantly downregulated the MTF-1 in the gill ($P < 0.05$). On the other hand, in the hepatopancreas, both doses of Cd elevated the MTF-1 mRNA levels in a dose-dependent fashion ($P < 0.05$). Although the 20-ppb Zn exposure exhibited the small increase of MTF-1 mRNA levels in the hepatopancreas, 100-ppb Zn did not show any significant difference as compared to the level observed in non-exposed control ($P > 0.05$). Unlike in other three tissues, Cd exposure was unable to induce the MTF-1 expression in muscle tissue. In the muscle, Zn exposure depressed the MTF-1 mRNA expression (more significant downregulation of MTF-1 in the 100-ppb exposed group than in the 20-ppb exposed group). For hemocytes, the group exposed with 100-ppb Cd displayed a small, but statistically significant, increase of MTF-1 expression, while again, Zn-exposure resulted in the rapid downregulation of MTF-1 irrespective of exposure doses ($P < 0.05$).

In contrast to the variable or opposite regulation of MTF-1 by Cd and Zn, the transcriptional response of MT (the main target of MTF-1) to the metal exposure treatments was relatively uniform. Further, unlike MTF-1 showing the downregulation upon Zn exposure in most instances, MT was found to be consistently upregulated by Zn in both 20 and 100-ppb exposure treatments ($P < 0.05$). Although the induced folds were variable among tissues, MT gene expression was unfailingly induced in all the four tissues by exposure treatments with 100-ppb Cd, 20-ppb Zn, and 100-ppb Zn, but not by 20-ppb Cd. Collectively, the modulation patterns upon metal exposure were apparently different between MTF-1 and MT genes, and the degree of Cd-mediated upregulation in each tissue was much higher for MT than MTF-1 (Fig. 5).

Previous studies have indicated that mammalian MTF-1 should be a constitutively expressed gene with a TATA-less promoter and that mammalian MTF-1s would not show any appreciable response to experimentally designed heavy metal and other stress factors. The plausible reason for the absence of metal (or stress) responsiveness has been explained by the lack of MRE motif in the mammalian (e.g., mouse and human) MTF-1 gene promoters (Auf der Maur et al. 2000; Bi et al. 2006). Hence, the MTF-1 activity in mammals is widely agreed to be largely induced at a post-translation level (Günther et al. 2012a; Smirnova et al. 2000; Saydam et al. 2001). However, in the present study, this abalone MTF-1 gene was proven to possess a putative MRE copy as well as a canonical TATA box in its promoter region. Accordingly, the present abalone MTF-1 gene displayed the Cd-mediated induction in multiple tissues (in gill, hepatopancreas and hemocytes, but not in muscles). Similarly, the Cd-induced expression of MTF-1 gene has also been reported in other cold-blooded animals such as zebrafish

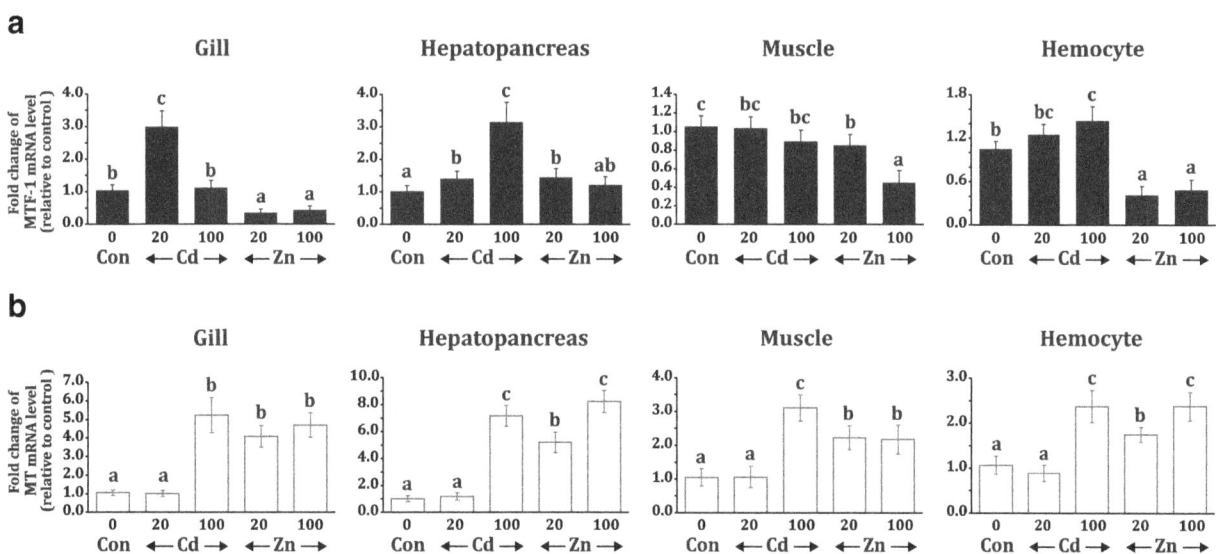

Fig. 5 Transcriptional responses of abalone MTF-1 (**a**) and MT (**b**) to experimental heavy metal exposures. Metal exposures were done with cadmium or zinc with the dose strengths of 20 and 100 ppb for 24 h. RT-qPCR data were presented as fold changes of MTF-1 mRNAs in expressed groups relative to that of non-exposed control (Con) based on the normalization against two internal control genes (*RPL5* and *PPL7*). Histograms with different letters indicate the significantly different means based on ANOVA followed by Duncan's multiple ranged tests at $P < 0.05$

(Dubé et al. 2011; Cheuk et al. 2008) and carp (Ferencz and Hermesz 2009), which is obviously different from the principle for the MTF-1 regulation in mammals (Bi et al. 2006; Saydam et al. 2001).

However, the transcription of abalone MTF-1 was not induced by zinc except only a minute increase in the hepatopancreas; furthermore, Zn exposure even down-regulated the mRNA expression of abalone MTF-1 in other tissues. This finding is similar with no induction of MTF-1 mRNA observed for zebrafish cells exposed to Zn (Cheuk et al. 2008), although Zn exposure in zebra-fish was reported to be able to activate the nuclear trans-location of MTF-1 from cytoplasm (Chen et al. 2007). On the other hand, a recent study with oyster *C. gigas* has shown that mRNA expression level of MTF-1 could be increased by Zn exposure, but the induced amount was only modest even treated with high dose of Zn (Meng et al. 2015). In the same report, authors have in-dicated that RNA interference of MTF-1 would result in the depression of Zn-mediated induction of MT (the known target of MTF-1), consequently confirming that MTF-1 is a superordinate regulator of the MT (Meng et al. 2015). However, in this study, the inducible pattern of MTF-1 was not in agreement with that of MT expression. Both Cd (inducer of abalone MTF-1) and Zn (non-inducer of abalone MTF-1) were found to be able to induce MT expression in all the tissues examined. Further, the quantitative relationship between MTF-1 and MT expression was not proportional. Currently, it has been widely agreed that Zn-mediated induction of MT could be achieved by direct binding of Zn to the MTF-1 fingers, and other metals such as Cd and Cu may not replace Zn in zinc finger binding (Schmittgen and Livak 2008; Chen et al. 1999; Zhang et al. 2001). Based on this, the Cd-mediated induction of MT gene might be considered as an indirect consequence. Pos-sibly, Cd load might give rise to the release of Zn from MTs (also from other metalloproteins), and the increase of cellular Zn levels may activate MTF-1 through the Zn-binding. This zinc pool hypothesis can also be applied to the MT induction upon exposed to diverse oxidative stresses (Günther et al. 2012a). Hence, the con-troversy between MTF-1 and MT responses to zinc should be challenged in future study by examining cellu-lar Zn concentrations (or tissue burden) under various Zn-exposure conditions.

Response to heat shock treatment
Abalone MTF-1 was proven an early phase, heat-shock responsive protein as evidenced by the rapid modulation upon thermal increases. The MTF-1 mRNA levels in the gill were rapidly increased (fourfold relative to 20 °C; $P < 0.05$) as early as when water temperature reached 25 °C (Fig. 6). The expression level was further elevated

when the temperature reached 30 °C (i.e., designated 30 °C+0 h) ($P < 0.05$). Afterward, the MTF-1 expression levels began to be decreased down with the continued incubation at 30 °C (i.e., 30 °C+12 h and 30 °C+24 h); however, the expression level at the end of thermal treat-ment was still significantly higher than that of 20 °C group ($P < 0.05$). On the other hand, in the hepatopancreas, the beginning of MTF-1 upregulation was more or less lagged as compared to that in the gill. Significant induction of MTF-1 in the hepatopancreas became evident from the 30 °C+12 h group, and the elevated expression level remained constant at 30 °C+24 h ($P < 0.05$). In muscles, the significant induction of MTF-1 was found at 30 °C+0 h and 30 °C+12 h but soon returned to the initial level observed at 20 °C group. Meanwhile, hemocyte dis-played the rapid induction of MTF-1 mRNA at 25 °C; however, the expression dropped sharply at 30 °C+0 h in which the decreased expression level was even lower than that of 20 °C group ($P < 0.05$). Afterward, the MTF-1 ex-pression rebounded at 30 °C+0 h group. Then, the re-bound was followed by a further increase at 30 °C+12 h group ($P < 0.05$) (Fig. 6).

As aforementioned, the zinc pool sensing mechanism by MTF-1 upon exposure to stress factors might be adopted for the involvement of MTF-1 in host defense pathways against oxidative stress, since the abrupt and substantial changes of water temperature might be a causative factor to generate oxidative stress in poikilo-thermal invertebrates (Kim et al. 2007; Attig et al. 2014; Banni et al. 2014). Accordant with this explanation, the potential involvement of MT protein in the heat shock response has already been reported in this abalone spe-cies and also in other aquatic animals (Lee and Nam 2016a; Negri et al. 2013; Jarque et al. 2014). Within this context, the cold shock-induced MTF-1 has been reported in the common carp (*Cyprinus carpio*) brain, with an ex-planation that sudden temperature drop gave rise to the alteration of physiologically accessible Zn concentrations in that tissue (Ferencz and Hermesz 2008).

Meanwhile, a series of mammalian studies has pro-posed that MT and heat shock protein (HSP) genes might work in a non-cooperative way in their transcrip-tional responses to stress treatments. Experimental evi-dences for this proposal may include that (1) HSF should boost the activity of HSP gene promoter but hardly affect an MRE-containing promoter of MT gene upon heat shock and metal exposure, (2) heat shock-induced nuclear translocation of MTF-1 has been shown to be insufficient to activate a MT gene promoter, (3) di-verse target gene searches for MTF-1 have been indica-tive of HSP genes as non-affected genes, and (4) MTF-1 has been likely to repress HSF-regulated genes through a direct protein-protein interaction (Lichtlen et al. 2001; Saydam et al. 2003; Uenishi et al. 2006). However, on the

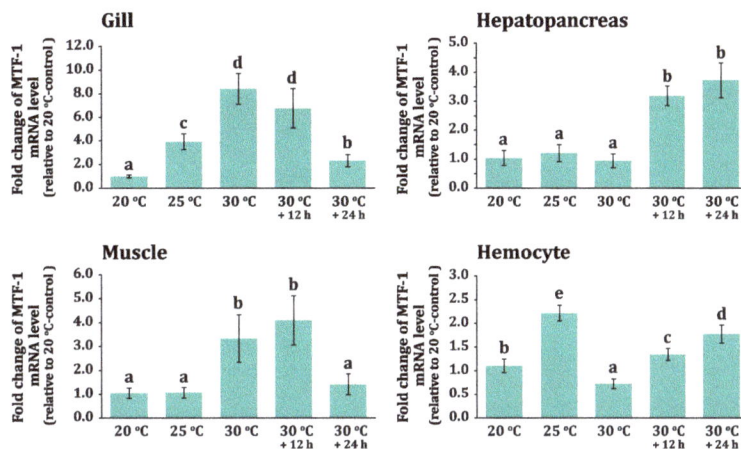

Fig. 6 Differential mRNA expression patterns of abalone MTF-1 in response to heat shock treatments. During heat elevation with an increase rate of 1 °C/h, abalones were sampled at 20 °C (just before the elevation), 25 °C, and 30 °C. After reaching 30 °C, abalones were exposed at the constant 30 °C and sampled after 12 h (30 °C+12 h) and 24 h (30 °C+24 h), respectively. RT-qPCR data were presented as fold changes of MTF-1 mRNAs in expressed groups relative to that of non-heat shocked group (i.e., 20 °C control) based on the normalization with $RPL5$ and $PPL7$ reference genes. Histograms with different letters indicate the significantly different means based on ANOVA followed by Duncan's multiple ranged tests at $P < 0.05$

contrary, the present study has shown that the transcription of abalone MTF-1 could be directly activated by heat shock. Hence, our finding suggests that the thermal stress-mediated activation of MTF-1 in abalone may be achieved not only at a post-translation level (i.e., indirectly based on cellular zinc sensing abovementioned) but also at a transcriptional level (i.e., de novo synthesis). The presence of potential HRE motifs in the abalone MTF-1 gene regulatory region is also in congruent with the present hypothesis. Possibly, the activation of MTF-1 from the dual routes is likely to be advantageous for permitting the poikilothermal animals to prepare more efficiently antioxidant components (i.e., antioxidant enzyme genes containing MREs in their promoters as well as MT) against the oxidative stress caused by thermal fluctuations (Kim et al. 2007; Cho et al. 2009). However, further efforts are needed to get direct evidence on the cooperativity between HSF and MTF-1 in this abalone species.

Conclusions

Novel MTF-1 was isolated and characterized from a commercially important marine mollusc species, Pacific abalone (*H. discus hannai*). The abalone MTF-1 was found to share a conserved feature in zinc finger, DNA-binding domain with its orthologs; however, it represented non-conservative features in remaining other parts. Bioinformatic analysis of the 5′-upstream region has predicted diverse transcription factor binding motifs that are potentially related with metal regulation, stress responses, and development. Abalone MTF-1 was ubiquitously detected in various tissues, in which the highest expression level was observed in the testes of

mature males. Abalone MTF-1 was expressed during the entire period of embryonic and early ontogenic development. From the heavy metal exposure, abalone MTF-1 was found to be Cd inducible (but not by Zn); however, the induced amounts were only modest as compared to that of MT. Abalone MTF-1 was highly modulated in responsive to heat shock potentially via both indirect zinc pool sensing and direct de novo transcription. Data from this study could be a useful basis to approach various researches regarding the stress responses in this abalone species particularly including detoxification of heavy metals and adaptation to thermal stresses.

Additional file

Additional file 1: Table S1. Summarized information on oligonucleotide primers used in this study. **Figure S1.** Full-length cDNA and deduced amino acid sequences of abalone *Haliotis discus hannai* metal responsive transcription factor-1 (MTF-1). In the nucleotide sequence, stop codon is indicated by an asterisk and putative polyadenylation signal is underlined. On the other hand, in the amino acid sequence (in a singlet code), presumed nuclear localization signal (NLS) is underlined (in the front of the first zinc finger), while six putative C_2H_2-zinc fingers in the DNA-binding domain are boxed. **Figure S2.** Multiple sequence alignments of abalone MTF-1 along with its representative orthologs. Putative NLS, six C_2H_2-zinc fingers, NES embedded in transactivation domain, and C-terminal Cys-cluster are indicated. **Figure S3.** Neighbor-joining phylogenetic tree based on amino acid sequences of zin-finger DNA-binding domain and NLS region in MTF-1 orthologs. Tree was computed by poisson correction method using MEGA 7.0 program. Confidence level of each clade was evaluated with bootstrap testing (1000 replicates). **Figure S4.** Putative transcription factor-binding motifs predicted in the 5′-flanking upstream region of the abalone MTF-1 gene. Transcription factor sites were predicted with perfect and imperfect matches to consensus core sequences. TRANSFAC® search (GeneXplain GmbH, Germany) was carried out with the cut-offs scores >0.99 (core score) and >0.95 (matrix score).

Acknowledgements
Not applicable.

Funding
This study was supported by the grant from the Golden Seed Project (GSP), Ministry of Oceans and Fisheries, Republic of Korea.

Authors' contributions
SYL carried out molecular cloning, bioinformatics, and expression analyses. YKN designed this study, evaluated data, and drafted the manuscript. Both authors read and approved the final manuscript.

Competing interests
The authors declare that they have no competing interests.

References
Attig H, Kamel N, Sforzini S, Dagnino A, Jamel J, Boussetta H, Viarengo A, Banni M. Effects of thermal stress and nickel exposure on biomarkers responses in *Mytilus galloprovincialis* (Lam). Mar Environ Res. 2014;94:65–71.

Auf der Maur A, Belser T, Wang Y, Günes C, Lichtlen P, Georgiev O, Schaffner W. Characterization of the mouse gene for the heavy metal-responsive transcription factor MTF-1. Cell Stress Chaperones. 2000;5:196–206.

Banni M, Hajer A, Sforzini S, Oliveri C, Boussetta H, Viarengo A. Transcriptional expression levels and biochemical markers of oxidative stress in *Mytilus galloprovincialis* exposed to nickel and heat stress. Comp Biochem Physiol C Pharmacol Toxicol Endocrinol. 2014;160:23–9.

Bi Y, Lin GX, Millecchia L, Ma Q. Superinduction of metallothionein I by inhibition of protein synthesis: role of a labile repressor in MTF-1 mediated gene transcription. J Biochem Mol Toxicol. 2006;20:57–68.

Bittel DC, Smirnova IV, Andrews GK. Functional heterogeneity in the zinc fingers of metalloregulatory protein metal response element-binding transcription factor-1. J Biol Chem. 2000;275:37194–201.

Chen X, Chu M, Giedroc DP. MRE-Binding transcription factor-1: weak zinc-binding finger domains 5 and 6 modulate the structure, affinity, and specificity of the metal-response element complex. Biochemistry. 1999;38:12915–25.

Chen W-Y, John JAC, Lin C-H, Chang C-Y. Molecular cloning and developmental expression of zinc finger transcription factor MTF-1 gene in zebrafish, *Danio rerio*. Biochem Biophys Res Commun. 2002;291:798–805.

Chen W, John JAC, Lin C, Chang C. Expression pattern of metallothionein, MTF-1 nuclear translocation, and its dna-binding activity in zebrafish (*Danio rerio*) induced by zinc and cadmium. Environ Toxicol Chem. 2007;26:110–7.

Cheuk WK, Chan PC-Y, Chan KM. Cytotoxicities and induction of metallothionein (MT) and metal regulatory element (MRE)-binding transcription factor-1 (MTF-1) messenger RNA levels in the zebrafish (*Danio rerio*) ZFL and SJD cell lines after exposure to various metal ions. Aquat Toxicol. 2008;89:103–12.

Cheung AP-L, Au CY-M, Chan WW-L, Chan KM. Characterization and localization of metal-responsive-element-binding transcription factors from tilapia. Aquat Toxicol. 2010;99:42–55.

Cho YS, Lee SY, Bang IC, Kim DS, Nam YK. Genomic organization and mRNA expression of manganese superoxide dismutase (Mn-SOD) from *Hemibarbus mylodon* (Teleostei, Cypriniformes). Fish Shellfish Immunol. 2009;27:571–6.

De SK, Enders GC, Andrews GK. High levels of metallothionein messenger RNAs in male germ cells of the adult mouse. Mol Endocrinol. 1991;5:628–36.

Dubé A, Harrisson J-F, Saint-Gelais G, Séguin C. Hypoxia acts through multiple signaling pathways to induce metallothionein transactivation by the metal-responsive transcription factor-1 (MTF-1). Biochem Cell Biol. 2011;89:562–77.

Ferencz Á, Hermesz E. Identification and characterization of two mtf-1 genes in common carp. Comp Biochem Physiol C Toxicol Pharmacol. 2008;148:238–43.

Ferencz Á, Hermesz E. Identification of a splice variant of the metal-responsive transcription factor MTF-1 in common carp. Comp Biochem Physiol C Toxicol Pharmacol. 2009;150:113–7.

Giedroc DP, Chen X, Apuy JL. Metal response element (MRE)-binding transcription factor-1 (MTF-1): structure, function, and regulation. Antioxid Redox Signal. 2001;3:577–96.

Günes C, Heuchel R, Georgiev O, Müller K-H, Lichtlen P, Blüthmann H, Marino S, Aguzzi A, Schaffner W. Embryonic lethality and liver degeneration in mice lacking the metal-responsive transcriptional activator MTF-1. EMBO J. 1998;17:2846–54.

Günther V, Lindert U, Schaffner W. The taste of heavy metals: Gene regulation by MTF-1. Biochim Biophys Acta, Mol Cell Res. 2012a;1823:1416–25.

Günther V, Davis AM, Georgiev O, Schaffner W. A conserved cysteine cluster, essential for transcriptional activity, mediates homodimerization of human metal-responsive transcription factor-1 (MTF-1). Biochim Biophys Acta, Mol Cell Res. 2012b;1823:476–83.

Hahn ME, Karchner SI, Evans BR, Franks DG, Merson RR, Lapseritis JM. Unexpected diversity of aryl hydrocarbon receptors in non-mammalian vertebrates: insights from comparative genomics. J Exp Zool A Ecol Genet Physiol. 2006;305:693–706.

Jackson DP, Joshi AD, Elferink CJ. Ah receptor pathway intricacies; signaling through diverse protein partners and DNA-motifs. Toxicol Res. 2015;4:1143–58.

Jarque S, Prats E, Olivares A, Casado M, Ramón M, Piña B. Seasonal variations of gene expression biomarkers in *Mytilus galloprovincialis* cultured populations: temperature, oxidative stress and reproductive cycle as major modulators. Sci Total Environ. 2014;499:363–72.

Jenny MJ, Warr GW, Ringwood AH, Baltzegar DA, Chapman RW. Regulation of metallothionein genes in the American oyster (*Crassostrea virginica*): ontogeny and differential expression in response to different stressors. Gene. 2006;379:156–65.

Kang J, Lee YG, Jeong DU, Lee JS, Choi YH, Shin YK. Effect of abalone farming on sediment geochemistry in the Shallow Sea near Wando, South Korea. Ocean Sci J. 2015;50:669–82.

Kim K-Y, Lee SY, Cho YS, Bang IC, Kim KH, Kim DS, Nam YK. Molecular characterization and mRNA expression during metal exposure and thermal stress of copper/zinc-and manganese-superoxide dismutases in disk abalone, *Haliotis discus discus*. Fish Shellfish Immunol. 2007;23:1043–59.

Laity JH, Andrews GK. Understanding the mechanisms of zinc-sensing by metal-response element binding transcription factor-1 (MTF-1). Arch Biochem Biophys. 2007;463:201–10.

Lee SY, Nam YK. Transcriptional responses of metallothionein gene to different stress factors in Pacific abalone (*Haliotis discus hannai*). Fish Shellfish Immunol. 2016a;58:530–41.

Lee SY, Nam YK. Evaluation of reference genes for RT-qPCR study in abalone *Haliotis discus hannai* during heavy metal overload stress. Fish Aquat Sci. 2016b;19:21.

Li Y, Kimura T, Laity JH, Andrews GK. The zinc-sensing mechanism of mouse MTF-1 involves linker peptides between the zinc fingers. Mol Cell Biol. 2006;26:5580–7.

Li Y, Kimura T, Huyck RW, Laity JH, Andrews GK. Zinc-induced formation of a coactivator complex containing the zinc-sensing transcription factor MTF-1, p300/CBP, and Sp1. Mol Cell Biol. 2008;28:4275–84.

Lichtlen P, Schaffner W. Putting its fingers on stressful situations: the heavy metal-regulatory transcription factor MTF-1. BioEssays. 2001;23:1010–7.

Lichtlen P, Wang Y, Belser T, Georgiev O, Certa U, Sack R, Schaffner W. Target gene search for the metal-responsive transcription factor MTF-1. Nucleic Acids Res. 2001;29:1514–23.

Mao H, Wang D-H, Yang W-X. The involvement of metallothionein in the development of aquatic invertebrate. Aquat Toxicol. 2012;110:208–13.

Meng J, Zhang L, Li L, Li C, Wang T, Zhang G. Transcription factor CgMTF-1 regulates CgZnT1 and CgMT expression in Pacific oyster (*Crassostrea gigas*) under zinc stress. Aquat Toxicol. 2015;165:179–88.

Negri A, Oliveri C, Sforzini S, Mignione F, Viarengo A, Banni M. Transcriptional response of the mussel *Mytilus galloprovincialis* (Lam.) following exposure to heat stress and copper. PLoS One. 2013;8:e66802.

O'Shields B, McArthur AG, Holowiecki A, Kamper M, Tapley J, Jenny MJ. Inhibition of endogenous MTF-1 signaling in zebrafish embryos identifies novel roles for MTF-1 in development. Biochim Biophys Acta, Mol Cell Res. 2014;1843:1818–33.

Park C-J, Kim SY. Abalone aquaculture in Korea. J Shellfish Res. 2013;32:17–9.

Qiu J, Liu Y, Yu M, Pang Z, Chen W, Xu Z. Identification and functional characterization of MRE-binding transcription factor (MTF) in *Crassostrea gigas* and its conserved role in metal-induced response. Mol Biol Rep. 2013;40:3321–31.

Roesijadi G, Hansen KM, Unger ME. Cadmium-induced metallothionein expression during embryonic and early larval development of the mollusc *Crassostrea virginica*. Toxicol Appl Pharmacol. 1996;140:356–63.

Saydam N, Georgiev O, Nakano MY, Greber UF, Schaffner W. Nucleo-cytoplasmic trafficking of metal-regulatory transcription factor 1 is regulated by diverse stress signals. J Biol Chem. 2001;276:25487–95.

Saydam N, Steiner F, Georgiev O, Schaffner W. Heat and heavy metal stress synergize to mediate transcriptional hyperactivation by metal-responsive transcription factor MTF-1. J Biol Chem. 2003;278:31879–83.

Schmittgen TD, Livak KJ. Analyzing real-time PCR data by the comparative CT method. Nat Protoc. 2008;3:1101–8.

Smirnova IV, Bittel DC, Ravindra R, Jiang H, Andrews GK. Zinc and cadmium can promote rapid nuclear translocation of metal response element-binding transcription factor-1. J Biol Chem. 2000;275:9377–84.

Stoytcheva ZR, Vladimirov V, Douet V, Stoychev I, Berry MJ. Metal transcription factor-1 regulation via MREs in the transcribed regions of selenoprotein H and other metal-responsive genes. Biochim Biophys Acta, Gen Subj. 2010;1800:416–24.

Uenishi R, Gong P, Suzuki K, Koizumi S. Cross talk of heat shock and heavy metal regulatory pathways. Biochem Biophys Res Commun. 2006;341:1072–7.

Wang Y, Wimmer U, Lichtlen P, Inderbitzin D, Stieger B, Meier PJ, Hunziker L, Stallmach T, Forrer R, Rulicke T, Georgiev O, Schaffner W. Metal-responsive transcription factor-1 (MTF-1) is essential for embryonic liver development and heavy metal detoxification in the adult liver. FASEB J. 2004;18:1071–9.

Zhang B, Egli D, Georgiev O, Schaffner W. The Drosophila homolog of mammalian zinc finger factor MTF-1 activates transcription in response to heavy metals. Mol Cell Biol. 2001;21:4505–14.

16

Oxidative stress and non-specific immune responses in juvenile black sea bream, *Acanthopagrus schlegelii*, exposed to waterborne zinc

Jun-Hwan Kim[1], Hee-Ju Park[2], Kyeong-Wook Kim[2] and Ju-Chan Kang[2*]

Abstract

Juvenile black sea bream, *Acanthopagrus schlegelii*, were exposed to waterborne zinc (Zn) at concentrations of 0, 200, and 400 µg/L, at temperatures of 18 or 26 °C for 4 weeks. Superoxide dismutase (SOD) activities in the liver and gill of *A. schlegelii* significantly increased following exposure to waterborne Zn. Significant reduction in glutathione S-transferase (GST) activity in the liver and gill was observed following exposure to waterborne Zn. Glutathione (GSH) concentrations in the liver and gill also significantly decreased following exposure to waterborne Zn. Phagocytosis and lysozyme in the plasma and kidney were significantly increased following exposure to waterborne Zn. High water temperature increased alterations in the antioxidant and immune responses. The results of the present study suggest that waterborne Zn induced significant alterations in oxidative stress, increased immune responses and high temperature that trigger Zn toxicity.

Keyword: *Acanthopagrus schlegelii*, Sea bream, Zinc, Oxidative stress, Non-specific immune response

Background

Zinc (Zn) is an essential trace metal in fish, providing a vital structural and catalytic function to more than 300 proteins integral to piscine growth, reproduction, development, and immune function (Bury et al. 2003). However, Zn is released into aquatic environments as a result of industrial activities and exposure to high concentrations of Zn that induces toxicity in fish (Hogstrand et al. 2002).

Waterborne Zn induces histological aberrations in fish, such as leukocyte infiltration, epithelial cell proliferation, necrosis, hypertrophy, and mucus secretion (Wood 2001). In addition, Zn exposure is particularly toxic to fish as it inhibits calcium uptake, which creates an internal imbalance and results in hypocalcemia and disturbance of acid-base regulation (Santore et al. 2002).

Exposure to metal stimulates the production of reactive oxygen species (ROS), leading to oxidative metabolism injury and oxidative stress (Lushchak 2011). Exposure to Zn specifically leads to ROS generation, oxidant injury, excitation of inflammation, and cell death (Xia et al. 2008). The influence of metal exposure on ROS generation differs according to the exposure period and route of uptake (Coteur et al. 2005). ROS produced by aerobic organisms readily reacts with most biological molecules; thus, high concentrations of ROS production induce pathophysiological damage, such as arthritis, diabetes, inflammation, cancer, and genotoxicity (Senevirathne et al. 2006).

Non-specific immune responses in fish are considered a response to stressors, and metal exposure is associated with changes in the fish immune system (Kim and Kang 2016a). The non-specific immune system is considered a first line of defense against toxicants (Saurabh and Sahoo 2008). Given that immune parameters are influenced by aquatic toxins, metals, pesticides, hydrocarbons, and other chemicals, they can be used as important indicators to evaluate toxic effects in fish exposed to toxic substances.

Black sea bream, *Acanthopagrus schlegelii*, is a member of the family Sparidae. This species inhibits the inshore shelf of East Asia countries, such as South Korea

---

* Correspondence: jckang@pknu.ac.kr
[2]Department of Aquatic Life Medicine, Pukyong National University, Busan 48513, Korea
Full list of author information is available at the end of the article

and Japan at depths of 50 m. The species is omnivorous and protandrous, and breeds from February to May, dopositing eggs in the demersal zone. *A. schlegelii* is one of the most commonly cultured fish species and highly sought after in South Korea. The optimum temperature for growth and development of *A. schlegelii* is 18 °C; however, during the Korean summer, seawater temperature reaches 26 °C. The degree of waterborne Zn-induced toxicity in aquatic animals depends on water temperature and water chemistry, as well as the life stages of individuals (McGeer et al. 2000). However, information on the relationship between Zn toxicity and water temperature is scarce. Therefore, the present study evaluated the toxic effects of exposure to waterborne Zn and the influence of water temperature on Zn-induced toxicity using the antioxidant and non-specific immune responses of *A. schlegelii*.

Methods

Experimental fish and conditions

Black sea breams, *Acanthopagrus schlegelii*, were obtained from a local fish farm in Tongyeong, Korea. Fish were acclimatized for 3 weeks under laboratory conditions (Table 1). Fish were fed a commercial diet twice daily (Woosungfeed, Daejeon City, Korea). After acclimatization, 72 fish (body length, 17.8 ± 1.5 cm; body weight, 96.7 ± 6.8 g) were selected for study. Waterborne Zn exposure took place in 20-L glass tanks, containing 6 fish per treatment group. Water temperature was adjusted from ambient at a rate of ± 1 °C/day until a final temperature of 26 °C was reached. The acclimation period commenced once the final temperature had been sustained for 24 h. Zinc oxide (Sigma, St.Louis, MO, USA) solution was dissolved in respective glass tanks. Zn concentrations were 0, 200, and 400 µg/L (at 18 and 26 °C). An extremely high concentration of 400 µg/L Zn is much higher than that in nature, but this exposure experiment focused on Zn toxicity in experimental fish. Glass tank water was completely replaced once per 2 days and made the same concentration in respective glass tank. At the end of each period (at 2 and 4 weeks), fish were anesthetized in buffered 3-aminobenzoic acid ethyl ester methanesulfonate (Sigma Chemical, St. Louis, MO).

Waterborne Zn analysis

Seawater samples were digested in 65% (v/v) HNO3 and dried at 120 °C. The procedure was repeated until total digestion. The entirely digested samples were diluted in 2% (v/v) HNO3. The samples were filtered through a 0.2-µm membrane filter (Advantec mfs, Ins.) under pressure for analysis. For determination of total Zn concentrations, the digested and extracted solutions were analyzed by ICP-MS. The ICP-MS measurements were performed using an ELAN 6600DRC ICP-MS instrument with argon gas (Perkin-Elmer). Total Zn concentrations were determined by external calibration. ICP multi-element standard solution VI (Merck) was used for standard curve. The Zn concentrations were expressed as microgram per liter (Table 1).

Antioxidant response analysis

Liver and gill tissues were excised and homogenized with 10 volumes of ice-cold homogenization buffer using Teflon-glass homogenizer (099CK4424, Glass-Col, Germany). The homogenate was centrifuged at $10,000g$ for 30 min under refrigeration, and the obtained supernatants were stored at −80 °C for analysis.

Superoxide dismutase (SOD) activity was measured with 50% inhibitor rate about the reduction reaction of WST-1 using SOD Assay kit (Dojindo Molecular Technologies, Inc.). One unit of SOD is defined as the amount of the enzyme in 20 µl of sample solution that inhibits the reduction reaction of WST-1 with superoxide anion by 50%. SOD activity was expressed as unit mg protein^{-1}.

* WST-1 = 2-(4-lodophenyl)-3-(4-nitrophenyl)-5-(2,4-disulfophenyl)- $2H$ -tetrazolium, monosodium salt Glutathione-S-transferase(GST) activity was measured according to the method of modified Habig et al. (1974). The reaction mixture consisted of 0.2 M phosphate buffer (pH 6.5), 10 mM glutathione (GSH) (Sigma), and 10 mM 1-chloro-2,-dinitrobenzene, CDNB (Sigma). The change in absorbance at 25 °C was recorded at 340 nm, and the enzyme activity was calculated as 340 nm, and the enzyme activity was calculated as nmol min^{-1} mg protein^{-1}.

Reduced glutathione was measured following the method of Beutler (1984). Briefly, 0.2 ml fresh supernatant was added to 1.8 ml distilled water. Three milliliters of the precipitating solution (1.67 g metaphosphoric acid, 0.2 g EDTA, and 30 g NaCl in 100 ml distilled water) was mixed with supernatants. The mixture was centrifuged at $4500g$ for 10 min. 1.0 mL of supernatant was added to 4.0 ml of 0.3 M NaHPO4 solution, and 0.5 mL DTNB (5,5′-dithiobis-2-nitrobenzoic acid) was then added to this solution. Reduced glutathione was

Table 1 Analyzed waterborne zinc concentration (mg/kg) from each source

Exposure seawater (µg/L)						
Concentrations (µg/L)	0 (18 °C)	0 (26 °C)	200 (26 °C)	200 (26 °C)	400 (18 °C)	400 (26 °C)
Zinc concentrations	0	0	200	200	400	400
Actual zinc levels	8.6	9.8	212.5	208.7	421.3	418.5

measured as the difference in the absorbance values of samples in the presence and the absence of DTNB at 412 nm. GSH value was calculated as μmol mg protein^{-1} in the tissues.

Non-specific immune response analysis

Blood samples were collected within 35–40 s through the caudal vein of the fish in 1-ml disposable heparinized syringes. The blood samples were centrifuged to separate serum from blood samples at 3000g for 5 min at 4 °C. Kidney tissues were obtained using sterilized dissecting tools, and the excised tissues were homogenizing with 10 volumes of ice-cold homogenization buffer (0.004 M phosphate buffer, pH 6.6) using Teflon- glass homogenizer (099CK4424, Glass-Col, Germany). Homogenate was centrifuged at 10,000g for 10 min under refrigeration, and the obtained supernatant was stored at –70 °C for analysis. Protein content was determined by the Bio-Rad Protein Assay Kit (Bio-Rad Laboratories GmbH, Munich, Germany) based on Bradford dye-binding procedure, using bovine serum albumin as standard.

Phagocytosis was measured with phagocytosis assay kit (Cell biolabs, Inc.). Briefly, we added 10 μl of *Escherichia coli* suspension in 100 μl plasma sample of 96-well plate. The samples were mixed well, and we immediately transferred the plate to a cell culture incubator at 37 °C for 3–6 h. Each sample including a negative control without *E. coli* particles was assayed in duplicate. After, we added 200 μl of cold 1× PBS to each well and promptly removed PBS solution by centrifugation and gentle aspiration. We added 100 μl of fixation solution by centrifugation and gentle aspiration and then washed the sample twice with 1× PBS. We added 100 μl of prediluted 1× blocking solution to each well, incubated the sample for 30 min at room temperature on an orbital shaker, and promptly removed blocking solution by centrifugation and gentle aspiration. The sample was washed three times with 1× PBS. We added 100 μl of prediluted 1× permeabilization solution to each well, and incubated the sample 5 min at room temperature. We promptly removed permeabilization solution by centrifugation and gentle aspiration, and the sample was washed twice with 1× PBS. We promptly removed the PBS by centrifugation and gentle aspiration, initiated the reaction by adding 100 μl of substrate, and incubated the sample for 10–30 min at room temperature. We then stopped the reaction by adding 100 μl of the stop solution and mixed it by placing the plate on an orbital plate shaker for 30 s. Finally, we read the absorbance of each well at 450 nm.

Lysozyme activity was determined by a turbidimetric method (Ellis 1990) using *Micrococcus lysodeikticus* (Sigma) as substrate (0.2 mg/ml 0.05 M phosphate buffer, pH 6.6 for kidney sample and pH 7.4 for plasma). A standard curve was made with lyophilized hen egg white lysozyme (Sigma), and the rate of change in turbidity was measured at 0.5-min and 4.5-min intervals at 530 nm. The results were expressed as microgram per milliliter and microgram per gram equivalent of hen egg white lysozyme activity (Anderson and Siwicki 1994).

Statistical analysis

The experiment was conducted in exposure period for 4 weeks and performed triplicate. Statistical analyses were performed using the SPSS/PC+ statistical package (SPSS Inc, Chicago, IL, USA). Significant differences between groups were identified using one-way ANOVA and Tukey's test for multiple comparisons or Student's t test for two groups. The significance level was set at $P < 0.05$.

Results
Antioxidant responses

Antioxidant responses such as SOD activity, GST activity, and GSH concentration were analyzed to assess the oxidative stress by the waterborne Zn exposure depending on water temperature. Liver and gill SOD activity of the *A. schlegelii* is presented in Fig. 1. The liver SOD activity was significantly increased in 400 μg/L Zn at 18 °C and in 100 and 200 μg/L Zn at 26 °C after 2 weeks. After 4 weeks, a significant SOD activity in the liver was observed in the fish exposed to waterborne Zn greater than 200 μg/L at 18 and 26 °C. The gill SOD activity was substantially increased in 400 μg/L Zn at 26 °C after 2 weeks, and there was no significant alteration at 18 °C after 2 weeks. After 4 weeks, the gill SOD activity was notably increased in 400 μg/L Zn at 18 and 26 °C.

Liver and gill GST activity of the *A. schlegelii* is presented in Fig. 2. A significant decrease in the liver GST activity was observed in 400 μg/L at 18 °C and in 200, 400 μg/L at 26 °C after 2 weeks. After 4 weeks, the liver GST was significantly decreased in 400 μg/L at 18 and 26 °C. In the gill tissue, the GST activity was notably decreased in 400 μg/L at 18 and 26 °C after 2 and 4 weeks.

Liver and gill GSH activity of the *A. schlegelii* is demonstrated in Fig. 3. The liver GSH concentration was substantially decreased in 400 μg/L Zn at 18 and 26 °C after 2 weeks. After 4 weeks, a significant decrease in the liver GSH concentration was observed in the concentration of 400 μg/L Zn at 18 °C and 200, 400 μg/L Zn at 26 °C. The gill GSH concentration was notably decreased in 400 μg/L Zn at 18 and 26 °C. After 4 weeks, the gill GSH concentration was significantly decreased in 400 μg/L Zn at 18 °C and 200, 400 μg/L Zn at 26 °C.

Non-specific immune responses

Non-specific immune responses such as phagocytosis and lysozyme activity were analyzed to evaluate the effects on the immune responses by the waterborne Zn

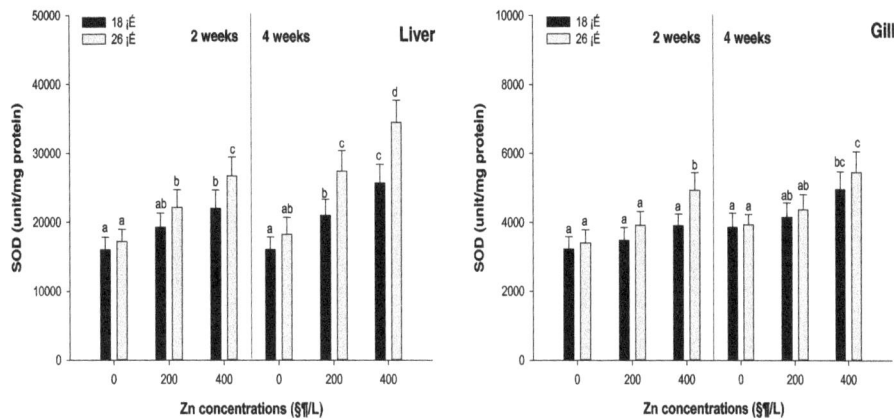

Fig. 1 Changes of SOD activity in the liver and gill of black sea bream, *A. schlegelii*, exposed to two different concentrations. *Vertical bar* denotes a standard error. *Values with different superscript* are significantly different (*P* < 0.05) as determined by Tukey's multiple range test

exposure depending on water temperature. Plasma and kidney phagocytosis activity of the *A. schlegelii* is shown in Fig. 4. The phagocytosis activity in the plasma was significantly increased in 400 µg/L at 18 and 26 °C after 2 weeks. After 4 weeks, the phagocytosis activity was substantially increased in 400 µg/L at 18 and 200, and 400 µg/L at 26 °C. The phagocytosis activity in the kidney was significantly increased in 400 µg/L at 18 and 200, and 400 µg/L at 26 °C after 2 weeks. After 4 weeks, a notable increase in the phagocytosis activity was observed in 400 µg/L at 18 and 26 °C.

Plasma and kidney lysozyme of the *A. schlegelii* is demonstrated in Fig. 5. The lysozyme activity in the plasma was significantly increased in 400 µg/L at 26 °C after 2 weeks. But, there was no notable change at 18 °C after 2 weeks. After 4 weeks, a significant increase in the lysozyme activity was observed in 400 µg/L at 18 and 26 °C. The lysozyme activity in the kidney was also significantly increased in 400 µg/L at 18 and 26 °C after 2 and 4 weeks.

Discussion

Oxygen is an essential component of various metabolic processes in aerobic organisms. However, organisms that are reliant on oxygen must also resist its toxicity, as concentrations of ROS cause substantial damage to cell structures (Ahmad et al. 2004). Similarly, high concentrations of metal exposure cause redox reactions, free radical production, and ROS in fish tissues (Brucka-Jastrzebska 2010). Liver and gill tissues are generally used to assess antioxidant responses for oxidative stress (Kim and Kang 2016b; Kim et al. 2017; Kim and Kang 2017).

Several antioxidant responses are observed in fish, for example, superoxide dismutase (SOD), which catalyzes the transformation of superoxide anion to hydrogen peroxide (H_2O_2) (Ozturk-Urek et al. 2001). In the present study, exposure to waterborne Zn resulted in significant increases in SOD activities in the liver and gill of *A. schlegelii*. Farombi et al. (2007) also reported significant increases in the SOD activity in African cat fish, *Clarias gariepinus*, exposed to metals and a metalloid, including

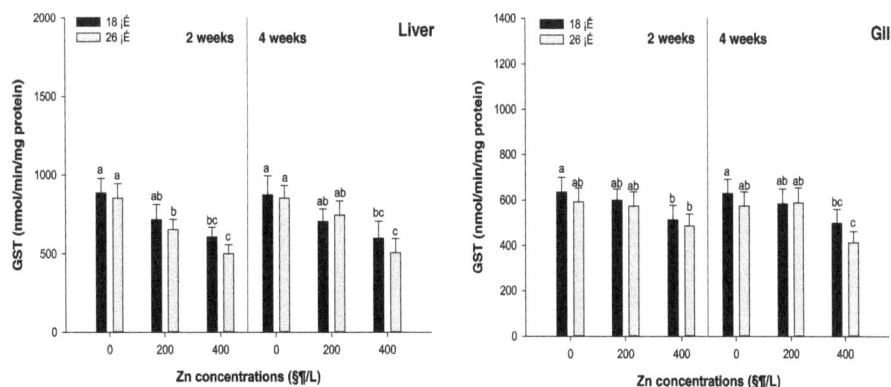

Fig. 2 Changes of GST activity in the liver and gill of black sea bream, *A. schlegelii*, exposed to two different concentrations. *Vertical bar* denotes a standard error. *Values with different superscript* are significantly different (*P* < 0.05) as determined by Tukey's multiple range test

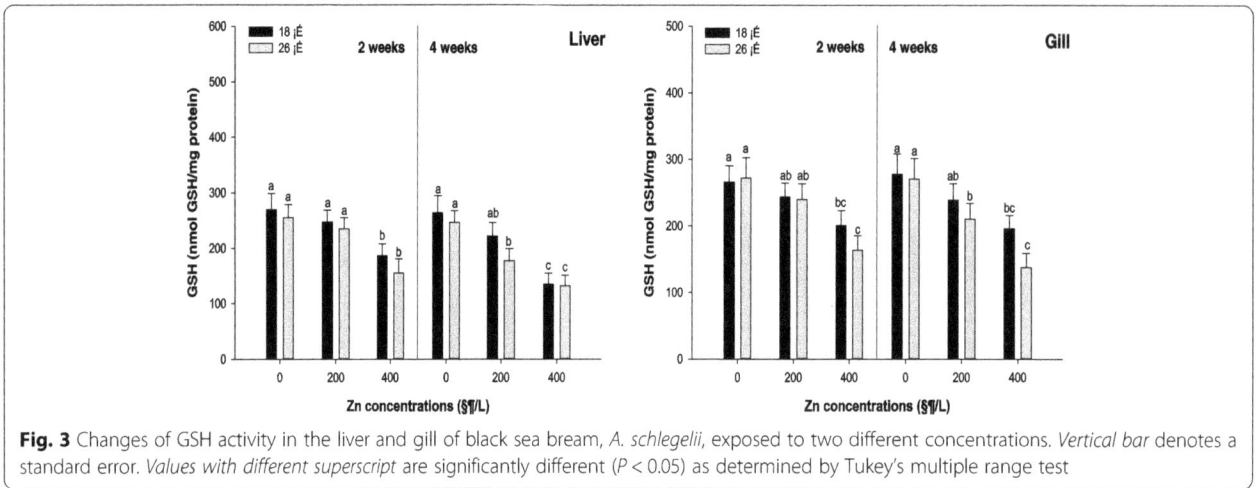

Fig. 3 Changes of GSH activity in the liver and gill of black sea bream, *A. schlegelii*, exposed to two different concentrations. *Vertical bar* denotes a standard error. *Values with different superscript* are significantly different ($P < 0.05$) as determined by Tukey's multiple range test

zinc, copper, cadmium, arsenic, and lead. Accumulation of metal may generate superoxide anions, which activates SOD to scavenge superoxide radicals. Glutathione-S transferase (GST) catalyzes the conjugation of glutathione (GSH) to various electrophiles and functions as a critical defense mechanism against ROS and xenobiotics (White et al. 2003). In this study, exposure to waterborne Zn significantly decreased GST activity in *A. schlegelii*. Significant decreases in the GST activity have also been reported in Nile tilapia, *Orechromis niloticus*, exposed to copper, with the removal of ROS by other enzymes in the antioxidant system possibly compensating for GST (Kanak et al. 2014). GSH, a thiol-containing peptide associated with cellular defense against the toxic effects of xenobiotics, such as metals, is a known substrate for GST activity (Lange et al. 2002). Pandey et al. (2008) reported a significant decrease in GSH levels in spotted snakehead, *Channa punctate*, that were exposed to multiple metals. Decreased GSH concentrations were related to decreases in GSH-dependent enzymes, such

as GST, glutathione reductase (GR), and glutathione peroxidase (GPx). Sanchez et al. (2005) suggested that GSH concentrations are reduced by a cellular response that chelates and detoxifies metals, protecting cells from metal exposure. Loro et al. (2012) reported that exposing killifish, *Fundulus heteroclitus*, to waterborne Zn induced oxidative stress and changes in antioxidant enzymes; the release of Zn ions triggered the increased expression of genes coding for antioxidant enzymes. Similar to previous studies, exposure to waterborne Zn induced significant changes in antioxidant responses in *A. schlegelii* in this study. Based on these results, exposure to waterborne Zn should manifest in the experimental fish as oxidative stress.

Temperature is an important factor affecting oxidative stress, and a higher temperature can amplify oxidative stress (Lushchak and Bagnyukova 2006). Kim et al. (2007) reported that thermal stress, combined with metal exposure, induced significant increases in SOD activity in the disk abalone, *Haliotis discus discus*. Similarly, thermal stress

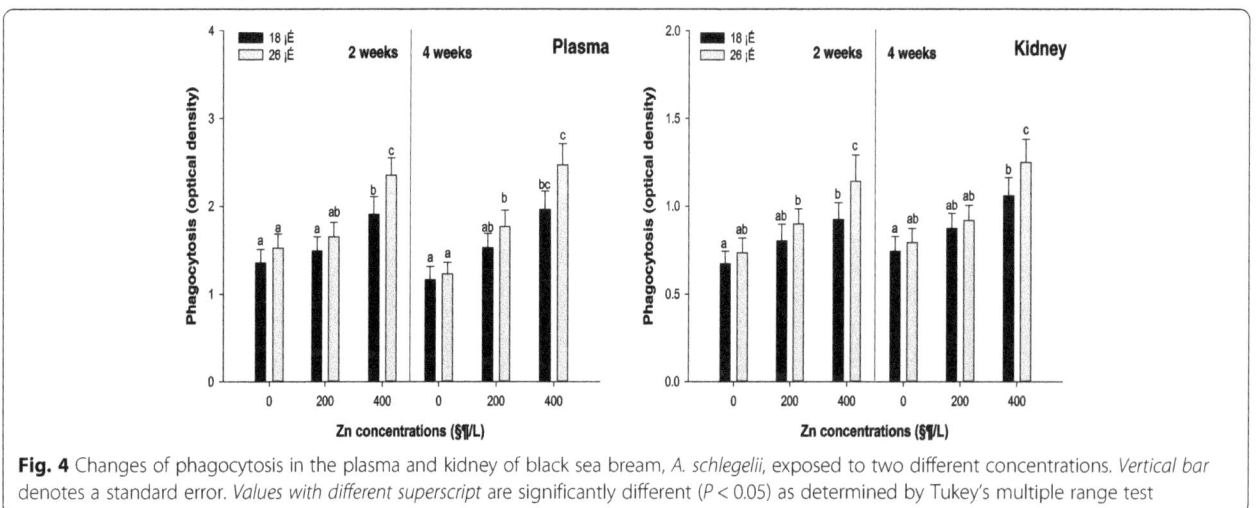

Fig. 4 Changes of phagocytosis in the plasma and kidney of black sea bream, *A. schlegelii*, exposed to two different concentrations. *Vertical bar* denotes a standard error. *Values with different superscript* are significantly different ($P < 0.05$) as determined by Tukey's multiple range test

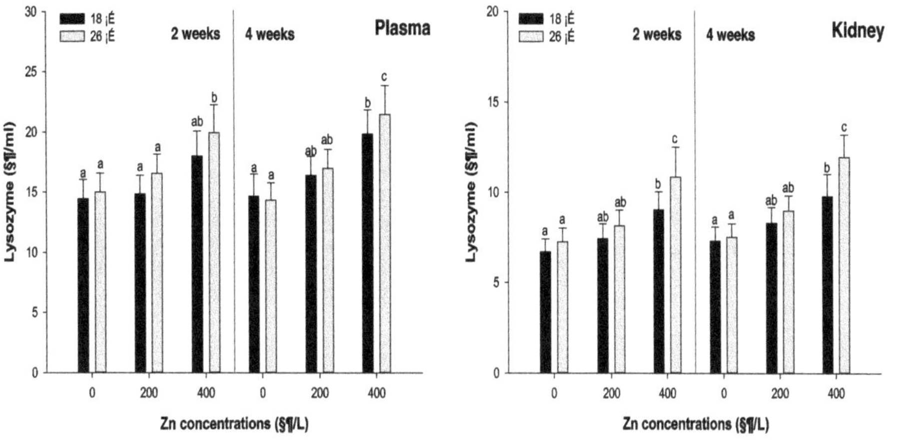

Fig. 5 Changes of lysozyme activity in the plasma and kidney of black sea bream, *A. schlegelii*, exposed to two different concentrations. *Vertical bar* denotes a standard error. *Values with different superscript* are significantly different (P < 0.05) as determined by Tukey's multiple range test

significantly decreased GST in *Channa punctata*, as the elevated temperature decreased GSH concentrations (Kaur et al. 2005). Moreover, in *O. niloticus*, waterborne arsenic significantly decreased gill GSH, and this effect was more pronounced at a higher temperature (Min et al. 2014).

Several studies have indicated that metal exposure affects various parameters in the host immune system, increasing susceptibility to infection and allergy (Bernier et al. 1995). In addition, Arunkumar et al. (2000) suggested that metal exposure induces immune responses in fish, either directly, by binding the tertiary structures of biologically active molecules, or indirectly, by acting as stressors that influence corticosteroid concentrations. Kidney and plasma function in immune systems, and the samples are generally used to assess immune responses(-Kim and Kang 2015; 2016c; 2016d; 2016e).

Phagocytosis is an important immune response wherein pathogenic particles are engulfed by intracellular vacuoles, and, removed. Therefore, it is a critical immunological parameter for evaluating the health status and immunity of fish exposed to toxicants (Risjani et al. 2014; Nagasawa et al. 2015). In this study, phagocytosis activity in *A. schlegelii* significantly increased with exposure to waterborne Zn. Pillet et al. (2000) also reported substantial increases in phagocytic activity in harbor seals, *Phoca vitulina*, and grey seals, *Halichoerus grypus*, exposed to Zn. Of the various non-specific immune responses, lysozyme is a key component of the innate immune response in fish and is stimulated by exposure to aquatic toxins (Bols et al. 2001). In this study, exposure to waterborne Zn caused a significant increase in lysozyme activity in *A. schlegelii*. Sanchez-Dardon et al. (1999) also reported notable elevations in the serum lysozyme in rainbow trout, *Oncorhynchus mykiss*, exposed to Zn, cadmium, mercury, and metal mixtures. Celik et al. 2012 reported increases in both phagocytic and lysozyme activity in Mozambique tilapia,

Oreochromis mossambicus, exposed to Zn. Given that lysozyme is a reliable parameter for monitoring the influence of environmental changes in innate immunity in fish (Bols et al. 2001), our results suggest that exposure to waterborne Zn can be considered an environmental stressor for *A. schlegelii*.

In the present study, a combination of high temperature (26 °C) and Zn concentration significantly affected the immune responses of *A. schlegelii*. As fish are ectothermic, their physiology and immune function is directly affected by water temperature (Morvan et al. 1998). Parry and Pipe (2004) also reported a significant increase in phagocytosis in the blue mussel, *Mytilus edulis*, at a higher temperature. Furthermore, high water temperatures were associated with increased lysozyme activity in Pacific abalone, *Haliotis discus hannai*, exposed to nickel.

Conclusions

In conclusion, exposure to waterborne Zn significantly affected antioxidant responses (SOD, GST, and GSH) in *A. schlegelii*. Non-specific immune responses such as phagocytosis and lysozyme activity were also substantially influenced by exposure to waterborne Zn. Rising water temperatures from global warming may exacerbate the seasonal increases in water temperature in Korea, and changes in water temperature are an important aspect of waterborne Zn toxicity in these coastal waters. The results of the present study indicate that exposure to waterborne Zn affects fish such as *A. schlegelii*, and that water temperature is a critical factor in the toxicity of waterborne Zn.

Abbreviations
GSH: Glutathione; GST: Glutathione S-transferase; SOD: Superoxide dismutase; Zn: Zinc

Acknowledgements
This work was supported by a Research Grant of Pukyong National University (year 2016).

Authors' contributions
HJ and KW carried out the environmental toxicity studies and manuscript writing. JH participated in the design of the study and data analysis. JC participated in its design and coordination and helped to draft the manuscript. All authors read and approved the final manuscript.

Competing interests
The authors declare that they have no competing interests.

Disclosure
The dataset(s) supporting the conclusions of this article is not included in the article.

Author details
[1]West Sea Fisheries Research Institute, National Institute of Fisheries Science, Incheon 22383, Korea. [2]Department of Aquatic Life Medicine, Pukyong National University, Busan 48513, Korea.

References
Ahmad I, Pacheco M, Santos MA. Enzymatic and nonenzymatic antioxidants as an adaptation to phagocyte-induced damage in *Anguilla anguilla* L. following in situ harbor water exposure. Ecotoxicol Environ Saf. 2004;57:290–302.

Anderson DP, Siwicki AK. Simplified assays for measuring nonspecific defense mechanisms in fish. In: Fish Health Section. Am. Fisheries Soc. Meeting, Seattle, WA. 1994. p. 26.

Arunkumar RI, Rajasekaran P, Michael RD. Differential effect of chromium compounds on the immune response of the African mouth breeder *Oerochromis mossambicus* (Peters). Fish Shellfish Immunol. 2000;10:667–76.

Bernier J, Brousseau P, Krzystyniak K, Tryphonas H, Fournier M. Immunotoxicity of heavy metals in relation to Greak Lakes. Environ Health Perspect. 1995;103:23–34.

Beutler E. A manual of biochemical methods. In: Beuter E, editor. Red cell metabolism. Philadelphia: Greene and Straton; 1984. p. 72–136.

Bols NC, Brubacher JL, Ganassin RC, Lee LEJ. Ecotoxicology and innate immunity in fish. Dev Comp Immunol. 2001;25:853–73.

Brucka-Jastrzebska E. The effect of aquatic cadmium and lead pollution on lipid peroxidation and superoxide dismutase activity in freshwater fish. Pol J Environ Stud. 2010;19:1139–50.

Bury NR, Walker PA, Glover CN. Nutritive metal uptake in teleost fish. J Exp Biol. 2003;206:11–23.

Celik ES, Kaya H, Yilmaz S, Akbulut M, Tulgar A. Effects of zinc exposure on the accumulation, haematology and immunology of Mozambique tilapia, *Oreochromis mossambicus*. Afr J Biotechnol. 2012;12:744–53.

Coteur G, Danis B, Dubois P. Echinoderm reactive oxygen species (ROS) production measured by peroxidase, luminol-enhanced chemiluminescence (PLCL) as an immunotoxicological tool. Echinodermata. 2005;39:71–83.

Ellis AE. Lysozyme assay. In: Stolen JS, Fletcher TC, Anderson DP, Roberson BS, editors. Techniques in fish immunology. Fair Haven, NJ: SOS Publications; 1990. p. 101–3.

Farombi EO, Adelowo OA, Ajimoko YR. Biomarkers of oxidative stress and heavy metal levels as indicators of environmental pollution in African cat fish (*Clarias gariepinus*) from Nigeria Ogun River. Int J Environ Res Public Health. 2007;4:158–65.

Habig WH, Pabst MJ, Jakoby WB. Glutathione S-transferases the first enzymatic step in mercapturic acid formation. J Biol Chem. 1974;249:7130 9.

Hogstrand C, Balesaria S, Glover CN. Application of genomics and proteomics for study of the integrated response to zinc exposure in a non-model fish species, the rainbow trout. Comp Biochem Physiol B. 2002;133:523–35.

Kanak EG, Dogan Z, Eroglu A, Atli G, Canli M. Effects of fish size on the response of antioxidant systems of *Oreochromis niloticus* following metal exposures. Fish Physiol Biochem. 2014;40:1083–91.

Kaur M, Atif F, Ali M, Rehman H, Raisuddin S. Heat stress-induced alterations of antioxidant in the freshwater fish *Channa punctata* Bloch. J Fish Biol. 2005;67:1653–65.

Kim JH, Kang JC. Oxidative stress, neurotoxicity, and non-specific immune responses in juvenile red sea bream, *Pagrus major*, exposed to different waterborne selenium concentrations. Chemosphere. 2015;135:46–52.

Kim JH, Kang JC. Effects of sub-chronic exposure to lead (Pb) and ascorbic acid in juvenile rockfish: antioxidant responses, MT gene expression, and neurotransmitters. Chemosphere. 2017;171:520–7.

Kim KY, Lee SY, Cho YS, Bang IC, Kim KH, Kim DS, Nam YK. Molecular characterization and mRNA expression during metal exposure and thermal stress of copper/zinc- and manganese- superoxide dismutases in disk abalone, *Haliotis discus discus*. Fish Shellfish Immunol. 2007;23:1043–59.

Kim JH, Kang JC. The toxic effects on the stress and immune responses in juvenile rockfish, *Sebastes schlegelii* exposed to hexavalent chromium. Environ Toxicol Pharmacol. 2016a;43:128-133.

Kim JH, Kang JC. Oxidative stress, neurotoxicity, and metallothionein (MT) gene expression in juvenile rock fish *Sebastes schlegelii* under the different levels of dietary chromium (Cr^{6+}) exposure. Ecotoxicol Environ Saf. 2016b;125:78-84.

Kim JH, Kang JC. The toxic effects on the stress and immune responses in juvenile rockfish, *Sebastes schlegelii* exposed to hexavalent chromium. Environ Toxicol Pharmacol. 2016c;43:128-133.

Kim JH, Kang JC. The immune responses in juvenile rockfish, Sebastes schlegelii for the stress by the exposure to the dietary lead (II). Environ Toxicol Pharmacol. 2016d;46:211-216.

Kim JH, Kang JC. The immune responses and expression of metallothionein (MT) gene and heat shock protein 70 (HSP 70) in juvenile rockfish, *Sebastes schelgelii*, exposed to waterborne arsenic (As^{3+}). Environ Toxicol Pharmacol. 2016e;47:136-141.

Kim JH, Park HJ, Kim KW, Hwang IK, Kim DH, Oh CW, Lee JS, Kang JC. Growth performance, oxidative stress, and non-specific immune responses in juvenile sablefish, Anoplopoma fimbria, by changes of water temperature and salinity. Fish Physiol Biochem. 2017. doi: 10.1007/s10695-017-0382-z.

Lange A, Ausseil O, Segner H. Alterations of tissue glutathione levels and metallothionein mRNA in rainbow trout during single and combined exposure to cadmium and zinc. Comp Biochem Physiol C. 2002;131:231–43.

Loro VL, Jorge MB, Silva KRD, Wood CM. Oxidative stress parameters and antioxidant response to sublethal waterborne zinc in a euryhaline teleost *Fundulus heteroclitus*: protective effects of salinity. Aquat Toxicol. 2012;110–111:187–93.

Lushchak VI. Environmentally induced oxidative stress in aquatic animals. Aquat Toxicol. 2011;101:13–30.

Lushchak VI, Bagnyukova TV. Temperature increase results in oxidative stress in goldfish tissues. 2. Antioxidant and associated enzymes. Comp Biochem Physiol C: Toxicol Pharmacol. 2006;143:36–41.

McGeer JC, Szebedinszky C, McDonald DG, Wood CM. Effects of chronic sublethal exposure to waterborne Cu, Cd or Zn in rainbow trout. 1: Iono-regulatory disturbance and metabolic costs. Aquat Toxicol. 2000;50:231–43.

Min E, Jeong JW, Kang JC. Thermal effects on antioxidant enzymes response in Tilapia, *Oreochromis niloticus* exposed Arsenic. J Fish Pathol. 2014;27:115–25.

Morvan CL, Troutaud D, Deschaux P. Differential effects of temperature on specific and nonspecific immune defences in fish. J Exp Biol. 1998;201:165–8.

Nagasawa T, Somamoto T, Nakao M. Carp thrombocyte phagocytosis requires activation factors secreted from other leukocytes. Dev Comp Immunol. 2015;52:107–11.

Ozturk-Urek R, Bozkaya LA, Tarhan L. The effects of some antioxidant vitamin- and trace element-supplemented diets on activities of SOD, CAT, GSH-Px and LPO levels in chicken tissues. Cell Biochem Funct. 2001;19:125–32.

Pandey S, Parvez S, Ansari RA, Ali M, Kaur M, Hayat F, Raisuddin S. Effects of exposure to multiple trace metals on biochemical, histological and ultrastructural features of gills of a freshwater fish, *Channa punctata* Bloch. Chem Biol Interact. 2008;174:183–92.

Parry HE, Pipe RK. Interactive effects of temperature and copper on immunocompetence and disease susceptibility in mussel (*Mytilus edulis*). Aquat Toxicol. 2004;69:311–25.

Pillet S, Lesage V, Hammil M, Bouquegneau J-M, Fournier M. *In vitro* exposure of seal reripheral blood leukocytes to different metals reveal a sex-dependent effect of zinc on phagocytic activity. Mar Pollut Bull. 2000;40:921–7.

Risjani Y, Yunianta, Couteau J, Minier C. Cellular immune responses and phagocytic activity of fishes exposed to pollution of volcano mud. Mar Environ Res. 2014;96:73e80.

Sanchez W, Palluel O, Meunier L, Coquery M, Porcher JM, Ait-Aissa S. Copper-induced oxidative stress in three-spined stickleback: relationship with hepatic metal levels. Environ Toxicol Pharmacol. 2005;19:177–83.

Sanchez-Dardon J, Voccia I, Hontela A, Chilmonczyk S, Dunier M, Boermans H, Blakley B, Fournier M. Immunomodulation by heavy metals tested individually or in mixtures in rainbow trout (*Oncorhynchus mykiss*) exposed in vivo. Environ Toxicol Chem. 1999;18:1492–7.

Santore RC, Mathew R, Paquin PR, DiToro D. Application of the biotic ligand model to predicting zinc toxicity to rainbow trout, fathead minnow, and Daphnia magna. Comp Biochem Physiol C: Toxicol Pharmacol. 2002;133:271–85.

Saurabh S, Sahoo PK. Lysozyme: an important defence molecule of fish innate immune system. Aquacult Res. 2008;39:223–39.

Senevirathne M, Kim SH, Siriwardhana N, Ha JH, Lee KW, Jeon YJ. Antioxidant potential of ecklonia cavaon reactive oxygen species scavenging, metal chelating, reducing power and lipid peroxidation inhibition. Food Sci Technol Int. 2006;12:27–38.

White CC, Viernes H, Krejsa CM, Botta D, Kavanagh TJ. Fluorescence-based microtiter plate assay for glutamate-cysteine ligase activity. Anal Biochem. 2003;318:175–80.

Wood CM. Toxic responses of the gill. In: Schlenk D, Benson WH, editors. Target Organ Toxicity in Marine and Freshwater Teleosts. London, UK: Taylor and Francis; 2001. p. 1–89.

Xia T, Kovochich M, Liong M, Madler L, Gilbert B, Shi H, Yeh JI, Zink JI, Nel AE. Comparison of the mechanism of toxicity of zinc oxide and cerium oxide nanoparticles based on dissolution and oxidative stress properties. ACS Nano. 2008;2:2121–34.

17

Postmortem changes in physiochemical and sensory properties of red snow crab (*Chionoecetes japonicus*) leg muscle during freeze storage

Joon-Young Jun[1], Min-Jeong Jung[1], Dong-Soo Kim[2], In-Hak Jeong[3] and Byoung-Mok Kim[1*]

Abstract

In order to evaluate the maximal storable period of the raw crab for a non-thermal muscle separation, the quality changes of the leg meat of red snow crab (*Chionoecetes japonicus*) during freeze storage were investigated. Fresh red snow crabs were stored at −20 °C for 7 weeks, and the leg muscle was separated by a no heating separation (NHS) method every week. During the storage, considerable loss of the leg muscle did not occur and microbiological risk was very low. In contrast, discoloration appeared at 2-week storage on around carapace and the leg muscle turned yellow at storage 3-week. In physiochemical parameters, protein and free amino acids gradually decreased with storage time, expected that proteolytic enzymes still activated at −20 °C. At 4-week storage, the sensory acceptance dropped down below point 4 as low as inedible and notable inflection points in pH and acidity were observed. The volatile base nitrogen was low, though a little increase was recorded. These results suggested that the maximal storable period at −20 °C of the raw material was within 2 weeks and it was depended on external factor such as the discoloration. The present study might be referred as basic data for approaches to solve quality loss occurred in non-thermal muscle separation.

Keywords: Red snow crab, *Chionoecetes japonicus*, Postmortem change, Leg muscle, Non-thermal muscle separation

Background

Red snow crab (*Chionoecetes japonicus*) belongings to Malacostraca, has an innate red color and inhabits a cold seawater in depth of 500 to 23,000 m in the East Sea of the Korean coast (Park et al. 2003). This crab is very popular in Korean market because of its unique flavor and taste (fresh, sweet, and umami). Since the red snow crab is caught all year round stably except a close season (July to August) with the total allowable catch (KFA 2015) and low priced, the crab may be a potential resource in the term of food industry. Unfortunately, industrial utilization is low.

In food industry, the red snow crab has been mainly produced and distributed into frozen meats (leg and body) or meat flake through muscle separation process after boiling with water (Kim et al. 2005). For the meat production, the boiling process is necessary as a pre-processing for preservation from the spoilage induced by enzymes and micro-organism, but in which some problems, such as waste water, loss of nutritional components and high energy cost have been raised. However, it is very difficult to separate the muscles from the shells of body and legs intactly without the boiling process (Ahn et al. 2006).

In recent, Kim et al. (2015) suggested a no heating separation (NHS) method for separating leg muscle using a miller machine equipped with multilayered roller, which can separate without the external damage of the leg muscle, and they investigated the effects of storage temperature and thawing condition on the meat quality with the NHS method. Although spoilage rate of fish muscle varies on their species, the postmortem bio-degradation of the red snow crab muscle is obviously occurred even in freeze storage. In other reports, the protein solubility of the muscles from hard and soft shell mud crabs (Scylla serrata) gradually decreased during storage at −20 °C and a notable pH variation was also found (Benjakul and Sutthipan 2009). Matsumoto and Yamanaka (1992) reported that ATP and

* Correspondence: bmkim@kfri.re.kr
[1]Division of Strategic Food Industry Research, Korea Food Research Institute, Seongnam 13539, Republic of Korea
Full list of author information is available at the end of the article

glycogen in the leg muscle from snow crab (*Chionoecetes opilio*) reduced quickly and the volatile base nitrogen reached the initial spoilage level (25 mg%) within 7 days when stored at −1 °C.

The quality loss of seafood is affected by many external factors, such as storage temperature and period, packaging, and rate of freezing and thawing (Srinivasan et al. 1997). For assessment of fish quality, Food and Agriculture Organization (FAO) of the United Nations recommended sensory, chemical, physical, and microbiological tests, in which appearance, odor, nitrogenous compounds, nucleotide catabolites, glyco-metabolites, pH, and spoilage, and pathogenic bacteria are included (Huss 1995). The purpose of this study is to suggest the maximal storable period for muscle separation from red snow crab (*C. japonicus*) during freeze storage through investigation of the postmortem quality changes in the leg muscle during storage at −20 °C with the NHS method. The present study might be referred as basic data for approaches to solve quality loss occurred in non-thermal muscle separation.

Methods

Materials and preparation of sample

About 30 of live male red snow crab (*C. japonicus*) with body weight and total length of 577.8 ± 118.1 g and 73.2 ± 6.4 cm were used in this study, which were purchased from the Jumunjin fishery market (Gangneung, Republic of Korea). The crab was transported in ice to our laboratory immediately, rinsed with a pure water to eliminate debris, and stored at −20 °C in a polystyrene icebox. For leg muscle separation, the no heating separation (NHS)

method of recommended by Kim et al. (2015) was employed. Briefly, the crab was thawed at 0 °C for 20 s, and the leg part was cut (Fig. 1) and passed through between the multilayered roller of miller machine (YMC-103; YongMa Machinery, Daegu, Republic of Korea) to separate the leg muscle.

Appearance and color

The appearance of the crab body and leg muscle collected in different storage period was photographed using a phone camera (Iphone 6S, Apple Inc., Cupertino, CA, USA). The color (Hunter's L, a, and b values) of the leg muscle was measured using a chroma meter (CR-300; Minolta Co. Ltd., Osaka, Japan).

Proximate composition and salinity

Moisture, ash, crude protein, and crude lipid of the leg muscle were measured according to the AOAC method (AOAC 2005). Carbohydrate was calculated as follows: "100% − (moisture% + ash% + crude protein% + crude lipid%)". For salinity determination, the sample was burned at 550 °C for 6 h to eliminate organic compounds and the ash was diluted in deionized water (DW). The salinity of the diluted sample was determined by volumetric titration with $AgNO_3$ using Mohr's method (Feng et al. 2012).

Free amino acid

For free amino acid analysis, 10 g of the sample was mixed with 100 mL of 75% ethanol, homogenized at 11,000 rpm for 30 s and centrifuged at 6000×g for 10 min. The supernatant was collected, and the residue

Fig. 1 Change in appearance of the crab body and leg muscle during 7-week storage at −20 °C

was mixed again with 100 mL of 75% ethanol to extract remain amino acid. After homogenization and centrifugation, the solvent in the collected supernatant was eliminated using a rotary evaporator. The final dried sample was dissolved in DW and filtrated using a 0.45-μm MCE syringe filter unit before analysis. The amino acids were analyzed using a high-speed amino acid analyzer (L-8800; Hitachi High-Technologies Co., Tokyo, Japan), according to the method of Kim et al. (2016).

pH, titratable acidity, and volatile base nitrogen

For pH determination, 5 g of the sample was mixed with 45 mL of DW, homogenized at 11,000 rpm for 30 s and centrifuged at $6000 \times g$ for 10 min. The pH of the supernatant was determined using a pH meter (SevenEasy S20K; Metteler Toledo International Inc., Columbus, OH, USA). After the pH determination, 20 mL of the supernatant was taken and titrated with 0.1 N NaOH until pH 8.3. The acidity was expressed as lactic acid%. The volatile base nitrogen (VBN) was determined by Conway's diffusion method (Choi et al. 2016).

Bacterial cell count

Aerobic and lactic acid bacterial cell in the sample were counted by plate count method using two different media. Ten grams of the sample was mixed with 90 mL of 0.1% sterile peptone buffered water (pH 7.2) and minced using a home blender sterilized with UV irradiation. The minced sample was transferred in a sterile plastic bag and homogenized using a stomacher. The homogenized sample was decimally diluted with 0.1% sterile peptone buffered water (pH 7.2). A 1 mL of the dilute was added onto petri-dish, and plate count agar (Difco; Becton Dickinson, Spark, MD, USA) for aerobic bacteria and lactobacilli MRS agar (Difco; Becton Dckinson) for lactic acid bacteria were poured, respectively. After incubation at 35 °C for 48 h, the viable cells (between 20 and 200 colonies) were counted and expressed as the logarithmic number of colony-forming units per gram sample.

Statistical analysis

A quantitative descriptive method (Lorentzen et al. 2014) with slight modification was conducted to test the freshness of the leg muscle with 20 non-trained panelists (ten men and ten women, age: 20–30). All panelists were asked to give freshness scores for appearance, texture (not eaten), flavor, and overall acceptance using a 7-point scale in which one represented "poorest quality" and seven represented "best quality."

Sensory test

All data except free amino acid were expressed as the mean ± standard deviation (SD) in triplicate. The values were statistically assed by a one-way ANOVA test; a

significant difference ($p < 0.05$) between means identified by least significant difference and Tukey's test using SPSS (IBM, Armonk, NY, USA).

Results and discussion

Appearance and color

Figure 1 shows the change in appearance of the crab body and leg muscle during 7-week storage at −20 °C. The carapace and legs at 0-week storage represented an innate white-red color, while of the carapace at 2-week storage changed to black color and it seemed to be expended to legs at 7-week storage. The leg muscle separated by the NHS method at storage 3-week yellowed compared to that of 0-week storage. From 5-week storage, surface peeling of the leg muscle was occurred and black color appeared from 6-week storage. On the Hunter's L, a, and b values of the leg muscle, L value gradually decreased with storage period except storage 3-week (Table 1). From 4-week storage, consistent decreases were found in a and b values. An enzymatic browning frequently appears in crustacean during storage, which is mainly related to phenol-oxidase (tyrosinase) that is responsible for the black discoloration called melanosis and physiologically important because the enzyme is contributed to natural wound healing in part (Kim et al. 2000).

Proximate composition and salinity

Changes in the proximate composition and salinity of the leg muscle during 7-week storage at −20 °C are listed in Table 2. The leg muscle (0-week storage) was composed of mostly protein (70.1%/dry basis, 12.20%/wet basis), followed by carbohydrate (16.4%/dry basis, 2.40%/wet basis), ash (11.6%/dry basis, 1.20%/wet basis), and lipid (1.8%/dry basis, 0.60%/wet basis). Although there are seasonal and regional variations, proximate composition of various species of crabs collected in the Republic of Korea have been reported in ranges of moisture 81.4–83.0%, protein 13.7–15.2%, lipid 0.5–0.8%, and ash 1.9–2.1% (NFRDI 2009).

By comparison with these values on dry basis, the red snow crab contained comparatively low protein and high carbohydrate than other crab species. The postmortem biochemical changes in fish muscle are very complex, but which is occurred generally with ATP degradation, glycolysis, rigor, tenderization, autolysis, and putrefaction continuously (Gill 2000). The first changes are initiated with endogenous enzymes promoting proteolysis of the muscle protein and lipid hydrolysis (Delbarre-Ladrat et al. 2006). In this study, no loss was occurred in moisture during 7-week storage at −20 °C, but the values varied in ranges of 80.5 to 85.7%. On dry basis, independent changes with storage period were found in ash and carbohydrate, whereas the crude protein gradually decreased with storage period except 5-week storage and the lipid slightly increased after 2-week storage.

Table 1 Change in color of the leg muscle during 7-week storage at −20 °C

Storage time (week)	L value (lightness)		a value (redness)		b value (yellowness)	
0	46.2	±1.8^{ab}	8.5	±1.2^{ab}	4.7	±1.0^{ab}
1	46.4	±2.6^{ab}	8.4	±1.0^{ab}	9.4	±3.0^{a}
2	43.8	±1.5^{bc}	10.4	±0.8^{a}	6.2	±2.9^{ab}
3	51.8	±3.0^{a}	6.5	±1.4^{ab}	7.2	±2.5^{ab}
4	42.9	±2.0^{bc}	8.2	±1.8^{ab}	8.0	±3.8^{a}
5	41.4	±1.2^{bc}	7.9	±3.2^{ab}	5.8	±2.0^{ab}
6	43.2	±2.2^{bc}	6.6	±1.8^{ab}	6.4	±1.4^{ab}
7	37.9	±2.6^{c}	4.6	±1.4^{b}	2.5	±1.7^{b}

Data expressed as the mean ± SD ($n = 3$). Different types of small superscript letters indicate significantly differences ($p < 0.05$)

Freeze storage is an important preservation method, and it slows enzyme activity and micro-organism growth, but the rate of the protein degradation relies on fish species (Srinivasan et al. 1997) and especially in case of crustacean, the proteolysis occurred in ice obviously. The salinities of the leg muscle during 7-week storage were in ranges of 1.6–2.4% (wet basis), and there were not much changes.

Free amino acid
Table 3 shows change in the free amino acid profile. In total, 13 types of composite amino acids were detected, including most essential amino acids except histidine. Amino acids below the detection limit were not indicated. The total content of free amino acid at 0-week storage was 2390.9 mg% (wet basis), which accounted for approximately 20% of the crude protein (11,500.0 mg%, wet basis).

The free amino acid composition of the *C. japonicus* is close to general crabs, which contain taurine, proline, glycine, alanine, and arginine as the major amino acids (NFRDI 2009). During 7-week storage, the total content decreased and consistent decreases were found in glycine, tyrosine, and arginine. In Matsumoto and Yamanaka (1992) report, total content of free amino acid in the leg muscle from *C. opilio* have been determined to 2261 mg% and it was decreased gradually with storage time when stored at 0 °C, but the composition was not similar to that of the *C. japonicus*. Decomposition of proteins and amino acids by enzymes increases NH_3 and amines, also increases VBN content and pH (Xu et al. 2008). Namely, it is considered that the free amino acid contents decreased due to the VBN content and pH were increased by excessive decomposition of amino acids at 7-day storage.

pH, acidity, and volatile base nitrogen
Figure 2 displays the changes in pH, acidity, and volatile base nitrogen (VBN). During 7-week storage, notable inflection points in both of pH and acidity were observed at 4-week storage. In more detail, the initial pH value (7.41) decreased until pH 7.05 at storage 3-week and tended to increase from 4-week storage. In contrast, the acidity increased during storage 3-week and subsequently decreased. pH change in fish occurs during storage or fermentation, which is affected by formation of organic acid and base nitrogen (Jun et al. 2016). In VBN, no considerable change was found and the values were in ranges of 11.9 to 18.3 mg% (wet basis). The increase of the pH in the leg muscle during freeze storage could be strongly associated with the formation of organic acid than VBN.

Miyagawa et al. (1990) studied that the changes in the free amino acid profile of *C. opilio* muscle during storage in ice for 28 days. In their report, the pH decreased until day 3 storage and increased in which term enzymes were

Table 2 Changes in proximate composition and salinity of the leg muscle during 7-week storage at −20 °C

Storage period (week)	Moisture (%, wet basis)	Ash (%, dry/wet)		Lipid (%, dry/wet)		Protein (%, dry/wet)		Carbohydrate (%, dry/wet)		Salinity (%, wet basis)
0	83.6 ± 0.3^{b}	11.6 ± 1.0^{a}	/1.20	1.8 ± 0.7^{bc}	/0.60	70.1 ± 0.7^{a}	/12.20	16.4 ± 2.0^{d}	/2.40	2.1 ± 0.2^{ab}
1	81.8 ± 0.4^{c}	8.3 ± 0.4^{b}	/1.51	0.8 ± 0.3^{c}	/0.15	70.5 ± 2.0^{a}	/12.84	20.4 ± 2.4^{bcd}	/3.72	1.7 ± 0.2^{b}
2	81.7 ± 0.1^{c}	8.0 ± 1.1^{b}	/1.47	0.7 ± 0.2^{c}	/0.12	70.6 ± 1.8^{a}	/12.91	20.7 ± 3.3^{bcd}	/3.79	1.6 ± 0.3^{b}
3	80.7 ± 0.2^{d}	8.0 ± 1.2^{b}	/1.54	1.0 ± 0.4^{c}	/0.20	69.4 ± 1.2^{a}	/13.39	21.6 ± 2.7^{bcd}	/4.16	2.0 ± 0.2^{ab}
4	81.5 ± 0.1^{c}	9.0 ± 0.2^{b}	/1.66	1.4 ± 0.1^{bc}	/0.26	66.5 ± 0.4^{a}	/12.29	23.2 ± 1.7^{abc}	/4.29	2.0 ± 0.3^{ab}
5	80.5 ± 0.4^{d}	7.9 ± 0.2^{b}	/1.55	1.5 ± 1.0^{bc}	/0.29	72.3 ± 2.3^{a}	/14.12	18.3 ± 1.5^{cd}	/3.58	2.0 ± 0.2^{ab}
6	85.1 ± 0.0^{a}	13.3 ± 0.6^{a}	/1.98	2.7 ± 0.2^{ab}	/0.40	55.9 ± 3.8^{b}	/8.31	27.5 ± 2.6^{a}	/4.16	2.1 ± 0.3^{ab}
7	84.8 ± 0.2^{a}	12.8 ± 0.1^{a}	/1.94	3.8 ± 0.1^{a}	/0.58	57.5 ± 3.7^{b}	/8.74	25.9 ± 2.4^{ab}	/3.94	2.4 ± 0.2^{a}

Data expressed as the mean ± SD ($n = 3$). Different types of small superscript letters indicate significantly differences ($p < 0.05$)

Table 3 Free amino acid profile of the leg muscle during 7-week storage at −20 °C

Amino acid	Storage period (week)				
	0	1	3	5	7
Nutritionally non-essential amino acid (mg 100 g^{-1}, wet base)					
Alanine	270.9	264.5	278.4	264.3	176.0
Aspartic acid	1.2	3.9	4.8	4.2	5.3
Glutamic acid	26.0	21.2	10.7	26.9	26.6
Glycine	657.0	582.8	562.8	450.1	359.3
Tyrosine	59.1	20.9	31.7	19.5	20.0
Sum	1014.2	893.3	888.4	765.0	567.2
Nutritionally essential amino acid (mg 100 g^{-1}, wet base)					
Threonine	1.3	7.0	18.3	5.1	21.5
Valine	58.1	46.2	59.4	56.2	53.2
Methionine	60.5	33.4	43.1	59.6	47.9
Isoleucine	53.3	37.3	36.8	50.6	39.1
Leucine	68.9	52.3	44.4	65.4	72.8
Phenylalanine	61.8	39.1	51.0	42.0	36.6
Lysine	91.7	83.0	93.2	121.1	71.7
Arginine	672.6	609.3	293.3	387.2	231.2
Sum	1068.2	907.6	639.5	787.2	573.7
Other amino acid (mg 100 g^{-1}, wet base)					
Phosphoserine	3.8	8.8	4.5	6.8	5.3
Taurine	210.0	172.7	192.9	182.6	135.0
Sarcosine	75.1	58.7	68.7	47.2	53.8
Ammonia	19.7	12.2	7.6	9.8	12.7
Sum	308.6	252.4	273.7	246.4	206.8
Total	2391.0	2053.3	1801.6	1798.6	1347.7

isolated from the leg muscle and of arginase was dominant. This might explain well the noticeable loss of arginine from storage 3-week in the present study. In a previous study for hard and soft mud crabs, the pH in the muscles decreased during 12-week storage at −20 °C, but according to the part of the muscles, a constant decrease was observed in claw muscle, while the pH in lump muscle temporary increased within 1- and 2-week storage (Benjakul and Sutthipan 2009).

Aerobic and lactic acid bacteria
During the entire storage at −20 °C, the both aerobic and lactic acid bacteria were detected at low levels with the maximal cell counts of 2.5 log CFU g^{-1} wet basis). After fish dies, micro-organism might proliferate freely, but microbial contamination or the population in muscle much less than intestine and grows slowly before tenderization (Delbarre-Ladrat et al. 2006). In addition, the growth of general spoilage bacteria is inhibited in freezing condition (Srinivasan et al. 1997). Although there was not statistical significance ($P < 0.05$), the lactic acid bacteria slightly increased between 1- and storage 3-week. These data matched with the increase of the acidity.

Sensory test
Appearance, flavor, and textural quality might be very important when consumer accept. Table 4 summarizes the change in sensory acceptability of the leg muscle during 7-week storage. The texture was measured by finger pressing, and the overall acceptance indicates a consumption possibility with in ranges of point 4 to 7. During the first 2-week storage, there were not changed in appearance, texture, flavor, and overall acceptance, but those decreased at storage 3-week. At 4-week storage, color, flavor, and overall acceptance was below 4 point, indicated that could not consume.

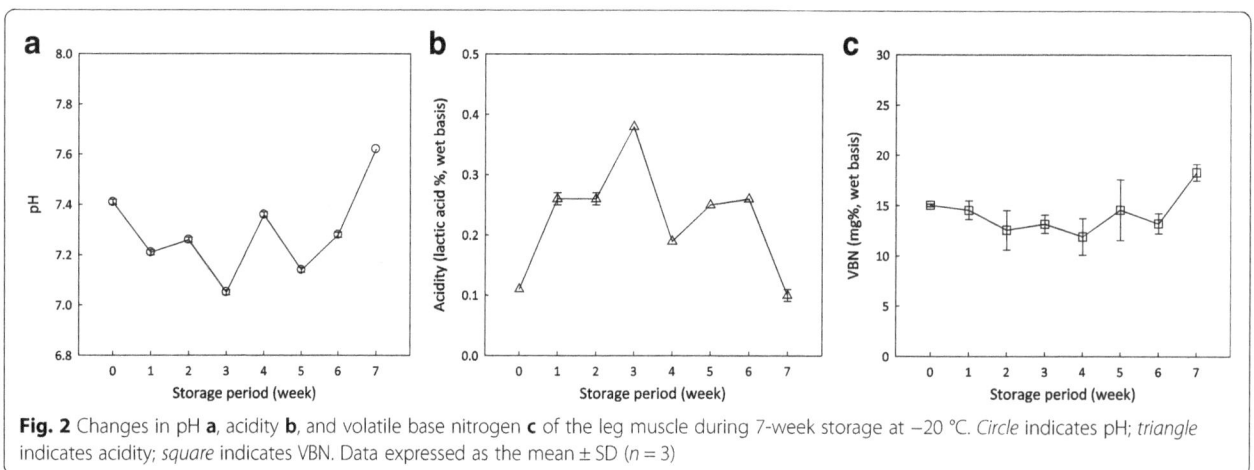

Fig. 2 Changes in pH **a**, acidity **b**, and volatile base nitrogen **c** of the leg muscle during 7-week storage at −20 °C. *Circle* indicates pH; *triangle* indicates acidity; *square* indicates VBN. Data expressed as the mean ± SD (n = 3)

Table 4 Change in sensory acceptability of the leg muscle during 7-week storage at −20 °C

Storage period (week)	Appearance	Texture	Flavor	Overall acceptance
0	6.5 ±0.6[a]	6.8 ±0.7[a]	6.5 ±0.6[a]	6.5 ±0.6[a]
1	6.4 ±1.1[a]	6.5 ±0.6[a]	6.3 ±0.5[a]	6.5 ±0.3[a]
2	6.5 ±0.9[a]	6.6 ±0.8[a]	6.3 ±0.9[a]	6.2 ±0.3[a]
3	4.4 ±1.1[ab]	4.8 ±0.2[ab]	4.3 ±0.2[ab]	4.7 ±1.1[ab]
4	3.4 ±1.1[bc]	4.3 ±0.6[b]	3.5 ±1.0[bc]	3.3 ±1.0[bc]
5	2.3 ±0.5[bc]	2.8 ±0.1[bc]	3.0 ±1.4[bc]	2.9 ±0.9[bc]
6	2.4 ±0.8[bc]	2.8 ±0.4[bc]	2.4 ±0.5[bc]	1.8 ±0.5[c]
7	1.3 ±0.5[c]	1.8 ±1.5[c]	1.8 ±1.0[c]	1.5 ±0.7[c]

Data expressed as the mean ± SD ($n = 3$). Different types of small superscript letters indicate significantly differences ($p < 0.05$)

From the results, the maximal storable period at −20 °C for muscle separation from the C. japonicus was suggested within 2 weeks and it was depended on the external factor such as discoloration than pH, VBN, and bacterial count. These results might be referred as basic data for approaches to solve quality loss occurred in non-thermal muscle separation.

Conclusions

In this study, the quality changes of the leg meat of red snow crab (C. japonicus) during freeze storage were investigated. During the storage, considerable loss of the leg muscle did not occur and microbiological risk was very low. In contrast, discoloration appeared at 2-week storage on around carapace and the leg muscle turned yellow at storage 3-week. In physiochemical parameters, protein and free amino acids gradually decreased with storage time, expected that proteolytic enzymes still activated at −20 °C. At 4-week storage, the sensory acceptance dropped down below point 4 as low as inedible and notable inflection points in pH and acidity were observed. The volatile base nitrogen was low, though a little increase was recorded. These results suggested that the maximal storable period at −20 °C of the raw material was within 2 weeks and it was depended on external factor such as the discoloration. The present study might be referred as basic data for approaches to solve quality loss occurred in non-thermal muscle separation.

Abbreviations
C. japonicus: Chionoecetes japonicus; C. opilio: Chionoecetes opilio; NHS: No heating separation

Authors' contributions
JYJ contributed to conduct the research and prepare the draft manuscript. MJJ contributed to conduct the physiochemical and sensory experiments. DSK and IHJ contributed to design the study and conduct the experiments. BMK contributed to monitor the experiments and finalize the manuscript. All authors read and approved the final manuscript.

Acknowledgements
This study was supported, in part, by grants from the KIMST (Korea Institute of Marine Science & Technology Promotion), Fishery Commercialization Technology Development Program (GA142600-03), and we thank the institute for the support.

Funding
Not applicable.

Competing interests
The authors declare that they have no competing interests.

Author details
[1]Division of Strategic Food Industry Research, Korea Food Research Institute, Seongnam 13539, Republic of Korea. [2]Jeonbuk Institute for Bioindustry, Jeonju 54810, Republic of Korea. [3]Department of Marine Food Science & Technology, Gangneung-Wonju National University, Gangneung 25457, Republic of Korea.

References
Ahn JS, Kim H, Cho WJ, Jeong EJ, Lee HY, Cha YJ. Characteristics of concentrated red snow crab Chionoecetes japonicus cooker effluent for making a natural crab-like flavorant. J Kor Fish Soc. 2006;39:431–6.
AOAC. Official methods of analysis, 18th ed. Methods 925.45, 923.03, 976.05, 991. 36. Gaithersburg: AOAC International Publishing; 2005.
Benjakul S, Sutthipan N. Muscle changes in hard and soft shell crabs during frozen storage. LWT-Food Sci Technol. 2009;42:723–9.
Choi YJ, Jang MS, Lee MA. Physicochemical changes in kimchi containing skate (Raja kenojei) pretreated with organic acids during fermentation. Food Sci Biotechnol. 2016;25:1369–77.
Delbarre-Ladrat C, Chéret R, Taylor R, Verrez-Bagnis V. Trends in postmortem aging in fish: understanding of proteolysis and disorganization of the myofibrillar structure. Crit Rev Food Sci Nutr. 2006;46:409–21.
Feng J, Zhan XB, Zheng ZY, Wang D, Zhang LM, Lin CC. A two-step inoculation of Candida ethchellsii to enhance soy sauce flavor and quality. Int J Food Sci Tech. 2012;47:2072–8.
Gill T. Chapter 2. Nucleotide-degrading enzymes, Seafood enzymes. New York: Marcel Dekker Inc; 2000. p. 37–68.
Huss HH. Quality and quality changes in fresh fish, FAO fisheries technical paper 348. Rome: Food and Agriculture Organization of the United Nations; 1995.
Jun JY, Lim YS, Lee MH, Kim BM, Jeong IH. Changes in the physiochemical quality of sailfin sandfish Arctoscopus japonicus sauces fermented with

soybean koji or rice koji during storage at room temperature. Korean J Fish Aquat Sci. 2016;49:101–8.

KFA. Korean fisheries yearbook. Seoul: Korea Fisheries Association; 2015.

Kim JM, Marshall MR, Wei CI. Chapter 10. Polyphenoloxidase, seafood enzymes. New York: Marcel Dekker Inc; 2000. p. 271–315.

Kim HS, Park CH, Choi SG, Han BW, Kang KT, Shim NH, Oh HS, Kim JS, Heu MS. Food component characteristics of red-tanner crab (*Chionoecets japonicus*) paste as food processing source. J Korean Soc Food Sci Nutr. 2005;34:1077–81.

Kim BM, Jeong JH, Jung MJ, Kim JC, Jun KH, Kim DS, Lee KP, Jun JY, Jeong IH. Effects of freezing storage temperature and thawing time on separation of leg meat from red snow crab *Chionoecetes japonicus*. Korean J Fish Aquat Sci. 2015;48:655–60.

Kim BM, Park JH, Kim DS, Kim YM, Jun JY, Jeong IH, Nam SY, Chi YM. Effects of rice koji inoculated with *Aspergillus luchuensis* on the biochemical and sensory properties of a sailfin sandfish (*Arctoscopus japonicus*) fish sauce. Int J Food Sci Tech. 2016;51:1888–99.

Lorentzen G, Skuland AV, Sone I, Johansen JO, Rotabakk BT. Determination of the shelf life of cluster of the red king crab (*Paralithodes camtschaticus*) during chilled storage. Food Control. 2014;42:207–13.

Matsumoto M, Yamanaka H. Post-mortem biochemical changes in the muscle of tanner crab during storage. Nippon Suisan Gakk. 1992;58:915–20.

Miyagawa M, Tabuchi Y, Yamane K, Matsuda H, Watabe S, Hashimoto K, Katakai R, Otsuka Y. Change in the free amino acid profile of snow crab *Chionoecetes opilio* muscle during storage in ice. Agric Biol Chem. 1990;54:359–64.

NFRDI. Chemical composition of marine products in Korea. 2nd ed. Busan: National fisheries Research and Development Institute; 2009.

Park JH, Min JG, Kim TJ, Kim JH. Composition of food components between red-tanner crab, *Chionoecetes japonicus* and Neodo-Daege, a new species of *Chionoecetes* sp. Caught in the East Sea of Korea. J Korean Fish Soc. 2003;36:62–4.

Srinivasan S, Xiong YL, Blanchard SP. Effects of freezing and thawing methods and storage time on thermal properties of freshwater prawns (*Macrobrachium rosenbergii*). J Sci Food Agric. 1997;75:37–44.

Xu W, Yu G, Xue C, Xue Y, Ren Y. Biochemical changes associated with fast fermentation of squid processing by-products for low salt fish sauce. Food Chem. 2008;107:1597–604.

18

Biochemical changes and drug residues in ascidian *Halocynthia roretzi* after formalin–hydrogen peroxide treatment regimen designed against soft tunic syndrome

Ji-Hoon Lee[1], Ju-Wan Kim[1], Yun-Kyung Shin[2], Kyung-Il Park[1] and Kwan Ha Park[1*]

Abstract

Soft tunic syndrome (STS) is a protozoal disease caused by *Azumiobodo hoyamushi* in the edible ascidian *Halocynthia roretzi*. Previous studies have proven that combined formalin–hydrogen peroxide (H_2O_2) bath is effective in reducing STS progress and mortality. To secure target animal safety for field applications, toxicity of the treatment needs to be evaluated. Healthy ascidians were bathed for 1 week, 1 h a day at various bathing concentrations. Bathing with 5- and 10-fold optimum concentration caused 100% mortality of ascidians, whereas mortality by 0.5- to 2.0-fold solutions was not different from that of control. Of the oxidative damage parameters, MDA levels did not change after 0.5- and 1.0-fold bathing. However, free radical scavenging ability and reducing power were significantly decreased even with the lower-than-optimal 0.5-fold concentration. Glycogen content tended to increase with 1-fold bathing without statistical significance. All changes induced by the 2-fold bathing were completely or partially restored to control levels 48 h post-bathing. Free amino acid analysis revealed a concentration-dependent decline in aspartic acid and cysteine levels. In contrast, alanine and valine levels increased after the 2-fold bath treatment. These data indicate that the currently established effective disinfectant regimen against the parasitic pathogen is generally safe, and the biochemical changes observed are transient, lasting approximately 48 h at most. Low levels of formalin and H_2O_2 were detectable 1 h post-bathing; however, the compounds were completely undetectable after 48 h of bathing. Formalin–H_2O_2 bathing is effective against STS; however, reasonable care is required in the treatment to avoid unwanted toxicity. Drug residues do not present a concern for consumer safety.

Keywords: Ascidians, Formalin–hydrogen peroxide combination, Toxicity, Biochemical parameters, Drug residues, Soft tunic syndrome

Background

Soft tunic syndrome (STS) in the ascidian *Halocynthia roretzi* has markedly reduced production of this edible invertebrate. Official figures indicate a gradual decrease in production to less than a half of the peak yearly production of approximately 22,500 t in 1995 (Kumagai et al. 2010).

The cause of STS is infection with a protozoal parasite *Azumiobodo hoyamushi*, which leads to softening of the rigid cellulose-protein tunic structure (Dache et al. 1992) without affecting the cellulose fiber structure itself (Kimura et al. 2015). Highly active protease enzymes are produced and excreted from *A. hoyamushi* cells (Jang et al. 2012). Although the disease spreads very rapidly, safe and effective measures have not been established to control the spread of STS in farms.

Chemical biocides are the first line of preventive measures against infective organisms in the absence of a practical method to deal with the infection. Different classes of biocidal agents have been tested, and formalin, H_2O_2, bronopol, povidone iodine, and NaOCl were found effective against the causative parasite (Park et al. 2014; Lee et al. 2016; Kumagai et al. 2016). The

* Correspondence: khpark@kunsan.ac.kr
[1]Department of Aquatic Life Medicine, College of Ocean Science and Technology, Kunsan National University, 558 Daehak-ro, Miryong-Dong, Gunsan City, Jeonbuk 54150, Republic of Korea
Full list of author information is available at the end of the article

combination of two anti-infective chemicals, formalin and H_2O_2, was the most promising choice of treatment owing to their synergistic efficacy (Park et al. 2014). Detailed results from treatment trials were published in a previous issue of this journal (Lee et al. 2016), and this paper thus constitutes an important counterpart companion to it.

Bathing ascidians with formalin and H_2O_2 suggested a possible use for the combination in treating infected ascidians; however, the possible side effects, except mortality, have not been examined. Certain side effects are expected, considering the non-selective mechanisms of action of these agents. The degree and recovery from toxicity should be considered in deciding the value of a treatment regimen. To determine the toxicity and safety of the formalin–H_2O_2 treatment, overall mortality, biochemical changes, and drug residue levels were assessed after a 1-week bathing treatment schedule in healthy ascidians.

Methods
Chemicals
Formalin and H_2O_2 were purchased from Sigma (St. Louis, MO, USA), and actual concentrations were assessed before use by HPLC–UV (Soman et al. 2008) and peroxidase–H_2O_2 analysis kit (Cell Biolabs, San Diego, CA), respectively. All other reagents were purchased from Sigma if not specified otherwise.

Test animals: *Halocynthia roretzi*
Healthy ascidians (114.7 ± 21.9 g, 90.9 ± 15.5 cm long) were obtained from a local dealer and acclimated to laboratory conditions for 1 week before commencing the experiment. The absence of *A. hoyamushi* was verified by polymerase chain reaction (Shin et al. 2014) with 10 randomly sampled ascidians. Animals were maintained at 15 °C, the temperature at which STS is most likely to occur and treatment administration is expected. The aquaria used were rectangular PVC tanks (L 1.0 m × W 0.65 m × H 0.3 m water level, 195 L). During experiments, feeding was not performed and water exchange was not needed.

Treatment procedures
Drug treatment was performed at 10:00 a.m. for 1 h in separate drug tanks (20 L acryl baths) kept at 15 °C that were artificially aerated. After drug bathing, the ascidians were returned to normal tanks. This treatment was repeated once daily for a week. Control groups were kept in normal seawater. In toxicity tests, recovery was checked again 48 h after termination of bathing (48 h post-bath group) when tunic signs were detected in the initial assessment.

Assessment of oxidative damage and oxido-reductive potential
To estimate the influence of the treatment on oxido-reductive potential in treated animals, three different parameters were assessed: malondialdehyde (MDA) content, free radical scavenging activity, and reducing power of ascidian soft tissues.

MDA content was assessed by the thiobarbituric acid-reactive substance (TBARS) method (Ohakawa et al. 1979), using 10 g of tissues after homogenization in 20% trichloroacetic acid (TCA) solution. For this, the whole soft tissue of one individual was homogenized and 10 g was taken. 2-Thiobarbituric acid (5 mL, 5 mM; Sigma) was added to an equal volume of tissue homogenates and stored in a refrigerator before absorbance measurements at 530 nm (Optozen POP UV/Vis spectrophotometer, Meacasys, Seoul, Korea). Free radical scavenging activity was determined according to the DPPH scavenging method (Blios 1958). For this, 6 g of soft tissues were homogenized in 100 mL methanol followed by addition of 1,1-diphenyl-2-picrylhydrazyl (DPPH) dissolved in methanol. The mixture was reacted for 10 min at room temperature, and absorbance was measured at 517 nm with a spectrophotometer.

Reducing power was determined (Oyaizu 1986) using the methanol-added homogenates described in the section describing the determination of free radical scavenging activity. Phosphate buffer (1 mL, 200 mM, pH 6.6) was added to 1 mL of tissue homogenate and mixed with 1% potassium ferricyanide solution (1 mL). After incubation at 50 °C for 20 min, 10% TCA was added to stop the reaction. Absorbance was measured at 700 nm using a spectrophotometer. Positive controls contained 10 μM ascorbic acid (vitamin C) instead of ascidian tissues.

Glycogen content
Glycogen content was analyzed according to the anthrone method (Roe and Dailey 1966). Minced soft tissues (2 g) were mixed with 30% KOH solution (1 mL) to hydrolyze glycogen to glucose. Final colored product obtained in reaction with anthrone was diluted with distilled water to appropriate concentrations before spectrophotometric measurements at 620 nm. The standard curve was prepared with D-glucose after identical processing.

Free amino acid composition
Free amino acids were analyzed according to the ninhydrin post-column derivatization method (Friedman 2004) optimized for the Hitachi amino acid analyzer (Hitachi L-8900, Hitachi, Tokyo, Japan). Soft tissues (5 g) were homogenized with distilled water (5 mL) and centrifuged at $3000 \times g$ for 10 min at 3 °C. Next, to 1 mL of supernatant, 5% TCA (0.9 mL) was added to precipitate

proteins, followed by centrifugation at $5000 \times g$ for 10 min at 3 °C. After 10-fold dilution of the supernatant with 0.02 N HCl, the samples were filtered through 0.2-μm membrane filters. The amino acids were separated with an ion exchange column (4.6×60 mm; Hitachi HPLC Packed Column No. 2622 Li type) installed in an amino acid analyzer and UV detector (Hitachi L-8900). The mobile phase was Wako buffer solution (L-8900 PF-1,2,3,4, Wako Pure Chemical Industries, Ltd., Osaka, Japan) run at a flow rate of 0.35 mL/min. Amino acid contents were quantified following a post-column nin-hydrin reaction on-line with 0.3 mL/min ninhydrin solu-tion flow. The separation column was kept at 30–70 °C, and the ninhydrin reaction was carried out at 135 °C. In-dividual amino acids were identified against the standard amino acid mixtures (Wako), with absorbance measured at 570 and 440 nm. The volume of the sample injection was 20 μL.

Analyses of formalin and H_2O_2

The bathing drug solutions and treated tissues were used for analyses of test drug concentrations. The bathing solution was analyzed directly after it was used for bathing without any further treatment. The ascidian tis-sues were homogenized in two volumes of distilled water and centrifuged to obtain supernatants. Formalin content was analyzed by HPLC–UV following complex formation with 2,4-dinitrophenylhydrazine (Soman et al. 2008). The limit of detection sensitivity was approxi-mately 500 nM.

H_2O_2 analysis was performed using the OxiSelect hydrogen peroxide colorimetric assay kit (Cell Biolabs, San Diego, CA) in accordance with the manufacturer's instructions. The detection sensitivity limit was ap-proximately 500 nM. Colored products were detected at 540 nm and quantified by comparison with the standard curve.

Statistical analysis

Data are expressed as mean ± standard deviation (SD). Statistical analyses performed on biochemical parame-ters were conducted by one-way analysis of variance followed by Duncan's multiple comparison tests. Sig-nificance in the difference of means was declared for p values <0.05.

Results

Mortality of ascidians

Figure 1 illustrates the mortality of ascidians following bathing treatment with formalin–H_2O_2 combination. Ascidians were treated for 1 h a day over a week at indi-cated concentrations, and survival was recorded. Since the optimum anti-parasitic treatment under identical conditions was formalin:H_2O_2 = 40:10 ppm (determined

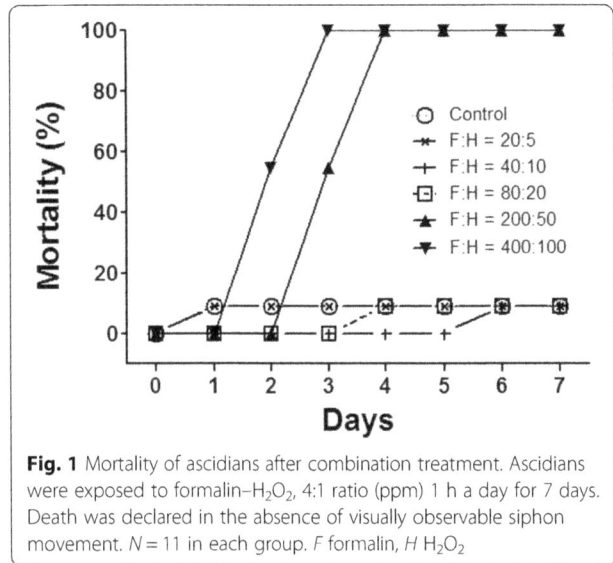

Fig. 1 Mortality of ascidians after combination treatment. Ascidians were exposed to formalin–H_2O_2, 4:1 ratio (ppm) 1 h a day for 7 days. Death was declared in the absence of visually observable siphon movement. $N = 11$ in each group. F formalin, H H_2O_2

in previous studies, see refs. (Park et al. 2014; Lee et al. 2016)), the treatments corresponded to exposure from 0.5- (20:5 ppm) to 10-fold (400:100 ppm) optimal treat-ment. The mortality was concentration-dependent.

Whereas 0.5- to 2.0-fold treatments caused 10% mor-tality, not different from that by the non-treated control, 5- and 10-fold bathing led to 100% ascidian mortality.

Oxidative damage and oxido-reductive potential

Oxidative damage and the effect of treatment on the oxido-reductive potential of ascidian soft tissues are shown in Fig. 2. These parameters were determined 24 h post-bathing after the termination of the 1 week expos-ure schedule. Bathing of ascidians with 2-fold optimum concentration caused a slight but significant elevation of lipid peroxide levels after 1 week exposure for 1 h a day (Fig. 2a). In addition, free radical scavenging activity was diminished by exposure to treatment: significant reduc-tion was noticed even after 0.5-fold exposure (Fig. 2b). Along with the reduction in free radical scavenging, a significant decrease in reducing power was observed in the same tissues (Fig. 2c). Addition of vitamin C to control tissues markedly elevated the reduction poten-tial, as indicated by elevated free radical scavenging capacity and reducing power (Figs. 2b, c). These alter-ations returned to pre-treatment levels after 48-h re-covery in fresh seawater.

Tissue glycogen content

A biphasic pattern in glycogen levels was observed (Fig. 3). The levels increased in a concentration-dependent manner after 0.5- and 1-fold treatments and returned to control levels after 2-fold exposure. Glyco-gen content stayed unchanged 48 h after exposure when kept in fresh seawater.

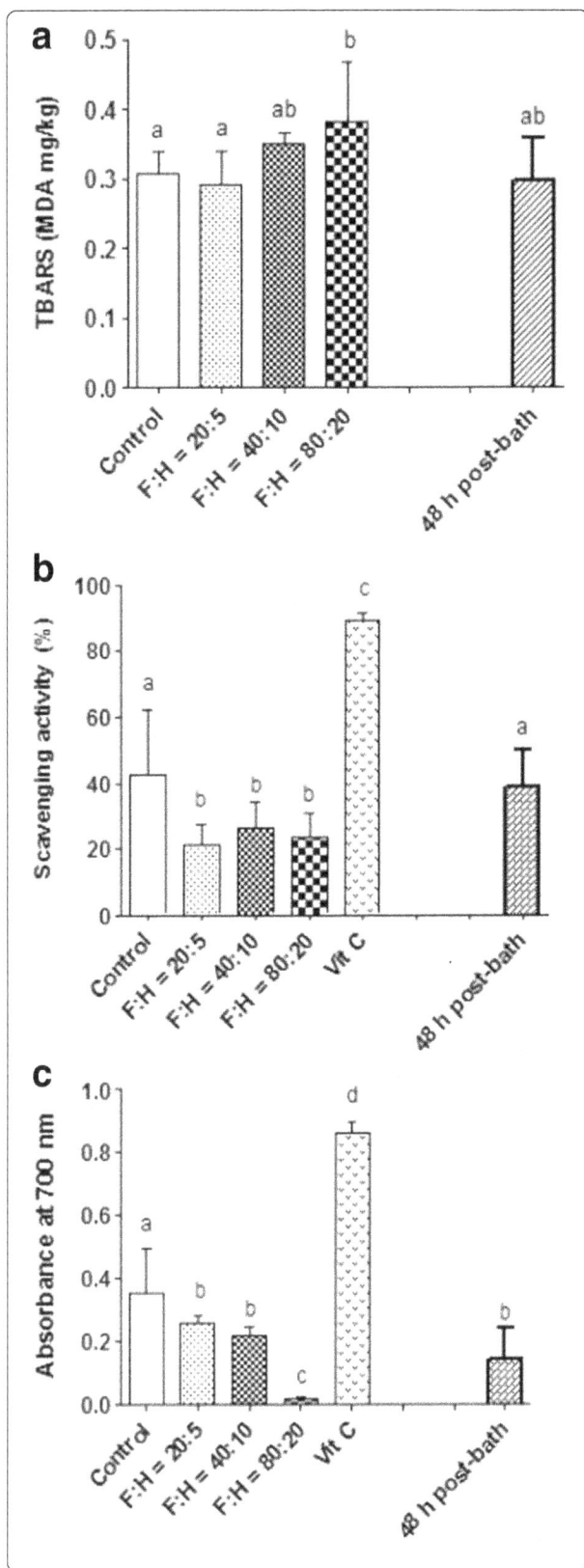

Fig. 2 Levels of oxidative damage in soft tissues of ascidians exposed to the combination treatment. Each damage parameter was assessed 24 h after the termination of the whole 7-day exposure scheme. Recovery (48 h post-bath column) was assessed in 2-fold exposure (F:H = 80:20 ppm) group after an additional 24 h in fresh seawater. a Malondialdehyde levels. b Free radical scavenging capacity. c Reduction power. F formalin, H H_2O_2., Vit C ascorbic acid (10 μM). $N = 7$. Superscripts over bars denote significant statistical difference by Duncan's multiple comparison tests at $p < 0.05$

Free amino acid composition

Free amino acid composition of the edible tissues following drug bathing is shown in Table 1. Taurine was the most prevalent amino acid-like substance, followed by amino acids proline, glutamic acid, glycine, and histidine. The most evident treatment-induced change was a concentration-dependent, significant decrease in aspartic acid concentration. In addition, significant decreases were noted for cysteine levels after 2-fold treatment and proline levels after 0.5-fold exposure. Notably, a significant increase in alanine and valine content was observed after 2-fold exposure.

Formalin and H_2O_2 residue concentrations

Drug residue concentrations in the bathing solution and treated ascidian tissues are shown in Fig. 4. Optimal, 1-fold treatment was used for the residue analysis experiment (40 ppm formalin and 10 ppm H_2O_2). Formalin concentration in the bathing solution was approximately 30 ppm after use for 1 h and declined slowly over the

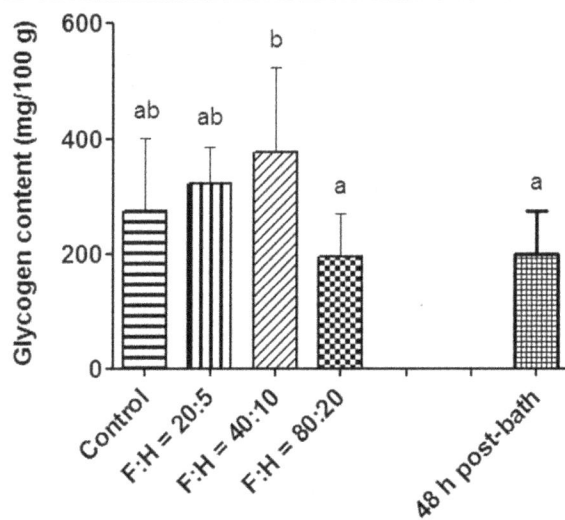

Fig. 3 Glycogen contents of soft tissues in ascidians exposed to combination treatment. Glycogen content was assessed 24 h after the termination of the whole 7-day exposure scheme. Recovery (48 h post-bath column) was assessed in 2-fold exposure (F:H = 80:20 ppm) group after an additional 24 h in fresh seawater. F formalin, H H_2O_2. $N = 7$. Superscripts over bars denote significant statistical difference by Duncan's multiple comparison tests at $p < 0.05$

Table 1 Free amino acid content of edible tissues exposed to the treatment

Amino acids	Amino acid content (mg/100 g muscle)			
	Control ($n = 7$)	$0.5 \times (n = 10)$	$1 \times (n = 10)$	$2 \times (n = 9)$
Taurine (Tau)	1860.36 ± 357.57	1526.89 ± 186.85	1816.14 ± 242.77	1601.10 ± 247.87
Aspartic acid (Asp)	98.34 ± 47.38[a]	75.68 ± 30.71[ab]	57.42 ± 16.54[bc]	37.61 ± 28.71[c]
Threonine (Thr)	72.56 ± 29.51	78.14 ± 24.18	70.97 ± 22.51	87.15 ± 31.46
Serine (Ser)	47.49 ± 13.97	47.24 ± 10.11	57.21 ± 14.36	61.54 ± 20.62
Glutamic acid (Glu)	305.64 ± 54.55	248.89 ± 67.42	309.87 ± 67.07	295.48 ± 70.03
Glycine (Gly)	238.45 ± 57.63	183.81 ± 30.83	207.38 ± 33.59	183.10 ± 48.06
Alanine (Ala)	97.86 ± 22.88[a]	89.12 ± 13.86[a]	98.40 ± 18.56[a]	142.61 ± 45.83[b]
Citrulline (Cit)	0.29 ± 0.76	0.83 ± 2.61	2.26 ± 3.04	0.90 ± 1.45
Valine (Val)	23.85 ± 7.56[a]	27.13 ± 10.06[ab]	22.67 ± 10.05[a]	36.07 ± 13.19[b]
Cysteine (Cys)	5.22 ± 3.00[a]	3.80 ± 3.80[ab]	5.38 ± 2.88[a]	1.45 ± 1.66[b]
Methionine (Met)	15.09 ± 6.93	12.47 ± 5.04	11.71 ± 4.42	12.10 ± 3.42
Isoleucine (Ile)	19.08 ± 6.26	20.28 ± 7.14	17.67 ± 6.44	25.44 ± 8.85
Leucine (Leu)	27.92 ± 7.88	34.92 ± 14.07	26.16 ± 9.72	39.49 ± 11.85
Tyrosine (Tyr)	44.50 ± 21.21	27.68 ± 13.18	35.48 ± 28.24	46.99 ± 20.80
Phenylalanine (Phe)	23.20 ± 5.38	32.51 ± 8.87	25.86 ± 6.54	29.99 ± 7.62
Ornithine (Orn)	2.61 ± 0.80	2.49 ± 0.75	2.58 ± 0.74	2.31 ± 0.70
Lysine (Lys)	34.67 ± 8.98	38.28 ± 12.12	36.35 ± 6.27	33.25 ± 5.63
Histidine (His)	144.42 ± 40.33	120.30 ± 37.19	151.46 ± 49.64	145.57 ± 29.00
Arginine (Arg)	15.26 ± 4.53	19.46 ± 7.67	13.99 ± 4.21	12.66 ± 4.80
Proline (Pro)	688.83 ± 286.60[a]	453.67 ± 139.60[b]	607.11 ± 116.11[ab]	558.56 ± 96.81[ab]

Values with different superscript letters are significantly different ($p < 0.05$)

next 24 h (Fig. 4a). The tissue formalin concentrations in the ascidians were approximately 1/3 of the bath concentration after 1 h and undetectable after 48 h (Fig. 4b). H_2O_2 concentrations exhibited a similar pattern to formalin; the agent was stable in the seawater bath and barely detectable in ascidian tissues after 24 h (Fig. 4c, d). The lowest concentrations of formalin and H_2O_2 were about 0.4 and 0.1 ppm, respectively. These concentrations apply for both ascidian tissues and culture water.

Discussion

This study was performed to assess the toxicity of combined formalin–H_2O_2 treatment in edible ascidians. Formalin–H_2O_2 combination is very effective against the tunic-infecting parasite *A. hoyamushi* (Park et al. 2014; Lee et al. 2016). The treatment concentrations tested here were based on concentrations exerting antiprotozoal effects and used for treating STS. Biochemical responses were monitored to evaluate the toxic effects of the formalin–H_2O_2 combination. Drug residue concentrations were analyzed to correlate toxicity with drug levels in the tissue.

Oxidative damaging effects of the combined agents were evaluated by examining lipid peroxidation, free radical scavenging activity, and reduction potential in edible tissues. Significant changes in these parameters were observed at optimal treatment concentrations of 40 ppm formalin and 10 ppm H_2O_2. H_2O_2 exposure stimulates lipid peroxidation, as H_2O_2 biocidal effects in living organisms are based on production of free radicals (Siddique et al. 2012; Cavaletto et al. 2002). In addition, formaldehyde causes lipid peroxidation (Gulec et al. 2006; Saito et al. 2005) directly and via a secondary mechanism involving the production of reactive oxygen species (Hancock et al. 2001). Although further studies are required, it is reasonable to assume that the combined formalin–H_2O_2 treatment stimulated lipid peroxidation at the 2-fold effective concentration in this marine invertebrate. It is known that reactive oxygen species deplete endogenous reducing biomaterials in cells (Lushchak 2014) and glutathione is the representative reducing agent in marine invertebrates (Conners 1998). Lipid peroxidation is postponed until reducing reserves of the cell are completely exhausted. The observed pronounced decline in free radical scavenging ability and reducing power compared to elevation of lipid peroxidation could indicate that some biochemical changes occur than others.

Major glycogen deposits in ascidians occur in the pyloric gland, which plays a homologous role in the liver

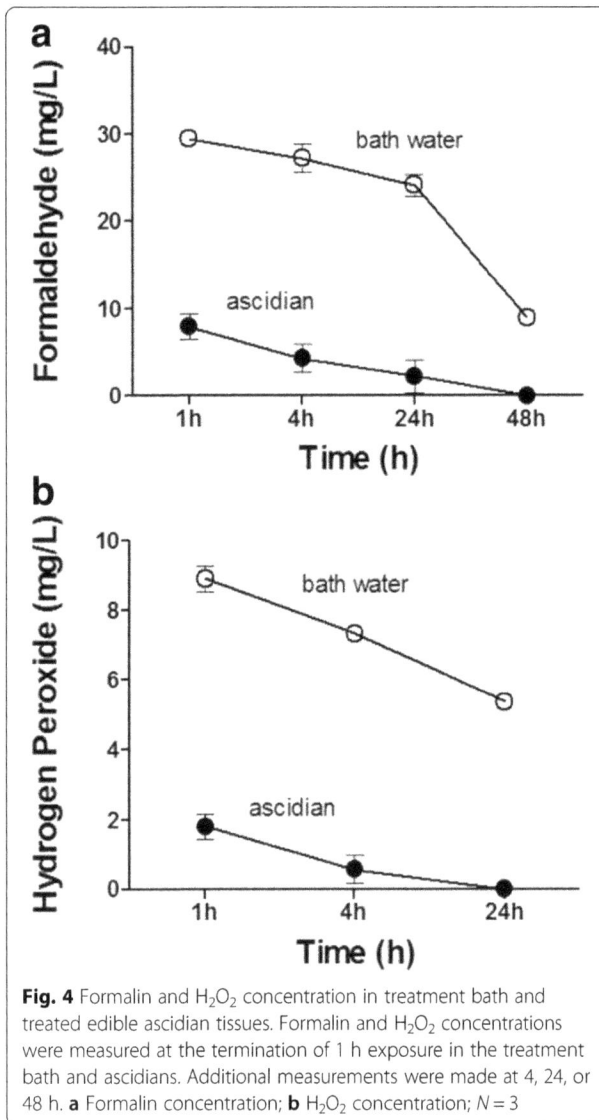

Fig. 4 Formalin and H_2O_2 concentration in treatment bath and treated edible ascidian tissues. Formalin and H_2O_2 concentrations were measured at the termination of 1 h exposure in the treatment bath and ascidians. Additional measurements were made at 4, 24, or 48 h. **a** Formalin concentration; **b** H_2O_2 concentration; $N = 3$

in other animals (Ermak 1977). Glycogen storage in the pyloric gland indicates disturbed metabolic activity (Gaill 1980), and thus, the increasing trend after 1-fold exposure reflects a perturbation in energy balance caused by the treatment. However, interpretation becomes complicated, as the 2-fold exposure did not increase glycogen content. It is known that reduction status induce changes in glycogen contents in mice (Nocito et al. 2015).

Free amino acid content is an indicator of toxic response in various aquatic invertebrate animals (Cook et al. 1972; Hosoi et al. 2003). Changes in free amino acid levels in tissues during stress occur because of altered amino acid utilization for protein synthesis (Kültz 2005). Amino acid changes in ascidians occur seasonally (Watanabe et al. 1983). However, changes in amino acid composition in response to chemical stress have not been studied in *H. roretzi*.

Observed amino acid patterns match the typical characteristics of edible tissues of this species: high content of taurine, proline, glutamic acid, and glycine, as described by Watanabe (Watanabe et al. 1983). Aspartic acid concentration was reduced in a concentration-dependent manner. Although aspartic acid levels were relatively high (taurine > proline > glutamic acid, glycine, histidine > aspartic acid, alanine, threonine>serine), the importance of this amino acid in physiology of ascidians is not known. Cysteine content was diminished after 2-fold exposure. Reduced cysteine content may reflect the changes in reducing potential because cysteine is used as a precursor in the synthesis of glutathione, which reactive compounds like formalin and H_2O_2 may deplete (Poole 2015). Cysteine protects against free radical damage caused by paraquat (Shoji et al. 1992), although the significance of cysteine in ascidians is not known.

A very interesting phenomenon observed in the free amino acid analysis was the elevation of alanine and valine levels. Alanine is important for intracellular osmolality regulation in Pacific oysters, with salinity changes inducing immediate elevation of alanine levels in mantle tissues (Hosoi et al. 2003). However, the importance of alanine and valine in ascidians in relation to stress requires further studies.

Biochemical toxic responses were observed 24 h post-bathing; however, associated residue levels of formalin and H_2O_2 were undetectably low. This finding implies that the exposed animals were recovering and further progression of toxicity is not expected. In addition, rapid elimination of treatment agents is ideal from the aspect of food safety. In contrast to the rapid decline of formalin and H_2O_2 residues in edible ascidian tissues, the compounds are reasonably stable in aquatic media, making daily 1 h bathing treatment possible (Jung et al. 2001; Yamamoto et al. 2011). In addition to their role in the main purpose of this study, which is examining toxic responses to formalin–H_2O_2 bath treatments, biochemical parameters assessed can be used to monitor the impact of these chemicals during treatment.

In view of toxicity form this study, formalin–H_2O_2 bathing sounds promising to disinfect ascidians against STS-causing parasites. The practice will be specifically useful before landing ascidian seedlings in Korean waters.

Conclusions

STS is a highly infectious protozoal disease that has severely affected ascidian industry in Asian countries. Bathing treatment with formalin–H_2O_2 combination solution is an effective method for reducing STS mortality. Bathing treatment with optimal drug concentrations induces a certain recoverable level of biochemical changes. Our results indicate that the two biocidal drugs studied possess inherent toxicity related to their mechanisms of action. However, the treatment toxicity is acceptable as long as excess bathing concentrations are avoided. After treatment, both drugs are eliminated relatively quickly from edible ascidian tissues.

Abbreviations
STS: Soft tunic syndrome; TCA: Trichloracetic acid; MDA: Malondialdehyde

Acknowledgements
We thank the Marine Biology Education Center of the Kunsan National University for providing the ascidian culture facilities during the experiment.

Funding
This work was financially supported by the National Fisheries Research and Development Institute of Korea (NFRDI, RP-2016-AQ-038).

Authors' contributions
JHL carried out the bathing treatments, sample preparations, and necessary chemical and biochemical analyses. KIP performed the tissue histology on the treated ascidian specimens. YKS, KIP, and KHP participated in the writing of the manuscript. All authors read and approved the final version of the manuscript.

Competing interests
The authors declare that they have no competing interests.

Author details
[1]Department of Aquatic Life Medicine, College of Ocean Science and Technology, Kunsan National University, 558 Daehak-ro, Miryong-Dong, Gunsan City, Jeonbuk 54150, Republic of Korea. [2]Southeast Sea Fisheries Institute, National Institute of Fisheries Science, Youngwun-Ri 361, Sanyang-Eup, Tongyeong City, Kyungnam, Republic of Korea.

References
Blios MS. Antioxidant determination by the use of a stable free radical. Nature. 1958;181:1199–200.

Cavaletto N, Ghezzi A, Burlando B, Evangelisti V, Cerrato N, Viarengo A. Effect of hydrogen peroxide on antioxidant enzymes and metallothionein level in the digestive gland of Mytilus galloprovincialis. Comp Biochem Physiol Part C. 2002;131:447–55.

Conners DE. Master's Thesis. Medial University of South Carolina/University of Chaleston-South Carolina, Environmental Studies Program. Charleston, SC, USA. 1998

Cook PA, Gabbott PA, Youngson A. Seasonal changes in the free amino acid composition of the adult barnacle, Balnus balanoides. Comp Biochem Physiol. 1972;42B:409–21.

Dache YB, Revol JF, Gail F, Goffinet G. Characterization and supramolecular architecture of the cellulose-protein fibrils in the tunic of the sea peach (Holocynthia papillosa, Ascidiacea, Urochordata). Biol Cell. 1992;76:87–96.

Ermak TH. Glycogen deposits in the pyloric gland of the ascidian Styela clava (Urochordata). Cell Tissue Res. 1977;176:47–55.

Friedman M. Applications of the ninhydrin reaction for analysis of amino acids, peptides, and proteins to agricultural and biomedical sciences. J Agric Food Chem. 2004;52:385–406.

Gaill F. Glycogen and degeneration in the pyloric gland of Dendrodos grossularia (Ascidiacea, Tunicata). Cell Tissue Res. 1980;208:197–206.

Gulec M, Gurel A, Armutcu F. Vitamin E protects against oxidative damage caused by formaldehyde in the liver and plasma of rats. Braz J Med Biol Res. 2006;34:639–43.

Hancock JT, Desikan R, Neill SJ. Role of reactive oxygen species in cell signaling pathways. Biochem Soc Trans. 2001;29:343–50.

Hosoi M, Kubota S, Toyohara M, Toyohara H, Hayashi I. Effect of salinity change on free amino acid content in Pacific oyster. Fish Sci. 2003;69:395–400.

Jang HB, Kim YK, del Castillo CS, Nho SW, Cha IS, Park SB, Ha MA, Hikima JI, Hong SJ, Aoki T, Jung TS. RNA-seq-based metatrasnscriptomic and microscopic investigation reveals novel metalloproteases of Neobodo sp. as potential virulence factors for soft tunic syndrome in Halocynthia roretzi. PLoS One. 2012;7:e52379.

Jung SH, Kim JW, Jeon IG, Lee YH. Formaldehyde residues in formalin-treated olive flounder (Paralichthys schlegeli), and seawater. Aquaculture. 2001;194:253–62.

Kimura S, Nakayama K, Wada M, Kim EJ, Azumi K, Ojima T, Nozawa A, Kitamura SI, Hirose E. Cellulose is not degraded in the tunic of the edible ascidian Halocynthia roretzi contracting soft tunic syndrome. Dis Aquat Org. 2015;116:143–8.

Kültz D. Molecular and evolutionary basis of the cellular stress response. Ann Rev Physiol. 2005;67:225–7.

Kumagai A, Suto A, Ito H, Tanabe T, Takahashi K, Kamaishi T, Miwa S. Mass mortality of cultured ascidians Halocynthia roretzi associated with softening of the tunic and flagellate-like cells. Dis Aquat Org. 2010;90:223–34.

Kumagai A, Tanabe T, Nawata A, Suto A. Disinfection of fertilized eggs of the edible ascidian Halocynthia roretzi for prevention of soft tunic syndrome. Dis Aquat Org. 2016;118:153–8.

Lee JH, Lee JG, Zeon SR, Park KI, Park KH. Methods to eradicate soft tunic syndrome (STS)-causing protozoa Azumiobodo hoyamushi, the highly infectious parasite from the edible ascidian (Halocynthia roretzi). Fish Aquat Sci. 2016;19:1–6.

Lushchak VI. Free radicals, reactive oxygen species, oxidative stress and its classification. Chem Biol Inter. 2014;224:164–75.

Nocito L, Kleckner AS, Yoo EY, Jones AR, Liesa M, Corkey BE. The extracellular redox state modulates mitochondrial function, gluconeogenesis, and glycogen synthesis in murine hepatocytes. PLoS One. 2015;10:1–17.

Ohakawa H, Ohishi N, Yagi K. Assay for lipid peroxides in animal tissues by thiobarbituric acid reaction. Anal Biochem. 1979;95:351–8.

Oyaizu M. Studies on product of browning reaction prepared from glucose amine. Jpn J Nutr. 1986;7:307–15.

Park KH, Zeon SR, Lee JG, Choi SH, Shin YK, Park KI. In vitro and in vivo efficacy of drugs against the protozoan parasite Azumiobodo hoyamushi that causes soft tunic syndrome in the edible ascidian Halocynthia roretzi (Drasche). J Fish Dis. 2014;37:309–17.

Poole BS. The basics of thiols and cysteines in redox biology and chemistry. Free Rad Biol Med. 2015;80:148–57.

Roe JH, Dailey RE. Determination of glycogen with the anthrone reagent. Anal. Biochem. 1966;15:245–50.

Saito Y, Nishio K, Yoshida Y, Niki E. Cytotoxic effect of formaldehyde with free radicals via increment of cellular reactive oxygen species. Toxicology. 2005;210:235–45.

Shin YK, Nam KW, Park KH, Yoon JM, Park KI. Quantitative assessment of Azumiobodo hoyamushi distribution in the tunic of soft tunic syndrome-affected ascidian Halocynthia roretzi using real-time polymerase chain reaction. Parasites Vectors. 2014;7:539–43.

Shoji K, Yuri M, Toshiya H, Moriyo K, Hideaki S, Tayayuki F. Effect of C-cystine on toxicity of paraquat in mice. Toxicol Lett. 1992;60:75–82.

Siddique YH, Ara G, Afzal M. Estimation of lipid peroxidation induced by hydrogen peroxide in cultured human lymphocytes. Dose-Response. 2012;10:1–10.

Soman A, Qiu Y, Li QC. HPLC-UV method development and validation for the determination of low level formaldehyde in a drug substance. J Chromatogr Sci. 2008;46:461–5.

Watanabe K, Maezawa H, Nakamura H, Konosu S. Seasonal variation of extractive nitrogen and free amino acids in the muscle of the ascidian Halocynthia roretzi. Bull Jpn Soc Sci Fish. 1983;49:1755–8.

Yamamoto A, Toyomura S, Saneyoshi M, Hatai K. Control of fungal infection of salmonid eggs by hydrogen peroxide. Fish Pathol. 2011;36:241–6 (in Japanese).

19

Physical and functional properties of tunicate (*Styela clava*) hydrolysate obtained from pressurized hydrothermal process

Hee-Jeong Lee, Sol-Ji Chae, Periaswamy Sivagnanam Saravana and Byung-Soo Chun[*] ⓘ

Abstract

In this study, marine tunicate *Styela clava* hydrolysate was produced by an environment friendly and green technology, pressurized hot water hydrolysis (PHWH) at different temperatures (125–275 °C) and pressure 50 bar. A wide range of physico-chemical and bio-functional properties such as color, pH, protein content, total carbohydrate content, reducing sugar content, and radical scavenging activities of the produced hydrolysates were evaluated. The appearance (color) of hydrolysates varied depending on the temperature; hydrolysates obtained at 125–150 °C were lighter, whereas at 175 °C gave reddish-yellow, and 225 °C gave dark brown hydrolysates. The L* (lightness), a* (red–green), and b* (yellow–blue) values of the hydrolysates varied between 35.20 and 50.21, −0.28 and 9.59, and 6.45 and 28.82, respectively. The pH values of *S. clava* hydrolysates varied from 6.45 (125 °C) to 8.96 (275 °C) and the values were found to be increased as the temperature was increased. The hydrolysis efficiency of *S. clava* hydrolysate was ranged from 46.05 to 88.67% and the highest value was found at 250 °C. The highest protein, total carbohydrate content, and reducing sugar content of the hydrolysates were found 4.52 mg/g bovine, 11.48 mg/g and 2.77 mg/g at 175, and 200 and 200 °C, respectively. Hydrolysates obtained at lower temperature showed poor radical scavenging activity and the highest DPPH, ABTS, and FRAP activities were obtained 10.25, 14.06, and 10.91 mg trolox equivalent/g hydrolysate (dry matter basis), respectively. Therefore, *S. clava* hydrolysate obtained by PHWH at 225–250 °C and 50 bar is recommended for bio-functional food supplement preparation.

Keywords: *S. clava*, Bio-functional, Hydrolysate, Pressurized hot water hydrolysis

Background

Marine resources are generally considered as health beneficial due to richness in wide range of bio-functional compounds. Recently, researchers investigated on the extraction of many bioactive compounds from various marine animals, including tunicate *Styela clava* (Donia and Hamann 2003). *S. clava* is a marine organism geographically distributed to Northwest Pacific (including Korea), northern Europe, North America, and Australia (Jumeri and Kim, 2011). It is an important mariculture species used as a popular food in Korea (Ko and Jeon 2015). The protein, lipid, carbohydrate and ash content of the flesh tissue of *S. clava* was reported 67.80, 6.54, 16.77, and 7.05% (dry basis), respectively (KO et al. 2012a). Several studies on *S. clava* have pointed out

various bioactivities including anti-oxidative (Lee et al. 2010), anti-inflammatory (Xu et al. 2008), anticancer (Kim et al. 2006), antihypertensive (KO et al. 2012a; KO et al. 2012b), and hepato-protective effects (Xu et al. 2008; Jumeri and Kim, 2011).

Oxidation reaction in living organisms is unavoidable which generates hydrogen peroxide (H_2O_2), singlet oxygen ($^1O_2$), superoxide radicals ($O_{2}^-\bullet$), and hydroxyl radicals (OH•), which are commonly known as reactive oxygen species (ROS) (Wang et al. 2006). Those ROS generated from energy metabolism, stress, exogenous chemicals, or in the food systems are able to oxidize biomolecules of the cell components and cause destructive and irreversible damages (Prasad et al. 2010). The primary target site of the degradation process is the DNA, proteins, cell membranes, and vital cellular constituents which induce fatal physiological disorders including atherosclerosis, muscular dystrophy, rheumatoid arthritis, neurological

* Correspondence: bschun@pknu.ac.kr
Department of Food Science and Technology, Pukyong National University, 45 Yongso-ro, Nam-Gu, Busan 48513, Republic of Korea

Physical and functional properties of tunicate (Styela clava) hydrolysate obtained from pressurized...

145

dysfunctions, cataracts, cancer, and aging (Valko et al. 2004). Imbalance between ROS and ingested antioxidant molecules severely cause oxidative stress (Kang et al. 2017). In taking of antioxidant compounds may neutralize those ROS and guard body system from the problems. Additionally, ROS leads to rancidity, breakdown, and toxicity of functional biomolecules of food components, thus rendering qualitative degradation of foods. Peptides obtained from the breakdown of animals and plant proteins have been found to exhibit various bio-functional activities. Usually peptides are inactive when staying intact with parent proteins (Matsui et al. 2002) and exhibit various bio-functional activities once liberated, depending on their compositional, structural, and sequential properties. Hydrolysis is important for chemical decomposition in which the compounds are spitted into smaller compounds by reacting with water. Hydrolysis of protein under controlled condition generates peptides, and thus the functional properties of a protein is improved (Fujimoto et al. 2012). So, a proper hydrolysis technique at suitable and optimum conditions plays an important role in industrial bio-functional food preparations.

Present hydrolysis methods used in the industries such as chemical (acid, alkali, or catalytic) hydrolysis and enzymatic hydrolysis have several drawbacks. Violent reaction conditions, removal of chemicals from the products, and environmental pollution are the primary disadvantages of chemical hydrolysis. High operation cost and long production cycle make enzymatic hydrolysis inconvenient in industrial application. On the contrary, pressurized hot water extraction can provide a new dimension in hydrolytic reactions. Pressurized hot water extraction is done by boiling water at 100 to 374 °C under pressure (10 to 60 bar) to keep the water in a liquid condition (Saravana et al. 2016a). It is considered as environment friendly green technique which offers high extraction yield (Özel and Göğüş 2014). A number of physical and chemical changes in water at sub-critical conditions, especially in hydrogen bond, dielectric constant, ion product, etc. facilitate reactions of organic compounds and generate many valuable materials (Yoshii et al. 2001; Laria et al. 2004; Tomita and Oshima 2004; Yagasaki et al. 2005). For example, the dielectric constant of water at room temperature is 80, which can be changed to 27 by heating at 250 °C, but the liquid state can be maintained by manipulating pressure (Carr et al. 2011). Hydrolysis in subcritical water is environment friendly technology as it is free of environmental pollution (Cheng et al. 2008). At present, sub-critical water hydrolysis attracted attention for hydrolysis and conversion of biomass to useful compounds (Kruse and Gawlik 2003; Bicker et al. 2005; Uddin et al. 2010). Sub-critical water hydrolysis does not use organic solvents which ensures great advantages as organic solvents used in any process operation must be recycled, deposited, or incinerated resulting in a non-aggressive waste to the environment. Moreover, sub-critical water hydrolysis does not require biomass pretreatment; it is fast, generates lower residue, and presents lesser corrosion than conventional methods (Zhao et al. 2012).

Several studies of the bioactivities of S. clava, have been carried out previously which employed chemical hydrolysis, enzymatic hydrolysis, and solvent extraction methods, but there is no report regarding the hydrolysates of S. clava obtained by pressurized hot water extraction. The main objective of the study was to produce S. clava hydrolysate by pressurized hot water extraction at different temperature and pressure to optimize suitable conditions of hydrolysate regarding physical parameters (color and pH) and bio-functional activities (reducing sugars, antioxidants, and antihypertensive).

Methods

Chemicals and reagents
2,2-azinobis-3ethylbenzothiazoline-6-sulphonic acid (ABTS), 2,2-diphenyl-1-picrylhydrazyl (DPPH), 2,4,6-tripyridyl-s-triazine (TPTZ), 6-hydroxy-2,5,7,8-tetramethylchroman-2-carboxylic acid (Trolox) were purchased from Sigma-Aldrich Chemical Co. (St. Louis, MI, USA). Iron (III) chloride 6-hydrate ($FeCl_3.6H_2O$), sodium acetate ($C_2H_3NaO_2.3H_2O$) acetic acid ($C_2H_4O_2$) were purchased from Merck (Darmstadt, Hessen Germany). All other reagents used in this study were of high-performance liquid chromatography (HPLC) or analytical grade.

Sample collection and preparation
Fresh, solitary Mideodeok (S. clava) was purchased from a local market in Gosung, Gyoungnam Province, Korea. Upon arrival at the laboratory the samples were immediately dissected and rinsed with tap water to remove the contaminants and then the tunics were separated from the muscle and frozen at −40 °C. The frozen samples were freeze dried at −113 °C for about 72 h then freeze-dried sample was milled using mechanical blender and sieved to pass 710 μm sieve, filled in airtight plastic bag and kept in refrigerator at −40 °C until needed for further analysis.

Pressurized hot water hydrolysis (PHWH)
The PHWH was performed in a 200-cm^3 batch reactor made of 276 Hastelloy with temperature control (Fig. 1). Freeze-dried S. clava powder and water (1:30 w/v) was loaded into the reactor. Then the reactor was closed and heated using an electric heater to the required temperature (125–275 °C) and pressures of 50 bar. The temperature and pressure in the reactor were controlled

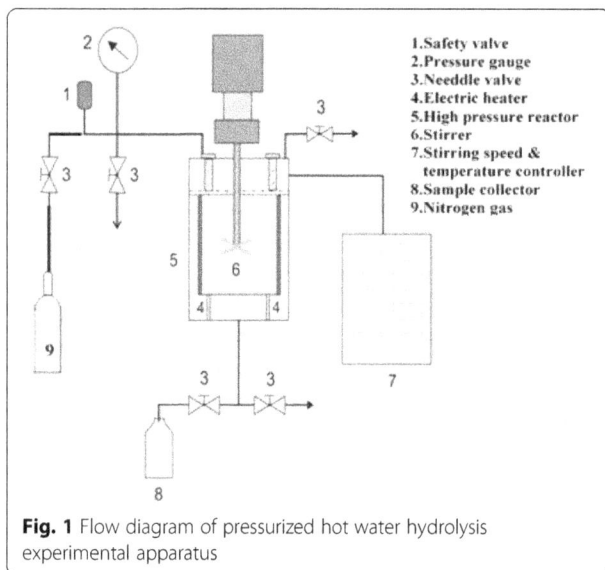

1. Safety valve
2. Pressure gauge
3. Needle valve
4. Electric heater
5. High pressure reactor
6. Stirrer
7. Stirring speed & temperature controller
8. Sample collector
9. Nitrogen gas

Fig. 1 Flow diagram of pressurized hot water hydrolysis experimental apparatus

using a temperature controller and pressure gauge, respectively. The sample was stirred using a four-blade stirrer at 150 rpm. The reaction time was recorded after the set temperature and pressers were achieved and it was 5 min for each condition. After the end of the reaction, the hydrolysate samples from the reactor were collected and filtered using Whatman nylon membrane filter (0.45 μm) lyophilized and stored at 4 °C until needed for analysis. The hydrolysis yield was calculated using the following reaction:

$$\text{Yield}(\%) = \frac{(\text{Mass of samle before hydrolysis}) - (\text{Mass of residue after hydrolysis})}{\text{Mass of sample before hydrolysis}} \times 100 \tag{1}$$

To compare the PHWH, a control was kept by having a room temperature extraction for 18 h with the same ratio and after the extraction process, same process was carried out as before for the PHWH extracts.

Protein content
The protein content of the soluble product was analyzed by Lowry's assay (Lowry et al. 1951), using bovine serum albumin (BSA) as a standard.

Total carbohydrate content
The total carbohydrate content was determined by using anthrone reagent following the method reported by Carroll et al. (1956).

Reducing sugar content
The reducing sugars content was analyzed by dinitrosalicylic (DNS) colorimetric method (Miller 1959), using D-glucose as a standard. For each of the 3 ml of the sample, 3 mL of DNS reagent was added. The mixture

was then heated in boiling water for 5 min until the red-brown color developed. Then, 1 mL of 40% potassium sodium tartrate (Rochelle salt) solution was added to stabilize the color, after which the mixture was cooled to room temperature in a water bath. The absorbance was then measured with a spectrophotometer at 575 nm.

Antioxidant activity analysis
DPPH radical scavenging activity
The stable free radical scavenging activity was determined by DPPH• assay according to the method of Thitilertdecha et al. (2008). One mL of 60 mM DPPH• solution in ethanol was mixed with 3 mL of sample at different concentrations. The control consisted of 1 mL of DPPH• solution and 3 mL of ethanol. The mixture was incubated at room temperature for 30 min, and the absorbance was measured at 517 nm. The ability to scavenge DPPH radicals was calculated as the DPPH• scavenging by the following equation:

$$\%\text{DPPH}\bullet\text{scavenging} = [(A0{-}A1)/A0] \times 100 \tag{2}$$

where A0 is absorbance of the control and A1 is the absorbance of the mixture containing the sample. Trolox were used as positive control.

ABTS$^+$ radical cation scavenging activity
ABTS$^{\bullet+}$ radical scavenging activity was determined according to the modified method of Re et al. (1999). ABTS• + was produced by reacting 7 mM aqueous ABTS• + solution with 2.45 mM potassium persulfate in the dark at room temperature for 16 h and was used within 2 days. The ABTS• + solution was diluted with ethanol to an absorbance of (0.70 ± 0.02) at 734 nm. One mL of diluted sample was mixed with 3 mL of diluted ABTS• + solution. The mixture was then allowed to stand for 20 min at room temperature, and the absorbance was immediately recorded at 734 nm. Standard curve was constructed using standard concentrations of Trolox.

The FRAP assay
The FRAP (ferric reducing antioxidant power) assay was done according to Benzie and Strain (1996) with some modifications. Briefly, acetate buffer 300 mM (pH 3.6) was prepared by mixing 3.1 g of $C_2H_3NaO_2.3H_2O$ and 16 mL $C_2H_4O_2$ in 1 L deionized water, 10 mM TPTZ (2, 4, 6-tripyridyls-triazine) in 40 mM HCl; 20 mM $FeCl_3.6H_2O$. Working FRAP reagent was prepared as required by mixing 25 mL acetate buffer, 2.5 mL TPTZ solution, and 2.5 mL $FeCl_3.6H_2O$ solution. The standard curve was linear between 5 to 500 μg/mL Trolox.

Statistical analysis

Statistical analysis was performed using SPSS (Version 20 for windows, IBM, Chicago, IL, USA).

Results and discussion

Color and pH of hydrolysates

The appearance of *S. clava* hydrolysates attained by PHWH process varied depending on the usage of various temperature (Fig. 2, Table 1). The brown color of extracts was more intense at higher treatment temperatures. Extracts obtained at 125 and 150 °C were light white, whereas extracts prepared at 175 °C were reddish-yellow. Hydrolysate obtained at higher temperatures progressively became brownish-yellow, and the 225 °C hydrolysate was dark brown, while the control (25 °C) showed a pale white color.

The pH of the obtained *S. clava* hydrolysate was measured and values were shown in the Table 1. The pH values were varied from 6.45 (125 °C) to 8.96 (275 °C) in the *S. clava* hydrolysate and the values were found to be increased as the temperature was increased. Initially, the pH was 6.45 for lower temperature, when the temperature increased the pH reduce to 5.04 at 200 °C. After that the pH gradually increased as the temperature increases. The pH is increased at the high temperature due to the formation of the salts and degradation of the all the matters. The low pH is a result of sugars degradation into organic acids and these organic acids react in chain, providing the acidity for speeding up the subsequent reactions as an autocatalytic process (Sasaki et al. 1998).

Color is a very important quality parameter in food industry. As respects to color coordinates (Table 1) lightness (L*) values ranged from 35.20 to 50.21 for different condition of the hydrolysates. High lightness is due to the pigments presence and some hygroscopic substances are increase, when a sample is treated thermally and its lightness value is increased. The coordinates a* (red-green) showed values ranged from −0.28 to 9.59 while for the coordinate b* (yellow-blue) the values ranged from 6.45 to 28.82.

Table 1 Color and pH of hydrolysates at different temperatures

Conditions (°C)	Color			pH
	L*	a*	b*	
Control (25)	41.26 ± 0.14^d	1.20 ± 0.08^f	5.94 ± 1.23^d	6.38 ± 1.21^d
125	42.39 ± 0.67^d	1.27 ± 0.13^f	6.45 ± 0.20^d	6.45 ± 0.07^d
150	50.21 ± 0.92^a	-0.28 ± 0.07^g	12.31 ± 0.06^c	5.51 ± 0.01^e
175	49.30 ± 0.04^b	1.83 ± 0.05^e	21.47 ± 0.07^b	5.35 ± 0.03^f
200	36.72 ± 0.28^d	7.28 ± 0.02^d	14.03 ± 0.24^c	5.04 ± 0.01^g
225	35.20 ± 0.37^f	9.59 ± 0.12^a	12.05 ± 0.35^c	6.74 ± 0.03^c
250	47.32 ± 0.10^c	9.22 ± 0.09^b	28.29 ± 0.08^a	8.58 ± 0.00^b
275	50.09 ± 0.15^{ab}	8.44 ± 0.06^c	28.82 ± 0.72^a	8.96 ± 0.02^a

Values are mean ± SD. Values with different superscript letters within rows indicate statistical significant difference ($P \leq 0.05$) as determined by Duncan's multiple range test

Hydrolysis efficiency

The conditions used in PHWH ranged from 125 to 275 °C with a reaction time of 5 min and pressure was maintained 50 bar for all the conditions. The product obtained after reaching room temperature was normally a mixture of water and solids sorted in a matrix consisting of two layers due to precipitation of particles after extraction. The upper layer comprised of a less viscous aqueous solution with very low turbidity, while the lower layer was predominantly wet *S. clava* residue. The aroma of the hydrolysate was somehow toasty for lower temperature conditions when the temperature is increased the aroma was turned to be a pungent. This change in aroma with increasing temperature was also reported by Saravana et al. (2016a) for hydrolysates of *Saccharina japonica* using PHWH. The hydrolysis efficiency of *S. clava* hydrolysate ranged from 46.05 to 88.67% (Fig. 3), while the control (25 °C) showed a extraction yield of 40%. In the present study, it is shown that hydrolysis efficiency for *S. clava* hydrolysate increased consistently with increasing temperature. This was considered by improved mass transfer rate, increase in solubility of the analytes, and decrease in solvent viscosity and surface tension that take place due to increasing temperature and pressure under subcritical conditions (Herrero et al. 2015). Previous work by Asaduzzaman and Chun (2014) with *Scomber japonicus* using the same PHWH apparatus reported similar results for hydrolysis efficiency. Thus, increasing temperature and pressure conditions facilitate deeper penetration of solvent into the sample matrix which enhanced greater surface contact and improved mass transfer to the solvent. PHWH technique has been adapted for various biomasses, including proteins, carbohydrates and fatty acids, and the yield and form of hydrolysate differ depending on the reaction conditions, including the original source, particle size, temperature, pressure, hydrolysis time, etc. (Rogalinski et al. 2008).

Fig. 2 Effect of temperature on the appearance of the hydrolysates

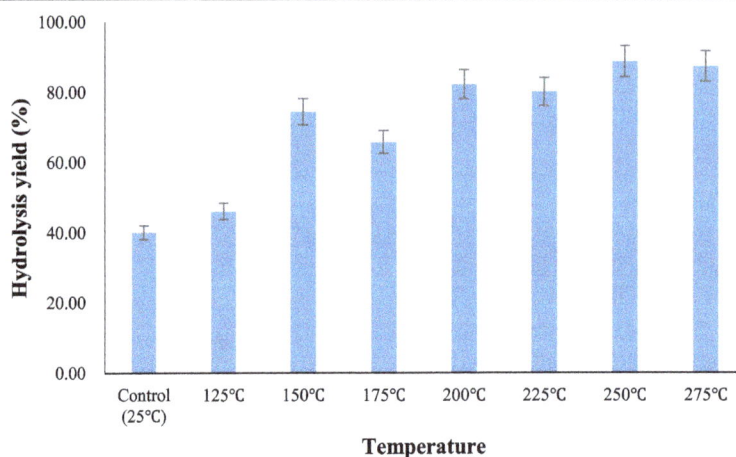

Fig. 3 Effect of temperature on hydrolysis yield of *S. clava*

While it has been reported that the hydrolysate yield and form change can be affected by temperature, pressure, and hydrolysis time, few studies have reported that the hydrolysate yield can be altered by use of different particle sizes (Toor et al. 2011).

Total protein, total carbohydrate, and reducing sugar content

Total protein for *S. clava* hydrolysate increased from 125 to 175 °C but decreased slightly as temperature increased further (Fig. 4). Watchararuji et al. (2008) stated that protein content improved with a raise in temperature up to 220 °C for rice bran by PHWH. Generally, the rise in protein content with increasing temperature in PHWH is due to a change in polarity of water in the subcritical region. According to Thiruvenkadam et al. (2015), the increase in dielectric constant and the decrease in density (1 g/cm^3 at 25 °C to 0.75 g/m^3 at 300 °C) compared to ambient conditions consequently enable hydrocarbons to become more water soluble. This is characterized by breakdown of hydrogen bonding in the water molecules which changes the polarity of water in the subcritical region from complete polarity to moderately non-polar. This condition tends to enhance the attraction of water towards non-polar hydrocarbons thus increases miscibility and enhances hydrocarbon solubility in water. However, the decrease of proteins after 250 °C was most probably due to denaturing as a result of exceedingly high temperatures. Actually, proteins can be denatured by heat when their resistance to thermal denaturation capacity is exceeded (Haque et al. 2016). Thus, this study showed that the best condition for utilizing proteins from *S. clava* by PHWH is around 175 °C with a high yield of 4.24 mg/g. These observations suggest that proteinaceous substances are the main components in the extracts obtained at higher temperatures. However, the decrease of protein content at 220 and 240 °C suggested that at these temperatures degrade the proteins and the generation of small components,

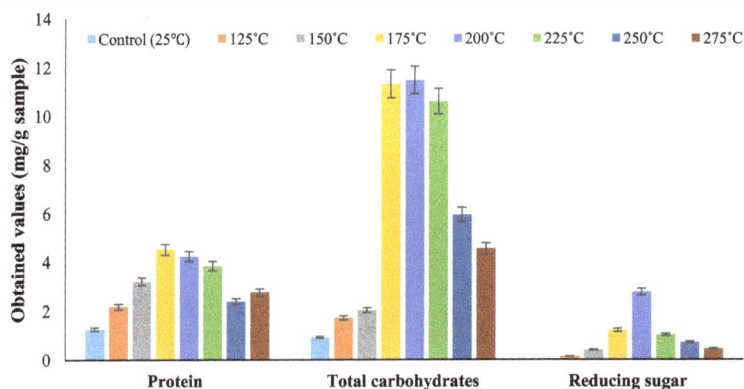

Fig. 4 Protein, total carbohydrate, and reducing sugar content of *S. clava* hydrolysates

such as organic acids, were produced (Saravana et al. 2016b).

The highest yield of total carbohydrate content was obtained at 200 °C, after that the composition was gradually decreased as the temperature increases. At 200 °C the total carbohydrate content was 11.48 mg/g (Fig. 4). Recent report says the total carbohydrate content in the *S. clava* can be altered throughout the season and it was found high in March with a range from 21.6 to 25.9% (Lee et al. 2006). The highest yields for reducing sugar were recovered at condition 200 °C (Fig. 4). All sugars were higher at milder conditions but decreased gradually as temperature and pressure increased. Quitain et al. (2002) stated that the reducing sugar content was dropped as the temperature and reaction time increases and this is perhaps due to the degradation into other products such as ketones/aldehydes, and it could lead to produce the organic acids. Therefore, PHWH treatment without acid or base catalyst is a promising step towards bioethanol production.

Antioxidant activities

DPPH radical scavenging activity

DPPH, can easily undergo reduction by an antioxidant and it is a stable radical with a maximum absorbance at 517 nm. Liu et al. (2010) and Peng et al. (2009) stated that DPPH a proton-donating can change color from purple to yellow by scavenging the substance (H⁺) and the absorbance is reduced. All hydrolysates effectively showed DPPH activity (Fig. 5a). The decline of DPPH in the incidence of the *S. clava* hydrolysates shows that mixed peptide/amino acids were capable of reducing DPPH apparently by combination the odd electron of the DPPH radicals. The DPPH assay was expressed in terms of trolox equivalent and the high antioxidant activity was found at 225 °C (Trolox equivalent 10.20 mg/g sample) after that the activity was decreased. From the obtained results, *S. clava* hydrolysate has the capability to efficiently reduce DPPH radical, which shows that the hydrolysates are good antioxidant compounds with radical scavenging activity. No DPPH activity was found for control extract. Wu et al. 2003 distinguished that for mackerel hydrolysates, DPPH scavenging activity enriched progressively with increasing hydrolysis time. During hydrolysis, a varied amount of smaller peptides and free amino acids are produced, depending upon the temperature conditions. The changes in size, composition of amino acids can have an effect in the antioxidant activity (Wu et al. 2003). Earlier studies have shown that high DPPH or other radical scavenging activities of protein hydrolysates or peptides are frequently related with vastly hydrophobic amino acids or overall hydrophobicity (Li et al. 2008).

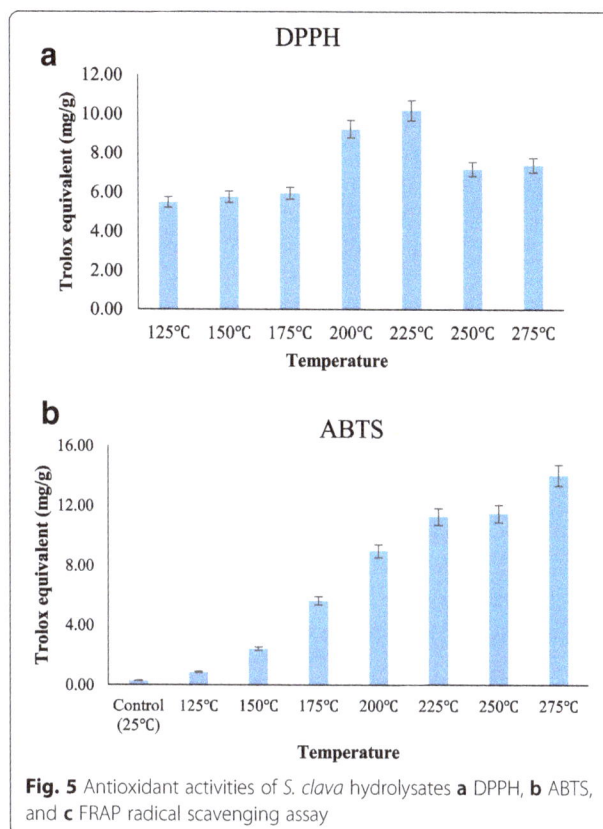

Fig. 5 Antioxidant activities of *S. clava* hydrolysates **a** DPPH, **b** ABTS, and **c** FRAP radical scavenging assay

ABTS•+

ABTS antioxidant activity was found to be increased as the temperature increases, *S. clava* hydrolysate showed high antioxidant activity at 275 °C and the amount of ABTS•+ antioxidant was ranged from 0.86–14.06 mg/g Trolox equi (Fig. 5b), for the control extract a very low activity was found. The alterations in ABTS•+ scavenging activity between the hydrolysates valor due to the changes in sequence length and amino acid composition. Normally, every hydrolysate which consists of proteins that can act as hydrogen donors and react with radicals, altering them into more stable products and thereby ending the radical chain reaction. Free radical reducing has been described to have the chief antioxidative mechanism of peptides due to amino acids such as Ala, Tyr, and Leu. Even though protein hydrolysates were the key point of this work, other substances in *S. clava*, such as phenolics, might have extracted together with protein and that could have contributed to the antioxidant activity of the crude hydrolysates (Jumeri and Kim, 2011).

FRAP

The FRAP assay is a rapid, simple, dependable and inexpensive method extensively used in most of the research laboratory where antioxidant capacity measurement is carried out (Apak et al. 2007). The FRAP of the *S. clava* hydrolysate was measured as the ability to reduce Fe^{3+} to Fe^{2+}, which indicated the capacity of antioxidants to donate an electron or hydrogen, and an increase in absorbance at 700 nm indicated greater reducing power. As showed in Fig. 5c, 225 °C hydrolysate showed a high FRAP absorbance of 10.91 mg/g. No antioxidant activity was found using FRAP test for control extract.

The FRAP result indicates that antioxidant activity increases as the temperature increases after 225 °C the activity decreased gradually. Many studies reported that protein hydrolysates from other sources possessed strong FRAP. Fish protein hydrolysates from smooth hound muscle protein and yellow stripe trevally were reported to have FRAP values of 0.60 at 2.0 mg/g and 0.52 at 3.6 mg/g, respectively (Bougatef et al. 2009; Klompong et al. 2007). Additionally, only one research for the FRAP of abalone viscera hydrolysates by alkali protease, papain, neutral protease, pepsin, and trypsin was reported by Zhou et al. (2012), and abalone viscera hydrolysates exhibited the FRAP value of below 0.90 at 10.0 mg/g.

Conclusions

The results of this study indicate that temperature has great influence on the physico-chemical and biofunctional properties on the PHWH of *S. clava*. The highest DPPH and FRAP activities were obtained in the hydrolysate at 225 °C, while ABTS activity showed highest activity at 275 °C. The highest protein, total carbohydrate, and reducing sugar content of the hydrolysates were found at 175, 150, and 200 °C, respectively. So, PHWH of *S. clava* at 225 °C and 50 bar can produce high antioxidant activity. Therefore, PHWH have the potential to produce functional compounds from *S. clava*, which can be used as antioxidant supplement in food industry.

Abbreviations
ABTS: 2,2-azinobis-3ethylbenzothiazoline-6-sulphonic acid; DPPH: 2,2-diphenyl-1-picrylhydrazyl; FRAP: Ferric reducing antioxidant power; PHWH: Pressurized hot water hydrolysis; TPTZ: 2,4,6-tripyridyl-s-triazine; Trolox: 6-hydroxy-2,5,7,8-tetramethylchroman-2-carboxylic acid

Acknowledgements
This work was supported by Business for Cooperative R&D between Academy and Research Institute funded by Korea Small and Medium Administration in 2016 project number (C0406103).

Funding
This study was funded by the Business for Cooperative R&D between Academy and Research Institute funded by Korea Small and Medium Administration in 2016 project number (C0406103).

Authors' contributions
HJL, SJC PSS, and BSC designed the study. PSS wrote the article. HJL, SJC PSS, and BSC performed the experiment. HJL and SJC analyzed and interpreted the data. All authors have read, commented upon, and approved the final article.

Competing interests
The authors declare that they have no competing interests.

References
Apak R, Güçlü K, Demirata B, Özyürek M, Çelik SE, Bektaşoğlu B. Comparative evaluation of various total antioxidant capacity assays applied to phenolic compounds with the CUPRAC assay. Molecules. 2007;12(7):1496–547.

Asaduzzaman A, Chun B-S. Hydrolyzates produced from mackerel Scomber japonicus skin by the pressurized hydrothermal process contain amino acids with antioxidant activities and functionalities. Fisheries Sci. 2014;80(2):369–80.

Benzie IF, Strain JJ. The ferric reducing ability of plasma (FRAP) as a measure of "antioxidant power": the FRAP assay. Anal Biochem. 1996;239(1):70–6.

Bicker M, Endres S, Ott L, Vogel H. Catalytical conversion of carbohydrates in subcritical water: A new chemical process for lactic acid production. J Mol Catal A Chem. 2005;239(1):151–7.

Bougatef A, Hajji M, Balti R, Lassoued I, Triki-Ellouz Y, Nasri M. Antioxidant and free radical-scavenging activities of smooth hound (*Mustelus mustelus*) muscle protein hydrolysates obtained by gastrointestinal proteases. Food Chem. 2009;114(4):1198–205.

Carr AG, Mammucari R, Foster N. A review of subcritical water as a solvent and its utilisation for the processing of hydrophobic organic compounds. Chem Eng J. 2011;172(1):1–17.

Carroll NV, Longley RW, Roe JH. The determination of glycogen in liver and muscle by use of anthrone reagent. J Biol Chem. 1956;220:583–93.

Cheng H, Zhu X, Zhu C, Qian J, Zhu N, Zhao L, et al. Hydrolysis technology of biomass waste to produce amino acids in sub-critical water. Bioresour Technol. 2008;99(9):3337–41.

Donia M, Hamann MT. Marine natural products and their potential applications as anti-infective agents. Lancet Infect Dis. 2003;3(6):338–48.

Fujimoto K, Yoshii N, Okazaki S. Free energy profiles for penetration of methane and water molecules into spherical sodium dodecyl sulfate micelles obtained using the thermodynamic integration method combined with molecular dynamics calculations. J Chem Phys. 2012;136(1):014511.

Jumeri, Kim SM. Antioxidant and anticancer activities of enzymatic hydrolysates of solitary tunicate (*Styela clava*). Food Sci Biotechnol. 2011;20(4):1075–85.

Haque MA, Timilsena BYP, Adhikari B. Food Proteins, Structure, and Function. 2016.

Herrero M, del Pilar S-CA, Cifuentes A, Ibáñez E. Plants, seaweeds, microalgae and food by-products as natural sources of functional ingredients obtained using pressurized liquid extraction and supercritical fluid extraction. TrAC Trend Anal Chem. 2015;71:26–38.

Kang N, Kim S-Y, Rho S, Ko J-Y, Jeon Y-J. Anti-fatigue activity of a mixture of seahorse (Hippocampus abdominalis) hydrolysate and red ginseng. Fish Aqua Sci. 2017;20(1):3.

Kim J-J, Kim S-J, Kim S-H, Park H-R, Lee S-C. Antioxidant and anticancer activities of extracts from *Styela clava* according to the processing methods and solvents. J Korean Soc Food Sci Nutr. 2006;35(3):278–83.

Klompong V, Benjakul S, Kantachote D, Shahidi F. Antioxidative activity and functional properties of protein hydrolysate of yellow stripe trevally (*Selaroides leptolepis*) as influenced by the degree of hydrolysis and enzyme type. Food Chem. 2007;102(4):1317–27.

Ko S-C, Jeon Y-J. Anti-inflammatory effect of enzymatic hydrolysates from Styela clava flesh tissue in lipopolysaccharide-stimulated RAW 264.7 macrophages and in vivo zebrafish model. Nutr Res Pract. 2015;9(3):219–26.

Ko S-C, Lee J-K, Byun H-G, Lee S-C, Jeon Y-J. Purification and characterization of angiotensin I-converting enzyme inhibitory peptide from enzymatic hydrolysates of Styela clava flesh tissue. Process Biochem. 2012a;47(1):34-40.

Ko S-C, Kim DG, Han C-H, Lee YJ, Lee J-K, Byun H-G. Nitric oxide-mediated vasorelaxation effects of anti-angiotensin I-converting enzyme (ACE) peptide from Styela clava flesh tissue and its anti-hypertensive effect in spontaneously hypertensive rats. Food Chem. 2012b;134(2):1141-5.

Kruse A, Gawlik A. Biomass conversion in water at 330–410 C and 30–50 MPa. Identification of key compounds for indicating different chemical reaction pathways. Ind Eng Chem Res. 2003;42(2):267–79.

Laria D, Martí J, Guardia E. Protons in supercritical water: a multistate empirical valence bond study. J Am Chem Soc. 2004;126(7):2125–34.

Lee J-S, Kang S-J, Choi B-D. Seasonal variation in the nutritional content of Mideodeok Styela clava. Fish Aqua Sci. 2006;9(2):49–56.

Lee D-W, You D-H, Yang E-K, Jang I-C, Bae M-S, Jeon Y-J. Antioxidant and ACE inhibitory activities of Styela clava according to harvesting time. J Korean Soc Food Sci Nutr. 2010;39(3):331–6.

Li Y, Jiang B, Zhang T, Mu W, Liu J. Antioxidant and free radical-scavenging activities of chickpea protein hydrolysate (CPH). Food Chem. 2008;106(2):444–50.

Liu Q, Kong B, Xiong YL, Xia X. Antioxidant activity and functional properties of porcine plasma protein hydrolysate as influenced by the degree of hydrolysis. Food Chem. 2010;118(2):403–10.

Lowry OH, Rosebrough NJ, Farr AL, Randall RJ. Protein measurement with the Folin phenol reagent. J Biol Chem. 1951;193(1):265–75.

Matsui T, Yukiyoshi A, Doi S, Sugimoto H, Yamada H, Matsumoto K. Gastrointestinal enzyme production of bioactive peptides from royal jelly protein and their antihypertensive ability in SHR. J Nutr Biochem. 2002;13(2):80–6.

Miller GL. Use oi Dinitrosalicylic Acid Reagent tor Determination oi Reducing Sugar. Anal Chem. 1959;31(3):426–8.

Özel MZ, Göğüş F. Subcritical water as a green solvent for plant extraction. In: Chemat F, Vian MA, editors. Alternative Solvents for Natural Products Extraction. Berlin: Springer-Verlag; 2014. p. 73-89.

Peng X, Xiong YL, Kong B. Antioxidant activity of peptide fractions from whey protein hydrolysates as measured by electron spin resonance. Food Chem. 2009;113(1):196–201.

Prasad KN, Xie H, Hao J, Yang B, Qiu S, Wei X, et al. Antioxidant and anticancer activities of 8-hydroxypsoralen isolated from wampee [Clausena lansium (Lour.) Skeels] peel. Food Chem. 2010;118(1):62–6.

Quitain AT, Faisal M, Kang K, Daimon H, Fujie K. Low-molecular-weight carboxylic acids produced from hydrothermal treatment of organic wastes. J Hazard Mater. 2002;93(2):209–20.

Re R, Pellegrini N, Proteggente A, Pannala A,Yang M, Rice-Evans C. Antioxidant activity applying an improved ABTS radical cation decolorization assay. Free Radic Biol Med. 1999; 26: 1231-1237.

Rogalinski T, Ingram T, Brunner G. Hydrolysis of lignocellulosic biomass in water under elevated temperatures and pressures. J Supercrit Fluids. 2008;47(1):54–63.

Saravana PS, Choi JH, Park YB, Woo HC, Chun BS. Evaluation of the chemical composition of brown seaweed (Saccharina japonica) hydrolysate by pressurized hot water extraction. Algal Res. 2016a;13:246-54.

Saravana PS, Getachew AT, Ahmed R, Cho Y-J, Lee Y-B, Chun B-S. Optimization of phytochemicals production from the ginseng by-products using pressurized hot water: Experimental and dynamic modelling. Biochem Eng J. 2016b;113:141-51.

Sasaki M, Kabyemela B, Malaluan R, Hirose S, Takeda N, Adschiri T, et al. Cellulose hydrolysis in subcritical and supercritical water. J Supercrit Fluids. 1998;13(1):261–8.

Thiruvenkadam S, Izhar S, Yoshida H, Danquah MK, Harun R. Process application of Subcritical Water Extraction (SWE) for algal bio-products and biofuels production. Appl Energy. 2015;154:815–28.

Thitilertdecha N, Teerawutgulrag A, Rakariyatham N. Antioxidant and antibacterial activities of Nephelium lappaceum L. extracts. LWT-Food Sci Technol. 2008;41(10):2029–35.

Tomita K, Oshima Y. Stability of manganese oxide in catalytic supercritical water oxidation of phenol. Ind Eng Chem Res. 2004;43(24):7740–3.

Toor SS, Rosendahl L, Rudolf A. Hydrothermal liquefaction of biomass: a review of subcritical water technologies. Energy. 2011;36(5):2328–42.

Uddin M, Ahn H-M, Kishimura H, Chun B-S. Production of valued materials from squid viscera by subcritical water hydrolysis. J Environ Biol. 2010;31(5):675–9.

Valko M, Izakovic M, Mazur M, Rhodes CJ, Telser J. Role of oxygen radicals in DNA damage and cancer incidence. Mol Cell Biochem. 2004;266(1-2):37–56.

Wang L, Tu Y-C, Lian T-W, Hung J-T, Yen J-H, Wu M-J. Distinctive antioxidant and antiinflammatory effects of flavonols. J Agric Food Chem. 2006;54(26):9798–804.

Watchararuji K, Goto M, Sasaki M, Shotipruk A. Value-added subcritical water hydrolysate from rice bran and soybean meal. Bioresour Technol. 2008;99(14):6207–13.

Wu H-C, Chen H-M, Shiau C-Y. Free amino acids and peptides as related to antioxidant properties in protein hydrolysates of mackerel (Scomber austriasicus). Food Res Int. 2003;36(9):949–57.

Xu C-X, Jin H, Chung Y-S, Shin J-Y, Woo M-A, Lee K-H, et al. Chondroitin sulfate extracted from the Styela clava tunic suppresses TNF-α-induced expression of inflammatory factors, VCAM-1 and iNOS by blocking Akt/NF-κB signal in JB6 cells. Cancer Lett. 2008;264(1):93–100.

Yagasaki T, Iwahashi K, Saito S, Ohmine I. A theoretical study on anomalous temperature dependence of p K w of water. J Chem Phys. 2005;122(14):144504.

Yoshii N, Miura S, Okazaki S. Free energy profiles for penetration of methane and water molecules into spherical sodium dodecyl sulfate micelles obtained using the thermodynamic integration method combined with molecular dynamics calculations. Chem Phys Lett J. 2001;345:195–200.

Zhao Y, Lu W-J, Wu H-Y, Liu J-W, Wang H-T. Optimization of supercritical phase and combined supercritical/subcritical conversion of lignocellulose for hexose production by using a flow reaction system. Bioresour Technol. 2012;126:391–6.

Zhou D-Y, Zhu B-W, Qiao L, Wu H-T, Li D-M, Yang J-F, et al. In vitro antioxidant activity of enzymatic hydrolysates prepared from abalone (Haliotis discus hannai Ino) viscera. Food Bioprod Process. 2012;90(2):148–54.

20

Circadian rhythm of melatonin secretion and growth-related gene expression in the tiger puffer *Takifugu rubripes*

Byeong-Hoon Kim[1†], Sung-Pyo Hur[5†], Sang-Woo Hur[2], Yuki Takeuchi[3], Akihiro Takemura[4] and Young-Don Lee[1,6*]

Abstract: Somatostatin (SS) and growth hormone-releasing hormone (GHRH) are primary factors regulating growth hormone (GH) secretion in the pituitary. To date, it remains unknown how this rhythm is controlled endogenously, although there must be coordination of circadian manners. Melatonin was the main regulator in biological rhythms, and its secretion has fluctuation by photic information. But relationship between melatonin and growth-related genes (*ghrh* and *ss*) is unclear. We investigated circadian rhythms of melatonin secretion, *ghrh* and *ss* expressions, and correlation between melatonin with growth-related genes in tiger puffer *Takifugu rubripes*. The melatonin secretion showed nocturnal rhythms under light and dark (LD) conditions. In constant light (LL) condition, melatonin secretion has similar patterns with LD conditions. *ss1* mRNA was high during scotophase under LD conditions. But *ss1* rhythms disappeared in LL conditions. *Ghrh* appeared opposite expression compared with melatonin levels or *ss1* expression under LD and LL. In the results of the melatonin injection, *ghrh* and *ss1* showed no significant expression compared with control groups. These results suggested that melatonin and growth-related genes have daily or circadian rhythms in the tiger puffer. Further, we need to know mechanisms of each *ss* and *ghrh* gene regulation.

Keywords: Photoperiod, Melatonin, Somatostatin, GHRH, Tiger puffer

Background

Biological rhythms are controlled by many environmental changes including light, temperature, universal gravitation, and weather conditions (Fraser et al. 1993; Forward et al. 1998; Wan et al. 2013; Guerra-Santos et al. 2017). In non-mammalian vertebrates, photic signals are transmitted via neural pathways from the retina to the pineal gland. These signals control the secretion of various hormones (Iigo et al. 1997; Ayson and Takemura 2006; Revel et al. 2006; Moore and Menaker 2011; Hur et al. 2011).

Body growth is enhanced by growth hormone (GH), which is released from the pituitary gland in vertebrates (McLean et al. 1997; Raven et al. 2012; Fuentes et al. 2013). The secretion of GH is primarily controlled by growth hormone-releasing hormone (GHRH) and somatostatin (SS), which are synthesized in the hypothalamus (Klein and Sheridan 2008; Luque et al. 2008). Moreover, IGF-1, which is secreted from the liver, is also known to control body growth through interactivity with GH (Wood et al. 2005). Studies show that GH secretion is controlled by various environmental factors such as temperature and photoperiod.

GHRH is a member of the glucagon superfamily, and its primary function is to stimulate GH synthesis and secretion by binding to GHRH receptors (GHRHR) in the anterior pituitary. Recent studies in fish show that GHRH and pituitary adenylate cyclase-activating peptide (PACAP) encodings differ by only a single gene, and GH induction in the pituitary gland by GHRH has been identified (Lee et al. 2007). In contrast, SS has been widely detected not only in the central nervous system but also in peripheral tissues. It inhibits GH secretion in the pituitary (Very et al. 2001). These two GH-regulating hormones, GHRH and SS, have opposing functions: GHRH enhances GH release from the pituitary in vivo and in vitro, while SS inhibits its secretion in many teleost fishes (Canosa et al. 2007).

* Correspondence: leemri@jejunu.ac.kr
†Equal contributors
[1]Marine Science Institute, Jeju National University, Jeju 6333, South Korea
[6]19-5, Hamdeok 5(o)-gill, Jocheon, Jeju Special Self-Governing Province 695-965, Republic of South Korea
Full list of author information is available at the end of the article

Melatonin is an indole-derived hormone that is synthesized in the retina and pineal gland. The secretion of this hormone is primarily controlled by environmental light conditions in vertebrates, including fish. The plasma melatonin level in Mozambique tilapia *Oreochromis mossambicus* increases during the night and decreases at day (Nikaido et al. 2009). Although it has been suggested that melatonin secretions affect physiological functions, including body growth in fish (Taylor et al. 2005; Herrero et al. 2007; De Pedro et al. 2008; Maitra et al. 2013), the effect of melatonin on the transcription of growth-related genes (*ghrh* and *ss*) is still unclear.

The tiger puffer *Takifugu rubripes* is a commercially valuable species in South Korea. The aim of the present study is to profile its growth-related gene expressions and to evaluate the effect of melatonin on expressions of these genes.

Methods

Animal

A total of 200 tiger puffer *T. rubripes* (body weight 128.4 ± 2.1 g, body length 18.7 ± 0.1 cm) were used in this study. Fish were obtained from the Tham-Ra Fishery located in Soegwipo, Jeju, South Korea, and transported to Marine Research Institute, Jeju National University. The fish were acclimatized under natural photoperiod and water temperature (20–21 °C). The fish were fed commercial pellets (Daehan co., MP3, Busan, South Korea) equivalent to 1 to 2% of body weight at 0900 and 1600 h daily.

Distribution of growth-related genes in the parts of brain were examined by RT-PCR and real-time quantitative RT-PCR (Real-time PCR). The fish ($n = 4$) brain were divided to five portions each, the telencephalon, optic tectum, diencephalon, cerebellum, and medulla

oblongata (Hur et al. 2011) (Fig. 1). Sampling was conducted at 1200 h during the daytime, and sampling methods was explained above.

For the circadian variation, fish were adapted under 12-h light and 12-h dark photoperiod condition (12L:12D, light on = 0800 h and light off = 2000 h) for 1 week. After adapted, fish were divided into two groups; 12L:12D group and 24-h light photoperiod condition (24L) group. The 12L:12D group fish ($n = 98$, BW 126.2 ± 4.1 g, TL 18.0 ± 0.2 cm) and 24L group fish ($n = 98$, BW 136.4 ± 3.1 g, TL 19.0 ± 0.2 cm) were reared for 3 days. Fish were anesthetized in MS-222 and killed by decapitation at 4-h intervals for 1 day at zeitgeber time (ZT) 2, ZT 6, ZT 10, ZT 14, ZT18, and ZT 22 for 12L:12D group and circadian time (CT) 2, CT 6, CT 10, CT 14, CT 18, and CT 22 for 24L group fish. The two experimental groups were sampled at the same time in different light conditions. The sampling at darkness time was conducted under red dim light, and blood plasma, diencephalon, and pituitary were collected as described above. The collected blood plasma were used in melatonin levels' analysis, and diencephalon tissues were used in *ghrh1*, *ghrh2*, and *ss* mRNA expression analysis by fluoroimmuno assay (TR-FIA) and real-time qPCR. The experimental procedures followed the guidance approved by the Animal Care and Use committees of Jeju National University, Jeju, South Korea.

Melatonin intraperitoneal injection

A total of 20 fish were used for the effects of melatonin with several growth-related genes. Fish were reared under natural photoperiod and water temperature. The one group of fish was melatonin (Sigma, 1 mg/kg, $n = 10$) treated by intraperitoneal injection (i.p.), and the other group was injected with the only vehicle solution (0.6% of saline, $n = 10$) at 1100 h. The fish of melatonin

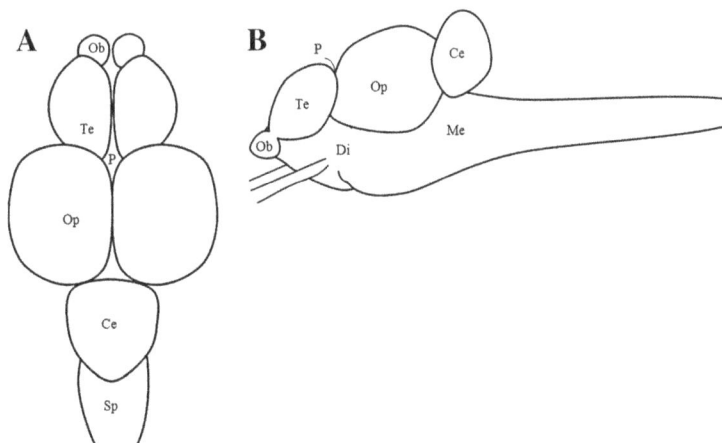

Fig. 1 Diagram showing dorsal view (**a**) and sagittal plane (**b**) of the puffer fish brain. *Ob* olfactory bulb, *Te* telencephalon, *Op* optic tectum, *Ce* cerebellum, *P* pineal gland, *PON* preoptic nucleus, *Sp* spinal cord

i.p. and vehicle groups were sampled melatonin i.p. after 1 h. Fish were anesthetized with MS-222 before sampling, and tissues were collected (diencephalon and pituitary) by decapitation. The collected samples were kept at −80 °C until the analysis.

Plasma melatonin measurement

The levels of melatonin were measured by time-resolved fluoroimmuno assay (TR-FIA) according to the previous report (Takemura et al. 2004). Briefly, a 96-well plate (AGC Techno Glass) was coated with 50 mM carbonate buffer, pH 9.6 (100 ll/well), containing a melatonin-bovine serum albumin (BSA) conjugate (5 ng/ml) for 2 h at 24 ± 0.5 °C in an incubator (Sanyo). After three washes with DELFIA wash buffer, 50 ll of samples/standards (7.8–4000 pg/ml) and 50 ll of anti-melatonin (1:200,000 in assay buffer) were placed in the wells. The plate was incubated overnight at 4 °C. After washing, 100 ll of the secondary antibody against rabbit immunoglobulin G labeled with europium (1:1000 in assay buffer) was added to the well and incubated at 24 °C for 1 h. After washing, 50 ll of DELFIA enhancement solution was added to the wells. The signal in each well was quantified using a time-resolved fluorometer (Arcus1234, Wallac, Oy, Finland). The composition of the assay buffer (pH 7.75) used in the present study was 0.05 M Tris, 0.9% NaCl, 0.5% BSA, (Sigma), 0.05% NaN3, 0.01% Tween 40, and 20 lM diethylenetriaminepentaacetic acid (DTPA, Kanto Chemicals, Tokyo, Japan). DELFIA washing buffer (pH 7.75) contained 0.05 M Tris, 0.9% NaCl, 0.1% NaN3, and 0.01% Tween 20.

RT-PCR and real-time quantitative PCR

Total RNA was extracted from the tissues of tiger puffer using the RNAiso Reagent (TaKaRa Bio, Japan) after absolutely homogenizing the samples. And 1 μg of total RNA was used for synthesis of first strand cDNA by Transcriptor First strand cDNA synthesis kit (Roche Diagnostics, Mannheim, Germany). Primer sets of each genes were designed by isolated ghrh1 (Gene bank, accession number; DQ659331), ghrh2 (DQ659332), ss1 (XM_003968318), and β-actin (U37499) of tiger puffer from National Center for Biotechnology Information (NCBI, Table 1). For the RT-PCR, each PCR reaction mix contained 50% of Emerald PCR Master mix (TaKaRa-Bio), 10 μm of each forward and reverse primer, and 50 ng of cDNA template. The RT-PCR cycling conditions were followed by 28 cycles of denaturation for 45 s at 94 °C, annealing for 45 s at 58 °C, and extension for 1 min at 72 °C. The real-time quantitative PCR was performed using the CFX™ Real-time System (Bio-Rad, Hercules, CA, USA) with 20 ng of cDNA using FastStart Universal SYBR Green Master (Roche Diagnostics). The real-time quantitative PCR amplification was performed by initial denaturation at 95 °C for 10 min, 40 cycles of 95 °C for 15 s, 60 °C for 1 min, and last 60 °C for

Table 1 Primer sets used in this study

Primers	Sequence (5'-3')	GeneBank Accession No.
ss1- F ssl-R	TCTGCT ACT GCAGT CCAACG CCTT CAGCCAGAGAGCAAGGTT	XM_003968318
ghrh 1-F	T ACAGCGT CAT CT GCT CACC	DQ659331
ghrh 1-R	GAAGGAGT AGAGGT CGT CAGC	
ghrhl-V	AGCGT CTT CT GCACACCT AT C	DQ659332
ghrh 1-R	AGT ACGCAT CGTCGTCTACC	
yS-actin-F	GCCAT CCTT CCTT GGT AT GGA	U37499
yS-actin-R	GT CGT ACT CCT GCTT GCT GA	

1 min. Growth-related gene expressions in diencephalon were normalized to amount of the internal control β-actin gene.

Statistical analysis

The plasma melatonin levels and growth-related genes were expressed as means ± SEM and considered significantly different at P < 0.05. Data were analyzed by one-way analysis of variance (ANOVA) followed by Tukey's test to assess statistically significant differences among the different time points in the daily and circadian variation experiments. The significant differences of growth-related gene expression after melatonin injection were tested by Student's t test. P < 0.05 was considered significant.

Results

The mRNA expression of growth-related genes in central nerves tissues

We profiled the distribution of the growth-related gene expression in the several parts of the tiger puffer brain by RT-PCR (Fig. 2a) and real-time qPCR (Fig. 2b–d). The ss1 expression was detected in the telencephalon (Te), optic tectum (Op), diencephalon (Di), and medulla oblongata (ME) but not in the cerebellum (Ce), while the ghrh1 and ghrh2 seemed to express in the all parts of the brain (Fig. 2a).

The ss1 was significantly highly expressed in the Te, Di, and ME than Op and Ce (Fig. 2b). The significantly highest expression of ghrh1 was observed in the ME, and the highest expression of ghrh2 was detected in the Te (Fig. 2c, d).

Daily and circadian rhythm of melatonin and growth-related genes

Plasma melatonin levels were significantly elevated at the middle of the night (ZT18), and its higher level was kept until the dark phase under 12L:12D conditions (Fig. 3a). When the fish were reared under 24L conditions, the plasma melatonin showed similar patterns with that under

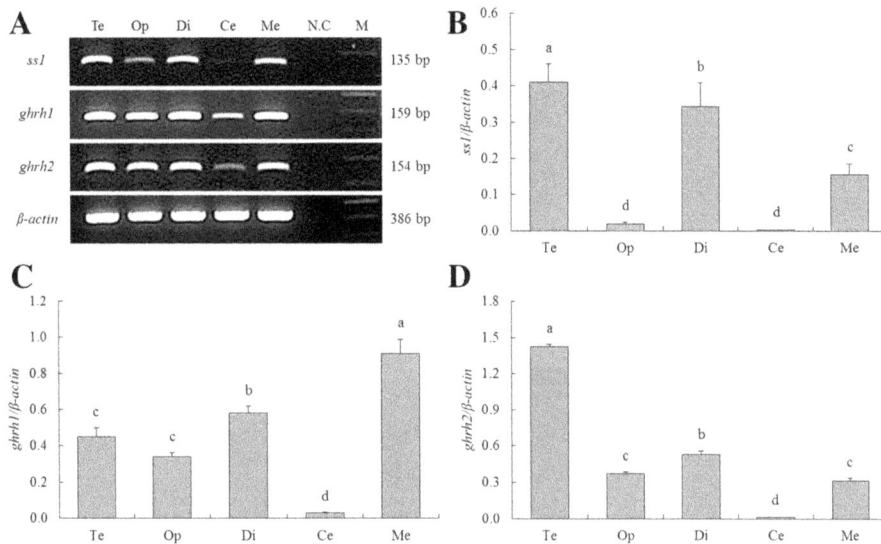

Fig. 2 Tissue-specific expression of growth-related genes in the portion of the brain of tiger puffer. The brain was further divided into five parts. They were analyzed by RT-PCR (**a**) and qPCR (**b–d**). The expression of β-actin mRNA was used as reference. *Te* telencephalon, *Op* optic tectum, *Di* diencephalon, *Ce* cerebellum, *Me* medulla oblongata, *N.C* negative control, *M* 100 bp DNA ladder marker. Means represented by different *letters* are significant ($P < 0.05$). Values are mean ± SEM

12L:12D conditions (Fig. 3b). Under 12L:12D conditions, *ss1* expression in the diencephalon was more increased during scotophase than photophase (Fig. 4a). *ss1* mRNA showed the highest expression at ZT18, and the lowest expression was detected at ZT2. However, this significant ($P < 0.05$) expression patterns disappear under 24L conditions (Fig. 4b).

Each *ghrh1* and *ghrh2* mRNA in the diencephalon showed alike expression patterns under 12L:12D and 24L conditions. Expression of *ghrh1* mRNA was more increased during photophase than scotophase under 12L:12D conditions, but its rapidly low expression was detected in ZT6 (Fig. 5a). Under 24L conditions, *ghrh1* mRNA showed different expression patterns compared with 12L:12D conditions, and this mRNA detected no sudden low expression in CT6 (Fig. 5b). *ghrh2* mRNA showed alike expression pattern with *ghrh1* mRNA under all experiment conditions (Fig. 6a, b). This gene revealed significant expressions under 12L:12D and 24 L conditions. But, this gene detected no significant expression between photophase and scotophase.

Effect of melatonin treatment on the expressions of growth-related genes

The expression of growth-related genes in the diencephalon and pituitary after melatonin i.p. were analyzed by qPCR. One hour after melatonin i.p., *ss1* mRNA showed higher expression levels in melatonin group than saline group (Fig. 7a). However, *ghrh1* and *ghrh2* mRNA after melatonin i.p. showed no different expression in the melatonin group with saline group (Fig. 7b, c).

Discussion and conclusions

Photoperiod is a well-known regulation factor for many physiological responses in fish, including growth and development (Norberg et al. 2001; Taylor et al. 2005; Biswas et al. 2008; Gunnarsson et al. 2012). In this study, we decided to evaluate the effect of light/dark cycle on the expression of growth-related genes in the tiger puffer.

The distribution of growth-related genes in the brain tissue was evaluated using RT-PCR and qPCR. The data revealed widespread distribution of these genes in the brain (Fig. 2), similar to that in the orange-spotted grouper *Epinephelus coioides* (Xing et al. 2005; Qian et al. 2012). In the orange-spotted grouper, the expression of *ss* mRNA was detected in all brain regions, excluding the epithalamus. *Ghrh* mRNA was also expressed in many brain regions. These data suggest that *ss* and *ghrh* expression may mediate not only GH secretion in the pituitary but also multiple biological functions in the tiger puffer.

The daily and circadian regulation of melatonin is well known in vertebrates, including fish. Plasma melatonin levels in the Senegalese sole *Solea senegalensis* showed typical daily rhythms consisting of low levels during photophase and high levels during scotophase (Bayarri et al. 2004). For tench *Tinca tinca* identified the circadian secretion of melatonin under continuous dark conditions as well as light and dark conditions (Oliveira et al. 2009). Our experimental data showed similar results; plasma melatonin secretion in tiger puffers increased during the dark phase in 12L:12D conditions (Fig. 3a). Similar melatonin secretion patterns were found during

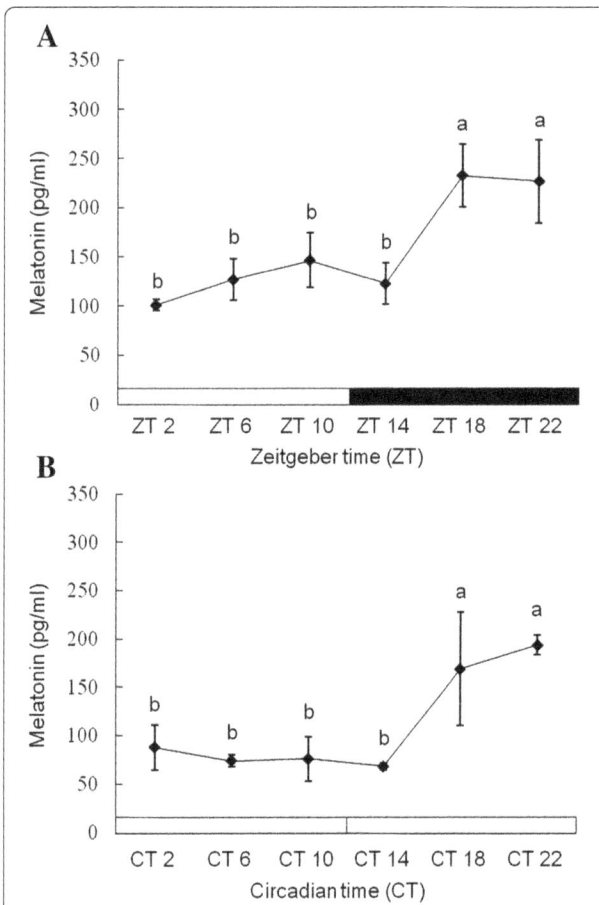

Fig. 3 Daily and circadian rhythms of melatonin secretion in the plasma of tiger puffer under 12L:12D (**a**) and continuous light (**b**) conditions. Means represented by different *letters* are significant ($P < 0.05$). Values are mean ± SEM

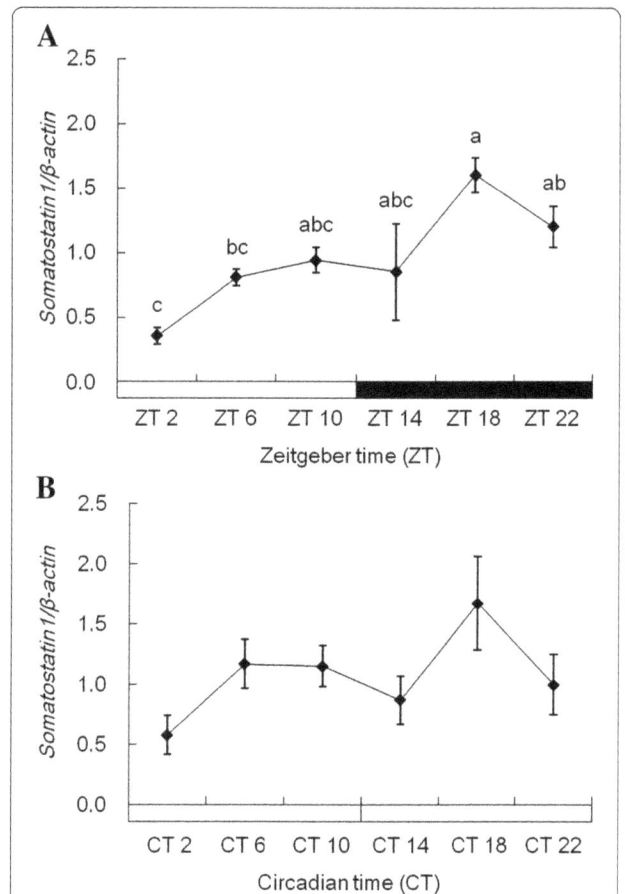

Fig. 4 Daily and circadian expressions of *ss1* mRNA in the diencephalon of tiger puffer under 12L:12D (**a**) and continuous light (**b**) conditions. The relative values of somatostatin1 mRNA expressions were normalized against β-actin and then averaged. Means represented by different *letters* are significant ($P < 0.05$). Values are mean ± SEM

continuous light conditions (Fig. 3b). In some fishes, the rise of melatonin secretion during scotophase is controlled using an endogenous clock. In pike *Esox* and zebrafish *Danio rerio*, the expression of the melatonin synthesis gene (arylalkylamine *N*-acetyltransferase (AANAT)) shows nocturnal rhythms in 12L:12D conditions, and this gene has similar expression rhythms during continuous light or dark conditions (Bégay et al. 1998). These fish have circadian rhythms of high secretion during scotophase caused by the circadian oscillation of the AANAT gene via an endogenous clock. Therefore, we suggest that melatonin have the circadian rhythm in the tiger puffer.

SS and GHRH regulate the synthesis and secretion of GH in the pituitary of vertebrates, including fish (Bertherat et al. 1995; Sheridan and Hagemeister 2010). However, SS and GHRH rhythms in fish are not well understood since the majority of research has focused on mammals. Circadian variation of SS levels in rat plasma showed more elevation during scotophase than

during photophase (Peinado et al. 1990; Ishikawa et al. 1997). We investigated biological rhythms of growth-related genes in the diencephalon of the tiger puffer. The expression patterns of *ss1* mRNA were similar to those shown in previous studies. The *ss1* mRNA expression in the diencephalon of tiger puffers increased during the dark phase in 12L:12D conditions (Fig. 4). However, *ss1* mRNA expression rhythm disappeared under 24L conditions. Therefore, *ss1*mRNA suggest to have the daily rhythm according to light and dark.

In our presents, *ghrh1* and *ghrh2* mRNA exhibited diurnal rhythms in tiger puffers (Figs. 5 and 6). In rats, *ghrh* mRNA expression also exhibited diurnal rhythms (Bredow et al. 1996). Peak expression of hypothalamic *ghrh* mRNA occurred during the early part of photophase in rats. The *ghrh* expression gradually decreased up to scotophase, and the mRNA expressions maintained low levels during scotophase. Diurnal rhythms of hypothalamic *ghrh* mRNA were also observed in rats (Gardi et al. 1999). In these

Fig. 5 Daily and circadian expressions of *ghrh1* mRNA in the diencephalon of tiger puffer under 12L:12D (**a**) and continuous light (**b**) conditions. The relative values of *ghrh1* mRNA expressions were normalized against β-actin and then averaged. Means represented by different *letters* are significant (*P* < 0.05). Values are mean ± SEM

Fig. 6 Daily and circadian expressions of *ghrh2* mRNA in the diencephalon of tiger puffer under 12L:12D (**a**) and continuous light (**b**) conditions. The relative values of *ghrh2* mRNA expressions were normalized against β-actin and then averaged. Means represented by different *letters* are significant (*P* < 0.05). Values are mean ± SEM

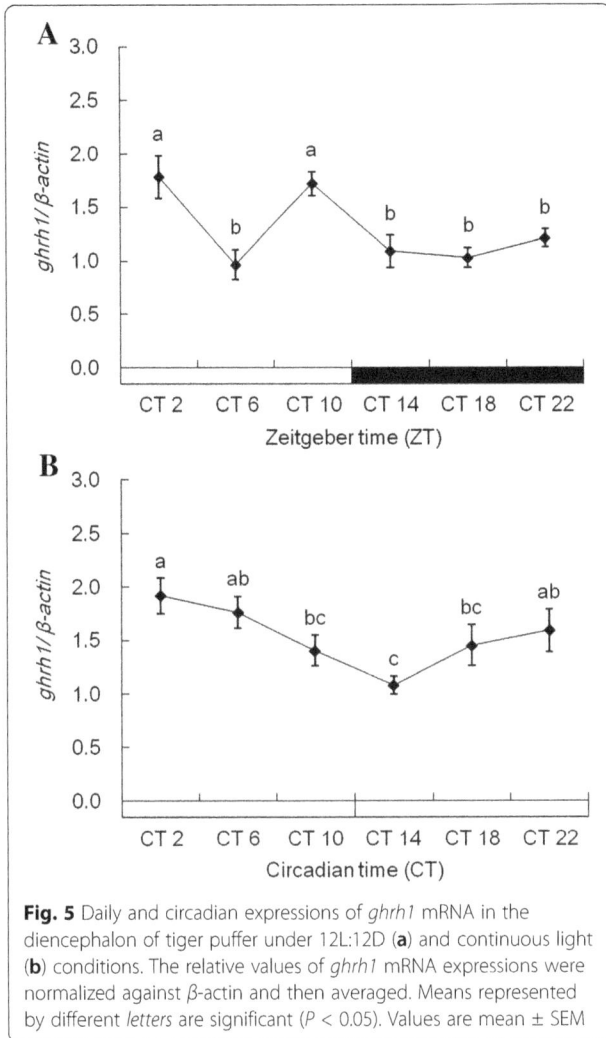

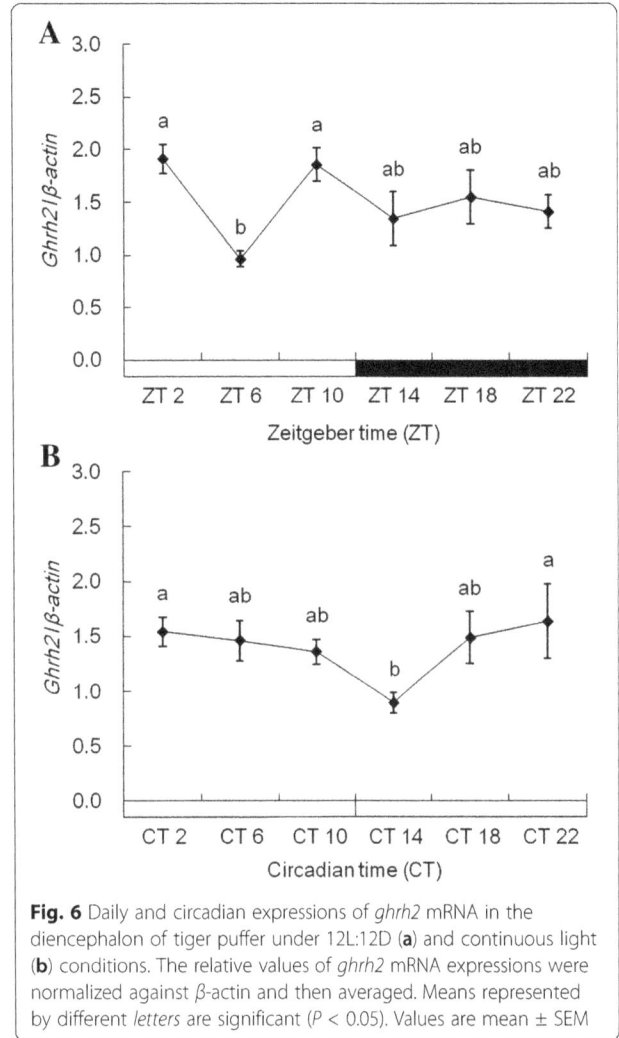

studies, hypothalamic *ghrh* levels increased rapidly in the first hour after light onset, then decreased for 4 h after light onset. The *ghrh* level gradually increased up to the beginning of scotophase and then steadily decreased during scotophase. The *ghrh1* and *ghrh2* mRNA rhythms in tiger puffers in this study were similar to the *ghrh* mRNA rhythms in rats. The expression of *ghrh1* mRNA in tiger puffers was high in the early part of photophase and then decreased toward the midpoint of photophase. Further, the *ghrh1* mRNA increased from the midpoint up to the end of photophase. During scotophase, *ghrh1* mRNA showed prolonged low expression levels. Therefore, we suggest that tiger puffers have daily rhythms of *ghrh* expression that vary with light and dark conditions.

Our results showed that the *ss1* mRNA expression pattern was similar to the pattern of melatonin secretion, but *ghrh1* and *ghrh2* mRNA expression patterns were opposite to those of melatonin secretion. We then examined the correlation of melatonin secretion and growth-related gene

expressions in tiger puffers. In our experiment, *ss1* mRNA in diencephalon of tiger puffer was showed inducing expression by melatonin i.p (Fig. 7a). But, no effect of melatonin on mRNA expressions of *ghrh1* and *ghrh2* was detected (Fig. 7b, c). This suggests that the daily fluctuation of *ss1* gene in the tiger puffer is controlled by melatonin, but not *ghrh* gene. Few studies have addressed the correlation of melatonin with SS or GHRH in vertebrates, including mammals. Our review of the literature found studies only on the relationship of melatonin to SS in the rat hippocampus (Izquierdo-Claros et al. 2004). This study showed a reduction in SS levels through decreasing somatostatinergic system activity, including inhibition of SS receptor activity by melatonin in the hippocampus. Although this finding seems to conflict with our results, we suggest that daily and circadian rhythms of growth-related genes are directly or indirectly controlled by melatonin in the tiger puffer.

Briefly, our results suggest that melatonin secretion and growth-related gene expressions follow daily and circadian rhythms in the tiger puffer. As a result, melatonin showed

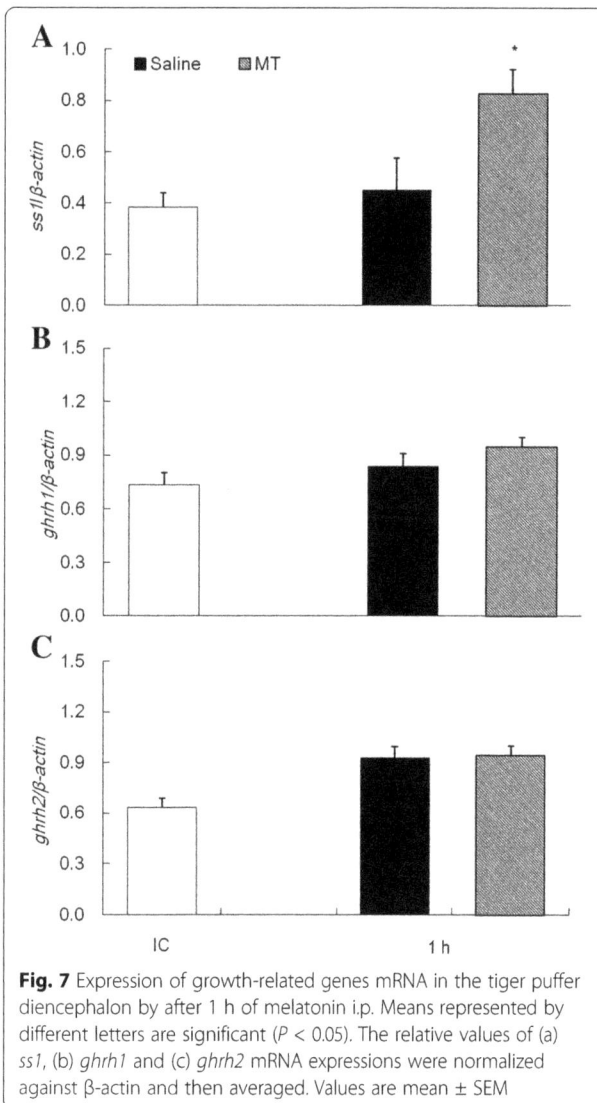

Fig. 7 Expression of growth-related genes mRNA in the tiger puffer diencephalon by after 1 h of melatonin i.p. Means represented by different letters are significant ($P < 0.05$). The relative values of (a) *ss1*, (b) *ghrh1* and (c) *ghrh2* mRNA expressions were normalized against β-actin and then averaged. Values are mean ± SEM

Acknowledgements

This research was supported by the Basic Science Research Program through the National Research Foundation of Korea (NRF) funded by the Ministry of Education, Science, and Technology (2012R1A6A3A04041089).

Funding

This research was supported by the Basic Science Research Program through the National Research Foundation of Korea (NRF).

Authors' contributions

BH and SP designed and carried out the data analysis and manuscript writing. SW participated in the fish sampling and data analysis. YT and AT participated in the plasma melatonin data analysis. YD participated in its design and coordination and helped to draft the manuscript. All authors read and approved the final manuscript.

Competing interests

The authors declare that they have no competing interests.

Author details

[1]Marine Science Institute, Jeju National University, Jeju 6333, South Korea. [2]Aquafeed Research Center, National Institute of Fisheries Science (NIFS), Pohang 37517, South Korea. [3]Department of Electrical Engineering and Bioscience, School of Advanced Science and Engineering, Waseda University, Shinjuku, Tokyo 162-8480, Japan. [4]Department of Chemistry, Biology and Marine Sciences, Faculty of Science, University of the Ryukyus, Nishihara, Okinawa 903-0213, Japan. [5]Jeju International Marine Science Research & Logistics Center, Korea Institute of Ocean Science & Technology, Jeju 63349, South Korea. [6]19-5, Hamdeok 5(o)-gill, Jocheon, Jeju Special Self-Governing Province 695-965, Republic of South Korea.

that nocturnal rhythm in 12L:12D condition and this rhythm was similarly showed in 24L condition compared with 12L:12D condition. Through these results, melatonin is suggested to have the circadian rhythm regardless of the day and night. The *ss1* expression appears to be induction by melatonin treatment. However, *ss1* expression rhythm showed different expression rhythm when compared with melatonin rhythm under 24L condition. Therefore, *ss1* mRNA suggest that directly or indirectly regulated by melatonin. The correlation between *ghrh* expression rhythms with melatonin was not confirmed. Therefore, we more need to further investigate the clarity of the mechanisms between melatonin and growth-related genes.

Abbreviations

AANAT: Arylalkylamine *N*-acetyltransferase; GH: Growth hormone; GHRH: Growth hormone-releasing hormone; LD: Light and dark; LL: Constant light; SS: Somatostatin

References

Ayson FG, Takemura A. Daily expression patterns for mRNAs of GH, PRL, SL, IGF-I and IGF-II in juvenile rabbitfish, *Siganus guttatus*, during 24-h light and dark cycles. Gen Comp Endocrinol. 2006;149:261–8.

Bayarri MJ, Munoz-Cueto JA, López-Olmeda JF, Vera LM, De Lama MR, Madrid JA, Sanchez-Vazquez FJ. Daily locomotor activity and melatonin rhythms in Senegal sole (*Solea senegalensis*). Physiol Behav. 2004;81:577–83.

Bégay V, Falcón J, Cahill GM, Klein DC, Coon SL. Transcripts encoding two melatonin synthesis enzymes in the Teleost pineal organ: circadian regulation in pike and Zebrafish, but not in trout 1. Endocrinology. 1998;139:905–12.

Bertherat J, Bluet-Pajot MT, Epelbaum J. Neuroendocrine regulation of growth hormone. Eur J Endocrinol. 1995;132:12–24.

Biswas AK, Seoka M, Ueno K, Yong AS, Biswas BK, Kim YS, Taki K, Kumai H. Growth performance and physiological responses in striped knifejaw, *Oplegnathus fasciatus*, held under different photoperiods. Aquaculture. 2008;279:42–6.

Bredow S, Taishi P, Obél F Jr, Guha-Thakurta N, Krueger JM. Hypothalamic growth hormone-releasing hormone mRNA varies across the day in rats. Neuroreport. 1996;7:2501–6.

Canosa LF, Chang JP, Peter RE. Neuroendocrine control of growth hormone in fish. Gen Comp Endocrinol. 2007;151:1–26.

De Pedro N, Martínez-Álvarez RM, Delgado M. Melatonin reduces body weight in goldfish (*Carassius auratus*): effects on metabolic resources and some feeding regulators. J Pineal Res. 2008;45:32–9.

Forward RB, Tankersley RA, Reinsel KA. Selective tidal stream transport of spot (*Leistomus xanthurus* Lacepede) and pinfish [*Lagodon rhomboides* (Linnaeus)] larvae: contribution of circatidal rhythms in activity. J Exp Mar Biol Ecol. 1998;226:19–32.

Fraser NH, Metcalfe NB, Thorpe JE. Temperature-dependent switch between diurnal and nocturnal foraging in salmon. Pro Royal Soc Lon B. 1993;252:135–9.

Fuentes EN, Valdés JA, Molina A, Björnsson BT. Regulation of skeletal muscle growth in fish by the growth hormone-insulin-like growth factor system. Gen Comp Endocrinol. 2013;192:136–48.

Gardi J, Obal F, Fang J, Zhang J, Krueger JM. Diurnal variations and sleep deprivation-induced changes in rat hypothalamic GHRH and somatostatin contents. Am J Phys. 1999;277:R1339–44.

Guerra-Santos B, López-Olmeda JF, de Mattos BO, Baião AB, Pereira DSP, Sánchez-Vázquez FJ, Cerqueira RB, Albinati RCBA, Fortes-Silva R. Synchronization to light and mealtime of daily rhythms of locomotor activity, plasma glucose and digestive enzymes in the Nile tilapia (Oreochromis niloticus). Com Biochem Physiol A Mol Integr Physiol. 2017;204:40–7.

Gunnarsson S, Imsland AK, Siikavuopio SI, Árnason J, Gústavsson A, Thorarensen H. Enhanced growth of farmed Arctic charr (Salvelinus alpinus) following a short-day photoperiod. Aquaculture. 2012;350:75–81.

Herrero MJ, Martínez FJ, Míguez JM, Madrid JA. Response of plasma and gastrointestinal melatonin, plasma cortisol and activity rhythms of European sea bass (Dicentrarchus labrax) to dietary supplementation with tryptophan and melatonin. J Comp Physiol B. 2007;177:319–26.

Hur SP, Takeuchi Y, Esaka Y, Nina W, Park YJ, Kang HC, Jeong HB, Lee YD, Kim SJ, Takemura A. Diurnal expression patterns of neurohypophysial hormone genes in the brain of the threespot wrasse Halichoeres trimaculatus. Comp Biochem Physiol A Mol Integr Physiol. 2011;158:490–7.

Iigo M, Furukawa K, Hattori A, Ohtani-Kaneko R, Hara M, Suzuki T, Tabata M, Aida K. Ocular melatonin rhythms in the goldfish, Carassius auratus. J Biol Rhythm. 1997;12:182–92.

Ishikawa M, Mizobuchi M, Takahashi H, Bando H, Saito S. Somatostatin release as measured by in vivo microdialysis: circadian variation and effect of prolonged food deprivation. Brain Res. 1997;749:226–31.

Izquierdo-Claros RM, MDC BLO-A, Arilla-Ferreiro E. Acutely administered melatonin decreases somatostatin-binding sites and the inhibitory effect of somatostatin on adenylyl cyclase activity in the rat hippocampus. J Pineal Res. 2004;36:87–94.

Klein SE, Sheridan MA. Somatostatin signaling and the regulation of growth and metabolism in fish. Mol Cell Endocrinol. 2008;286:148–54.

Lee LT, Siu FK, Tam JK, Lau IT, Wong AO, Lin MC, Vaudry H, Chow BK. Discovery of growth hormone-releasing hormones and receptors in nonmammalian vertebrates. Proc Natl Acad Sci. 2007;104:2133–8.

Luque RM, Park S, Kineman RD. Role of endogenous somatostatin in regulating GH output under basal conditions and in response to metabolic extremes. Mol Cell Endocrinol. 2008;286:155–68.

Maitra SK, Chattoraj A, Mukherjee S, Moniruzzaman M. Melatonin: a potent candidate in the regulation of fish oocyte growth and maturation. Gen Comp Endocrinol. 2013;181:215–22.

McLean E, Devlin RH, Byatt JC, Clarke WC, Donaldson EM. Impact of a controlled release formulation of recombinant bovine growth hormone upon growth and seawater adaptation in coho (Oncorhynchus kisutch) and chinook (Oncorhynchus tshawytscha) salmon. Aquaculture. 1997;156:113–28.

Moore AF, Menaker M. The effect of light on melatonin secretion in the cultured pineal glands of Anolis lizards. Comp Biochem Physiol A Mol Integr Physiol. 2011;160:301–8.

Nikaido Y, Ueda S, Takemura A. Photic and circadian regulation of melatonin production in the Mozambique tilapia Oreochromis mossambicus. Comp Biochem Physiol A Mol Integr Physiol. 2009;152:77–82.

Norberg B, Weltzien FA, Karlsen Ø, Holm JC. Effects of photoperiod on sexual maturation and somatic growth in male Atlantic halibut (Hippoglossus hippoglossus L.). Comp Biochem Physiol B Biochem Mol Biol. 2001;129:357–65.

Oliveira C, Garcia EM, López-Olmeda JF, Sánchez-Vázquez FJ. Daily and circadian melatonin release in vitro by the pineal organ of two nocturnal teleost species: Senegal sole (Solea senegalensis) and tench (Tinca tinca). Comp Biochem Physiol A Mol Integr Physiol. 2009;153:297–302.

Peinado MA, Fajardo N, Hernandez G, Puig-Domingo M, Viader M, Reiter RJ, Webb SM. Immunoreactive somatostatin diurnal rhythms in rat pineal, retina and harderian gland: effects of sex, season, continuous darkness and estrous cycle. J Neural Transm. 1990;81:63–72.

Qian Y, Yan A, Lin H, Li W. Molecular characterization of the GHRH/GHRH-R and its effect on GH synthesis and release in orange-spotted grouper (Epinephelus coioides). Comp Biochem Physiol B Biochem Mol Biol. 2012; 163:229–37.

Raven PA, Sakhrani D, Beckman B, Neregård L, Sundström LF, Björnsson BT, Devlin RH. Growth and endocrine effects of recombinant bovine growth hormone treatment in non-transgenic and growth hormone transgenic coho salmon. Gen Comp Endocrinol. 2012;177:143–52.

Revel FG, Saboureau M, Masson-Pévet M, Pévet P, Mikkelsen JD, Simonneaux V. Kisspeptin mediates the photoperiodic control of reproduction in hamsters. Curr Biol. 2006;16:1730–5.

Sheridan MA, Hagemeister AL. Somatostatin and somatostatin receptors in fish growth. Gen Comp Endocrinol. 2010;167:360–5.

Takemura A, Susilo ES, Rahman MD, Morita M. Perception and possible utilization of moonlight intensity for reproductive activities in a lunar-synchronized spawner, the golden rabbitfish. J Exp Zoolog A Comp Exp Biol. 2004;301:844–51.

Taylor JF, Migaud H, Porter MJR, Bromage NR. Photoperiod influences growth rate and plasma insulin-like growth factor-I levels in juvenile rainbow trout, Oncorhynchus mykiss. Gen Comp Endocrinol. 2005;142:169–85.

Very NM, Knutson D, Kittilson JD, Sheridan MA. Somatostatin inhibits growth of rainbow trout. J Fish Biol. 2001;59:157–65.

Wan X, Zhang X, Huo Y, Wang G. Weather entrainment and multispectral diel activity rhythm of desert hamsters. Behav Process. 2013;99:62–6.

Wood AW, Duan C, Bern HA. Insulin-like growth factor signaling in fish. Int Rev Cytol. 2005;243:215–85.

Xing Y, Wensheng L, Haoran L. Polygenic expression of somatostatin in orange-spotted grouper (Epinephelus coioides): molecular cloning and distribution of the mRNAs encoding three somatostatin precursors. Mol Cell Endocrinol. 2005;241:62–72.

21

Molecular detection of *Kudoa septempunctata* (Myxozoa: Multivalvulida) in sea water and marine invertebrates

Alagesan Paari[1,2,4], Chan-Hyeok Jeon[1,2,5], Hye-Sung Choi[3,6], Sung-Hee Jung[3] and Jeong-Ho Kim[1,7*]

Abstract

The exportation of cultured olive flounder (*Paralichthys olivaceus*) in Korea has been recently decreasing due to the infections with a myxozoan parasite *Kudoa septempunctata*, and there is a strong demand for strict food safety management because the food poisoning associated with consumption of raw olive flounder harbouring *K. septempunctata* has been frequently reported in Japan. The life cycle and infection dynamics of *K. septempunctata* in aquatic environment are currently unknown, which hamper establishment of effective control methods. We investigated sea water and marine invertebrates collected from olive flounder farms for detecting *K. septempunctata* by DNA-based analysis, to elucidate infection dynamics of *K. septempunctata* in aquaculture farms. In addition, live marine polychaetes were collected and maintained in well plates to find any possible actinosporean state of *K. septempunctata*. The level of *K. septempunctata* DNA in rearing water fluctuated during the sampling period but the DNA was not detected in summer (June–July in farm A and August in farm B). *K. septempunctata* DNA was also detected in the polychaetes *Naineris laevigata* intestinal samples, showing decreased pattern of 40 to 0%. No actinosporean stage of *K. septempunctata* was observed in the polychaetes by microscopy. The absence of *K. septempunctata* DNA in rearing water of fish farm and the polychaetes *N. laevigata* intestinal samples during late spring and early summer indicate that the infection may not occur during this period. *N. laevigata* was suspected as the possible alternate invertebrate host of *K. septempunctata*, but the actinosporean stage was not found by well plate method and further studies will be necessary. This research provides important baseline information for understanding the infection dynamics of *K. septempunctata* in olive flounder farms and further establishment of control strategies.

Keywords: Kudoa septempunctata, Myxozoa, Olive flounder, *Paralichthys olivaceus*, *Naineris laevigata*

Background

Myxozoans belong to the group of metazoan parasites of fish and act as a cause for several outbreaks in both fresh water and marine fish (Canning and Okamura 2003). Disease transmission by these myxozoan parasites can often have disastrous economic impact in aquaculture industries, although most of them are known to have insignificant or negligible effect in fish (Yokoyama et al. 2012). The genus *Kudoa* comprises more than 70 species reported from a broad range of fish host (Miller and

Adlard 2012). Most of the species are histozoic which develop symptoms of macroscopic whitish cyst or cause post-mortem myoliquefaction (Shirakashi et al. 2012). However, some *Kudoa* species do not cause any of the symptoms mentioned above and *Kudoa septempunctata*, a newly found myxosporean found in olive flounder (*Paralichthys olivaceus*) is probably the most well-known example of them (Yokoyama et al. 2004; Matsukane et al. 2010) .

Since 2011, food poisoning due to ingestion of farmed olive flounders in Japan has been reported (Kawai et al. 2012). Epidemiological studies have revealed that this outbreak is associated with the presence of *K. septempunctata* in the causative foods (Kawai et al. 2012) and the food-borne outbreaks associated with consumption

* Correspondence: jhkim70@gwnu.ac.kr
[1]Department of Marine Bioscience, Gangneung-Wonju National University, Gangneung, Gangwon 25457, South Korea
[7]Department of Marine Bioscience, Gangneung-Wonju National University, Gangneung, Gangwon 210-702, South Korea
Full list of author information is available at the end of the article

of raw olive flounder harbouring *K. septempunctata* are becoming a prominent public health concern in Japan. As the customs of consuming raw fish is spreading, the occurrence of this food-borne disease is predicted to increase (Harada et al. 2012). Although there have been outbreaks in Japan since 2011, the question of olive flounder in Korea acting as a host of *K. septempunctata* remains unanswered (Iwashita et al. 2013). Considering the commercial value of olive flounder and the public health concern, it is urgently needed to solve the negative impact of this parasite on public health and food safety, but almost nothing is known about its transmission biology, infection dynamics in aquatic environment.

Myxozoan parasites have been believed to be transmitted from fish to fish until Wolf and Markiw (1984) demonstrated that freshwater oligochaete was essential for the transmission of *Myxobolus cerebralis* and since then, many studies have confirmed that some myxozoans undergo two-host life cycle (Lom and Dykova 2006; Markussen et al. 2015). Currently, more than 30 freshwater myxozoans are known to have two-host life cycles (Yokoyama et al. 2012), but only 7 marine myxozoans are found to have marine invertebrates to complete their life cycles (Karlsbakk and Køie 2012; Køie et al. 2004, 2007, 2008, 2013; Rangel et al. 2015), and neither life cycle nor invertebrate alternate hosts have been elucidated in kudoid myxozoans.

Environmental water analysis is indispensable for investigating epidemiology of myxozoan infections because fish myxozoans occur in aquatic environment and transmission between two different hosts also occur in environmental water. Many studies revealed that the disease transmission occurs via water in endemic area and appropriate water treatments were effective for the management of several myxozoan infections (Cobcroft and Battaglene 2013; Nehring et al. 2003; Yanagida et al. 2006). Thus, the environmental water analysis would be the first step to clarify infection dynamics and develop further effective management strategy for *K. septempunctata* infection. As knowledge about the infection dynamics of this parasite is scarce, we carried out a monthly inspection of water samples to study the occurrence pattern of *K. septempunctata* in aquatic environment by molecular analysis. We also investigated the prevalence of *K. septempunctata* in marine polychaetes collected around the farms using both of well plate method described by Yokoyama et al. (1991) and molecular analysis, to speculate the possible life cycle of *K. septempunctata*.

Methods

Water sampling and DNA extraction

Water samples were directly collected from the inlet pumping units from two aquaculture farms where *K.* *septempunctata* infection occurs (farm A, B). Sampling was conducted on a monthly basis during May and November 2014 and approximately 2 l of water was collected in each sampling. Water samples were filtered through a fresh nitrocellulose membrane filter (5 µm pore size; ADVENTEC, Japan) using a suction pump (DOA-P704-AA, GAST, USA), and the membrane filter were placed in an individual micro centrifuge tube and stored at –20 °C until DNA extraction.

Acetone dissolution method was used for extracting DNA from the filter samples (Hallett et al. 2012). The membrane filter in microcentrifuge tube was air dried and dissolved by adding 2 ml of acetone (Cica reagent, Japan). The completely solubilized filter components by repeated vortexing were then centrifuged at 3000*g* for 15 min, and the supernatant was discarded. This step was repeated twice to ensure complete dissolution of filtrate particles from the dissolved materials. To the dissolved filtrate samples, 1 ml of 95% ethanol was added and thoroughly mixed. The suspended pellet after centrifugation was then air dried and used directly for DNA extraction.

DNA was extracted by using a QIAamp DNA Mini Kit (QIAGEN, USA) according to the manufacturer's instructions with slight modifications. Briefly, 180 µl of tissue lysis buffer (Buffer AE, QIAGEN, USA) was added to the air-dried pellet sample, and then 20 µl of Proteinase K (QIAGEN, USA) was added. Following overnight incubation, wash buffers (Buffer AW1, AW2, QIAGEN, USA) were added and eluted using elution buffer (Buffer AE, QIAGEN, USA). Extracted DNA was stored at –20 °C until used for PCR detection.

PCR and real-time PCR for detecting *K. septempunctata* in rearing water

PCR was carried out for detecting *K. septempunctata* in water samples using the following set of primer sets: *Ks*f-GTGTGTGATCAGACTTGATATG; *Ks*R-AAGCCA AAACTGCTGGCCATTT [25]. 0.5 µM of forward and reverse primer, 1 µl of template DNA was added to the PCR premix tube (Bioneer, Korea) and the total volume was made to 20 µl using ultra-pure distilled water (Invitrogen, USA). PCR cycling parameters followed the protocols of Grabner et al. (2012) with some minor modifications. PCR cycling parameters were an initial denaturation at 95 °C for 4 min, followed by 35 cycles at 95 °C for 35 s, 56 °C for 30 s and 72 °C for 30 s and ended with a final extension at 72 °C for 7 min.

Real-time PCR was carried out using the following sequence of primers and probe; F-CATGGGATTAGCCC GGTTTA; R-ACTCTCCCCAAAGCCGAAA; P-(FAM)-TCCAGGTTGGGCCCTCAGTGAAAA (Kawai et al. 2012). Real-time PCR was carried out in a 0.2-ml PCR strip tube containing 2× Premix Ex Taq (Takara, Japan)

10 µl, primer (0.4 uM, Bioneer, Korea), probe (0.25 µM, Bioneer, Korea), ROX II reference dye (Takara, Japan), 4 µl template DNA using ABI 7500 Fast Real-time PCR system (Applied Biosystems, USA). Cycling parameters were preheating at 95 °C for 10 min, followed by 45 cycles at 95 °C for 15 s, 60 °C for 1 min and the analyses were conducted twice.

Collection of marine invertebrates and species identification

Marine invertebrates from the sediments of fish tanks and coastal areas near fish farms were collected to investigate the prevalence of *K. septempunctata* infection. For collecting invertebrate samples, mud was collected from approximately 0.5 m depth from outflow waterway of the same fish farms where the water samples were collected and transferred to laboratory. Sediments were sieved through a mesh (0.5 mm) to separate marine polychaetes within a day of collection of sediment. Live polychaetes were collected, washed with sterile sea water several times, and then maintained in a 12-well plate at 15 °C, for observing the possible actinosporean stages of *K. septempunctata.*

For collecting invertebrate samples from coastal areas, quadrats and dredging devices were used by trained divers to obtain marine invertebrates near the olive flounder farms. Subsamples of invertebrate samples collected from the gravel materials were washed with sterile sea water and were fixed in 70% ethanol for taxonomic identification and molecular detection.

Species identification for all the collected invertebrate samples was conducted by morphological observations or PCR amplification of the mitochondrial cytochrome c oxidase subunit I (mt COI) gene as described by Maturana et al. (2011). For molecular identification, PCR primers targeting the partial mt COI gene described by Folmer et al. (1994) were used. LCO1490: 5′-GGTCAACAAATCAT-AAAGATATTGG-3′; HC02198: 5′-TAAACTTCA GGGTGACCAAAAAATCA-3′ DNA was extracted from the polychaete samples using the QIAamp DNA Mini Kit with the protocol described previously, and PCR was carried out in a 20 µl of reaction volume consisted of 10 µl of PCR premix (Bioneer, Korea), 1 µl of template DNA, 1 µl of 10 µM of each primers and 17 µl of double-distilled deionized water. PCR cycling parameters were an initial denaturation phase at 94 °C for 1 min, followed by 35 cycles at 94 °C for 30 s, 49 °C for 55 s, and 72 °C for 90 s, and a final extension at 72 °C for 10 min. Following amplification, PCR products were analysed in a 2% agarose gels and stained with ethidium bromide. PCR products of the expected size were purified using PCR gel purification kit (Bioneer, Korea). Gel-purified PCR amplicons were sequenced in both directions utilizing the same primers used for initial amplification. The obtained sequences of mt COI gene were aligned and identified based on the percentage identity of nucleotide sequences previously registered in NCBI.

Microscopic detection of *K. septempunctata* from polychaetes

Live polychaetes collected from the outflow waterway sediments of fish tank were observed for the occurrence of actinospores by well plate method (Yokoyama et al. 1991). Briefly, collected polychaetes were washed with sterile sea water several times, individually placed in a 12-well plate and observed microscopically using an inverted microscope (Leica, Germany) every day to find if the possible *K. septempunctata* actinospores were released. Sea water in the wells was replaced in a 2-day interval. Squash preparations were also made from the isolated polychaetes samples at every 2 days interval for observation of the actinosporeans; several posterior segments of the intestinal region of polychaetes were squashed between the slide and cover slip, fixed with methanol, stained with Giemsa solutions and examined using a light microscope (Leica, Germany).

Invertebrate samples collected around the coastal regions were not maintained in well plates but immediately processed for microscopic analysis using squash preparations and for PCR detection using the same methodology mentioned above.

PCR and real-time PCR detection of *K. septempunctata* in marine invertebrates

Polychaetes collected from the outflow waterway sediments of fish tank were examined for *K. septempunctata* by PCR and real-time PCR. DNA was extracted from the whole body of polychaetes using QIAamp DNA Mini Kit (QIAGEN, USA) following manufacturer's instructions. The PCR and real-time PCR primers and amplification conditions used in this analysis are mentioned above. Prevalence of *K. septempunctata* was calculated as the proportion of infected invertebrate host in the whole number of studied host. PCR for detecting *K. septempunctata* was also conducted for invertebrate samples collected from the coastal regions using the same protocol mentioned previously.

Results
Detection of *K. septempuncta* in rearing water samples by PCR and real-time PCR

During the sampling period, positive signals for *K. septempunctata* were not detected in any of the water samples from two farms by PCR (Table 1). The standard curve for real-time PCR was derived from 10-fold serial dilutions of different plasmid DNA concentrations ranging from 1×10^8 to 1×10^1 copies/µl, as described by Kawai et al. (2012). The assays were linear with R^2 values

Table 1 Detection of *K. septempunctata* DNA in rearing water samples from fish farms by molecular methods

| Sampling date | PCR | | Real-time PCR | | | |
| | | | Ct value | | rDNA copy number | |
	Farm A	Farm B	Farm A	Farm B	Farm A	Farm B
2014. May.	–	–	40.9, 41.1	40.1, 43.0	9.0, 7.7	16.3, 1.9
2014. Jun.	–	–	ND, ND	42.2, 40.4	ND, ND	3.4, 13.1
2014. Jul.	–	–	ND, ND	36.1, 38.3	ND, ND	3.2×10^2, 62.5
2014. Aug.	–	–	42.2, ND	ND, ND	3.5, ND	ND, ND
2014. Sept.	–	–	41.4, 39.3	40.7, ND	6.2. 29.7	10.4, ND
2014. Oct.	–	–	36.3, 34.5	37.0, 37.9	2.8×10^2, 1.1×10^3	1.6×10^2, 87.5
2014. Nov.	–	–	40.0, 39.5	39.3, 39.8	18.3, 26.2	29.7, 20.4

of 0.993 (Fig. 1). Relative *K. septempunctata* DNA concentration was calculated based on the Ct value. The amount of *K. septempunctata* DNA was inversely proportional to the Ct value obtained in this study, and the Ct value of the highest standard (10^8 copies/µl) was 14.8 and the lowest standard (10^1 copies/µl) was 37.1.

The level of rDNA copy number for all of the water samples fluctuated during the sampling period in both of the farms. The Ct value ranged from 36.3–42.2 in farm A and 36.1–42.2 in farm B (Table 1). The highest rDNA copy number (1.1×10^3) of *K. septempunctata* in Farm A was recorded in October. In Farm B, the highest rDNA copy number was recorded 3.2×10^2 in August. Interestingly, *K. septempuctata* DNA was not detectable during June–July in farm A and August in farm B.

Incidence of *K. septempunctata* in marine invertebrate samples collected from fish farms

Microscopic observation of the marine polychaetes collected was conducted on a daily basis until they die. The live polychaetes survived for 7 to 10 days in the well plate; however, no actinosporean release was observed from them during the incubation period. The squashed slide

specimens were made with the intestinal segments of randomly selected polychaetes, and microscopic observation was also conducted after Giesma and eosin staining, but any actinosporean-like stage of *K. septempunctata* was not found (data not shown).

All of the polychaetes were identified using PCR. Primers targeting mt COI gene amplified a PCR product of size 710 bp, and the amplified sequences represent the species *Naineris laevigata* (Polychaeta, Orbiniidae) with 99.0% homology (data not shown).

K. septempunctata DNA was detected in polychaetes by PCR and real-time PCR during the sampling period. PCR detection of *K. septempunctata* in polychaete intestinal sample showed the mean prevalence of 9.5% (55/578) (Table 2). The highest prevalence of *K. septempunctata* in polychaetes by PCR (40.0%) was recorded in May of 2014, then, gradually decreased to 0% in August.

Quantitative analysis of *K. septempunctata* DNA in polychaetes samples revealed the parasitic DNA was detectable only in May and June. The Ct value was 38.9–41.4 in May and 35.5–38.3 in June. Although the incidence of PCR-

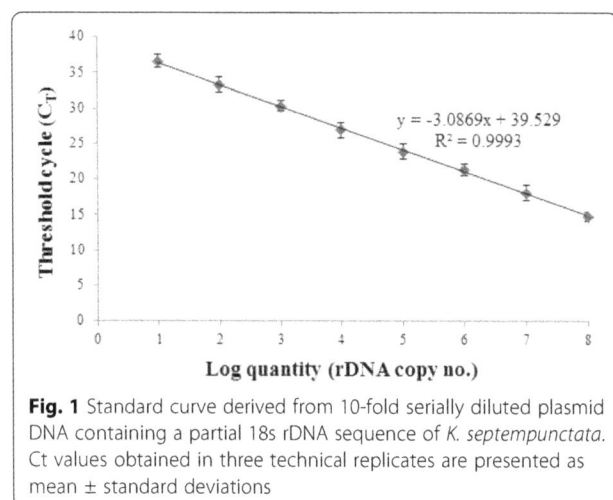

Fig. 1 Standard curve derived from 10-fold serially diluted plasmid DNA containing a partial 18s rDNA sequence of *K. septempunctata*. Ct values obtained in three technical replicates are presented as mean ± standard deviations

$y = -3.0869x + 39.529$
$R^2 = 0.9993$

Table 2 PCR and real-time PCR results for detecting *K. septempunctata* in orbiniid polychaete *N. laevigata* isolated from outflow waterway of fish tank

| | PCR Positive samples/ tested samples (%) | Real-time PCR | |
		Ct value	rDNA copy number/ body weight (g)
May	36/90 (40.0%)	38.9–41.4	2.5×10^3–1.6×10^3
June	14/62 (22.6%)	35.5–38.3	2.0×10^3–2.1×10^4
July	5/126 (4.0%)	nd	nd
August	0/84	nd	nd
September	0/82	nd	nd
October	0/62	nd	nd
November	0/72	nd	nd
	55/578 (9.5%)		

nd not detected

positive samples was higher in May, the rDNA copy number was higher in June (Table 2).

Incidence of *K. septempunctata* in marine invertebrate samples collected from coastal area

Invertebrate samples collected during May to November around the coastal area near olive flounder farm were identified by microscopic observation at the lowest taxon level, and the results are summarized in Table 3. All of the collected samples were negative for *K. septempunctata* by PCR and not detected by real-time PCR.

Discussion

Outbreaks due to consumption of raw olive flounder harbouring *K. septempunctata* have been reported in Japan since 2011 (Kawai et al. 2012; Harada et al. 2012).

Table 3 Detection of *K. septempunctata* in marine invertebrates collected from coastal waters in this study

Family name	Species name	Number of individuals examined	PCR
Arabellidae	*Arabella iricola*	4	–
Chrysopetalidae	*Chrysopetalum occidentale*	1	–
Cirratulidae	*Dodecaceria laddi*	1	–
Eunicidae	*Eunice antennata*	6	–
	Lysidice collaris	2	–
Euphrosinidae	*Euphrosine superba*	3	–
Nereidae	*Nereis multignatha*	4	–
	Nereis neoneanthes	1	–
	Perinereis cultrifera	5	–
	Platynereis bicanaliculata	5	–
Opheliidae	*Polyophalmus pictus*	3	–
Orbiniidae	*Naineris laevigata*	5	–
Phyllodocidae	*Eulalia virdis*	1	–
	Notophyllum sp.	2	–
Polynoidae	*Halosydna brevisetosa*	19	–
	Harmothoe imbricata	3	–
	Lepidonotus sp.	1	–
Serpulidae	*Hydroides exalata*	2	–
	Spirobranchus tetraceros	2	–
Syllidae	Unidentified 1	11	–
	Unidentified 2	4	–
	Trypanosyllis zebra	2	–
	Typosyllis ehlersioides	2	–
Terebellidae	Unidentified	3	–

However, information on the transmission biology of *K. septempunctata* is still lacking as we are not aware of the alternate polychaete host to complete its life cycle or the transmission dynamics of *K. septempunctata* in marine environment. In this study, we analysed rearing water samples from fish farms on a monthly basis for the presence of *K. septempunctata* by molecular methods.

Since the discovery of the myxozoan life cycle by Wolf and Markiw (1984), a lot of fresh water myxozoans have been known to use freshwater oligochaetes as alternate invertebrate hosts (Yokoyama et al. 2012). For marine myxozoans, however, polychaetes have been suggested as the best candidates for the alternative invertebrate hosts; seven marine myxozoan life cycles have been elucidated at present time, and all of them are known to use polychaetes as alternative invertebrate hosts (Karlsbakk and Køie 2012; Køie et al. 2004, 2007, 2008, 2013; Rangel et al. 2009), except for *Ortholinea auratae* using marine oligochaete as marine invertebrate host (Rangel et al. 2015). Thus, we exclusively sampled marine invertebrates from the sediment of outflow waterway of fish tank and around the fish farms, then investigated them by microscopic observation and molecular analysis, to find the possible invertebrate hosts they use for transmission. Our investigations on myxospore infection were restricted for few months due to adverse climate conditions.

Parasite density in aquatic environment is an important factor affecting the level of myxozoan outbreaks (Ray et al. 2012) because the transmission of actinospores to teleost hosts occurs in aquatic environment. Real-time PCR has successfully detected the actinospores to measure parasite density in freshwater environment (Hallet and Bartholomew 2009; Hallett et al. 2012; True et al. 2009) but less frequently in marine environment. Alma-Bermejo et al. (2013) and Ishimaru et al. (2014) developed real-time PCR assay for detecting marine myxosporeans, *Ceratomyxa puntazzi* and *Kudoa yasunagai* in environmental sea water, respectively. They found the seasonal changes in parasitic density and mentioned that this may reflect the infection dynamics of marine myxozoans. In a similar manner, *K. septempunctata* DNA was detected both in water samples of two farms examined in this study. Overall Ct value was 36.3–42.2 in farm A and 36.1–43.0 in farm B, corresponding 3.5 to 2.8×10^2 copies of *K. septempunctata* 18s rDNA from 2 l water samples in farm A and 3.0 to 3.2×10^2 copies in farm B, respectively (Table 1). These values are lower than those of *K. yasunagai* (Ishimaru et al. 2014) but higher than those of *C. puntazzi* (Alma-Bermejo et al. 2013). These differences are thought to be due to many factors including different infection dynamics between the parasites and their hosts, different aquaculture systems, many physical and chemical factors in the aquatic environments, as suggested by Ishimaru et al. (2014). In particular, a flow-through system pumping seawater directly from the open sea makes more difficult to understand the infection dynamics of *K. septempunctata* in olive flounder farms as because it has been suggested that the actinosporean stages are generally fragile and easy to be destroyed by strong water flow (Kerans and Zale 2002; Hoz Franco and Budy 2004; Hallet and Bartholomew 2007). Nevertheless, considering the sea water volume flowing into the fish tanks, the parasite density calculated in this study cannot be negligible and should be considered for elucidating infection dynamics of *K. septempunctata* in olive flounder farms.

The level of *K. septempunctata* DNA in rearing waters fluctuated during the experimental period, but *K. septempuctata* DNA was not detected in June–July (farm A) or August (farm B). Similar results were obtained by Ishimaru et al. (2014) with *K. yasunagai* and Alama-Bermejo et al. (2013) with *C. punctzaaii*, suggesting changes in parasite density in the water may be related to the water temperature. But it is not clear if the parasite DNA detected from water in this study is originated from actinospores from invertebrate hosts or from myxospores from fish hosts, which is also indicated as the major shortcoming of their study by the authors mentioned above. In addition, our data have some limitations because we analysed water samples only for a half year. Seasonality in the prevalence has been reported in many myxosporean infections (Al-Qahtani et al. 2015; Abdel-Baki et al. 2015), thus sentinel fish exposure to water from the endemic area throughout the year and discovery of actinospores with subsequent quantification of them in seawater are thought to be necessary, to prove the parasite DNA in seawater came from actinosporean stage of *K. septempunctata*. Recently, Yokoyama et al. (2015) described that the *K. septempunctata* predominantly invade juvenile olive flounder in July. In our study, *K. septempunctata* DNA in rearing water was not detected during June and July or in August, which is also indicating the infection may occur during summer season and can be helpful for avoiding *K. septempunctata* infection of olive flounder.

All of the live polychaetes from fish farm sediments were identified as *Naineris laevigata* (Polychaeta, Orbiniidae) by PCR amplification of mt COI gene and maintained in 12-well plates for approximately 2 weeks, but no actinosporean stage were observed. Freshwater actinospores from oligochaetes have been successfully observed by well plate method (Yokoyama et al. 1991, 2012). However, we could not find any actinospores released from *N. laevigata* maintained in well plates in this study. Most of the actinospores from marine polychaetes have been observed either by collecting coelomic fluid of marine polychaetes with syringe needles or squash preparations (Køie et al. 2008; Rangel et al. 2009, 2011) and our study was the first trial to observe the release of actinospores from marine polychaetes by well

plate method. Rangel et al. (2011) mentioned that the well-plate method may not be suitable for relatively big marine polychaetes. Otherwise, different mechanisms may work for the release of the actinospores from the invertebrate hosts; Køie (2002) mentioned that the actinospores are released via the gonopores of polychaetes, while Rangel et al. (2009) described that the actinospores are released together with the gametes by the rupture of the host's body wall. If this is the case, it will be necessary to exclusively make squash preparation or histological sections to observe possible actinosporean stages of *K. septempunctata*.

PCR could successfully detect *K. septempunctata* DNA in polychaetes in this study. The percentage of PCR positive individuals was the highest (40.0%) in May, then gradually decreased during the experimental periods and maintained 0% after August of 2014. Seasonal prevalence of actinospores in invertebrate hosts has been conducted in several experiments, but the seasonal variation patterns were different depending on the species examined (Rangel et al. 2009, 2011). These are thought to reflect the seasonality of vertebrate host life cycle or invertebrate host life cycle for at least some myxozoans. In case of *N. laevigata* in this study, seasonal pattern of PCR-positive rate was also observed. However, this should be carefully interpreted because PCR can detect both of mature and developmental stages of actinospores. In general, the prevalence of actinosporean infection in polychaetes is estimated by microscopic observation and known to be very low (Rangel et al. 2009, 2011). Thus, it would be helpful to detect any seasonal patterns in prevalence of actinosporean infection in polychaetes hosts by molecular method, but successful microscopic observation should be accompanied, which is also indispensable to make a clear conclusion whether *N. laevigata* is the alternate host of *K. septempunctata*.

Conclusions

Myxozoan infections in wild and farmed fish is becoming increasingly important as marine aquaculture expands to meet the resource demands and some of them actually cause economic loss in aquaculture industry by causing considerable mortality or losing market value of them. *K. septempunctata* does not belong to any of two types mentioned above because it does not cause any negative effect on host but may affect human beings. Thus, effective control methods for *K. septempunctata* infection in olive flounder are urgently needed. Based on the knowledge of the transmission biology, several methods have been suggested for controlling myxozoan infections and some of them have been proved to be effective. The information obtained in this study is thought to be helpful for establishing strategies to avoid *K. septempunctata* infection in olive flounder farms.

Abbreviations

Ct: Threshold cycle; mt COI: Mitochondrial cytochrome c oxidase subunit 1; PCR: Polymerase chain reaction

Acknowledgements

We thank Jong-Goo Choi for his help during the experiments.

Funding

This work was supported by grants from the National Institute of Fisheries Science (R2016065).

Author's contributions

JHK conceived and designed the experiments. JHK and AP prepared the manuscript. AP and CHJ carried out sample collections and analyses. JHK, AP, HSC, SHJ, and CHJ interpreted the results. All authors read and approved the final manuscript.

Competing interests

The authors declare that they have no competing interests.

Author details

[1]Department of Marine Bioscience, Gangneung-Wonju National University, Gangneung, Gangwon 25457, South Korea. [2]East Coast Life Science Institute, Gangneung-Wonju National University, Gangneung, Gangwon 25457, South Korea. [3]Pathology Division, National Institute of Fisheries Science (NIFS), Busan 46041, South Korea. [4]Present address: Department of Life science, Christ University, Bengaluru, Karnataka, India. [5]Present address: East Sea Life Resources Center, Korea Fisheries Resources Agency, 36340, Uljin, South Korea. [6]Present address: Inland Aquaculture Research Center, NIFS, Changwon 51688, South Korea. [7]Department of Marine Bioscience, Gangneung-Wonju National University, Gangneung, Gangwon 210-702, South Korea.

References

Abdel-Baki AA, Mansour L, Al-Qahtani HA, Al-Omar SY, Al-Quraishy S. Morphology, seasonality and phylogenetic relationships of Ceratomyxa husseini n. sp. from the gall-bladder of Cephalopholis hemistiktos (Rüppell) (Perciformes: Serranidae) in the Arabian Gulf of Saudi Arabia. Syst Parasitol. 2015;91:91–9.

Alama-Bermejo G, Sima R, Raga JA, Holzer AS. Understanding myxozoan infection dynamics in the sea: seasonality and transmission of Ceratomyxa puntazzi. Int J Parasitol. 2013;43:771–80.

Al-Qahtani HA, Mansour L, Al-Quraishy S, Abdel-Baki AS. Morphology, phylogeny and seasonal prevalence of Ceratomyxa arabica n. sp. (Myxozoa: Myxosporea) infecting the gallbladder of Acanthopagrus bifasciatus (Pisces: Sparidae) from the Arabian Gulf, Saudi Arabia. Parasitol Res. 2015;114:465–671.

Canning EU, Okamura B. Biodiversity and evolution of the myxozoa. Adv Parasitol. 2003;56:43–131.

Cobcroft JM, Battaglene SC. Ultraviolet irradiation is an effective alternative to ozonation as a sea water treatment to prevent Kudoa neurophila (Myxozoa: Myxosporea) infection of striped trumpeter Latris lineata (Forster). J Fish Dis. 2013;36:57–65.

Folmer O, Black M, Hoeh W, Lutz R, Vrijenhoek R. DNA primers for amplification of mitochondrial cytochrome C oxidase subunit I from diverse metazoan invertebrates. Mol Mar Biol Biotechnol. 1994;3:294–9.

Grabner DS, Yokoyama H, Shirakashi S, Kinami R. Diagnostic PCR assays to detect and differentiate Kudoa septempunctata, K. thyrsites and K. lateolabracis (Myxozoa, Multivalvulida) in muscle tissue of olive flounder (Paralichthys olivaceus). Aquaculture. 2012;338:36–40.

Hallet SL, Bartholomew JL. Development and application of a duplex qPCR for river water samples to monitor the myxozoan parasite Parvicapsula minibicornis. Dis Aquat Org. 2009;86:36–50.

Hallet SL, Bartholomew JL. Effects of water flow on the infection dynamics of Myxobolus cerebralis. Parasitology. 2007;135:371–84.

Hallett SL, Ray RA, Hurst CN, Buckles GR, Atkinson SD, Bartholomew JL. Density of the waterborne parasite Ceratomyxa shasta and its biological effects on salmon. Appl Environ Microbiol. 2012;78:3724–31.

Harada T, Kawai T, Jinnai M, Ohnishi T, Sugita-Konishi Y, Kumeda Y. Detection of Kudoa septempunctata 18S ribosomal DNA in patient fecal samples from novel food-borne outbreaks caused by consumption of raw olive flounder (Paralichthy solivaceus). J Clin Microbiol. 2012;50:2964–8.

Hoz Franco E, Budy P. Linking environmental heterogeneity to the distribution and prevalence of Myxobolus cerebralis: a comparison across sites in a Northern Utah water shed. Trans Am Fish Soc. 2004;133:1176–89.

Ishimaru K, Matsuura T, Tsunemoto K, Shirakashi S. Seasonal monitoring of Kudoa yasunagai from sea water and aquaculture water using quantitative PCR. Dis Aquat Org. 2014;108:45–52.

Iwashita Y, Kamijo Y, Nakahashi S, Shindo A, Yokoyama K, Yamamoto A, Omori Y, Ishikura K, Fujioka M, Hatada T, Takeda T, Maruyama K, Imai H. Food poisoning associated with Kudoa septempunctata. J. Emerg. Med. 2013; 44: 943–945.

Karlsbakk E, Køie M. The marine myxosporean Sigmomyxa sphaerica (Thélohan, 1895) gen. n., comb. n. (syn. Myxidium sphaericum) from garfish (Belone belone (L.)) uses the polychaete Nereis pelagica L. as invertebrate host. Parasitol. Res. 2012;110:211–218.

Kawai T, Sekizuka T, Yahata Y, Kuroda M, Kumeda Y, Iijima Y, Kamata Y, Sugita-Konishi Y, Ohnishi T. Identification of Kudoa septempunctata as the causative agent of novel food poisoning outbreaks in Japan by consumption of Paralichthys olivaceus in raw fish. Clin. Infect. Dis. 2012;54:1046–1052.

Kerans BL, Zale AV. The ecology of Myxobolus cerebralis. In Whirling Disease: Reviews and Current Topics (ed. Bartholomew, J. L. and Wilson, J. C.). American Fisheries Society Symposium. 29, Bethesda, Maryland, USA. 2002. p. 145–166.

Køie M. Spirorchid and serpulid polychaetes are candidates as invertebrate hosts for Myxozoa. Folia Parasitol. 2002;49:160–162

Køie M. Whipps CM, Kent ML. Ellipsomyxa gobii (Myxozoa: Ceratomyxidae) in the common goby Pomatoschistus microps (Teleostei: Gobiidae) uses Nereis spp. (Annelida: Polychaeta) as invertebrate hosts. Folia Parasitol. 2004;51:14–18.

Køie M, Karlsbakk E, Nylund A. A new genus Gadimyxa with three new species (Myxozoa, Parvicapsulidae) parasitic in marine fish (Gadidae) and the two host life cycle of Gadimyxa atlantica n. sp. J. Parasitol. 2007;93: 1459–1467

Køie M, Karlsbakk E, Nylund A. The marine herring myxozoan Ceratomyxa auerbachi (Myxozoa: Ceratomyxidae) uses Chone infundibuliformis (Annelida: Polychaeta: Sabellidae) as invertebrate host. Folia Parasitol. 2008;55:100–104.

Køie M, Karlsbakk E, Einen ACB, Nylund A. A parvicapsulid (Myxozoa) infecting Sprattus sprattus and Clupea harengus (Clupeidae) in the Northeast Atlantic uses Hydroides norvegicus (Serpulidae) as invertebrate host. Folia Parasitol. 2013;60:149–154.

Lom J, Dykova I. Myxozoan genera: definition and notes on taxonomy, life-cycle terminology and pathogenic species. Folia Parasitol. 2006;53:1–36.

Markussen T, Agusti C, Karlsbakk E, Nylund A, Brevik O, Hansen H. Detection of the myxosporean parasite Parvicapsula pseudobranchicola in Atlantic salmon (Salmo salar L.) using in situ hybridization (ISH). Parasite Vectors. 2015;8:1–6.

Matsukane Y, Sato H, Tanaka S, Kamata Y, Sugita-Konishi Y. Kudoa septempunctata n. sp. (Myxosporea: Multivalvulida) from an aquacultured olive flounder (Paralichthys olivaceus) imported from Korea. Parasitol. Res. 2010;107:865–872.

Maturana CS, Moreno RA, Labra FA, González CA, Rozbaczylo N, Carrasco FD, Poulin E, DNA barcoding of marine polychaetes species of southern Patagonian fjords. Rev. Biol. Mar. Oceanogr. 2011;46:35–42.

Miller TL, Adlard RD. Brain infecting kudoids of Australia's coral reefs, including a description of Kudoa lemniscati n. sp (Myxosporea: Kudoidae) from Lutjanus lemniscatus (Perciformes: Lutjanidae) off Ningaloo Reef, Western Australia. Parasitol. Int. 2012;61:333–342.

Nehring RB, Thompson KG, Taurman K, Atkinson W. Efficacy of passive sand filtration in reducing exposure of salmonids to the actinospore of Myxobolus cerebralis. Dis. Aquat. Org. 2003;57:77–83.

Ray RA, Holt RA, Bartholomew JL. Relationship between temperature and Ceratomyxa shasta induced mortality in Klamath river salmonids. J. Parasitol. 2012;95:561–569.

Rangel LF, Santos MJ, Cech G, Székely C. Morphology, Molecular Data, and
        Development of *Zschokkella mugilis* (Myxosporea, Bivalvulida) in a polychaete
        alternate Host, *Nereis diversicolor*. J. Parasitol. 2009;95:561–569.
Rangel LF, Cech G, Székely C, Santos MJ. A new actinospore type
        Unicapsulactinomyxon (Myxozoa), infecting the marine polychaete, *Diopatra
        neapolitana* (Polychaeta: Onuphidae) in the Aveiro Estuary (Portugal).
        Parasitology 2011; 138:698–712.
Rangel LF, Rocha S, Castro R, Severino R, Casal G, Azevedo C, Cavaleiro F, Santos
        MJ. The life cycle of *Ortholinea auratae* (Myxozoa: Ortholineidae) involves an
        actinospore of the triactinomyxon morphotype infecting a marine
        oligochaete. Parasitol. Res. 2015;114: 2671–2678.
Shirakashi S, Morita A, Ishimaru K, Miyashita S. Infection dynamics of *Kudoa
        yasunagai* (Myxozoa: Multivalvulida) infecting brain of cultured yellowtail
        Seriola quinqueradiata in Japan. Dis. Aquat. Org. 2012;101:123–130.
True K, Purcell MK, Foott JS. Development and validation of a quantitative PCR to
        detect *Parvicapsula minibicornis* and comparison to histologically ranked
        infection of juvenile Chinook salmon, *Oncorhynchus tshawytscha* (Walbaum),
        from the Klamath River, USA. J. Fish Dis. 2009;32:183–192.
Wolf K, Markiw ME. Biology contravenes taxonomy in the myxozoa: new
        discoveries show alternation of invertebrate and vertebrate hosts. Science.
        1984;225:1449–1452.
Yanagida T, Sameshima M, Nasu H, Yokoyama H, Ogawa K. Temperature effects
        on the development of *Enteromyxum* spp. (Myxozoa) in experimentally
        infected tiger puffer *Takifugu rubripes* (Temminck & Schlegel). J. Fish Dis.
        2006;29:561–567.
Yokoyama H, Ogawa K, Wakabayashi H. A new collection method of
        actinosporeans-a probable infective stage of myxosporeans to fishes-from
        tubificids and experimental infection of goldfish with the actinosporean,
        Raabeia sp. Fish Pathol. 1991;26:133–138.
Yokoyama H, Whipps CM, Kent ML, Mizuno K, Kawakami H. *Kudoa thyrsites* from
        Japanese flounder and *Kudoa lateolabracis* n. sp. from Chinese sea bass:
        causative myxozoans of post-mortem myoliquefaction. Fish Pathol. 2004;39:
        79–85.
Yokoyama H, Grabner D, Shirakashi S. Transmission biology of the Myxozoa. In:
        Carvalho ED, David GS, Silva RJ (eds) Health and environment in aquaculture,
        In Tech, Rijeka, 2012.p. 3–42.
Yokoyama H, Lu M, Mori K, Satoh J, Mekata T, Yoshinaga T. Infection dynamics of
        Kudoa septempunctata (Myxozoa: Multivalvulida) in hatchery-produced olive
        flounder *Paralichthys olivaceus*. Fish Pathol. 2015;50:60–67.

22

Preparation and characterization of protein isolate from Yellowfin tuna *Thunnus albacares* roe by isoelectric solubilization/precipitation process

Hyun Ji Lee[1], Gyoon-Woo Lee[1], In Seong Yoon[1], Sung Hwan Park[1], Sun Young Park[2], Jin-Soo Kim[2] and Min Soo Heu[1*]

Abstract

Isoelectric solubilization/precipitation (ISP) processing allows selective, pH-induced water solubility of proteins with concurrent separation of lipids and removal of materials not intended for human consumption such as bone, scales, skin, etc. Recovered proteins retain functional properties and nutritional value. Four roe protein isolates (RPIs) from yellowfin tuna roe were prepared under different solubilization and precipitation condition (pH 11/4.5, pH 11/5.5, pH 12/4.5 and pH 12/5.5). RPIs contained 2.3–5.0 % moisture, 79.1–87.8 % protein, 5.6–7.4 % lipid and 3.0–3.8 % ash. Protein content of RPI-1 and RPI-2 precipitated at pH 4.5 and 5.5 after alkaline solubilization at pH 11, was higher than those of RPI-3 and RPI-4 after alkaline solubilization at pH 12 ($P < 0.05$). Lipid content (5.6–7.4 %) of RPIs was lower than that of freeze-dried concentrate (10.6 %). And leucine and lysine of RPIs were the most abundant amino acids (8.8–9.4 and 8.5–8.9 g/100 g protein, respectively). S, Na, P, K as minerals were the major elements in RPIs. SDS-PAGE of RPIs showed bands at 100, 45, 25 and 15 K. Moisture and protein contents of process water as a 2'nd byproduct were 98.9–99.0 and 1.3–1.8 %, respectively. Therefore, yellowfin tuna roe isolate could be a promising source of valuable nutrients for human food and animal feeds.

Keywords: Protein isolate, Roe, Yellowfin tuna, Chemical composition, Isoelectric solubilization/precipitation process

Background

Processing of raw fish into food products generates large quantities of byproducts such as scales, head, viscera and roes. Byproducts utilization will improve the economic aspects of processing industry and further their nutritional beneficiation through valuable essential amino acid and fatty acid components (Narsing Rao et al. 2012). It has been estimated that the value addition of human food developed from the byproduct will increase significantly in the future (Kristinsson and Rasco 2000; Tahergorabi et al. 2012).

Yellowfin tuna *Thunnus albacares* is a large epipelagic species widely distributed in the tropical and subtropical waters of the major oceans (Collette and Nauen 1983; Zudaire et al. 2013). Due to its high demand, yellowfin is harvested widely, and many types of fishing gear are used. Yellowfin tuna is widely used in raw fish dishes. And yellowfin tuna roe, a byproduct generated from fish processing (1.5–3.0 % of total weight), is generally used as animal feed or pet food preparation (Chalamaiah et al. 2013; Klomklao et al. 2013; Intarasirisawat et al. 2011). In our previous study, we fractionated inhibitors from fish roes, and confirmed the distribution of protease inhibitory activity in crude extracts from fish roes (Ji et al. 2011; Kim et al. 2013a; 2013b). In the present study, yellowfin tuna roe was used as a model for fish processing byproduct, and it was the starting material for isoelctric solubilization/precipitation (ISP). ISP process is to solubilize the muscle protein at low or high pH to separate soluble proteins from bone, skin, connective

* Correspondence: heu1837@dreamwiz.com
[1]Department of Food and Nutrition/Institute of Marine Industry, Gyeongsang National University, Jinju 52828, Korea
Full list of author information is available at the end of the article

tissue, cellular membranes, and neutral storage lipids through the centrifugation (Nolsøe and Undeland 2009). The solubilized proteins are recovered by isoelectric precipitation to give a highly functional and stable protein isolate (Kristinsson and Ingadottir 2006; Chanarat and Benjakul 2013). The ISP processing has been applied to beef and fish processing byproducts (Chen and Jaczynski 2007a; 2007b; Mireles DeWitt et al. 2002). Various methods of protein isolate preparation have been reported for different protein sources, including legumes (Horax et al. 2004), oilseeds (Horax et al. 2011), cereals (Agboola et al. 2005; Ju et al. 2001; Paraman et al. 2007) and fish protein (Azadian et al. 2012; Chanarat and Benjakul 2013) based on solubility behavior of their proteins. Proteins isolates are the basic functional components of various high protein processed food products and thus determine the textural and nutritional properties of the foods. These properties contribute to the quality and sensory attributes of food systems (Foh et al. 2012). The roe of marine sources are the most underutilized fish by-products, which have considerable chance for value-addition to produce food and feed. Roes are easily decomposed with short shelf-life and hence, the roes should be processed immediately or converted into value added foods to enhance their shelf-life (Narsing Rao et al. 2012).

No scientific information is available on the protein isolate preparation from yellowfin tuna roe. The aims of this study was to investigate the chemical compositions, amino acid profile, mineral profile, color from yellowfin tuna protein roe isolate and second by-products with isoelectric solubilization/precipitation using basic and acidic pH treatments.

Methods
Raw material
Yellowfin tuna *Thunnus albacares* (YT) roe was obtained from Dongwon F&B Co., Ltd. (Changwon, Korea). Roe was stored at −70 °C in sealed polyethylene bags, and transferred to the laboratory. Frozen roe was partially thawed for 24 h at 4 °C and then cut into small pieces with an approximate thickness of 1.5–3 cm and minced with food grinder (SFM-555SP, Shinil Industrial Co., Ltd., Seoul Korea). Minced roe was frozen at −20 °C until used.

Chemicals
Bovine serum albumin (BSA), casein, hemoglobin, β-Mercaptoethanol (β-ME), glycerol, N,N,N′,N′-tetramethyl ethylene diamine (TEMED), sodium carbonate, sodium hydroxide, sodium L-tatarate, and potassium hydroxide were purchased from Sigma-Aldrich Co., LLC. (St. Louis, MO, USA). Copper (II) sulfate pentahydrate was purchased from Yakuri Pure Chemicals Co., Ltd. (Kyoto, Japan). Bromophenol blue and Folin-Ciocalteu's reagent

were purchased Junsei Chemical Co., Ltd. (Tokyo, Japan). All reagents used analytical grade.

Preparation of roe protein isolates (RPIs)
The frozen minced roe was homogenized with distilled deionized water (DDW) at a ratio of 1:6 (w/v) using a homogenizer (POLYTRON® PT 1200E, KINEMATICA AG, Luzern, Switzerland). The homogenate was adjusted to pH 11 and 12 with 2 N NaOH, respectively. Alkaline solubilization was solubilized protein and inactivated endogenous enzymes in the homogenate. Once the desired pH was reached, the solubilization reaction was allowed to take place at 4 °C for 1 h, followed by centrifugation at 12,000 g and 4 °C for 30 min using a refrigerator centrifuge (Supra 22 K, Hanil Science Industrial Co., Ltd., Incheon, Korea). After centrifugation, two alkaline solubles (pH 11 and 12) in the supernatant fraction, and two alkaline insolubles (pH 11 and 12) as processing 2'nd byproduct in the precipitate fraction were separated. First, to prepare the protein isolates from alkaline solubles through acid precipitation, those of pH were readjusted by addition of 2 N HCl to pH 4.5 and 5.5, respectively, a value near the isoelectric point of fish proteins. The suspensions were centrifuged at 12,000 g and 4 °C for 30 min. After centrifugation, the supernatants were collected as roe process waters (RPWs) and referred to as RPW-1 (pH 11/4.5), RPW-2 (pH 11/5.5), RPW-3 (pH 12/4.5) and RPW-4 (pH 12/5.5), respectively. Precipitates by alkali solubilization and acid precipitation were additionally washed with DDW by centrifugation at 12,000 g and 4 °C for 30 min to remove the NaCl. After centrifugation, the washed roe protein isolates (RPIs) were lyophilized and referred to as RPI-1 (pH 11/4.5), RPI-2 (pH 11/5.5), RPI-3 (pH 12/4.5), and RPI-4 (pH 12/5.5), respectively. The roe protein isolate washed water (RPI-WW) as the 2'nd byproduct referred to as RPI-WW1 (pH 11/4.5), RPI-WW2 (pH 11/5.5), RPI-WW3 (pH 12/4.5) and RPI-WW4 (pH 12/5.5), respectively. Alkaline insolubles were resuspended with DDW, then readjusted to pH 6.5 with 2 N HCl, and centrifuged at 12,000 g and 4 °C for 30 min. The supernatants were collected and referred to as roe alkaline insoluble washed water (RAI-WW11 and RAI-WW12, respectively). The precipitate was lyophilized and referred to as roe alkaline insolubles (RAI-1 and RAI-2, respectively). All samples were stored at −20 °C until further experiments. A flow chart for the preparation of roe protein isolates (RPIs) is shown in Fig. 1.

Proximate composition
The proximate composition was determined according to the AOAC method (AOAC 1995). Moisture content was determined by oven-drying method at 105 °C until a constant weight was reached. The ash content was obtained

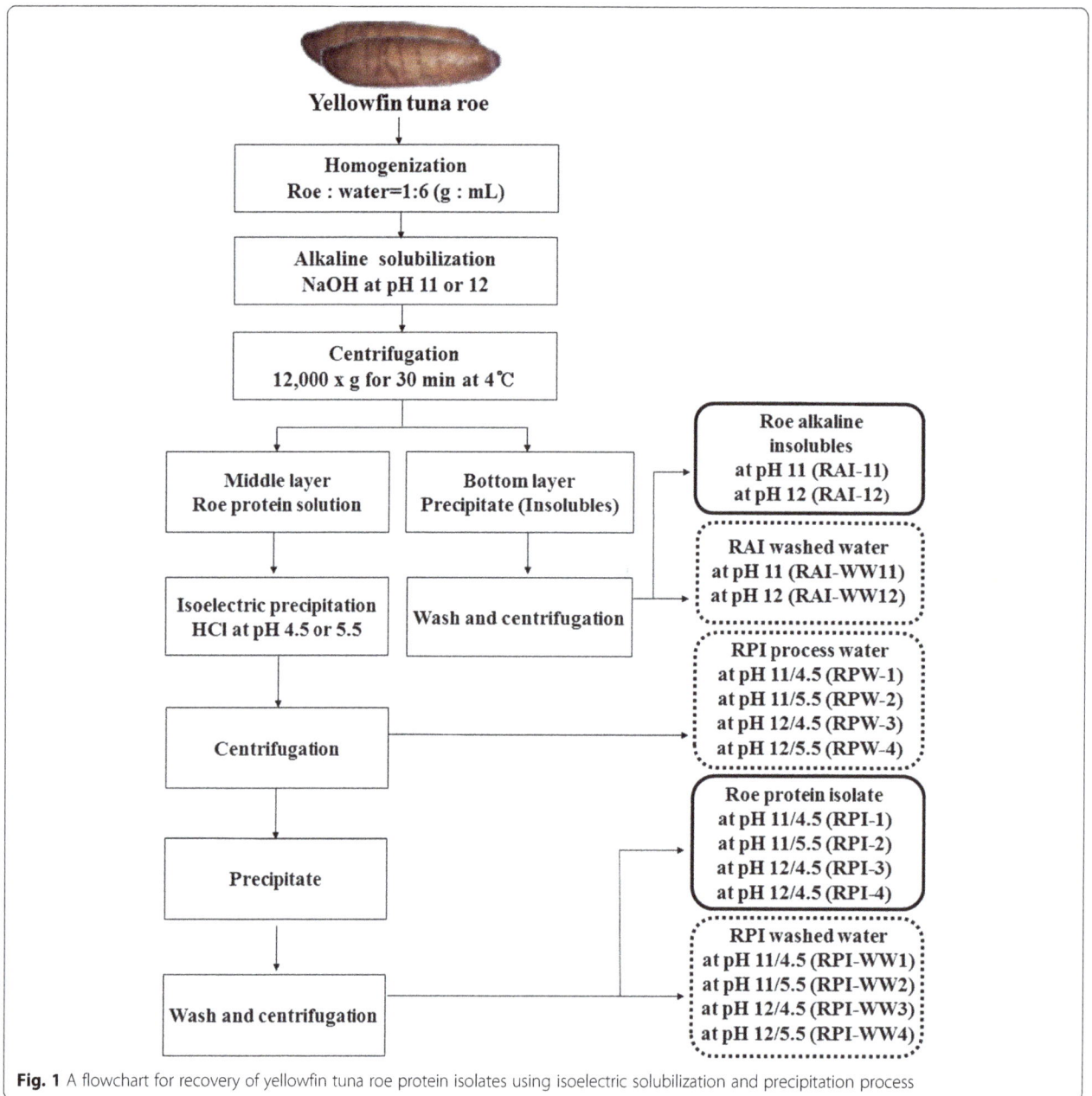

Fig. 1 A flowchart for recovery of yellowfin tuna roe protein isolates using isoelectric solubilization and precipitation process

by ashing a sample in a muffle furnace (Thermolyne 10500 furnace, a subsidiary of Sybron Co., Dubuque, IA, USA) at 550 °C until a constant final weight for ash was achieved. The total crude protein (N × 6.25) content of samples was determined using the semi-micro Kjeldahl method. Total lipid content was determined according to the Soxhlet extraction method.

Protein concentration

Soluble protein concentration of sample was determined by the method of Lowry et al. (1951) using bovine serum albumin as a standard.

Total amino acid

Total amino acid analysis was conducted according to AOAC method (AOAC 1995). The sample (20 mg) was hydrolyzed with 2 mL of 6 N HCl at 110 °C for 24 h in heating block (HF21; Yamoto Science Co, Tokyo, Japan) and filtered out using vacuum filtrator (ASPIRATOR A-3S, EYELA, Tokyo, Japan). Amino acids were quantified using the amino acid analyzer (Biochrom 30, Biochrom Ltd., Cambridge, United Kingdom) employing sodium citrate buffers (pH 2.2) as step gradients. The data are reported as mg of amino acid per 100 g of protein.

Mineral

Analysis of iron (Fe), copper (Cu), manganese (Mn), cadmium (Cd), nickel (Ni), lead (Pb), zinc (Zn), chromium (Cr), magnesium (Mg), sodium (Na), phosphorus (P), potassium (K), calcium (Ca), and sulfur (S) contents in sample was carried out using the inductively coupled plasma optical emission spectrophotometry (OPTIMA 4300 DV, Perkin Elmer, Shelton, Conn., USA). Briefly, teflon digestion vessel was washed overnight in a solution of 2 % nitric acid (v/v) prior to use.

Sample was dissolved in 10 mL of 70 % nitric acid. The mixture was heated on the hot plate until digestion was completed. The digested samples were added in 5 mL of 2 % nitric acid and filtered using filter paper (Advantec No. 2, Toyo Roshi Kaisha, Ltd., Tokyo, Japan). Sample was massed up to 100 mL with 2 % nitric acid in a volumetric flask

Hunter color

Hunter color properties of samples were equilibrated to room temperature for 2 h prior to the color measurement. Colors were determined using color meter (ZE-2000 Nippon Denshoku Inc., Japan). The colorimeter was calibrated by using a standard plate (L* = 96.82, a* = –0.35, b* = 0.59) supplied by the manufacturer. The values for the CIE (Commission Internationale d'Eclairage of France) color system using tristimulus color values, L* (lightness), a* (redness), and b* (yellowness) were determined. The whiteness was calculated by the following equation:

$$\text{Whiteness} = 100 - \sqrt{(100 - L^*)^2 + a^{*2} + b^{*2}}$$

SDS-PAGE

The molecular weight distribution of protein isolates and their 2'nd byproducts was observed by sodium dodecyl sulfate polyacrylamide gel electrophoresis (SDS–PAGE) according to the method of Laemmli (1970). Briefly, 10 mg of sample was solubilized in 1 mL of 8 M urea solution containing 2 % β-mercaptoethanol and 2 % sodium dodecyl sulfate (SDS) solution. Protein solution was mixed at 4:1 (v/v) ratio with the SDS-PAGE sample treatment buffer (62.5 mM Tris–HCl (pH 6.8), 2 % SDS (w/v), 10 % glycerol, 2 % β-mercaptoethanol and 0.002 % bromophenol blue) and boiled at 100 °C for 3 min. The sample (20 μg protein) was loaded on the Any KD™ Mini-PROTEAN® TGX™ Precast gel (Bio-Rad Lab., Inc., Hercules, CA, USA) and subjected to electrophoresis at a constant current of 10 mA per gel using a Mini-PROTEAN® Tetra cell (Bio-Rad Lab. Inc., Hercules, CA, USA). Electrophoresed gel was stained in 0.125 % Coomassie brilliant blue R-250 and destained in 25 % methanol and 10 % acetic acid until background was clear. Molecular weight of protein bands

was estimated using Precision Plus Protein™ standards (10–250 K, Bio-Rad Lab., Inc., Hercules, CA, USA).

Statistical analysis

All experiments were conducted in triplicates. The average and standard deviation were calculated. Data were analyzed using analysis of variance (ANOVA) procedure by means of the statistical software of SPSS 12.0 KO (SPSS Inc., Chicago, IL, USA). The mean comparison was made using the multiple range Duncan's test ($P < 0.05$).

Results and discussions

Proximate composition

Roe protein isolates (RPIs) from yellowfin tuna roe were prepared according to previously described ISP process. Proximate compositions of RPIs and positive controls (casein and hemoglobin) are shown in Table 1. Yields of RPIs prepared from YTR by ISP process were in range of 11.6–14.1 % with slight difference. Moisture content of the RPIs ranged 2.3 to 5.0 %. RPI-1 (87.8 %) had the highest protein content than RPI-2 (83.2 %), RPI-4 (79.6 %) and RPI-3 (79.1 %), respectively ($P < 0.05$). protein content of RPIs was but lower than hemoglobin (94.4 %, $P < 0.05$). Protein content of RPIs were higher than that reported for fish protein powders obtained from different raw materials (Sathivel et al. 2004, 2005, 2006; Sathivel and Bechtel 2006; Shaviklo et al. 2011). Protein powder prepared from fish roe or surimi (Sathivel et al. 2009; Huda et al. 2001) had similar to protein content. Lipid (5.6–7.4 %) and ash content (3.0–3.8 %) of RPIs were lower than those of FDC ($P < 0.05$). Pires et al. (2012) reported that mineral and fat were eliminated in the supernatant obtained after ISP process. Protein content of RPI-1 and –2 precipitated at pH 4.5 and 5.5 after alkaline solubilization at pH 11, was significantly higher than those of RPI-3 and –4 after alkaline solubilization at pH 12 ($P < 0.05$). Whereas, yield of RPI-1 and –2 was lower than that of RPI-3 and –4. In this result, total protein yield of all RPIs was not significantly different ($P > 0.05$). Final supernatant referred to roe process water (RPW) as 2'nd byproduct was generated through alkaline solubilization and acid precipitation. Their moisture and protein contents are shown in Table 2. Moisture content of the RPWs ranged from 98.9 to 99.0 % and protein concentration (mg/mL) of RPWs ranged from 2.6 to 3.0 mg/mL. RPW-1 and RPW-3 (1478.5 and 1420.9 mg, respectively) had lower total protein than RPW-2 and RPW-4 (1861.2 and 1605.3 mg, respectively) in this experiment. Lower total protein of supernatant (RPW-1 and –3) obtained by acid precipitation at pH 4.5 could be more efficient to raise the yield of RPI as precipitate, compared with protein isolate at pH 5.5. Roe alkaline insoluble-washed waters (RAI-WWs) and roe protein isolate-washed waters (RPI-WWs) were obtained through washing process to

Table 1 Proximate composition of FDC, roe alkaline insolubles (RAIs) as 2'nd byproducts and roe protein isolates (RPIs) by isoelectric solubilization and precipitation (ISP) process

Sample	Yield[a] (g)	Protein yield[b] (g)	Moisture (%)	Protein (%)	Lipid (%)	Ash (%)
FDC	25.3	18.3	4.3 ± 0.1de	72.3 ± 0.4h	10.6 ± 0.1b	5.7 ± 0.5a
RAI-11	8.7	7.1	6.5 ± 0.1b	82.5 ± 0.2d	10.2 ± 0.1bc	2.0 ± 0.0d
RAI-12	8.0	5.9	10.1 ± 0.2a	73.7 ± 0.2g	13.1 ± 0.5a	2.3 ± 0.1d
RPI-1	12.6	10.1	4.7 ± 0.1cd	87.8 ± 0.7b	6.1 ± 0.2d	3.8 ± 0.2b
RPI-2	11.6	9.7	3.4 ± 0.1e	83.2 ± 0.7e	7.4 ± 0.1c	3.2 ± 0.1c
RPI-3	14.1	11.1	2.3 ± 1.5ef	79.1 ± 0.5f	5.6 ± 0.1d	3.2 ± 0.3c
RPI-4	13.4	10.6	5.0 ± 1.8c	79.6 ± 0.7f	6.3 ± 1.0d	3.0 ± 0.1c
Casein	-	-	4.0 ± 0.2d	85.5 ± 0.0c	-	-
Hemoglobin	-	-	2.0 ± 0.0f	94.4 ± 0.4a	-	-

FDC freeze-dried concentrate; RAI-11 and RAI-12, roe alkaline insolubles after alkaline solubilization at pH 11 or 12 respectively. RPI-1 and RPI-2, roe protein isolate adjusting at pH 4.5 and 5.5, respectively, after alkaline solubilization at pH 11; RPI-3 and RPI-4, roe protein isolate adjusting at pH 4.5 and 5.5, respectively, after alkaline solubilization at pH 12
Data are given as mean values ± SD ($n = 3$)
Means with different letters within the same column are significantly different at $P < 0.05$ by Duncan's multiple range test
-; not determined
[a]Yield is weight (g) of roe protein isolate obtained from 100 g of raw YTR
[b]Protein yield (g) = yield x protein (%)

reduce remaining salt in RAI and RPI. Their moisture and protein content are shown in Table 3. Moisture content of all washed water was ranged from 99.5 to 99.7 % with no difference. Total protein of RAI-WWs in range of 315.0–340.3 mg was higher than that of RPI-WWs in range of 133.5–192.4 mg). During ISP process, it was possible to recover more than 90 % of the process water (RPWs, RAI-WWs and RPI-WWs) contained 0.50-3.4 mg/mL protein.

Total amino acid
Total amino acid composition (g/100 g protein, %) of RPIs, RAIs, and positive controls (casein and hemoglobin) are shown in Table 4. Protein content of all samples ranged from 81.6 to 96.3 % on a dry base. From the result, RPIs had a EAAs/NEAAs acid ratio in range of 1.08 to 1.17. These were higher than that (0.92) of casein, but slightly lower than that (1.33) of hemoglobin. Leucine

(8.8–9.4 %), lysine (8.5–8.9 %) and isoleucine content (5.8–6.3 %) of RPIs were significantly higher than those of RAIs ($P < 0.05$). It is indicated that total essential amino acid content of RPIs in range of 51.9–53.9 % were higher than that (47.1–49.0 %) of RAIs. Intarasirisawat et al. (2011) reported that leucine (8.28–8.64 %) and lysine (8.24–8.30 %) were the predominant essential amino acids in defatted tuna roe from skipjack, tongol and bonito. Lysine is often considered a first limiting amino acid for cereal food. Therefore, it needs to be emphasized that the RPIs had a higher content of lysine than egg white (8.2 %) ($P < 0.05$). The lysine content of RPIs was higher than that reported for Channa (6.94 %) and Lates (6.86 %) roe protein concentrates (Narsing Rao et al. 2012). The major non–essential amino acids (NEAAs) of RAIs as 2'nd byproduct by alkaline solubilization were glutamic acid (13.7–13.8 %), aspartic acid (8.8–8.9 %) and arginine (6.4–6.6 %),

Table 2 Moisture and protein contents of roe process waters (RPWs) obtained from yellowfin tuna roe during isoelectric solubilization and precipitation (ISP) process

Sample	Moisture (%)	Protein (%)	Volume (mL/100 g roe)	Protein[a] (mg/mL)	Total protein (mg)
RPW-1	99.0 ± 0.2a	1.5 ± 0.1b	547.6 ± 0.0	2.7bc	1478.5
RPW-2	98.9 ± 0.1a	1.8 ± 0.0a	547.4 ± 0.0	3.4a	1861.2
RPW-3	98.9 ± 0.0a	1.3 ± 0.0c	546.5 ± 0.0	2.6c	1420.9
RPW-4	98.9 ± 0.0a	1.6 ± 0.1b	535.1 ± 0.4	3.0b	1605.3

RPW-1 and RPW-2, roe process water adjusting pH 4.5 and 5.5, respectively, after alkaline solubilization at pH 11; RPW-3 and RPW-4, roe process water adjusting pH 4.5 and 5.5, respectively, after alkaline solubilization at pH 12
Values are means ± standard deviation of triplicate determinations
Means with different letters within the same column are significantly different at $P < 0.05$ by Duncan's multiple range test
[a]Based on Lowry's method (1951)

Table 3 Moisture and protein contents of roe alkaline insoluble washed waters (RAI-WWs) and roe protein isolate washed waters (RPI-WWs) by a washing process

Sample	Moisture (%)	Protein (%)	Volume (mL/100 g roe)	Protein[a] (mg/mL)	Total protein (mg)
RAI-WW11	99.5 ± 0.1a	0.3 ± 0.0a	262.5	1.21 ± 0.0a	315.0
RAI-WW12	99.5 ± 0.0a	0.3 ± 0.2a	283.6	1.16 ± 0.0b	340.3
RPI-WW1	99.7 ± 0.1a	0.2 ± 0.0a	320.6	0.58 ± 0.0c	192.4
RPI-WW2	99.7 ± 0.1a	0.2 ± 0.0a	352.3	0.47 ± 0.0e	176.2
RPI-WW3	99.7 ± 0.1a	0.2 ± 0.0a	317.1	0.48 ± 0.0de	158.6
RPI-WW4	99.7 ± 0.2a	0.2 ± 0.3a	266.9	0.53 ± 0.0d	133.5

RAI-WW11 and RAI-WW12, roe alkaline insoluble washed waters of RAI-11 and RAI-12, respectively
RPI-WW1-4, roe protein isolate washed water of RAIs (1–4)
Values are means ± standard deviation of triplicate determinations
Means with same letters within the moisture (%) and protein (%) are not significantly different at $P > 0.05$, and means with different letters within the protein concentration are significantly different $P < 0.05$ by Duncan's multiple range test
[a]Based on Lowry method (1951)

Table 4 Total amino acid (g/100 g protein) composition of FDC, RAIs and RPIs prepared by ISP process and positive controls

Amino acid	FDC	RAI-11	RAI-12	RPI-1	RPI-2	RPI-3	RPI-4	Casein	Hb
Protein content (%)[a]	81.6d	87.2c	82.8d	89.9b	86.9c	90.1b	89.5b	89.1b	96.3a
Asp	8.7d	8.9cd	8.8cd	9.3b	9.1bc	9.3b	9.2b	8.2e	11.2a
Thr[b]	5.0e	5.1ab	5.0d	5.1abc	5.1ab	5.1a	5.0cd	3.9g	4.7f
Ser	5.6a	5.3c	5.4b	5.2cd	5.2d	5.4b	5.0e	4.0g	4.4f
Glu	13.1d	13.8b	13.7c	12.7e	12.4h	12.4g	12.5f	22.1a	9.3i
Pro[c]	6.1c	7.7b	7.9b	5.7d	5.1e	4.7g	4.9ef	10.0a	4.0de
Gly[c]	4.9c	6.0b	6.4a	4.5ef	4.2f	4.3ef	4.3ef	2.4g	4.6de
Ala[c]	6.6c	6.1d	6.2d	6.7bc	6.9b	6.8b	6.8bc	3.8e	9.0a
Cys	0.7b	0.8b	1.1a	0.8b	0.9b	0.3d	0.2d	0.5c	0.1d
Val[bc]	6.3d	6.7c	6.3cd	6.5cd	6.5cd	6.6c	6.6cd	7.3b	10.2a
Met[bc]	2.9c	2.5d	2.5d	2.9b	3.0b	3.1a	3.1a	1.3e	0.0f
Ile[bc]	5.4d	4.9e	4.5f	5.8c	6.2ab	6.1b	6.3a	5.7c	0.8g
Leu[bc]	8.6f	8.0g	7.5h	8.8e	9.3c	9.3bc	9.4b	9.1d	13.3a
Tyr	3.4a	2.6c	3.4a	3.2b	3.1b	3.4a	3.2b	1.2d	0.3e
Phe[bc]	4.4h	4.6f	4.3i	4.5g	4.7e	4.8d	4.9c	5.0b	7.6a
His[b]	3.4b	3.2c	3.1c	3.2c	3.1c	3.1c	3.2c	2.9d	6.4a
Lys[b]	8.5c	7.6e	7.3f	8.5c	8.9b	8.8b	8.9b	8.3d	10.3a
Arg[b]	6.6a	6.4a	6.6a	6.6a	6.4a	6.4a	6.5a	4.4b	3.8c
TAA	100.2	100.2	100.0	100.0	100.1	99.9	100.0	100.1	100.0
EAA	51.1	49.0	47.1	51.9	53.2	53.3	53.9	47.9	57.1
EAA/NEAA	1.04	0.96	0.89	1.08	1.14	1.14	1.17	0.92	1.33
Hydrophobic amino acid	45.2	46.5	45.6	45.4	45.9	45.7	46.3	44.6	49.5

TAA total amino acid, *EAA* essential amino acids, *NEAA* non essential amino acids
FDC freeze-dried concentrate; RAI-11 and RAI-12, roe alkaline insolubles after alkaline solubilization at pH 11 or 12 respectively. RPI-1 and RPI-2, roe protein isolate adjusting at pH 4.5 and 5.5, respectively, after alkaline solubilization at pH 11; RPI-3 and RPI-4, roe protein isolate adjusting at pH 4.5 and 5.5, respectively, after alkaline solubilization at pH 12; *Hb* hemoglobin
Values are means ± standard deviation of duplicate determinations
Means with different letters within the same row are significantly different at $p < 0.05$ by Duncan's multiple range test
[a]Based on dry weight
[b]Essential amino acids for infant
[c]Hydrophobic amino acids

respectively. Amino acids namely glutamic acid (13.14 g), aspartic acid (8.08 g) and arginine (5.76 g) were reported per 100 g roe protein of Channa roe protein concentrate (Narsing Rao et al. 2012).

NEAAs of RAIs were relatively higher than those data of FDC. Predominant essential amino acids (EAAs) of RAI-11 were leucine (8.0 %), lysine (7.6 %) and valine (6.7 %), respectively. These were similar to those of RAI-12 which had leucine (7.5 %), lysine (7.3 %) and valine (6.3 %). EAAs/NEAAs ratio of RAI-11 (0.96) was higher than that of RAI-12 (0.89). From the result, total essential amino acid (isoleucine, leucine, lysine, methionine, phenylalanine, threonine, valine) content of RAI-11 (49.0 %) was higher than that of RAI-12 (47.1 %). However, EAAs content of RAIs (47.1–49.0 %) were lower than that of FDC (51.1 %). Thus, EAAs/NEAAs ratio of RAIs (0.89–0.96) were lower than that (1.04) of FDC. RPIs were rich in glutamine, asparagine, leucine, and lysine which accounted for 12.4–12.7 %, 9.1–9.3 %, 8.8–9.4 % and 8.3–8.9 % of total amino acid, respectively. The hydrophobic amino acid content of RAIs and RPIs were similar in range of 45.4–46.3 %. However, RAI was richer in proline and glycine than RPIs ($P < 0.05$). In the case of proline and glycine the difference may be due to the elimination of collagenous material during the protein recovery by the alkaline solubilization. Lysine content of RAIs and RPIs was similar to that (6.1 to 9.7 %) of pollock protein samples reported by Sathivel et al. (2006). RPIs and RAIs can be used as a nutritional supplement due to the content of essential amino acids composition.

Mineral

The mineral contents (mg/100 g) of FDC and RAIs and RPIs are given in Table 5. The main functions of essential minerals include skeletal structure, maintenance of colloidal system and regulation of acid–base equilibrium, and mineral also constitute important components of hormones, enzymes and enzyme activators (Belitz and Grosch 2001). The major mineral contents of RPIs and RAIs were S (591.4–715.7 mg/100 g) K (36.2–130.6 mg/100 g), Na (78.2–303.3 mg/100 g), and Ca (7.8–39.4 mg/100 g). The K content of RPIs was found to be in ranged of 36.2–68.5 mg/100 g, respectively. Among RPIs, the RPI-4 (pH 12/5.5) had the highest content of Na (244.4 mg/100 g) and K (68.5 mg/100 g) than RPI-3 (pH 12/4.5), RPI-2 (pH 11/5.5) and RPI-1 (pH 11/4.5) ($P < 0.05$). However, Na content of RPIs was lower than that of crab (Gokoglu and Yerlikaya 2003), rainbow trout (Gokoglu et al. 2004) and fish based dishes (Martinez-Valverde et al. 2000). P content of RPIs (215.3–254.6 mg/100 g) was lower than that (337.8 mg/100 g) of rainbow trout reported for Gokoglu et al. (2004) and higher than that (215.0–231.0 mg/ 100 g) of European perch reported Orban et al. (2007). Mg content of RPI-2 and RPI-4 (32.3–21.4 mg/100 g) were higher than those (6.6-4.6 mg/100 g) of RPI-1 and RPI-3 ($P < 0.05$). S and Ca content of RPIs were no significant difference ($P > 0.05$). Na, K and Mg content of RAI-11 (213.0, 108.9 and 23.4 mg/100 g, respectively) were lower than those (303.3, 130.6 and 51.7 mg/100 g, respectively) of RAI-12 ($P < 0.05$). These values were higher than RPIs ($P < 0.05$). This result is indicated that Na, K and Mg content of RPIs could be eliminated during the alkaline solubilization process. Mg of RAIs was similar to that (33.0–34.0 mg/100 g) of sea bream (Orban et al. 2000) and that (25.1–33.6 mg/100 g) of Baltic herring (Tahvonen et al. 2000). But, P content of RAIs contained in range of 84.3–97.1 mg/100 g was lower than those of RPIs ($P < 0.05$) because of solubilization of P content for RAIs during alkaline solubilization process. Ca content of RAIs (39.4-37.2 mg/100 g, respectively) was

Table 5 Mineral contents (mg/100 g of sample) of FDC, RAIs and RPIs prepared by ISP process and positive controls

Sample	FDC	RAI-11	RAI-12	RPI-1	RPI-2	RPI-3	RPI-4	Casein	Hb
Moisture (%)	4.3	6.5	10.1	4.7	3.4	2.3	5.0	4.0	2.0
K	1179.9 ± 8.3a	108.9 ± 3.9d	130.6 ± 0.6c	36.2 ± 1.7g	57.5 ± 1.3f	51.6 ± 1.7f	68.5 ± 4.9e	912.0 ± 12.3b	71.4 ± 3.0e
S	992.3 ± 92.6b	591.4 ± 33.4e	627.9 ± 40.9cde	609.2 ± 40.9cde	715.7 ± 26.9c	664.1 ± 75.0cde	698.9 ± 73.9cd	1984.1 ± 3.2a	442.0 ± 46.2f
Na	376.2 ± 2.1b	213.0 ± 3.6e	303.3 ± 3.7c	78.2 ± 0.1h	123.7 ± 1.0g	188.0 ± 2.0f	244.4 ± 4.2d	706.3 ± 10.9a	212.6 ± 1.8e
P	257.7 ± 2.8a	84.3 ± 1.5f	97.1 ± 1.4e	254.6 ± 2.5a	246.8 ± 7.8b	235.7 ± 1.2c	215.3 ± 1.6d	34.7 ± 0.4g	29.1 ± 0.3h
Mg	66.8 ± 0.4a	23.4 ± 0.2d	51.7 ± 0.5b	6.6 ± 0.1f	32.3 ± 0.6c	4.6 ± 0.1g	21.4 ± 0.0e	-	-
Zn	45.0 ± 0.3b	42.8 ± 0.3c	44.4 ± 0.2b	23.2 ± 0.4d	52.1 ± 0.4a	23.0 ± 0.2d	52.6 ± 0.8a	-	-
Ca	33.5 ± 0.3b	39.4 ± 1.0b	37.2 ± 0.2b	7.8 ± 0.1d	15.3P ± 0.5cd	12.4 ± 0.1cd	16.5 ± 0.3cd	987.2 ± 17.3a	21.2 ± 0.2c
Fe	9.8 ± 0.1d	5.8 ± 0.1e	6.6 ± 0.1e	12.9 ± 0.3bc	11.9 ± 0.1c	13.5 ± 0.2b	13.2 ± 0.3b	4.8 ± 0.0f	250.1 ± 1.8a
Mn	0.1 ± 0.0c	0.1 ± 0.0c	0.0 ± 0.0d	−0.1 ± 0.0e	0.0 ± 0.0d	−0.1 ± 0.0e	0.0 ± 0.0d	4.0 ± 0.0a	0.4 ± 0.0b

FDC freeze-dried concentrate; RAI-11 and 12, roe alkaline insolubles after alkaline solubilization at pH 11 and 12, respectively; RPI-1, 2, 3 and 4, roe protein isolate adjusting at pH 4.5 and 5.5, respectively after alkaline solubilization at pH 11 and 12; Hb hemoglobin
Values are means ± standard deviation of triple determinations
Means with different letter within the same row are significantly different at $P < 0.05$ by Duncan's multiple range test
-; not determined

Table 6 L*, a* and b* color values, whiteness of FDC, RAIs and RPIs by ISP process

Hunter color	FDC	RAI-11	RAI-12	RPI-1	RPI-2	RPI-3	RPI-4
L*	59.2 ± 0.1d	64.3 ± 0.1a	63.7 ± 0.1b	59.6 ± 0.2c	57.2 ± 0.2f	57.8 ± 0.2e	57.7 ± 0.2e
a*	6.5 ± 0.1a	5.7 ± 0.1c	5.9 ± 0.0b	4.6 ± 0.1f	5.6 ± 0.1c	5.0 ± 0.1e	5.5 ± 0.1d
b*	18.6 ± 0.0c	19.5 ± 0.4b	20.0 ± 0.0a	16.4 ± 0.1e	17.3 ± 0.0d	16.2 ± 0.0e	17.4 ± 0.1d
ΔE	42.3 ± 0.1b	38.2 ± 0.6e	38.9 ± 0.1d	40.7 ± 0.2c	43.3 ± 0.2a	42.4 ± 0.2d	43.0 ± 0.1a
Whiteness	54.7 ± 0.1d	58.9 ± 0.3a	58.1 ± 0.1b	56.2 ± 0.2c	53.5 ± 0.2f	54.5 ± 0.2b	53.9 ± 0.1e

FDC freeze-dried concentrate; RAI-11 and 12, roe alkaline insolubles after alkaline solubilization at pH 11 and 12, respectively; RPI-1, 2, 3 and 4, roe protein isolate adjusting at pH 4.5 and 5.5, respectively after alkaline solubilization at pH 11 and 12; *Hb* hemoglobin
Values are means ± standard deviation of triplicate determinations
Means with different letters within the same row are significantly different at $p < 0.05$ by Duncan's multiple range test

significantly higher ($P < 0.05$) than those found in RPIs. During the ISP process, K, Na, Ca and S content of RPIs and RAIs were significantly lower than those (912.0, 706.3, 987.2 and 1984.1 mg/100 g, respectively) of casein ($P < 0.05$). In case of hemoglobin, similar results were observed in RAIs and RPIs except for Fe and S. Potentially, the RAIs as recovered insoluble may be useful in animal feeds as a mineral additive due to the relatively high concentration of minerals (Ca, Mg, and K).

Hunter color

The color properties of FDC, RAIs and RPIs are presented in Table 6. L*, a*, and b* of RAIs were higher than RPIs ($P < 0.05$). Among the RPIs, RPI-1 was the lightest with L* value (59.6) followed by RPI-3 (57.8) RPI-4 (57.7) and, RPI-2 (57.2), respectively ($P < 0.05$). RPI-1 and RPI-3 had the lower b* value of 16.4 and 16.2, respectively than those (17.3 and 17.4, respectively) of RPI-2 and RPI-4 ($P < 0.05$). RPI-2 and RPI-4 had higher

a* value (5.6 and 5.5, respectively) than that (4.6 and 5.0, respectively) of RPI-1 and RPI-3 ($P < 0.05$). However, RPI-1 and RPI-3 had higher whiteness values (56.2 and 54.5, respectively) than the other RPIs (53.5 for RPI-2 and 53.9 for RPI-4). RAI-11 was lightest ($P < 0.05$) with L* value of 64.3 than that (63.7) of RAI-12. a*, b* and ΔE values of RAI-11 (5.7, 19.5 and 38.2 respectively) were slightly lower than those of RAI-12. Whiteness value of RAI-11 (58.9) was higher than that (58.1) of RAI-12 ($P < 0.05$). This different in color could be due to the separation pigment content caused by ISP process. Among the RAIs and RPIs, RAI-11 could be more useful as a food ingredient because of high value of whiteness.

SDS-PAGE

SDS-PAGE patterns of FDC, RAIs, RPIs, RPWs, RAI-WWs and RPI-WWs obtained by ISP process are shown in Fig. 2. Protein with a molecular weight (MW) ranged from 37 to 50 K was dominant in RAI-11 and RAI-12.

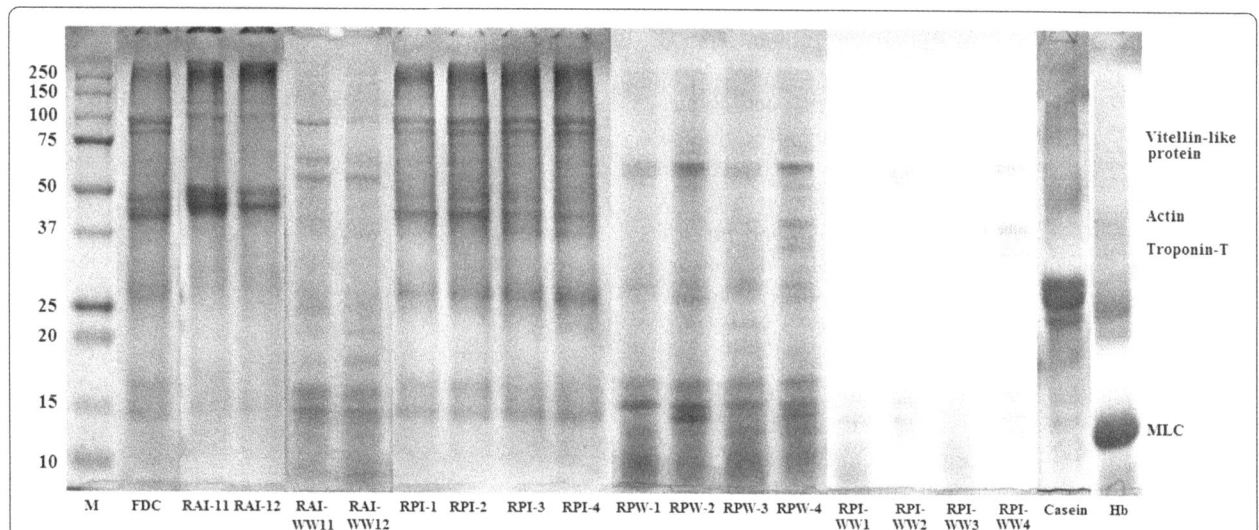

Fig. 2 SDS-PAGE patterns of FDC, RAIs, RPIs, RPWs, RAI-WWs and RPI-WWs prepared by ISP process and positive controls. M, protein maker; *FDC*, freeze-dried concentrate; RAI-11 and 12, roe alkaline insolubles after alkaline solubilization at pH 11 and 12, respectively; RPI-1, 2, 3 and 4, roe protein isolate adjusting at pH 4.5 and 5.5, respectively after alkaline solubilization at pH 11 and 12; RPW-1, 2, 3 and 4, roe process water adjusting pH 4.5 and 5.5, respectively after alkaline solubilization at pH 11 and 12; RAI-WW11 and 12, washed waters of RAI-11 and RAI-12, respectively. RPI-WW1, 2, 3 and 4, washed water of RAI-1, 2, 3 and 4; *Hb* hemoglobin

Other protein bands in range of 100–150 K, 25–37 K and 15–20 K, respectively were also observed. There was no difference between RAIs except for protein band in range of 37–50 K where RAI-11 was clearer than that of RAI-12. The protein with a MW of 97 kDa might be a vitellin-like protein, which was found in salmon *Oncorhynchus keta* and sturgeon *Acipenser transmontanus* roes (Al-Holy and Rasco 2006). Similar proteins with MW of 32.5, 29 and 32.5 K were found in skipjack *Kasuwonous pelamis*, tonggol *Thunnus tonggol* and bonito *Euthynnus affinis* roe, respectively reported Intarasirisawat et al. (2011). Those proteins might be ovomucoid (Al-Holy and Rasco 2006) or phosvitin (Losso et al. 1993). Generally, different roe samples showed different electrophoretic patterns, indicating the differences in protein compositions among all samples (Intarasirisawat et al. 2011). Compared with FDC, low molecular protein bands (0–15 K) of RAIs were more faint than those of FDC that because a little protein compounds and others were transferred to RAI-WWs. Washed soluble proteins were moved on RAI-WW11 and RAI-WW12. Overall, RAI-WW11 had lower molecular protein bands than RAI-WW12 because of difference on solubilization condition during alkaline solubilization and precipitation (ISP). Prominent protein bands of RPI-1-4 were in range of 75–100 K, 37–50 K, 15–20 K, and 10 K respectively. A similar SDS-PAGE pattern of lower molecular weight protein (10 K) was observed in meriga roe hydrolysate (Chalamaiah et al. 2010). This result was also similar to hemoglobin which protein band observed in range of around 25 K and 10–15 K, respectively. Protein band of casein with molecular weight ranged 25–37 K was clearer than that of RPIs. Protein band of RPIs in range from 50 to 75 were faint because those proteins already been washed and moved to RPI-WWs. Protein band of RPWs with molecular weight of 10 K were was clearer than that of RPI. This result is indicated that acid was affected to degradation on aggregation and association of low molecular weight. Al-Holy and Rasco (2006) reported that three prominent proteins of salmon caviar had MW of 96, 20 and 10 K, which could be vitellin and possibly lysozyme or phosvitin. Protein with a MW of approximately 27 K in the soluble fraction of sturgeon caviar may possibly represent ovomucoid, a glycoprotein, which normally has a MW of 27–29 K (Al-Holy and Rasco 2006). For RAIs and RPIs, new protein bands with molecular weight over the 250 K were formed instead of FDC. Azadian et al. (2012) reported that this can be explained by the dissociation of high molecular weight myosin and association of the protein to form a high molecular weight.

Conclusion

The roe of marine sources are the most underutilized fish by-products, which have considerable chance for value-addition to produce food and feed. This study was to investigate the chemical compositions, amino acid profile, mineral profile, color from yellowfin tuna protein roe isolate and second by-products with isoelectric solubilization/precipitation using basic and acidic pH treatments. RPIs and RAIs can be used as a nutritional supplement due to the content of essential amino acids composition. The RAIs as recovered insoluble may be useful in animal feeds as a mineral additive due to the relatively high concentration of minerals (Ca, Mg, and K). Therefore, yellowfin tuna roe isolate could be a promising source of valuable nutrients for human food and animal feeds.

Acknowledgements
This research was supported by Basic Science Research Program through the National Research Foundation of Korea(NRF) funded by the Ministry of Education(NRF-2014R1A1A4A01008620).

Authors' contributions
HJL and SHP carried out the preparation of protein isolates, participated in the analysis of chemical compositions and drafted the manuscript. ISY, G-WL and SYP participated in searching and screening references and performed the statistical analysis and carried out the SDS-PAGE. J-SK and MSH conceived of the study, and participated in its design and coordination and helped to draft the manuscript. All authors read and approved the final manuscript.

Competing interests
The authors declare that they have no competing interests.

Author details
[1]Department of Food and Nutrition/Institute of Marine Industry, Gyeongsang National University, Jinju 52828, Korea. [2]Department of Seafood Science and Technology/Institute of Marine Industry, Gyeongsang National University, Tongyeong 53064, Korea.

References
Agboola S, Ng D, Mills D. Characterization and functional properties of Australia rice protein isolates. J Cereal Sci. 2005;41:283–90.
Al-Holy MA, Rasco BA. Characterisation of salmon (*Oncorhynchus keta*) and sturgeon (*Acipenser transmontanus*) caviar proteins. J Food Biochem. 2006;30:422–8.
AOAC (Association of Analytical Chemists). Official Methods of Analysis. 16th ed. Washington DC: Association of Analytical Chemists; 1995. p. 69–74.
Azadian M, Nasab MM, Abedi E. Comparison of functional properties and SDS-PAGE patterns between fish protein isolate and surimi produced from silver carp. Eur Food Res Technol. 2012;235:83–90.
Belitz HD and Grosch W. 2001. Schieberle, P. Lehrbuch der Lebensmittelchemie, ISBN 3-540-41096-1 5. Aufl. Springer-Verlag Berlin, Heidelberg, New York
Chalamaiah M, Narsing Rao G, Govardhana Rao D, Jyothirmayi T. Protein hydrolysates from meriga (*Cirrhinus mrigala*) egg and evaluation of their functional properties. Food Chem. 2010;120:652–7.
Chalamaiah M, Balaswamy K, Narsing Rao G, Prabhakara Rao PG, Jyothirmayi T. Chemical composition and functional properties of mrigal (*Cirrhinus mrigala*) egg protein concentrates and their application in pasta. J Food Technol. 2013;50:514–20.
Chanarat S, Benjakul S. Impact of microbial transglutaminase on gelling properties of Indian mackerel fish protein isolates. Food Chem. 2013;136:929–37.
Chen YC, Jaczynski J. Gelation of protein recovered from Antarctic krill (*Euphausia superba*) by isoelectric solubilization/precipitation as affected by function additives. J Agric Food Chem. 2007a;55:1814–22.
Chen YC, Jaczynski J. Protein recovery from rainbow trout (*Oncorhynchus mykiss*) processing by-products via isoelectric solubilization/precipitation and its

gelation properties as affected by functional additives. J Agric Food Chem. 2007b;55:9079–88.

Collette BB, Nauen CE. FAO Species Catalogue. Vol. 2 Scombrids of the World: An Annotated and Illustrated Catalogue of Tunas, Mackerels, Bonitos, and Related Species Known to Date. Rome: FAO; 1983. p. 137. Fish. Synop., 125.

Foh MBK, Wenshui X, Amadou I, Jiang Q. Influence of pH shift on functional properties of protein isolated of tilapia (Oreochromis niloticus) muscles and of soy protein isolate. Food Bioprocess Technol. 2012;5:2192–200.

Gokoglu N, Yerlikaya P. Determination of proximate composition and mineral contents of blue crab (Callinectes sapidus) and swim crab (Portunus pelagicus) caught off the Gulf of Antalya. Food Chem. 2003;80:495–8.

Gokoglu N, Yerlikaya P, Cengiz E. Effect of cooking methods on the proximate composition and mineral contents of rainbow trout (Oncorhynchus mykiss). Food Chem. 2004;84:19–22.

Horax R, Hettiarachchy NS, Chen P, Jalaluddin M. Preparation and characterization of protein isolate from cowpea (Vigna unguiculata L. Walp.). J Food Sci. 2004;69:FCT114–8.

Horax R, Hettiarachchy N, Kannan A, Chen P. Protein extraction ptimisation, characterisation, and functionalities of protein isolate from bitter melon (Momordica charantia) seed. Food Chem. 2011;124:545–50.

Huda N, Abdullah A, Babji AS. Functional properties of surimi powder from three Malaysian marine fish. Int J Food Sci Technol. 2001;36:401–6.

Intarasirisawat R, Benjakul S, Visessanguan W. Chemical compositions of the roes from skipjack, tongol and bonito. Food Chem. 2011;124:1328–34.

Ji SJ, Lee JS, Shin JH, Park KH, Kim JS, Kim KS, Heu MS. Distribution of protease inhibitors from fish eggs as seafood processing byproducts. Kor J Fish Aquat Sci. 2011;44:8–17.

Ju ZY, Hettiarachchy NS, Rath N. Extraction, denaturation, and hydrophobic properties of rice flour proteins. J Food Sci. 2001;66:229–32.

Kim HJ, Kim KH, Song SM, Kim IY, Park SH, Gu EJ, Lee HJ, Kim JS, Heu MS. Fractionation and characterization of protease inhibitors from fish eggs based on protein solubility. Kor J Fish Aquat Sci. 2013a;46:119–28.

Kim JS, Kim KH, Kim HJ, Kim MJ, Park SH, Lee HJ, Heu MS. Chromatographic fractionation of protease inhibitors from fish eggs. Kor J Fish Aquat Sci. 2013b;46:351–8.

Klomklao S, Benjakul S, Kishimura H. Optimum extraction and recovery of trypsin inhibitor from yellowfin tuna (Thunnus albacores) roe and its biochemical properties. Int J Food Sci Technol. 2013;49:168–73.

Kristinsson HG, Ingadottir B. Recovery and properties of muscle proteins extracted from tilapia (Oreochromis niloticus) light muscle by pH shift processing. J Food Sci. 2006;71:132–41.

Kristinsson HG, Rasco BA. Biochemical and functional properties of Atlantic salmon (Salmo salar) muscle proteins hydrolyzed with various alkaline proteases. J Agric Food Chem. 2000;48:657–66.

Laemmli UK. Cleavage of structural proteins during the assembly of the head of bacteriophage T4. Nature. 1970;227:680–5.

Losso JN, Bogumil R, Nakai S. Comparative studies of phosvitin from chicken and salmon egg yolk. Comp Biochem Physiol B. 1993;106:919–23.

Lowry OH, Rosebrough NJ, Farr AL, Randall RJ. Protein measurement with the Folin phenol reagent. J Biol Chem. 1951;193:265–75.

Martinez-Valverde I, Periago MJ, Santaella M, Ros G. The content and nutritional significance of minerals on fish flesh in the presence and absence of bone. Food Chem. 2000;71:503–9.

Mireles DeWitt CA, Gomez G, James JM. Protein extraction from beef heart using acid solubilization. J Food Sci. 2002;67:3335–41.

Narsing Rao G, Balaswamy K, Satyanarayana A, Prabhakara RP. Physico-chemical, amino acid composition, functional and antioxidant properties of roe protein concentrates obtained from Channa striatus and Lates calcarifer. Food chem. 2012;132:1171–6.

Nolsøe H, Undeland I. The acid and alkaline solubilization process for the isolation of muscle proteins. Food Bioprocess Technol. 2009;2:1–27.

Orban E, Di Lena G, Ricelli A, Paoletti F, Casini I, Gambelli L, Caproni R. Quality characteristics of sharpsnout sea bream (Diplodus puntazzo) from different intensive rearing systems. Food Chem. 2000;70:27–32.

Orban E, Nevigato T, Masci M, Di Lena G, Casini I, Caproni R, Gambelli L, De angelis P, Rampacci M. Nutritional quality and safety of European perch (Perca fluviatilis) from three lakes of central Italy. Food Chem. 2007;100:482–90.

Paraman I, Hettiarachchy NS, Schaefer C, Beck MI. Hydrophobicity, solubility, and emulsifying properties of enzyme-modified rice endosperm protein. Cereal Chem. 2007;84:343–9.

Pires C, Costa S, Batista AP, Nunes MC, Raymundo A, Batista I. Properties of protein powder prepared from Cape hake by-products. J Food Eng. 2012;108:268–75.

Sathivel S, Bechtel PJ. Properties of soluble protein powders from Alaska pollock (Theragra chalcogramma). Int J Food Sci Technol. 2006;41:520–9.

Sathivel S, Bechtel PJ, Babbitt JK, Prinyawiwatkul W, Negulescu II, Reppond KD. Properties of protein powders from arrowtooth flounder (Atheresthes stomias) and herring (Clupea harengus) byproducts. J Agric Food Chem. 2004;52:5040–6.

Sathivel S, Bechtel PJ, Babbitt JK, Prinyawiwatkul W, Patterson M. Functional, nutritional, and rheological properties of protein powders from arrowtooth flounder and their application in mayonnaise. J Food Sci. 2005;70:E57–63.

Sathivel S, Bechtel PJ, Prinyawiwatkul W. Physicochemical and rheological properties of salmon protein powders. Int J Food Eng. 2006;2:2.

Sathivel S, Yin H, Bechtel PJ, King JM. Physical and nutritional properties of catfish roe spray dried protein powder and its application in an emulsion system. J Food Eng. 2009;95:76–81.

Shaviklo GR, Thorkelsson G, Arason S, Sveinsdottir K. Characteristics of freeze-dried fish protein isolated from saithe (Pollachius virens). J Food Sci Technol. 2011;49:309–18.

Tahergorabi R, Beamer SK, Matak KE, Jaczynski J. Isoelectric solubilization/precipitation as a means to recover protein isolate from Striped Bass (Morone saxatilis) and Its physicochemical properties in a nutraceutical seafood product. J Agric Food Chem. 2012;60:5979–87.

Tahvonen R, Aro T, Nurmi J, Kallio H. Mineral content in Baltic herring and Baltic herring products. J Food Compos Anal. 2000;13:893–903.

Zudaire I, Murua H, Grande M, Korta M, Arrizabalaga H, Areso JJ, et al. Fecundity regulation strategy of the yellowfin tuna (Thunnus albacares) in the Western Indian Ocean. Fish Res. 2013;138:80–8.

23

Establishment and long-term culture of the cell lines derived from gonad tissues of Siberian sturgeon (*Acipenser baerii*)

Jun Hyung Ryu[1], Yoon Kwon Nam[1,2] and Seung Pyo Gong[1,2,3*]

Abstract

To culture germline stem cells in vitro, establishment of the cell lines that can be used as the feeder cells is a prerequisite. In this study, we tried to establish gonad-derived cell lines in Siberian sturgeon (*Acipenser baerii*). Five 1-year-old *A. baerii* were used as a donor of gonad tissues, and gonad-dissociated cells were cultured in vitro. Subsequently, determination of growth conditions, long-term culture, characterization, and cryopreservation of the cell lines were also conducted. Five gonad-derived cell lines were stably established and cultured continuously over at least the 73th passage and 402 culture days under the media containing 20 % fetal bovine serum at 28 °C. All cell lines consisted of two main cell types based on morphology even if the ratio of the two cell types was different depending on cell lines. Despite long-term culture, all cell lines maintained diploid DNA contents and expression of several genes that are known to express in the *A. baerii* gonad. After freezing and thawing of the cell lines, post-thaw cell viabilities between 57.6 and 92.9 % depending on cell lines were indentified, suggesting that stable cryopreservation is possible. The results and the cell lines established in this study will contribute to the development of an in vitro system for *A. baerii* germline stem cell culture.

Keywords: Sturgeon, *Acipenser baerii*, Feeder cells, Long-term culture, Germline stem cells

Background

Germline stem cells are a very important cell type taking charge of gamete production (Brinster, 2007; Spradling et al., 2011; Lehmann, 2012). Due to their high application possibilities to animal transgenic research as a mediator conveying new traits to the next generation, lots of trials for in vitro culture and manipulation of them have been conducted in many mammalian (Guan et al., 2006; Aponte et al., 2008; Lee et al., 2013; Tiptanavattana et al., 2013; Lee et al., 2014) and some avian species (Park et al., 2008; Song et al., 2014). In fish, similar studies have been performed in small fish models (Sakai, 2002; Hong et al., 2004; Fan et al., 2008; Kawasaki et al., 2012; Wong and Collodi, 2013; Wong et al., 2013; Li et al., 2014), but the related ones dealing with large farmed fish have been rarely conducted (Shikina et al., 2008; Shikina and

Yoshizaki, 2010; Lacerda et al., 2013). Therefore, establishment of an in vitro culture system for germline stem cells derived from the economically valuable fish is one of the upcoming challenges in the field of fish transgenic research. As one of the important aquaculture fish, the Siberian sturgeon (*Acipenser baerii*) provides valuable food resources including a luxury food and caviar and has a merit in culture due to its ability to tolerate the changes of environmental factors such as temperature and low O_2 concentration (Gisbert and Ruban, 2003). Thus, to fulfill the improvement of the breed of this species, applying transgenic technology is a worthwhile work in regard to commercial aspect. Moreover, establishment of germline stem cells can contribute to species preservation of this endangered fish species (Ruban and Zhu, 2010; Hong et al., 2011; Lacerda et al., 2014).

To establish germline stem cell culture system in *A. baerii*, developing appropriate feeder cell lines that can support germline stem cells in vitro is a top priority. In general, cell lines derived from gonads are known to possess a capacity to support survival, maintenance,

* Correspondence: gongsp@pknu.ac.kr
[1]Department of Marine Biomaterials and Aquaculture, Pukyong National University, Busan 608-737, South Korea
[2]Department of Fisheries Biology, Pukyong National University, Busan 608-737, South Korea
Full list of author information is available at the end of the article

growth, and differentiation of germline stem cells during in vitro culture (Hofmann et al., 2003; Nagano et al., 2003; Hong et al., 2004). It has been reported that mouse spermatogonial stem cells formed colonies and maintained its characteristics for 5 months on testis-derived feeder cells during in vitro culture (Kanatsu-Shinohara et al., 2012). In fish, trout spermatocytes showed longer survival on Sertoli cells from testis than those without somatic cells in culture (Loir, 1989). Moreover, zebrafish ovarian-somatic feeder cells supported cell growth, survival, and germline competency of female germline stem cells in culture (Wong et al., 2013).

Therefore, in this study, we tried to establish *A. baerii* gonad-derived cell lines as a first step toward culturing *A. baerii* germline stem cells. Subsequently, we evaluated optimal growth conditions of the established cell lines and characterized them within the framework of the morphology, DNA contents, and expression of genes that are known to express in *A. baerii* gonad. In addition, considering the legal and ethical limitation about killing this endangered species repeatedly to secure a large amount of feeder cells, we tested the feasibility of long-term culture and cryopreservation of the established cell lines.

Methods

Fish

Siberian sturgeons (*A. baerii*) were reared by a water recycling system in hatchery of Pukyong National University (Busan, Korea) at ambient temperature and fed an artificial feed (Millennium Plus; Woosung Feed Corp., Daejeon, Korea). Five 1-year-old fish were used in this study, and the average body length and weight were 39.6 ± 3.3 cm and 192.3 ± 54.8 g, respectively. All procedures for animal management, euthanasia, and surgery were complied with the guidelines of Institutional Animal Care and Use Committee (IACUC) of Pukyong National University and the ethical guidelines published by International Council for Laboratory Animal Science (ICLAS).

Cell isolation and culture

Gonad tissues were dissected from the body with sterile scissors and tweezers following disinfection with 70 % ethanol (SK Chemicals, Sungnam, Korea). Each gonad was washed twice with Dulbecco's Phosphate-Buffered Saline (DPBS; Gibco, Grand Island, NY, USA) containing 1 % (v/v) penicillin and streptomycin (P/S; Gibco) in petri dishes (SPL Life Sciences, Pocheon, Korea) and placed in a 35-mm petri dish (SPL Life Science) filled with digestive solution consisting of a Leibovitz's L-15 Medium (L15; Gibco) supplemented with 500 U/mL collagenase type I (Worthington Biochemical Corporation, Lakewood Township, NJ, USA) and 0.05 % trypsin EDTA (Gibco). Then, the tissues were chopped using a

surgical blade and incubated for 30 min at 28 °C. After digestion, enzyme was inactivated by adding 10 % (v/v) fetal bovine serum (FBS; Gibco)-containing L15. All the tissue derivatives were filtered on 40-µm cell strainers (BDFalcon™, San Jose, CA, USA), and the isolated cells were retrieved by centrifugation at $400 \times g$ for 4 min. Viable cells were counted with a hemocytometer (Paul Marienfeld GmbH & Co. KG, Lauda-Königshofen, Germany) after trypan blue (Gibco) staining, and 5×10^5 live cells were seeded in 24-well culture plates (SPL Life Sciences) with L15 containing 20 % (v/v) FBS and 1 % (v/v) P/S. The cells were cultured at 28 °C with an air atmosphere, and the culture media were changed every 2 to 3 days. Subcultures were conducted when the cells reached 90 to 100 % confluency. For subculture, the cells were washed twice using DPBS supplemented with 1 % (v/v) P/S and detached by 0.05 % trypsin EDTA at room temperature for 5 min. After trypsin inactivation by adding one volume of 10 % (v/v) FBS-containing L15, the detached cell suspension was centrifuged at $400 \times g$ for 4 min. Then, the harvested cells were resuspended with culture media and subcultured at a 1:3 ratio up to the 7th passage and since then at a 1:5 ratio.

Measurement of growth rate

To investigate the growth rates under different culture conditions, 2.5×10^3 cells from each of three cell lines (designated as ABG1, ABG3, and ABG5) at passages 24 to 29 were seeded in 96-well microplates (Thermo Scientific, Vernon Hills, IL, USA) filled with L15 containing 20 % (v/v) FBS and 1 % (v/v) P/S and cultured under four different temperatures of 24, 26, 28, and 30 °C. Cell viability of each group was measured at days 1, 3, 5, and 7 after cell seeding using Cell Counting Kit-8 (CCK-8; Dojindo, Kushu, Japan) according to the manufacturer's instructions. After determination of optimal culture temperature, a test for determining optimal FBS concentration for cell growth was conducted under the optimal culture temperature. The same protocols with temperature test were used, but the cells were cultured in five different media consisting of L15 containing different FBS concentrations of 0, 5, 10, 15, and 20 % under a fixed culture temperature. This experiment was conducted in triplicate.

Analysis of DNA contents

From each cell line, 1×10^7 cells at passages 54 to 55 were harvested by trypsinization and centrifugation. Cell pellets were suspended in 1 mL DPBS at room temperature and were transferred into 4 mL absolute ethanol at −20 °C. After overnight at −20 °C, the cells were harvested by centrifugation and rehydrated in 5 mL DPBS for 15 min at room temperature. After that, RNase A (Bioneer, Daejeon, Korea) was treated with a final concentration of 200 μg/mL in DPBS for 30 min at room temperature and, subsequently, stained by propidium iodide (Invitrogen, Carlsbad, CA, USA) with a final concentration of 10 μg/mL in the dark for 1 h at room temperature. Finally, DNA content was measured by an Accuri C6 flow cytometer (BD Biosciences, San Jose, CA, USA). A previously established *A. baerii* heart-derived cell line, the ploidy of which was confirmed (Kim et al., 2014), was used as a control of normal diploidy.

RT-PCR

Total RNA was extracted from each cell line using the RNeasy Plus Mini Kit (Qiagen, Valencia, CA, USA) at passages 15 to 16 and passages 54 to 55. Extracted total RNA was treated by DNase I (Sigma-Aldrich, St. Louis, MO, USA) to eliminate genomic DNA, and first-strand complementary DNA (cDNA) was synthesized from 1 μg total RNA using a M-MLV cDNA Synthesis Kit (Enzynomics, Daejeon, Korea). PCR was conducted with the primers specific for six genes including *ar*, *lh*, *cyp17a1*, *sox9*, *star*, and *β-actin* under the following condition: initial denaturation step (94 °C for 5 min), cycling step (30 cycles of 94 °C for 30 s, 60 °C for 30 s, and 72 °C for 45 s), and the final extension step (72 °C for 10 min). The PCR products mixed with LoadingSTAR (DyneBio, Seongnam, Korea) were size fractionated by 1.2 % (*w/v*) agarose gel (Lonza, Rockland, ME, USA) electrophoresis and visualized with a UV transilluminator. The primer sequences are listed in Table 1.

Measurement of viability of frozen-thawed cells

To investigate cell viability after freezing and thawing, 2×10^5 cells from each cell line at passages 52 to 57 were suspended with freezing solution consisting of L15 containing 10 % (*v/v*) dimethyl sulfoxide (DMSO; Sigma-Aldrich) and 20 % (*v/v*) FBS, and the cell suspensions were transferred into 1.2-mL cryovials (Corning, Corning, NY, USA). Subsequently, the cryovials were frozen in pre-chilled freezing containers (Nalgene, Rochester, NY, USA) with the cooling rate of –1 °C/min. After 12 h in a deep freezer at –75 °C, the cryovials were stored in liquid nitrogen (–196 °C) for 24 to 27 days. The cells were thawed in a 37 °C water bath for 2 min, 30 s and harvested by centrifugation. The post-thaw cells were suspended with culture media, and 1×10^4 cells were seeded in 96-well microplates. Cell viability was measured using CCK-8. Non-frozen cells were used as a control, and cell viability was calculated as absorbance$_{sample}$/absorbance$_{control}$ × 100. This experiment was conducted in triplicate.

Statistical analysis

The Statistical Analysis System (SAS Institute, Cary, NC, USA) program was used to analyze the effect of each treatment. When analysis of variance (ANOVA) detected a significant main effect, treatments were analyzed subsequently by Duncan's method. Significant differences among treatments were defined by a *p* value <0.05.

Results

Establishment of *A. baerii* gonad-derived cell lines

Five gonad-dissociated cells derived from five individuals were subjected to in vitro culture. All cell populations were attached to substrata after 1 day of culture and reached 90 to 100 % confluency in 4 to 6 days through continuous proliferation. They were subcultured continuously, and 1×10^6 cells of each were frozen at least once before the 10th passage. Each cell line was named

Table 1 Primer sequences used in this study

Genes	Primer sequences (5' > 3')	Product size (bp)	Accession number
ar	Forward, TGAAGAAGATGAAGGGAGCAGAAGAT	231	DQ388357.1
	Reverse, TCTCCCCCAGTTCATTCAAGC		
cyp17a1	Forward, TCACACACTCCAGTATTGGTG	92	HQ026486.1
	Reverse, CCATTCCTTTTCATCGTGATG		
lh	Forward, CTGCAGAGAAGGAGGAATGT	140	AJ251656.1
	Reverse, GCGAAGATCCTTATAGGTGCA		
sox9	Forward, AGCAGCAAAAACAAGCCTCA	113	EU241882.1
	Reverse, AGCTCCGCGTTGTGAAGAT		
star	Forward, CAGAAGTCAATCAGCATCCT	79	FJ205610.1
	Reverse, TCAGCACCTTGTCTCCATTG		
β-actin	Forward, CCCTGTTCCAGCCATCCTTC	155	JX027376.1
	Reverse, GTCTGCAATGCCAGGGTACA		

Primer sequences for *ar*, *cyp17a1*, *lh*, *sox9*, and *star* genes were referred from Berbejillo et al. (2012)
ar androgen receptor, *cyp17a1* cytochrome P450, family 17, subfamily a, polypeptide 1, *lh* luteinizing hormone, *sox9* sry-box containing gene 9, *star* steroidogenic acute regulatory protein

as ABG1, ABG3, ABG5, ABG6, and ABG7 (Table 2). Up to date, all the five cell lines have been cultured stably and constantly over at least the 73th passage and 402 culture days. For preservation and future use, all cell lines were cryopreserved in liquid nitrogen with a sufficient number of vials containing each cell lines at an interval of about 5 passages (Table 2).

Determination of optimal growth condition

To decide optimal growth conditions for *A. baerii* gonad-derived cell lines, the effects of temperature and FBS concentration on cellular growth were evaluated. As shown in Fig. 1, similar effects of temperature on cellular growth were observed in the three cell lines tested. Significant high cell growth was detected at both 28 and 30 °C ($p < 0.05$), and it was decreased as temperature goes down. In 24 °C, the lowest cell growth was observed, but the cells kept attaching on the substrata, maintained normal morphology, and grew slowly. To test the effects of FBS concentration on cellular growth, culture temperature was fixed at 28 °C according to the result of temperature test. As expected, absence of FBS in the culture did not induce both cell growth and attachment in all the cell lines tested and growth rate was increased in a dose-dependent manner. The ABG3 and ABG5 lines grew better significantly in the media containing 15 and 20 % FBS relative to the other media while the ABG1 line showed maximum growth in the media containing 20 % FBS ($p < 0.05$).

Characterization of *A. baerii* gonad-derived cell lines

During the culture period, all the cell lines were composed of mainly two types of cells based on morphological criteria: wide-spread cells that occupy a wide surface area and small polygonal cells that grow densely. At an early passage (passage nos. 15 to 16), a ratio of wide-spread cells was relatively higher than that of small polygonal cells in all cell lines and a certain difference among cell lines was not observed visually. As the cell lines were subcultured consistently, a ratio of small polygonal cells was relatively increased and

some differences among cell lines appeared at a late passage (passage nos. 54 to 55). The ABG1 and ABG5 lines have maintained a high ratio of wide-spread cells, whereas the ABG3 and ABG6 lines contained dominantly small polygonal cells. In the ABG7 line, two cell types were mixed at a similar ratio (Fig. 2a). To evaluate ploidy of the cell lines after long-term culture, DNA contents of each cell line at late passage were measured by flow cytometry. As shown in Fig. 2b, all the five cell lines showed diploid DNA contents similar to the control cell line despite long-term culture. Two peaks within diploid peak of the ABG7 line are seemed to be caused by the property of the ABG7 line in which two cell types existed at the same ratio. Next, RT-PCR analysis was conducted to identify the expression of a set of genes that are known to be expressed in *A. baerii* gonad tissue in the established cell lines at both early and late passages. As the results, the expression of *ar*, *lh*, and *sox9* genes was maintained at late passage, more or less, in all the cell lines, but the *cyp17a1* gene was not expressed at all in both early and late passages of all the cell lines. In case of the *star* gene, overall expression was weak with no expression in the ABG1 line and its expression in the ABG5 line was lost at late passage (Fig. 2c).

Post-thaw cell viability of *A. baerii* gonad-derived cell lines

To evaluate the feasibility of cryopreservation of the established cell lines, the viabilities of post-thaw cells which were frozen and stored in liquid nitrogen (–196 °C) for 24 to 27 days were measured in each cell line. High post-thaw cell viabilities of more than 79.6 % were detected in four cell lines including ABG1, 3, 5, and 7 (92.9, 79.6, 86.4, and 90.0 %, respectively), but the ABG6 line showed significant low post-thaw cell viability of 57.6 % (Fig. 3; $p < 0.05$).

Discussion

In this study, we successfully established five *A. baerii* gonad-derived cell lines from five trials (success rate, 5/5 = 100 %) and all established cell lines could grow stably during a long period of more than 1 year without

Table 2 Information of *Acipenser baerii* gonad-derived cell lines established in this study

Name of cell line	Culture[a]		Cryopreservation[b]		
	Passage no. in current	Culture period (day)	First passage no. cryopreserved	Final passage no. cryopreserved	Total no. of frozen vials stored
ABG1	80	504	9	80	46
ABG3	73	434	7	73	58
ABG5	80	405	8	80	39
ABG6	78	525	5	70	54
ABG7	83	402	6	83	62

[a]Composition of culture media was Leibovitz's L-15 medium supplemented with 20 % (*v/v*) fetal bovine serum and 1 % (*v/v*) penicillin and streptomycin
[b]All cell lines were frozen and stored in liquid nitrogen (–196 °C)

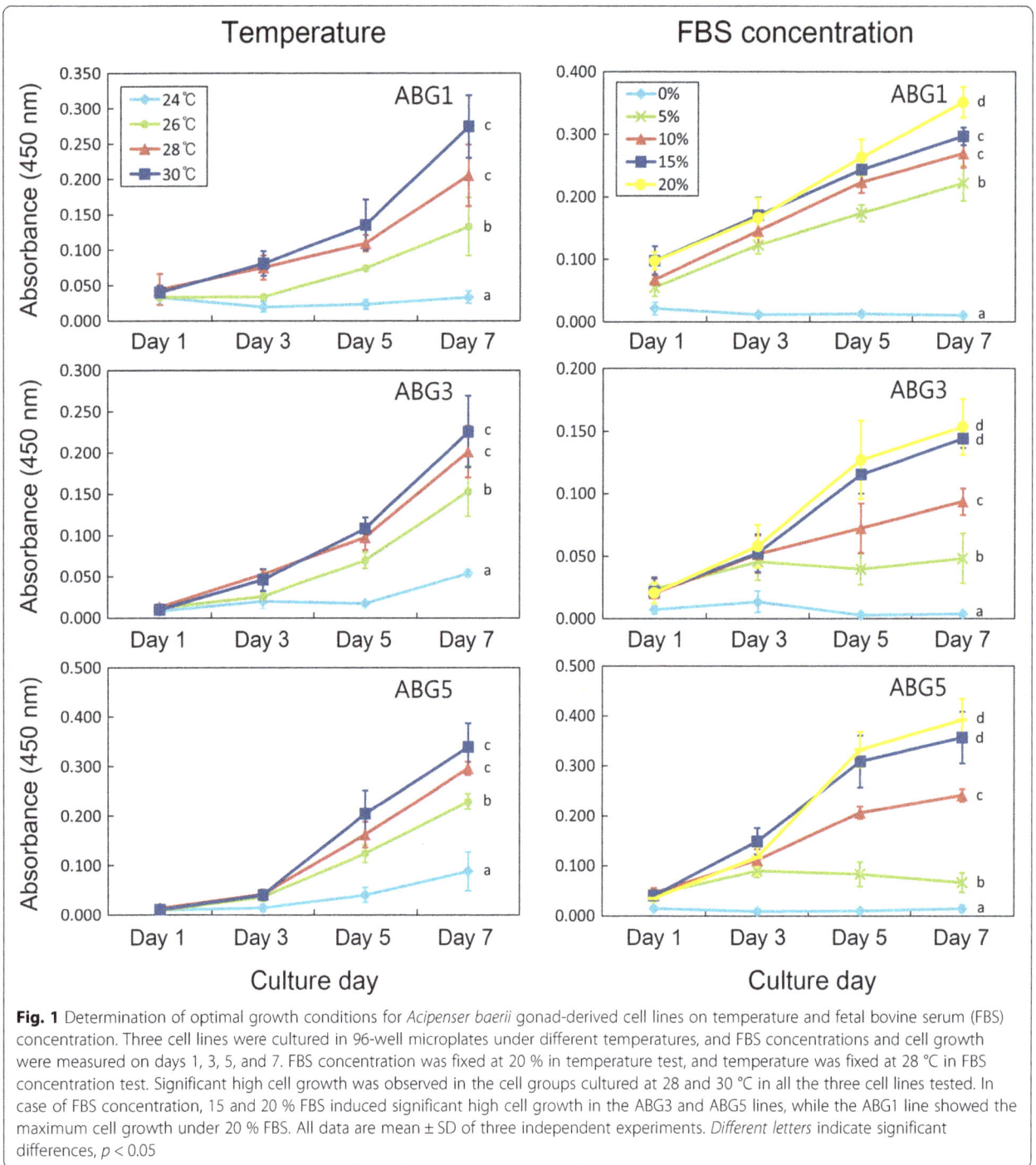

Fig. 1 Determination of optimal growth conditions for *Acipenser baerii* gonad-derived cell lines on temperature and fetal bovine serum (FBS) concentration. Three cell lines were cultured in 96-well microplates under different temperatures, and FBS concentrations and cell growth were measured on days 1, 3, 5, and 7. FBS concentration was fixed at 20 % in temperature test, and temperature was fixed at 28 °C in FBS concentration test. Significant high cell growth was observed in the cell groups cultured at 28 and 30 °C in all the three cell lines tested. In case of FBS concentration, 15 and 20 % FBS induced significant high cell growth in the ABG3 and ABG5 lines, while the ABG1 line showed the maximum cell growth under 20 % FBS. All data are mean ± SD of three independent experiments. *Different letters* indicate significant differences, $p < 0.05$

any significant growth retardation and marked deterioration in culture. Furthermore, these cell lines showed high post-thaw cell viabilities of more than 79.6 % except one line that showed 57.6 % post-thaw cell viability suggesting the feasibility of stable cryopreservation. These advantages of easiness in cell line derivation, long-term maintenance, and cryopreservation can maximize the availability of these cell lines as the feeder cells for culturing *A. baerii* germline stem cells. Culture of most stem cells requires

continuous supply of feeder cells (Evans and Kaufman 1981; Matsui et al., 1992; Kanatsu-Shinohara et al., 2003; Takahashi and Yamanaka, 2006; Jing et al., 2010; Pacchiarotti et al., 2010). Thus, in case of the feeder cells that have a relatively short lifespan like mouse embryonic fibroblasts that are feeder cells for culturing human and mouse embryonic stem cells, sacrifice of animals and labor-intensive work for cell preparation should be conducted repeatedly (Amit et al., 2003).

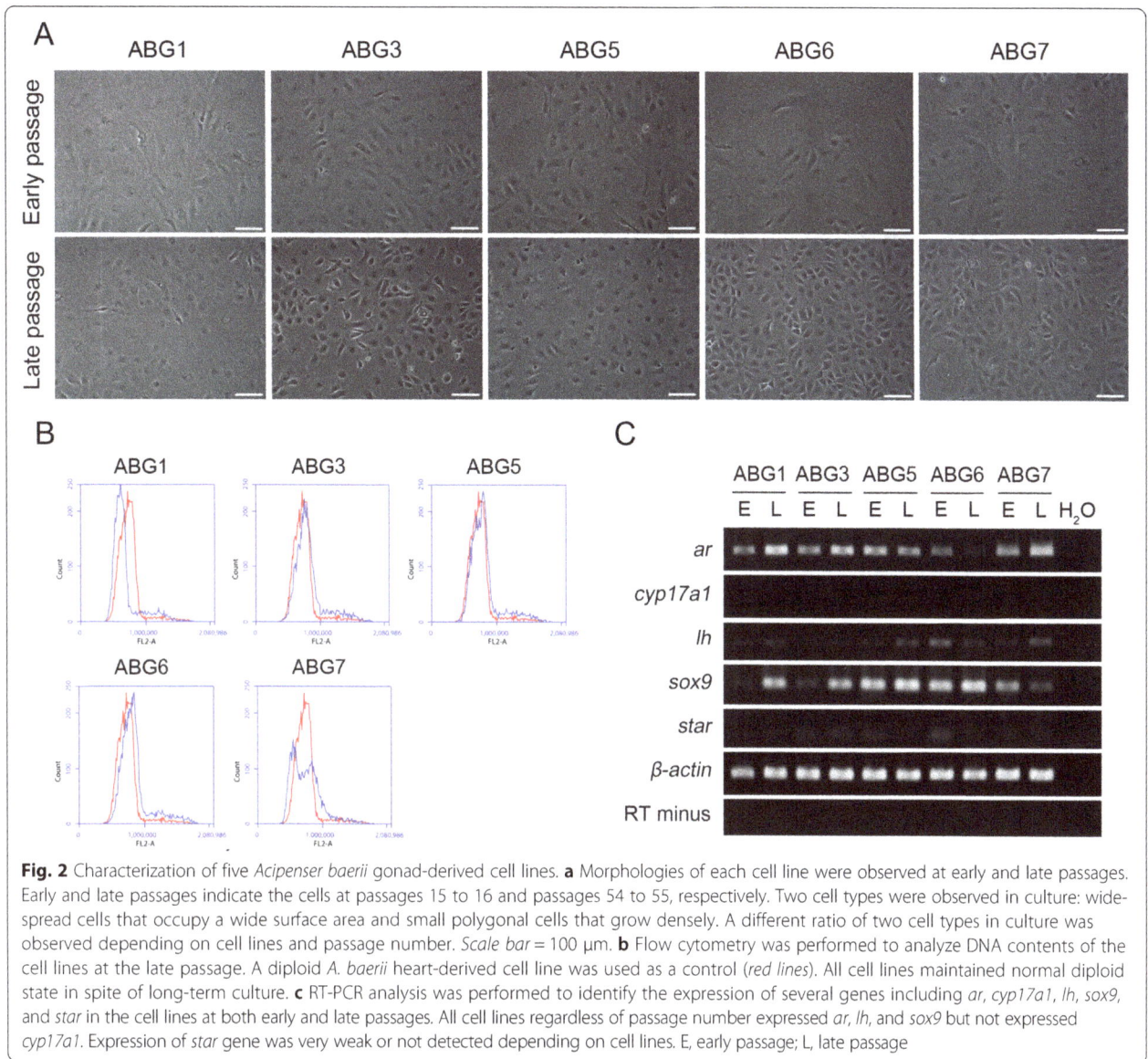

Fig. 2 Characterization of five *Acipenser baerii* gonad-derived cell lines. **a** Morphologies of each cell line were observed at early and late passages. Early and late passages indicate the cells at passages 15 to 16 and passages 54 to 55, respectively. Two cell types were observed in culture: widespread cells that occupy a wide surface area and small polygonal cells that grow densely. A different ratio of two cell types in culture was observed depending on cell lines and passage number. *Scale bar* = 100 μm. **b** Flow cytometry was performed to analyze DNA contents of the cell lines at the late passage. A diploid *A. baerii* heart-derived cell line was used as a control (*red lines*). All cell lines maintained normal diploid state in spite of long-term culture. **c** RT-PCR analysis was performed to identify the expression of several genes including *ar*, *cyp17a1*, *lh*, *sox9*, and *star* in the cell lines at both early and late passages. All cell lines regardless of passage number expressed *ar*, *lh*, and *sox9* but not expressed *cyp17a1*. Expression of *star* gene was very weak or not detected depending on cell lines. E, early passage; L, late passage

Unlike this, the cell lines that can grow long-term and be stored stably are free to such limitations. In addition, limitation of using *A. baerii* as an endangered species also can be overcome by the advantages. On the other side, these cell lines are able to play a role in universal biological studies such as toxicology and drug discovery, functional studies for genes and proteins, and study for cell-pathogen interaction (Lakra et al., 2011). Indeed, there is a report that conducted virus susceptibility test on Chinese sturgeon *A. sinensis* tail fin cell line to verify the feasibility of fish virus culture and isolation using a cultured cell line (Zhou et al., 2008).

For cell culture, we used 28 °C culture temperature and the media containing 20 % FBS based on our previous report that cultured *A. baerii* heart-derived cell lines (Kim et al., 2014) and the results from the experiment in this study. However, other conditions also can be attractive based on our results. We found that the established cell lines could be cultured in a wide range of culture temperatures from 24 to 30 °C even though significant high cell growth was observed in 28 and 30 °C conditions. Culture in various temperatures may have a potential merit as feeder cells because co-culture with germline stem cells may require a different optimal culture temperature as in the case of mammalian species (Lee et al., 2013). In addition, the results showed that 15 % FBS also induced significant high cell growth at a similar level with 20 % FBS in the ABG3 and ABG5 lines, suggesting that the application of a cell line-specific protocol can save overall cost by diminishing the use of high-cost FBS.

In general, tissue-derived primary cell populations in culture undergo a selection process by various factors

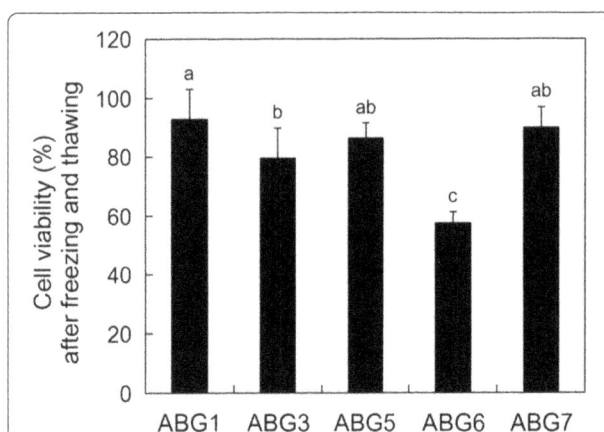

Fig. 3 Cell viability after freezing and thawing of five *Acipenser baerii* gonad-derived cell lines. Cell viability was measured immediately after thawing of frozen cells that were stored for 24 to 27 days in liquid nitrogen. More than 79.6 % post-thaw cell viability was observed in all the cell lines except ABG6 that showed 57.6 % cell viability. Non-frozen cells were used as a control. All data are mean ± SD of three the independent experiments. *Different letters* indicate significant differences, *p* < 0.05

such as cellular damage in isolation, cell attachment, limitations in nutrient or substrates, and growth competition between different cells. Thus, one or two cell types are eventually remained when a cell line forms in most case (Freshney, 2010). Likewise, we observed that two cell types were dominant in the culture of *A. baerii* gonad-dissociated cells in morphological criteria. At the initial stage of culture, it looked that wide-spread cells were more predominant than small polygonal cells but the proportion of one of the two cell types was increased as the culture was progressed in all the cell lines except the ABG7 line that showed a similar ratio of the two cell types. This indicates that both cell types can be easily adapted in the culture environment used in this study and they are the major targets for further study. In another aspect, two cell types may represent each cell population at different differentiation stages in a common cell lineage.

All cell lines showed diploid DNA contents at passages 54 to 55, indicating that all cell lines maintained an original normal state without cell transformation featured by aneuploidy or heteroploidy (Freshney, 2010). Moreover, the genes expressed at early passage were still expressed at late passage in all except the *star* gene of the ABG5 line. Although the data is very limited in terms of the number of genes tested, this result implies that the cell lines maintained their original characteristics in some degree at least up to passages 54 to 55. Taken together, these findings suggest that the *A. baerii* gonad-derived cell lines might be able to maintain their primary supportive role in germline stem cells in spite of long-term culture.

To establish gonad-derived cell lines, we used a 1-year-old *A. baerii* of which sex cannot be indentified externally (Keyvanshokooh and Gharaei, 2010). Unfortunately, we could not carry out histological inspection of gonad tissues from which the cell lines were originated due to a lack of available sample. Moreover, because there is no marker to discriminate sex at a cellular level in this species, sex of the cell lines established in this study is not clear. Therefore, it remains to be demonstrated if which germline stem cells, female or male, the established cell lines can support better. In previous reports, in vitro culture of germline stem cells requires several growth factors such as fibroblast growth factor (FGF), glial cell-derived neurotrophic factor (GDNF), leukemia inhibitory factor (LIF), or stem cell factor (SCF) (Matsui et al., 1992; Kubota et al., 2004; Zhang et al., 2011 in mammalian; Hong et al., 2004; Wong et al., 2013; Li et al., 2014 in fish). These growth factors can be supplied by feeder cells or exogenously. However, because commercially available growth factors are usually of human or mouse origin, supply of them by feeder cells seems to be more appropriate in fish germline stem cell culture (Wong et al., 2013). In this study, we did not identify the expression of the growth factors in the cell lines due to the absence of available information about them. Considering the importance of the growth factors in supporting germline stem cells in vitro, identification of their expression should be checked in future studies. Otherwise, the cell lines can be manipulated artificially so as to express stably such growth factors.

Conclusions

This study reports the establishment of five *A. baerii* gonad-derived cell lines that stable long-term culture and cryopreservation were possible. These cell lines can be utilized as a basic material to develop the culture system for germline stem cells in sturgeon species as well as a tool for general biological studies.

Acknowledgements
This work was supported by a Research Grant of Pukyong National University (year 2014).

Authors' contributions
JHR carried out the experiments and participated to write the manuscript. YKN conceived of the study and participated in its design. SPG conceived and designed the study, analyzed the data, and wrote the manuscript. All authors read and approved the final manuscript.

Competing interests
The authors declare that they have no competing interests.

Author details
[1]Department of Marine Biomaterials and Aquaculture, Pukyong National University, Busan 608-737, South Korea. [2]Department of Fisheries Biology,

Pukyong National University, Busan 608-737, South Korea. [3]Laboratory of Cell Biotechnology, Department of Marine Biomaterials and Aquaculture, College of Fisheries Science, Pukyong National University, Busan 608-737, South Korea.

References

Amit M, Margulets V, Segev H, Shariki C, Laevsky I, Coleman R, et al. Human feeder layers for human embryonic stem cells. Biol Reprod. 2003;68:2150–6.

Aponte PM, Soda T, Teerds KJ, Mizrak SC, van de Kant HJ, de Rooij DG. Propagation of bovine spermatogonial stem cells in vitro. Reproduction. 2008;136:543–57.

Berbejillo J, Martinez-Bengochea A, Bedo G, Brunet F, Volff JN, Vizziano-Cantonnet D. Expression and phylogeny of candidate genes for sex differentiation in a primitive fish species, the Siberian sturgeon, Acipenser baerii. Mol Reprod Dev. 2012;79:504–16.

Brinster RL. Male germline stem cells: from mice to men. Science. 2007;316:404–5.

Evans MJ, Kaufman MH. Establishment in culture of pluripotential cells from mouse embryos. Nature. 1981;292:154–6.

Fan L, Moon J, Wong TT, Crodian J, Collodi P. Zebrafish primordial germ cell cultures derived from vasa::Rfp transgenic embryos. Stem Cells Dev. 2008;17:585–97.

Freshney RI. Culture of animal cells: a manual of basic technique and specialized applications. 6th ed. Hoboken: Wiley-Blackwell; 2010.

Gisbert E, Ruban GI. Ontogenetic behavior of Siberian sturgeon, Acipenser baerii: a synthesis between laboratory tests and field data. Environ Biol Fish. 2003; 67:311–9.

Guan K, Nayernia K, Maier LS, Wagner S, Dressel R, Lee JH, et al. Pluripotency of spermatogonial stem cells from adult mouse testis. Nature. 2006;440:1199–203.

Hofmann MC, van Der Wee KS, Dargart JL, Dirami G, Dettin L, Dym M. Establishment and characterization of neonatal mouse Sertoli cell lines. J Androl. 2003;24:120–30.

Hong Y, Liu T, Zhao H, Xu H, Wang W, Liu R, et al. Establishment of a normal medakafish spermatogonial cell line capable of sperm production in vitro. Proc Natl Acad Sci U S A. 2004;101:8011–6.

Hong N, Li Z, Hong Y. Fish stem cell cultures. Int J Biol Sci. 2011;7:392–402.

Jing D, Fonseca AV, Alakel N, Fierro FA, Muller K, Bornhauser M, et al. Hematopoietic stem cells in co-culture with mesenchymal stromal cells—modeling the niche compartments in vitro. Haematologica. 2010;95:542–50.

Kanatsu-Shinohara M, Ogonuki N, Inoue K, Miki H, Ogura A, Toyokuni S, et al. Long-term proliferation in culture and germline transmission of mouse male germline stem cells. Biol Reprod. 2003;69:612–6.

Kanatsu-Shinohara M, Inoue K, Takashima S, Takehashi M, Ogonuki N, Morimoto H, et al. Reconstitution of mouse spermatogonial stem cell niches in culture. Cell Stem Cell. 2012;11:567–78.

Kawasaki T, Saito K, Sakai C, Shinya M, Sakai N. Production of zebrafish offspring from cultured spermatogonial stem cells. Genes to Cells. 2012;17:316–25.

Keyvanshokooh S, Gharaei A. A review of sex determination and searches for sex-specific markers in sturgeon. Aquac Res. 2010;41:e1–7.

Kim MS, Nam YK, Park C, Kim HW, Ahn J, Lim JM, et al. Establishment condition and characterization of heart-derived cell culture in Siberian sturgeon (Acipenser baerii). In Vitro Cell Dev Biol Anim. 2014;50:909–17.

Kubota H, Avarbock MR, Brinster RL. Growth factors essential for self-renewal and expansion of mouse spermatogonial stem cells. Proc Natl Acad Sci U S A. 2004;101:16489–94.

Lacerda SM, Costa GM, da Silva MA, Campos-Junior PH, Segatelli TM, Peixoto MT, et al. Phenotypic characterization and in vitro propagation and transplantation of the Nile tilapia (Oreochromis niloticus) spermatogonial stem cells. Gen Comp Endocrinol. 2013;192:95–106.

Lacerda SM, Costa GM, de França LR. Biology and identity of fish spermatogonial stem cell. Gen Comp Endocrinol. 2014;207:56–65.

Lakra WS, Swaminathan TR, Joy KP. Development, characterization, conservation and storage of fish cell lines: a review. Fish Physiol Biochem. 2011;37:1–20.

Lee WY, Park HJ, Lee R, Lee KH, Kim YH, Ryu BY, et al. Establishment and in vitro culture of porcine spermatogonial germ cells in low temperature culture conditions. Stem Cell Res. 2013;11:1234–49.

Lee KH, Lee R, Lee WY, Kim DH, Chung HJ, Kim JH, et al. Identification and in vitro derivation of spermatogonia in beagle testis. PLoS ONE. 2014;9: e109963.

Lehmann R. Germline stem cells: origin and destiny. Cell Stem Cell. 2012;10:729–39.

Li Z, Li M, Hong N, Yi M, Hong Y. Formation and cultivation of medaka primordial germ cells. Cell Tissue Res. 2014;357:71–81.

Loir M. Trout Sertoli cells and germ cells in primary culture: morphological and ultrastructural study. Gamete Res. 1989;24:151–69.

Matsui Y, Zsebo K, Hogan BL. Derivation of pluripotential embryonic stem cells from murine primordial germ cells in culture. Cell. 1992;70:841–7.

Nagano M, Ryu BY, Brinster CJ, Avarbock MR, Brinster RL. Maintenance of mouse male germ line stem cells in vitro. Biol Reprod. 2003;68:2207–14.

Pacchiarotti J, Maki C, Ramos T, Marh J, Howerton K, Wong J, et al. Differentiation potential of germ line stem cells derived from the postnatal mouse ovary. Differentiation. 2010;79:159–70.

Park TS, Kim MA, Lim JM, Han JY. Production of quail (Coturnix japonica) germline chimeras derived from in vitro-cultured gonadal primordial germ cells. Mol Reprod Dev. 2008;75:274–81.

Ruban G, Zhu B. Acipenser baerii, The IUCN Red List of Threatened Species 2010: e.T244A13046607. 2010.

Sakai N. Transmeiotic differentiation of zebrafish germ cells into functional sperm in culture. Development. 2002;129:3359–65.

Shikina S, Yoshizaki G. Improved in vitro culture conditions to enhance the survival, mitotic activity, and transplantability of rainbow trout type a spermatogonia. Biol Reprod. 2010;83:268–76.

Shikina S, Ihara S, Yoshizaki G. Culture conditions for maintaining the survival and mitotic activity of rainbow trout transplantable type A spermatogonia. Mol Reprod Dev. 2008;75:529–37.

Song Y, Duraisamy S, Ali J, Kizhakkayil J, Jacob VD, Mohammed MA, et al. Characteristics of long-term cultures of avian primordial germ cells and gonocytes. Biol Reprod. 2014;90:15.

Spradling A, Fuller MT, Braun RE, Yoshida S. Germline stem cells. Cold Spring Harb Perspect Biol. 2011;3:a002642.

Takahashi K, Yamanaka S. Induction of pluripotent stem cells from mouse embryonic and adult fibroblast cultures by defined factors. Cell. 2006;126:663–76.

Tiptanavattana N, Thongkittidilok C, Techakumphu M, Tharasanit T. Characterization and in vitro culture of putative spermatogonial stem cells derived from feline testicular tissue. J Reprod Dev. 2013;59:189–95.

Wong TT, Collodi P. Dorsomorphin promotes survival and germline competence of zebrafish spermatogonial stem cells in culture. PLoS ONE. 2013;8:e71332.

Wong TT, Tesfamichael A, Collodi P. Production of zebrafish offspring from cultured female germline stem cells. PLoS ONE. 2013;8:e62660.

Zhang Y, Yang Z, Yang Y, Wang S, Shi L, Xie W, et al. Production of transgenic mice by random recombination of targeted genes in female germline stem cells. J Mol Cell Biol. 2011;3:132–41.

Zhou GZ, Gui L, Li ZQ, Yuan XP, Zhang QY. Establishment of a Chinese sturgeon Acipenser sinensis tail-fin cell line and its susceptibility to frog iridovirus. J Fish Biol. 2008;73:2058–67.

24

Recovery of serine protease inhibitor from fish roes by polyethylene glycol precipitation

Hyun Ji Lee[1], Hyung Jun Kim[2], Sung Hwan Park[1], In Seong Yoon[1], Gyoon-Woo Lee[1], Yong Jung Kim[2], Jin-Soo Kim[2] and Min Soo Heu[1*]

Abstract

The fractionation of serine protease inhibitor (SPI) from fish roe extracts was carried out using polyethylene glycol-4000 (PEG4000) precipitation. The protease inhibitory activity of extracts and PEG fractions from Alaska pollock (AP), bastard halibut (BH), skipjack tuna (ST), and yellowfin tuna (YT) roes were determined against target proteases. All of the roe extracts showed inhibitory activity toward bromelain (BR), chymotrypsin (CH), trypsin (TR), papain-EDTA (PED), and alcalase (AL) as target proteases. PEG fractions, which have positive inhibitory activity and high recovery (%), were the PEG1 fraction (0–5 %, *w/v*) against cysteine proteases (BR and PA) and the PEG4 fraction (20–40 %, *w/v*) against serine proteases (CH and TR). The strongest specific inhibitory activity toward CH and TR of PEG4 fractions was AP (9278 and 1170 U/mg) followed by ST (6687 and 2064 U/mg), YT (3951 and 1536 U/mg), and BH (538 and 98 U/mg). The inhibitory activity of serine protease in extracts and PEG fractions from fish roe was stronger than that of cysteine protease toward common casein substrate. Therefore, SPI is mainly distributed in fish roe and PEG fractionation effectively isolated the SPI from fish roes.

Keywords: Polyethylene glycol, Roe, Serine protease inhibitor, Recovery

Background

Protease inhibitors commonly accumulate in high quantities in plant and animal tissues (Sangorrin et al. 2001), plant seeds, bird eggs, and various body fluids. Protease inhibitors are also found in poultry (Lopuska et al. 1999), blood plasma (Rawdkuen et al. 2005; Rawdkuen et al. 2007), fish roe (Kim et al. 2013a,b; Choi et al. 2002; Klomklao et al. 2014), and viscera (Kishimura et al. 2001).

These inhibitors play a significant role in the regulation of proteolysis, whether the target enzymes are of exogenous or endogenous origin. Protease inhibitors permit the regulation of the rate of proteolysis in the presence of the active enzyme (Barret 1986; Knight 1986; Cherqui et al. 2001). The presence of protease inhibitors has been demonstrated in the blood and muscle of rainbow trout (Clereszko et al. 2000), chum salmon (Yamashita and Konagaya 1991), white croaker (Sangorrin et al. 2001),

hake skeletal (Martone et al. 1991), and the roe of Alaska pollock, bastard halibut, skipjack tuna, yellowfin tuna (Kim et al. 2015; Ji et al. 2011), herring (Oda et al. 1998), and carp (Tsai et al. 1996).

In industries of surimi-based product, commercial protease inhibitors are used to prevent the modori (gel softening) phenomenon and to maximize the gel strength of surimi. The most commonly used inhibitors are bovine plasma protein (BPP), chicken egg white, potato powder, and whey protein concentrate (Hamann et al. 1990; Weerasinghe et al. 1996; Kim et al. 2015). However, the use of BPP has been prohibited, due to the occurrence of mad cow disease. Egg white is expensive and has an undesirable egg-like odor, while off-color problems may be encountered when potato powder is used (Akazawa et al. 1993). Therefore, alternative food-grade proteinase inhibitors from marine resources for surimi production are still needed.

Fish roe, a byproduct generated from fish processing (3.0–30.0 % depend on fish species), is a highly nutritious material rich in essential fatty acids and amino acids

---
* Correspondence: heu1837@dreamwiz.com
[1]Department of Food and Nutrition/Institute of Marine Industry, Gyeongsang National University, Jinju 52828, South Korea
Full list of author information is available at the end of the article

(Narsing Rao et al. 2012). Protease inhibitors in fish roe can have a major impact on nutritional value as they inhibit pancreatic serine proteases, thus impairing protein digestion. However, fish roe can be used as a potential source of proteinase inhibitor and can be for a variety of applications such as medicine, agriculture, and food technology (Klomklao et al. 2014).

Protein fractionation methods may be divided into those based on differential solubility, differential interaction with solid media, and differential interaction with physical parameters (Rawdkuen et al. 2005). In our previous study (Kim et al. 2013a), the protease inhibitor was fractionated from fish eggs using methods based on protein solubility using organic solvent and ammonium sulfate (AS). AS fractionation in isolating the protease inhibitor was more effective than organic solvent precipitation (Kim et al. 2013a). However, AS fractionation methods have the disadvantage of either requiring a high concentration or cooling to avoid denaturation (Rawdkuen et al. 2007). In the case of organic solvent fractionation, the component obtained by fractionation has a notable capacity for use, as a result of the denaturation of the protein during the process (Kim et al. 2014; Rawdkuen et al. 2007).

In order to avoid the disadvantages of these techniques, polyethylene glycol (PEG) is an alternative precipitating agent for protein fractionation. Chicken plasma was fractionated into the protease inhibitor by PEG precipitation (Rawdkuen et al. 2005; Rawdkuen et al. 2007). PEG has several advantages over other precipitants, including the least denaturation of proteins at ambient temperatures, negligible temperature control required in the range 4–30 °C, relatively small amount of precipitant required compared with AS or organic solvents, and low residual PEG concentration in the precipitate since most of the PEG is retained in the supernatant (Sharma and Kalonia 2004).

The objectives of this study were to find the best conditions for the polyethylene glycol fractionation of protein inhibitor and characterize the roe protease inhibitor from Alaska pollock and bastard halibut as white-fleshed fish and skipjack tuna and yellowfin tuna as dark-fleshed fish roes.

Methods
Materials
Alaska pollock (AP, *Theragra chalcogramma*) roe was obtained from Blue Ocean Co. (Busan, Korea). Bastard halibut (BH, *Paralichthys olivaceus*) was purchased from the fish market (Tongyoung, Korea) and immediately brought to the laboratory. Roe was separated from BH and stored at –70 °C in sealed polyethylene bags. Skipjack tuna (ST, *Katsuwonus pelamis*) and yellowfin tuna (YT, *Thunnus albacares*) roes were obtained from Dongwon F&B Co., Ltd. (Changwon, Gyungnam, Korea).

Fish roes were stored at –70 °C in sealed polyethylene bags until needed for inhibitor extraction.

Chemicals
Polyethylene glycol-4000 (PEG4000), which is a chemical used for fractionation, was obtained from the Yakuri Pure Chemicals Co., Ltd. (Kyoto, Japan). Trypsin, chymotrypsin, bromelain, and papain were from Sigma-Aldrich Chemical Co. (St. Louis, MO, USA). Alcalase 2.5 type DX, Neutrase 0.8 L, Flavourzyme 500 MG, and Protamex were purchased from Novozymes (Bagsvaerd, Denmark). Aroase AP-10 and Pancidase NP-2 were from Yakult Pharmaceutical Co., Ltd. (Tokyo, Japan). Protease-NP was purchased from Amorepacific Co., Ltd. (Seoul, Korea). Casein and Nα-benzoyl-DL-arginine-2-naphthylamide hydrochloride (BANA) as substrates were purchased from Sigma-Aldrich Chemical Co. (St. Louis, MO, USA). The buffer solutions (0.1 M sodium phosphate buffer, pH 6.0; 0.1 M Tris-HCl buffer, pH 9.0) for enzyme reaction were prepared according to the method of Dawson et al. (1986). Sodium dodecyl sulfate (SDS) and glycine were purchased from Bio Basic Inc. (Ontario, Canada). Coomassie brilliant blue R-250 was purchased from Bio-Rad Laboratories, Inc. (Hercules, CA, USA). Glycerol and β-mercaptoethanol were purchased from Sigma-Aldrich Chemical Co. (St. Louis, MO, USA). Bromophenol blue was purchased from Junsei Chemical Co., Ltd. (Tokyo, Japan).

All chemicals used were analytical grade.

Preparation of the CE
Crude extracts (CEs) were prepared according to the modified method of Kim et al. (2013a). For extraction of CE from fish roes, the frozen roes were partially thawed and homogenized with 3 volumes (*w/v*) of deionized distilled water. The homogenates were incubated at 20 °C for 6 h, stirring every 1 h, and then centrifuged at 12,000×*g* for 20 min at 4 °C. The supernatant was used as "crude extracts" for further study.

Fractionation of protease inhibitor from CE with PEG
Four CEs from fish roes were continuously fractionated using PEG4000 in the range of 0–5 % (PEG1), 5–10 % (PEG2), 10–20 % (PEG3), and 20–40 % (*w/v*, PEG4), and these fractions were collected by centrifugation (15,000×*g*, for 30 min at 4 °C) and dissolved in a minimum quantity of cold deionized water. The fractions were stored at –25 °C until further analysis.

Protein concentration
The protein concentration of CE and PEG fractions from fish roes was determined according to the method of Lowry et al. (1951) by bovine serum albumin as a standard protein.

Determination of inhibitory activity of CE and PEG fractions toward target proteases

Enzyme activities against 0.1 % (w/v) chymotrypsin (CH) and trypsin (TR) as serine protease; 0.1 % (w/v) papain-EDTA (PED) and bromelain (BM) as cysteine protein; and 1 % (v/v) Alcalase (AL) and Neutrase (NE) and 1 % (w/v) Protease-NP (PN), Pancidase NP-2 (NP), Protamex (PR), Aroase AP-10 (AP-10), and Flavourzyme (FL) as commercial food-grade protease were measured using casein as a substrate according to the methods of Ji et al. (2011).

The CE and PEG fractions were examined for inhibitory activity against commercial proteases as mentioned above. Protease inhibitory activity was measured using casein and BANA as substrates.

When casein was used as a substrate, 50 µL of the inhibitor solution (CE and PEG fractions) was mixed with enzymes (10–100 µL) in 1.5 mL of 0.1 M sodium phosphate buffer (pH 6.0) or 0.1 M Tris-HCl buffer (pH 9.0). After incubation for 10 min at room temperature, 0.5 mL of 2 % casein was added and mixed thoroughly. The mixture was incubated for 1 h at 40 °C. The enzymatic reaction was terminated by adding 2 mL of 5 % TCA and then centrifuged at 1910×g for 15 min at 4 °C. The liberated soluble peptides in the supernatant were estimated by measuring the absorbance at 280 nm to determine the residual protease activity.

Protease activities against 0.1 % TR and 0.1 % PED were measured using BANA as the substrate according to the methods of Rawdkuen et al. (2007) with a slight modification. The 0.1 % TR (50 µL) and 0.1 % PED (100 µL) were added to 50 µg of inhibitor solution in 1.5 mL of 0.1 M Tris-HCl buffer (pH 9.0) and 0.1 M sodium phosphate buffer (pH 6.0), respectively. The mixture was incubated for 10 min at room temperature. Then, 50 µL of 10 mM BANA was added and vortexed immediately to start the enzyme reaction. After incubating for 1 h at 40 °C, 0.5 mL of 2 % HCl/ethanol was added to terminate the reaction. The reaction mixture was centrifuged at 1910×g for 15 min. The residual activity of enzymes was measured by the absorbance at 540 nm (U-2900, UV-VIS spectrophotometer, Hitachi, Tokyo, Japan).

One unit of enzyme activity was defined as an increase of 0.1 absorbance per 1 h.

One unit of inhibitory activity was defined as the amount of an inhibitor that reduced 1 unit/mg of target protease activity for 1 h.

Relative inhibitory activity (RIA) was calculated as follows:

$$RIA\ (\%) = [(C-A)/C] \times 100$$

C = enzyme activity of control (without inhibitor), A = enzyme activity of sample (with inhibitor)

SDS-PAGE and native PAGE gel electrophoresis

Sodium dodecyl sulfate-polyacrylamide gel electrophoresis (SDS-PAGE) was carried out for the determination of the purity and molecular weight of the samples, as described by Laemmli (1970), using a 10 % Mini-PROTEAN® TGX™ Precast gel (Bio-Rad Laboratories, Inc., Hercules, CA, USA). Samples were prepared by mixing the CE and PEG fractions at a 4:1 (v/v) ratio with the SDS-PAGE sample treatment buffer (62.5 mM Tris-HCl (pH 6.8), 2 % SDS (w/v, pH 8.3), 10 % glycerol, 2 % β-mercaptoethanol, and 0.002 % bromophenol blue). The samples were heated in a boiling water bath at 100 °C for 5 min and loaded (20 µg protein) on the SDS-polyacrylamide gel, and electrophoresis was performed at constant amperage (10 mA/gel) using a Mini-PROTEAN® Tetra cell (Bio-Rad Laboratories Inc., Hercules, CA, USA). After electrophoresis, the gel was stained in a staining solution containing Coomassie brilliant blue R-250. De-staining was carried out using a solution containing acetic acid, methanol, and water (1:2:7, v/v/v). The molecular weight of samples was estimated using Precision Plus Protein™ standards (10–250 K) from Bio-Rad Laboratories, Inc., (Hercules, CA, USA). Native PAGE was performed according to the procedure of Kim et al. (2015), except that the sample was not heated and the SDS and reducing agent were left out.

Zymography

Casein zymography was performed on native PAGE. Briefly, after electrophoresis, the gel was flooded with 3 mL of 0.1 % chymotrypsin. The gel was incubated for 60 min at 40 °C to allow the protease to diffuse into the gel and then washed with distilled water. The gel was immersed in 0.1 M Tris-HCl buffer, (pH 9.0) with 2 % casein (v/v) for 2 h. The gel was then rinsed with distilled water, fixed, and stained with Coomassie brilliant blue R-250 to develop inhibitory zones detected as a dark band on a clear background.

Statistical analysis

All experiments were conducted in triplicates. The average and standard deviations were calculated. Data were analyzed using the analysis of variance (ANOVA) procedure by means of the statistical software SPSS 12.0 KO (SPSS Inc., Chicago, IL, USA). The mean comparison was made using Duncan's multiple range test ($P < 0.05$).

Results and discussion
Inhibitory activity of CEs

Commercial protease inhibitory activities of the crude extract (CE) from fish roes (AP, BH, ST, and YT) are shown in Fig. 1. Inhibitory activities against 11 commercial proteases were measured using casein as a substrate. The highest relative inhibitory activity (RIA, %) was found in all CEs for CH as a serine protease. Of the CEs,

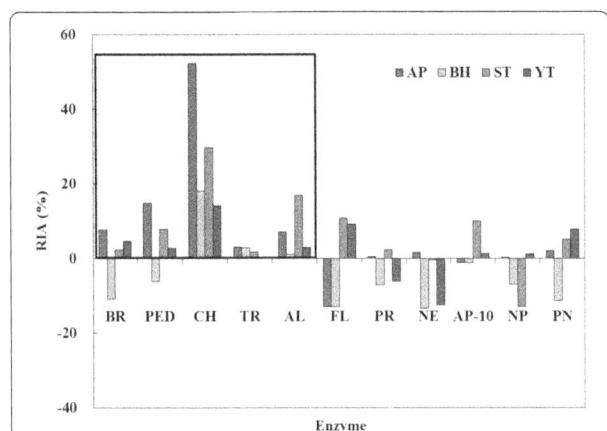

Fig. 1 Commercial protease inhibitory activity of the crude extract from fish roes toward casein as a substrate. *RIA (%)* relative inhibitor activity

AP showed the highest RIA (52.2 %), followed by ST (29.7 %), BH (18.1 %), and YT (14.0 %). RIAs (0.1–3.1 %) for TR as a serine protease were lower than those of CH. RIAs of BR and PED as a cysteine protease were observed for AP, ST, and YT except for BH. Among the commercial food-grade proteases, RIAs in all CEs were observed for AL. The other proteases, such as FL, PR, NE, AP-10, and PN, showed no effect on the inhibitory activity. Therefore, these results suggested that the CE from fish roes belongs to the serine protease inhibitor family and is also more sensitive to reaction with chymotrypsin than trypsin.

The protease inhibitory activities for trypsin (TR) and papain-EDTA (PED) of the CE from fish roes are shown in Fig. 2. Inhibitory activities were measured using BANA as a specific substrate for trypsin and papain. RIA for trypsin was the highest in AP (23.0 %), followed by ST (12.1 %), BH (8.4 %), and YT (8.0 %). Whereas, when PED as a cysteine protease was used, the CEs of all fish roes showed no effect on the inhibitory activity. Therefore,

these results confirmed that the CE from fish roes belongs to the serine protease inhibitor family.

Ji et al. (2011) confirmed the distribution of protease inhibitory activity in CEs from fish roes. ST (Choi et al. 2002) and YT (Klomklao et al. 2014) were reported to possess high trypsin inhibitory activity. The protease inhibitor from chum salmon egg (Kim et al. 2006), AP egg (Ustadi et al. 2005a), and glassfish egg (Ustadi et al. 2005b) inhibited the cysteine proteases such as papain and cathepsin L, but not trypsin, a serine protease.

Protein content of PEG fractions

The protein contents of CE and PEG fractions from fish roes are shown in Fig. 3. The protein contents of the CE of AP, BH, ST, and YT were 5655.0, 4183.0, 2849.6, and 3711.0 mg/100 g roe, respectively. The highest protein content of PEG fraction by PEG precipitation was found in PEG1 (0–5 % fraction) for AP and BH. The protein recovered in the PEG1 fraction of AP and BH represented 55.1 and 46.8 % of the total protein content of PEG fractions, respectively. Among the PEG fractions obtained from the CE of ST, the PEG4 fraction had the highest protein content (350.8 mg/100 g roe), which constituted approximately 38.8 % of the total protein content of PEG fractions, followed by PEG4 (349.4 mg/100 g roe), PEG2 (177.5 mg/100 g roe), and PEG1 fraction (26.3 mg/100 g roe). The protein content recovered from the PEG3 and PEG4 fractions of YT represented 42.3 and 40.5 % of the total protein content of PEC fractions. From the result, greater protein in the PEG fraction suggested that a higher amount of protease inhibitors was precipitated. Bovine blood plasma (Lee et al. 1987) and chicken plasma (Rawdkuen et al. 2005; Rawdkuen et al. 2007) were fractionated into proteins and protease inhibitor by PEG precipitation with high separation efficiencies.

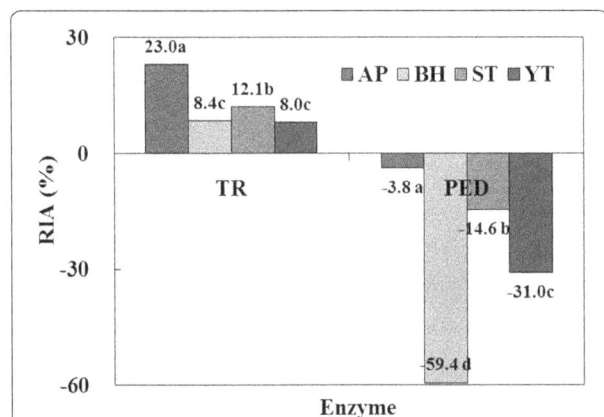

Fig. 2 Commercial protease inhibitory activity of the crude extracts from fish roes by the polyethylene glycol toward BANA as a substrate. Means with different letters within the sample are significantly different at $P < 0.05$ by Duncan's multiple range test. *RIA (%)* relative inhibitor activity

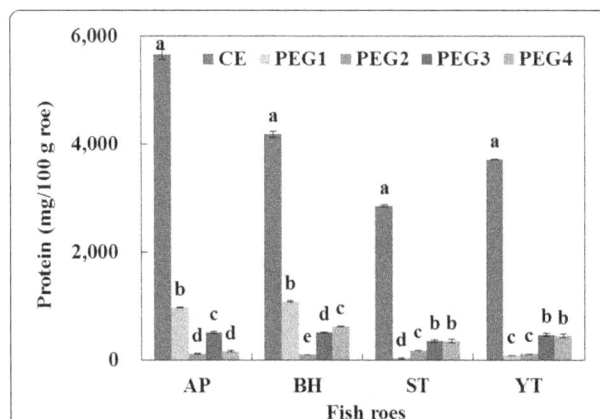

Fig. 3 Protein content (mg/100 g roe) of PEG fractions obtained from the crude extracts of fish roes by the polyethylene glycol precipitation. Means with different letters within the sample are significantly different at $P < 0.05$ by Duncan's multiple range test

Inhibitory activity of PEG fractions

Commercial protease inhibitory activity and the recovery of the CE and PEG fractions from fish roes are shown in Table 1. Inhibitory activities against 1 % AL, 0.1 % BR, 0.1 % PED, 0.1 % CH, and 0.1 % TR were measured using casein as a substrate.

All PEG fractions obtained from CE of AP and BH showed no effect on the specific inhibitory activity (SIA) for AL as a commercial food-grade protease. The SIA of 210.3 and 209.3 U/mg with recovery of 0.2 and 3.2 % were obtained for the PEG1 fraction of ST and YT, respectively. Among the PEG fractions of AP, the highest SIA (17.9 U/mg) and recovery (18.4 %) was found in the PEG1 fraction for BR, while the PEG2 fraction gave the highest SAI (220.8 U/mg) and recovery (11.8 %) for PED. However, all the PEG fractions of BH showed no effect on

the inhibitory activity for BR and PED. Of the PEG fractions, the PEG1 fraction of ST and YT showed the highest SIA for BR (72.6 and 45.7 U/mg, respectively) and PED (618.6 and 566.2 U/mg, respectively). From this result, it can be stated that the cysteine inhibitor from the PEG fraction of AP, ST, and YT is more concentrated in the PEG1 fraction (0–5 %). The highest SIA for CH was observed in the PEG4 fractions of AP, ST, and YT except for BH. The SIA of 9278.3, 6687.0, and 3951.1 U/mg with recoveries of 12.0, 49.1, and 68.7 % were obtained for AP, ST, and YT, respectively. The SIA and recovery for TR were highest in the PEG4 fraction of the four fish species. The SIA and recovery for TR in the PEG4 fraction were 1170.9 U/mg and 45.2 % for AP, 98.2 U/mg and 19.8 % for BH, 2064.2 U/mg and 312.4 % for ST, and 1536.2 U/mg and 419.2 % for YT. From the result, the greater SIA and

Table 1 Commercial protease inhibitory activities of PEG fractions obtained from the crude extracts of fish roes by the polyethylene glycol precipitation toward casein as a substrate

	Fraction	AP		BH		ST		YT	
		SIA (U/mg)	Recovery (%)	SIA (U/mg)	Recovery (%)	SIA (U/mg)	Recovery (%)	SIA (U/mg)	Recovery (%)
AL	CE	235.0	100.0	40.4	100.0	925.6	100.0	144.0	100.0
	PEG1	−44.9	−3.3	−537.2	−347.3	210.3	0.2	209.3	3.2
	PEG2	−65.9	−0.6	−42.7	−2.5	180.6	1.2	85.8	1.8
	PEG3	−23.4	−0.9	−4.7	−1.4	−160.1	−2.1	37.2	3.3
	PEG4	−65.3	−0.8	28.4	10.6	−26.7	−0.4	18.5	1.6
BR	CE	16.8	100.0	−58.7	100.0	9.9	100.0	16.2	100.0
	PEG1	17.9	18.4	−93.1	41.4	72.6	6.8	45.7	6.2
	PEG2	5.9	0.7	−23.0	0.9	−32.1	−20.2	−23.3	−4.2
	PEG3	17.3	9.5	−30.1	6.3	−28.6	−35.5	−26.4	−20.6
	PEG4	−14.1	−2.4	−19.4	5.0	−50.2	−62.5	−21.1	−15.7
PED	CE	38.4	100.0	−54.5	100.0	35.6	100.0	112.5	100.0
	PEG1	0.9	0.4	−66.2	31.7	618.6	16.1	566.2	11.0
	PEG2	220.8	11.8	20.3	−0.9	50.6	8.9	137.0	3.6
	PEG3	55.0	13.2	−42.9	9.6	9.3	3.2	50.5	5.7
	PEG4	137.0	10.1	−23.2	6.4	15.8	5.5	22.4	2.4
CH	CE	2183.9	100.0	672.5	100.0	1678.6	100.0	695.3	100.0
	PEG1	673.4	5.3	−1477.5	−57.4	5768.6	3.2	2475.6	7.8
	PEG2	1776.3	1.7	668.7	2.4	903.2	3.4	322.8	1.4
	PEG3	767.7	3.2	288.4	5.2	1311.0	9.6	511.2	9.3
	PEG4	9278.3	12.0	537.5	12.0	6687.0	49.1	3951.1	68.7
TR	CE	73.1	100.0	74.5	100.0	81.4	100.0	44.3	100.0
	PEG1	23.7	5.6	−990.1	−347.2	1999.7	22.7	−298.7	−14.8
	PEG2	−259.4	−7.3	−416.9	−13.3	−513.2	−39.3	−299.0	−19.9
	PEG3	−255.4	−32.2	−190.1	−31.2	302.8	45.7	118.9	33.9
	PEG4	1170.9	45.2	98.2	19.8	2064.2	312.4	1536.2	419.2

Minus (−) values are no protease inhibitory activity
Recovery (%) = (total inhibitory activity of fraction/total inhibitory activity of CE) × 100
PEG1–PEG4, 0–5, 5–10, 10–20, 20–40 % fractions obtained from polyethylene glycol-4000 precipitation
AL alcalase, *BM* bromelain, *PED* papain-EDTA, *CH* chymotrypsin, *TR* Trypsin, *CE* crude extract, *SIA* specific inhibitory activity, *RIA* (%) relative inhibitor activity

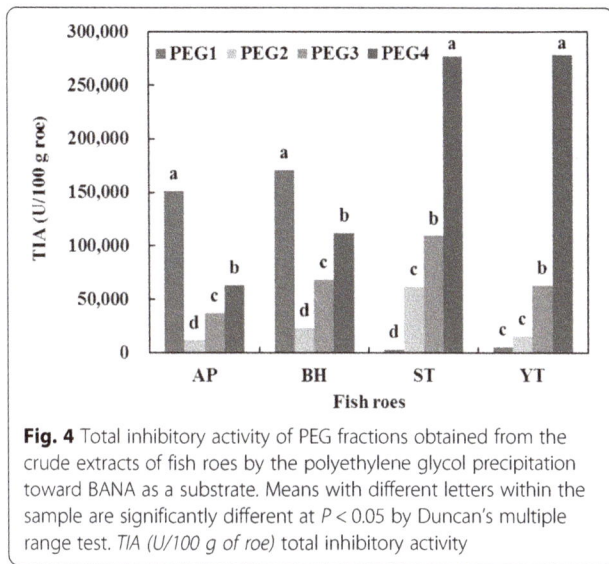

Fig. 4 Total inhibitory activity of PEG fractions obtained from the crude extracts of fish roes by the polyethylene glycol precipitation toward BANA as a substrate. Means with different letters within the sample are significantly different at $P < 0.05$ by Duncan's multiple range test. *TIA (U/100 g of roe)* total inhibitory activity

Whereas, it was observed that PEG precipitation for ST and YT gave maximum recovery of the inhibitor in a 20–40 % fraction (PEG4). Approximately 61.4 and 77.1 % of the total inhibitory activity of all PEG fractions were recovered in the PEG4 fraction of ST and YT, respectively. From the results, the serine protease inhibitor from four fish roes was more likely concentrated in the PEG1 (for AP and BH) and PEG4 fraction (for ST and YT).

Fractionation was commonly selected as a first step of purification, because the fractionation significantly reduced the volume of the solution and effectively removed contaminated proteins (Burnouf 1995). Rawdkuen et al. (2007) reported that PEG fractionation was more effective than AS fractionation. PEG might induce the conformational changes in the way which favored the inhibition of protease (Rawdkuen et al. 2005). Hao et al. (1980) reported that a variety of protease inhibitors were found in the 0–20 % PEG4000 fraction of plasma.

recovery of the PEG4 fraction suggested that a higher amount of serine protease inhibitor was precipitated in the PEG concentration range of 20–40 %.

Total inhibitory activity (TIA, U/100 g roe) of PEG fractions for trypsin using BANA as a specific substrate is shown in Fig. 4. Among all precipitates obtained from AP and BH, the PEG1 fraction had the highest inhibitory activity, followed by PEG4, PEG3, and PEG2 fraction. TIAs of 151,206.6 and 170,464.7 U/100 g roe were recovered in the PEG1 fraction toward AP and BH, respectively.

Native PAGE and SDS-PAGE

The native PAGE of the PEG fractions is shown in Fig. 5a. The PEG1, PEG2, and PEG3 fractions from AP contained protein bands similar to those of CE. A weakly cationic protein band which appeared in the PEG4 fraction of AP was rarely found in other fractions. In the CE of BH, protein bands with cationic proteins, weakly cationic protein, and weakly anionic protein were observed. After fractionation, increase in the weakly cationic protein bands was observed with increasing PEG concentration.

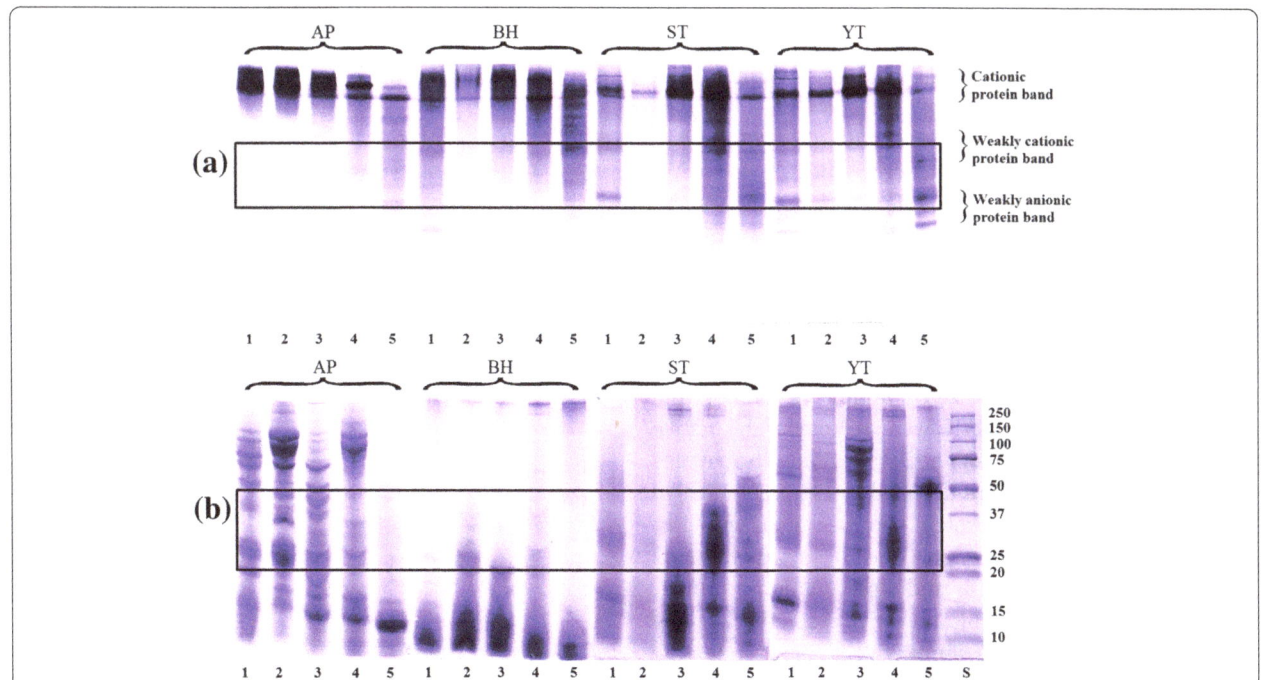

Fig. 5 Native PAGE (**a**) and SDS-PAGE (**b**) of PEG fractions obtained from the extracts of fish roes by the polyethylene glycol precipitation. Lane 1, CE; lane 2, PEG1; lane 3, PEG2; lane 4, PEG3; lane 5, PEG4. *S* standard maker

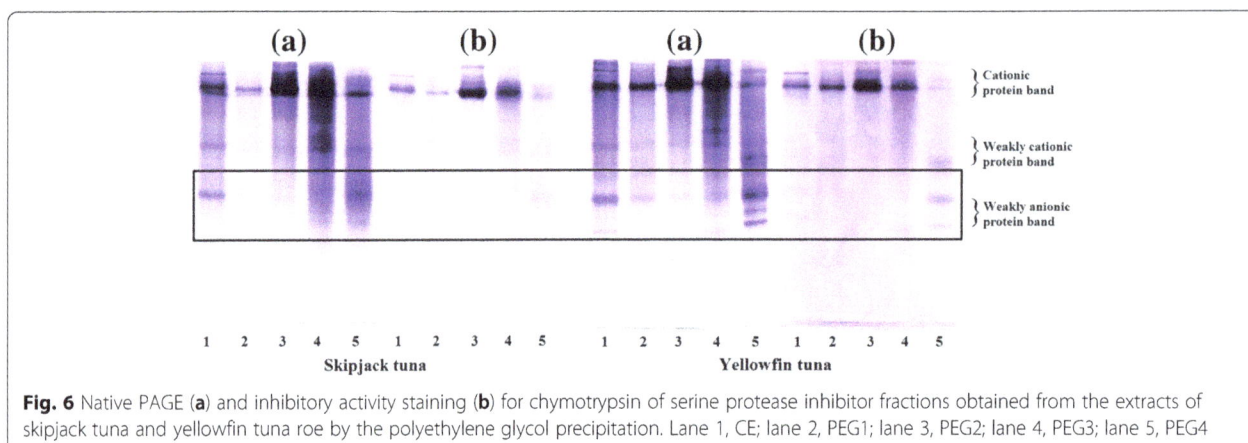

Fig. 6 Native PAGE (**a**) and inhibitory activity staining (**b**) for chymotrypsin of serine protease inhibitor fractions obtained from the extracts of skipjack tuna and yellowfin tuna roe by the polyethylene glycol precipitation. Lane 1, CE; lane 2, PEG1; lane 3, PEG2; lane 4, PEG3; lane 5, PEG4

The CE from ST and YT showed a similar protein pattern with cationic protein, weakly cationic protein, and weakly anionic protein bands. The PEG4 fraction from ST and YT consisted of bands with weakly cationic protein and weakly anionic protein as the major components.

The molecular weight distributions of the PEG fractions estimated from the mobility in SDS-PAGE are shown in Fig. 5b. The CE of AP contained a variety of proteins with different high and low molecular weights. Protein bands in the ranges of 150–75, 50, 25–20, and 15–10 K were observed. The PEG1, PEG2, and PEG3 fractions also had a pattern similar to that of CE from AP. Whereas, the PEG4 fraction showed only a low molecular band in the range of 15–10 K. Similar protein patterns were observed among the CE and PEG fractions from BH, in which low molecular proteins were predominant. The CE from ST had protein bands in the ranges of 25–20 and 15–10 K. The PEG1 and PEG2 fractions showed low molecular protein bands (25–10 K). The PEG3 and PEG4 fractions showed bands with higher molecular weight protein than those of PEG1 and PEG2 fractions. The CE from YT contained protein bands with a different molecular weight. After fractionation, the molecular band in range of 15–10 K was retained in the PEG4 fraction.

Native PAGE and detection of protease inhibitory activity by zymography
Due to the high serine protease inhibitory activity, the PEG fractions of ST and YT were selected. The native PAGE patterns and inhibitory activity staining for chymotrypsin of PEG fractions are depicted in Fig. 6. For native PAGE (Fig. 6a), a similar protein pattern was observed in CE (lane 1) and PEG4 fraction (lane 5), in which bands with weakly cationic protein and weakly anionic protein were dominant. The inhibitory activity staining of the PEG fractions from ST was similar to that of YT (Fig. 6b). All PEG fractions showed a dark major band with cationic protein bands observed. Whereas, inhibitory activity staining revealed that the weakly anionic proteins are the

predominant proteins in PEG4. From the result, using 20–40 % PG fractionation was found to be an effective method to fractionate the serine protease inhibitor from ST and YT roes.

Conclusions
The protease inhibitor from fish roes was successfully fractionated by using 200–400 g PEG/L. The PEG fractions from fish roes obtained showed high inhibitory activity against trypsin and chymotrypsin as serine protease. PEG is commonly exploited in large-scale protease inhibitor preparation or purification from fish roes for both seafood and surimi industry use.

Acknowledgements
This research was supported by the National Research Foundation of Korea (NRF) funded by the Ministry of Education, Science and Technology (2010–0009921).

Authors' contributions
HJL, HJK, and SHP carried out the enzymatic inhibitory activity analysis, participated in the PEG fractionation, and drafted the manuscript. ISY, GWL, and YJK participated in searching and screening references and performed the statistical analysis. JSK and MSH conceived of the study and participated in its design and coordination and helped to draft the manuscript. All authors read and approved the final manuscript.

Competing interests
The authors declare that they have no competing interests.

Author details
[1]Department of Food and Nutrition/Institute of Marine Industry, Gyeongsang National University, Jinju 52828, South Korea. [2]Department of Seafood Science and Technology/Institute of Marine Industry, Gyeongsang National University, Tongyeong 53064, South Korea.

References
Akazawa H, Miyauchi Y, Sakurada K, Wasson DH, Reppond KD. Evaluation of protease inhibitors in Pacific whiting surimi. J Food Sci. 1993. doi:10.1300/J030v02n03_06.
Barret AJ. An introduction to the proteinase. In: Barret AJ, Salvesen G, editors. Proteinase inhibitors. Amsterdam: Elsevier Science BV; 1986. p. 3–22.
Burnouf T. Chromatography in plasma fractionation: benefits and future trends. J Chromatogr B Biomed Sci Appl. 1995. doi:10.1016/0378-4347(94)00532-A.

Cherqui A, Cruz N, Simoes N. Purification and characterization of two serine protease inhibitors from the hemolymph of Mythimna unipuncta. Insect Biochem Mol Biol. 2001. doi:10.1016/S0965-1748(00)00172-7.

Choi JH, Park PJ, Kim SW. Purification and characterization of a trypsin inhibitor from the egg of skipjack tuna Katsuwonus pelamis. Fish Sci. 2002. doi:10.1046/j.1444-2906.2002.00576.x.

Clereszko A, Kwasnik M, Dabrowski K, Poros B, Glogowski J. Chromatographic separation of trypsin-inhibitory activity of rainbow trout blood and seminal plasma. Fish Shellfish Immunol. 2000. doi:10.1006/fsim.1999.0223.

Dawson RMC, Elliot DC, Elliot WH, Jones KM. Data for biochemical research. 3rd ed. Oxford, UK: Oxford Univ. Press; 1986. p. 417–41.

Hamann DD, Amato PM, Wu MC, Foegeding EA. Inhibition of modori (gel weakening) in surimi by plasma hydrolysate and egg white. J Food Sci. 1990. doi:10.1111/j.1365-2621.1990.tb05202.x.

Hao YL, Ingham KC, Wickerhauser M. Fractional precipitation of proteins with polyethylene glycol. In: Curling JM, editor. Methods of plasma protein fractionation. London, U.K.: Academic; 1980. p. 57–76.

Ji SJ, Lee JS, Shin JH, Park KH, Kim JS, Kim KS, et al. Distribution of protease inhibitors from fish eggs as seafood processing byproducts. Kor J Fish Aquat Sci. 2011. doi:10.5657/KFAS.2011.0008.

Kim HJ, Kim KH, Song SM, Kim IY, Park SH, Gu EJ, et al. Fractionation and characterization of protease inhibitors from fish eggs based on protein solubility. Kor J Fish Aquat Sci. 2013. doi:10.5657/KFAS.2013.0119.

Kim HJ, Lee HJ, Park SH, Jeon YJ, Kim JS, Heu MS. Recovery and fractionation of serine protease inhibitors from bastard halibut, Paralichthys olivaceus, roe. Kor J Fish Aquat Sci. 2015. doi:10.5657/KFAS.2015.0178.

Kim JS, Kim HS, Lee HJ, Park SH, Kim KH, Kang SI, et al. Lowering the bitterness of enzymatic hydrolysate using aminopeptidase-active fractions from the common squid (Todarodes pacificus) hepatopancreas. Korean J Food Sci Technol. 2014. doi:10.9721/KJFST.2014.46.6.716.

Kim JS, Kim KH, Kim HJ, Kim MJ, Park SH, Lee HJ, et al. Chromatographic fractionation of protease inhibitors from fish eggs. Kor J Fish Aquat Sci. 2013. doi:10.5657/KFAS.2013.0351.

Kim KY, Ustadi U, Kim SM. Characteristics of the protease inhibitor purified from chum salmon (Oncorhynchus keta) eggs. Food Sci Biotechnol. 2006;15:28–32.

Kishimura H, Saeki H, Hayashi K. Isolation and characteristics of trypsin inhibitor from the hepatopancreas of a squid (Todarodes pacificus). Comp Biochem Physiol Part B: Biochem Mol Biol. 2001. doi:10.1016/S1096-4959(01)00415-8.

Klomklao S, Benjakul S, Kishimura H. Optimum extraction and recovery of trypsin inhibitor from yellowfin tuna (Thunnus albacores) roe and its biochemical properties. Int J Food Sci Technol. 2014. doi:10.1111/ijfs.12294.

Knight CG. The characterization of enzyme inhibition. In: Barret AJ, Salvesen G, editors. Proteinase inhibitors. Amsterdam: Elsevier Science BV; 1986. p. 23–48.

Laemmli UK. Cleavage of structural proteins during the assembly of the head of bacteriophage T4. Nature. 1970. doi:10.1038/227680a0.

Lee YZ, Aishima T, Nakai S, Sim JS. Optimization for selective fractionation of bovine blood plasma proteins using poly(ethylene glycol). J Agric Food Chem. 1987. doi:10.1021/jf00078a024.

Lopuska A, Polanowska J, Wilusz T, Polanowski A. Purification of two low-molecular-mass serine proteinase inhibitors from chicken liver. J Chromatogr A. 1999. doi:10.1016/S0021-9673(99)00372-6.

Lowry OH, Rosebrough NJ, Farr AL, Randall RJ. Protein measurement with folin phenol reagent. J Biol Chem. 1951;193:256–75.

Martone CB, Busconi L, Folco E, Sanchez JJ. Detection of a trypsin-like serine protease and its endogenous inhibitor in hake skeletal muscle. Arch Biochem Biophys. 1991. doi:10.1016/0003-9861(91)90433-J.

Narsing Rao G, Prabhakara Rao P, Satyanarayana A, Balaswamy K. Functional properties and in vitro antioxidant activity of roe protein hydrolysates of Channa striatus and Labeo rohita. Food Chem. 2012. doi:10.1016/j.foodchem.

Oda S, Igarashi Y, Manaka KI, Koibuchi N, Sakai-Sawada M, Sakai M, et al. Sperm-activating proteins obtained from the herring eggs are homologous to trypsin inhibitors and synthesized in follicle cells. Dev Biol. 1998;204:55–63.

Rawdkuen S, Benjakul S, Visessanguan W, Lanier TC. Fractionation and characterization of cysteine proteinase inhibitor from chicken plasma. J Food Biochem. 2005. doi:10.1111/j.1745-4514.2005.00027.x.

Rawdkuen S, Benjakul S, Visessanguan W, Lanier TC. Cysteine proteinase inhibitor from chicken plasma: fractionation, characterization and autolysis inhibition of fish myofibrillar proteins. Food Chem. 2007. doi:10.1016/j.foodchem.

Sangorrin MP, Folco EJ, Martone CM, Sanchez JJ. Purification and characterization of a protease inhibitor from white croaker skeletal muscle (Micropogon opercularis). Intl J Biochem Cell Biol. 2001;33:691–9.

Sharma VK, Kalonia DS. Polyethylene glycol-induced precipitation of interferon alpha-2a followed by vacuum drying: development of a novel process for obtaining a dry, stable powder. AAPS PharmSci. 2004. doi:10.1208/ps060104.

Tsai YJ, Chang GD, Haung CJ, Chang YS, Haung FL. Purification and molecular cloning of carp ovarian cystatin. Comp Biochem Physiol B. 1996. doi:10.1016/0305-0491(95)02070-5.

Ustadi U, Kim KY, Kim SM. Characteristics of protease inhibitor purified from the eggs of Alaska pollock (Theragra chalcogramma). J Kor Fish Soc. 2005. doi:10.5657/kfas.2005.38.2.083.

Ustadi U, Kim KY, Kim SM. Purification and identification of a protease inhibitor from glassfish (Liparis tanakai) eggs. J Agric Food Chem. 2005;53:7667–72.

Weerasinghe VC, Morrissey MT, An H. Characterization of active components in food grade proteinase inhibitor for surimi manufacture. J Food Chem. 1996. doi:10.1021/jf950589z.

Yamashita M, Konagaya S. A comparison of cystatin activity in various tissues of chum salmon Oncorhynchus keta between feeding and spawning migrations. Comp Biochem Physiol A. 1991. doi:10.1016/0300-9629(91)90402-X.

25

Chemical composition of protein concentrate prepared from Yellowfin tuna *Thunnus albacares* roe by cook-dried process

Hyun Ji Lee[1], Sung Hwan Park[1], In Seong Yoon[1], Gyoon-Woo Lee[1], Yong Jung Kim[2], Jin-Soo Kim[2] and Min Soo Heu[1*]

Abstract

Roe is the term used to describe fish eggs (oocytes) gathered in skeins and is one of the most valuable food products from fishery sources. Thus, means of processing are required to convert the underutilized yellowfin tuna roes (YTR) into more marketable and acceptable forms as protein concentrate. Roe protein concentrates (RPCs) were prepared by cooking condition (boil-dried concentrate, BDC and steam-dried concentrate, SDC, respectively) and un-cooking condition (freeze-dried concentrate, FDC) from yellowfin tuna roe. The yield of RPCs was in the range from 22.2 to 25.3 g/100 g of roe. RPCs contained protein (72.3–77.3 %), moisture (4.3–5.6 %), lipid (10.6–11.3 %) and ash (4.3–5.7 %) as the major constituents. The prominent amino acids of RPCs were aspartic acid, 8.7–9.2, glutamic acid, 13.1–13.2, and leucine, 8.5–8.6 g/100 g of protein. Major differences were not observed in each of the amino acid. K, S, Na, and P as minerals were the major elements in RPCs. No difference noted in sodium dodecyl sulfate polyacrylamide gel electrophoresis protein band (15–100 K) possibly representing partial hydrolysis of myosin. Therefore, RPCs from YTR could be use potential protein ingredient for human food and animal feeds.

Keywords: Protein concentrate, Roe, Yellowfin tuna, Chemical composition, Cook-dried process

Background

Yellowfin tuna *Thunnus albacares* is a large tuna species found in the Pacific, Indian and Atlantic oceans. It is an important component of tuna fisheries worldwide and is one of the major target species for the tuna fishery in the major oceans, and popularly catched marine fish with annual availability of 44,013 t on overseas fishery in Korea during 2013 (Ministry of Ocean and Fisheries 2014). It is used extensively in raw cuisine such as sushi and sashimi. Byproducts such as scales, heads, skin, fat, visceral, and roe are generated increasingly and discarded as waste, without any attempt to recover the essential nutrients (Chalamaiah et al. 2010). Among byproducts, roes are highly nutritious material rich in essential fatty acids and amino acids (Heu et al. 2006;

Narsing Rao et al. 2012b). Fish roes are produced in large quantities during the spawning seasons, which constitute about 1–3 % of the weight of fish (Chalamaiah et al. 2013; Klomklao et al. 2013; Intarasirisawat et al. 2011). Currently roe obtained from fish such as salmon, cod, and pollock have a potential commercial market, especially they have a higher demand in Asian countries (Sathivel et al. 2009). Yellowfin tuna roe is an abundant and underutilized byproduct that can be used as a unique protein source (Heu et al. 2006). The roe can be used to recover protein that may be converted into a higher value food ingredient suitable for use as an emulsifier in food and feed systems (Sathivel et al. 2009).

Protein concentrates are widely used as ingredients in food industry because of their high nutritional quality, functional properties, high protein level and low content of antinutritional factors (Cordero-de-los-Santos et al. 2005). Using fish proteins in powder form presents some advantages since they do not require special storage

* Correspondence: heu1837@dreamwiz.com
[1]Department of Food and Nutrition/Institute of Marine Industry, Gyeongsang National University, Jinju 52828, Korea
Full list of author information is available at the end of the article

conditions and they can also easily be used as an ingredient in foods (Pires et al. 2012). Drying preserves fish by inactivation enzymes and removing the moisture necessary for bacterial and mold growth (Bellagha et al. 2002; Bala and Mondol 2001; Duan et al. 2004). Recently, protein concentrate preparation have been reported for different protein sources such as roe of *Channa striatus* and *Lates calcarifer* (Narsing Rao et al. 2012a), byproducts of Alaska pollock (Sathivel and Bechtel 2006), mrigal egg (Chalamaiah et al. 2013) and roes of hake and ling (Rodrigo et al. 1998).

Before the drying process, boiling and steaming of fish improve their digestibility, enhances palatability, and provide a safe eating by killing harmful bacteria, other micro-organisms and parasites. Thus, means of processing are required to convert the underutilized yellowfin tuna roes into more marketable and acceptable forms as protein concentrate. There is no information about protein concentrate by cook-dried process. The objectives of this study were to investigate chemical characteristic for protein concentrate from yellowfin tuna roe prepared by cook-dried process and to identify possibility on utilization of 2'nd byproduct. Hence, we expected that full utilization for fish roe concentrate will be possible by reducing the 2'nd byproduct wastage and environmental pollution.

Methods

Raw material

Yellowfin tuna *Thunnus albacares* roe was obtained from Dongwon F&B Co., Ltd. (Changwon, Korea). Roe was stored at −70 °C in sealed polyethylene bags, and transferred to the laboratory. Frozen roe was partially thawed for 24 h at 4 °C and then cut into small pieces with an approximate thickness of 1.5–3 cm and minced with food grinder (SFM-555SP, Shinil Industrial Co., Ltd., Seoul Korea). Minced roe was frozen at −20 °C until used.

Chemicals

Sodium dodecyl sulfate (SDS) and glycine were purchased from Bio Basic Inc., (Ontario, Canada). Coomassie Brilliant Blue R-250 was purchased from Bio-Rad Laboratories (Hercules, CA, USA). Bovine serum albumin (BSA), β-Mercaptoethanol (β-ME), egg white, glycerol, N,N,N′,N′-tetramethyl ethylene diamine (TEMED), sodium carbonate, sodium hydroxide, sodium L-tatarate, and potassium hydroxide were purchased from Sigma-Aldrich Co., Ltd. (St. Louis, MO, USA). Copper (II) sulfate pentahydrate was purchased from Yakuri Pure Chemicals Co., Ltd. (Kyoto, Japan). Bromophenol blue and Folin-Ciocalteu's reagent were purchased Junsei Chemical Co., Ltd. (Tokyo, Japan). All reagents used analytical grade.

Preparation of roe protein concentrates (RPCs)

300 g of frozen minced roe was put into pouch type tea bag (polyethylene polypropylene, 16 × 14.5 cm) for cook-dried process. Freeze-dried concentrate (FDC) as an un-cooked sample was prepared using freeze dryer (PVTFD50A, ilShinbiobase Co., Ltd., Dongducheon, Korea). To prepare the boil-dried concentrate (BDC), sample immerged in five volume of distilled deionized water (DDW) and was boiled for 20 min after sample core temperature was reached to 80 °C. In case of steam-dried concentrate (SDC), sample was steamed for 20 min after sample core temperature was reached to 80 °C. Cooked (boiled or steamed) samples were dried at 70 ± 1 °C for 15 h using an incubator (VS-1203P3V, Vision Scientific, Co., Ltd., Daejeon, Korea). Boil or steam-dried samples were ground to powder using a food grinder and sieved to pass through a 180 mesh to obtained dehydrated RPCs. Grounded powders were referred to as BDC and SDC, respectively. During these processes, cooking process drips such as boil-process drip (BPD) and steam-process drip (SPD) were generated as the 2'nd byproduct. They were packed in plastic bottles and stored at −20 °C for further experiments. A flow chart for the preparation of roe protein concentrates (RPCs) is shown in Fig. 1.

Proximate composition

The proximate composition was determined according to the AOAC method (AOAC 1995). Moisture content

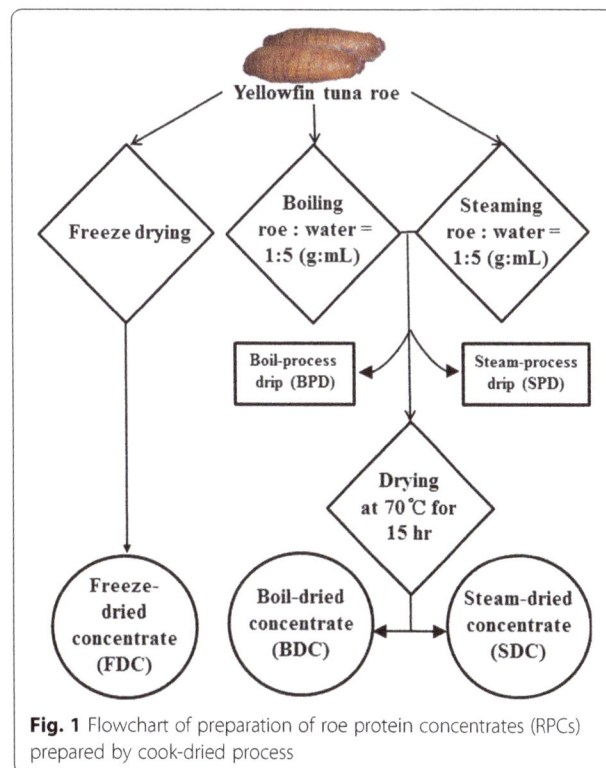

Fig. 1 Flowchart of preparation of roe protein concentrates (RPCs) prepared by cook-dried process

was determined by placing accurately weighed 0.2 g of sample into an aluminum pan. Sample was dried in a forced-air convection oven (WFO-700, EYELA, Tokyo, Japan) at 105 °C until a constant weight was reached. The ash content was determined by charring approximately 0.1 g of sample in a ceramic crucible over a hot plate and then heating in a muffle furnace (Thermolyne 10500 furnace, a subsidiary of Sybron Co., Dubuque, IA, USA) at 550 °C until a constant final weight for ash was achieved. The total crude protein (nitrogen x 6.25) content of samples was determined using the semi-micro Kjeldahl method. Protein concentration of cooking process drips was determined by the method of Lowry et al. (1951) using bovine serum albumin as a standard. Total lipid content was determined according to the Soxhlet extraction method. 5 g of sample was extracted with dimethyl ether and performed for 30 min at a drip rate of 10 mL/min. Total lipid content was determined on a gravimetric basis and expressed as percent.

Total amino acid
Total amino acid analysis was conducted according to AOAC method (AOAC 1995). The sample (20 mg) was hydrolyzed with 2 mL of 6 N HCl at 110 °C for 24 h in heating block (HF21, Yamoto Science Co, Tokyo, Japan) and filtered out using vacuum filtrator (ASPIRATOR A-3S, EYELA, Tokyo, Japan). Amino acids were quantified using the amino acid analyzer (Biochrom 30, Biochrom Ltd., Cambridge, United Kingdom) employing sodium citrate buffers (pH 2.2) as step gradients. The data are reported as g of amino acid per 100 g of protein. Asparagine is converted to aspartic acid and glutamine to glutamate during acid hydrolysis, so the reported values for these amino acids (Asp and Glu) represent the sum of the respective amine and amino acid in the proteins.

Mineral
Analysis of iron (Fe), copper (Cu), manganese (Mn), cadmium (Cd), nickel (Ni), lead (Pb), zinc (Zn), chromium (Cr), magnesium (Mg), sodium (Na), phosphorus (P), potassium (K), calcium (Ca), and sulfur (S) contents in sample was carried out using the inductively coupled plasma optical emission spectrophotometry (OPTIMA 4300 DV, Perkin Elmer, Shelton, Conn., USA). Briefly, teflon digestion vessel was washed overnight in a solution of 2 % nitric acid (v/v) prior to use.

Sample was dissolved in 10 mL of 70 % nitric acid. The mixture was heated on the hot plate until digestion was completed. The digested samples were added in 5 mL of 2 % nitric acid and filtered using filter paper (Advantec No. 2, Toyo Roshi Kaisha, Ltd., Tokyo, Japan). Sample was massed up to 100 mL

with 2 % nitric acid in a volumetric flask. Sample was run in triplicate. The concentration of mineral was calculated and expressed as mg/100 g sample.

Hunter color
Hunter color properties of samples were equilibrated to room temperature for 2 h prior to the color measurement. Colors were determined using color meter (ZE-2000, Nippon Denshoku Inc., Japan). The colorimeter was calibrated by using a standard plate (L* = 96.82, a* = −0.35, b* = 0.59) supplied by the manufacturer. The values for the CIE (Commission Internationale d'Eclairage of France) color system using tristimulus color values, L* (lightness), a* (redness), and b* (yellowness) were determined. The whiteness was calculated by the following equation:

$$\text{White index} = 100 - \sqrt{(100-L)^2 + a^2 + b^2}$$

SDS-PAGE
The molecular weight distribution of YTR and RPCs was observed by sodium dodecyl sulfate polyacrylamide gel electrophoresis (SDS–PAGE). It was performed according to the method of Laemmli (1970). Briefly, 10 mg of sample was solubilized in 1 mL of 8 M urea solution containing 2 % β-mercaptoethanol and 2 % sodium dodecyl sulfate (SDS) solution. Protein solution was mixed at 4:1 (v/v) ratio with the SDS-PAGE sample treatment buffer (62.5 mM Tris–HCl (pH 6.8), 2 % SDS (w/v), 10 % glycerol, 2 % β-mercaptoethanol and 0.002 % bromophenol blue) and boiled at 100 °C for 3 min. The sample (20 μg protein) was loaded on the 10 % Mini-PROTEAN® TGX™ Precast gel (Bio-Rad Lab., Inc., Hercules, CA, USA) and subjected to electrophoresis at a constant current of 10 mA per gel using a Mini-PROTEAN tetra cell (Bio-Rad Lab., Inc., Hercules, CA, USA).

Electrophoresed gel was stained in 0.125 % Coomassie brilliant blue R-250 and destained in 25 % methanol and 10 % acetic acid until background was clear. Molecular weight of protein bands was estimated using Precision Plus Protein™ standards (10–250 K, Bio-Rad Lab., Inc., Hercules, CA, USA).

Statistical analysis
All experiments were conducted in triplicates. The average and standard deviation were calculated. Data were analyzed using analysis of variance (ANOVA) procedure by means of the statistical software of SPSS 12.0 KO (SPSS Inc., Chicago, IL, USA). The mean comparison was made using the Duncan's multiple range test ($P < 0.05$).

Results and discussions

Proximate composition

Proximate compositions of yellowfin tuna roe (YTR) and roe protein concentrates (RPCs) named freeze-dried concentrate (FDC), boil-dried concentrate (BDC) and steam-dried concentrate (SDC) are shown in Table 1. YTR as a raw material contained 77.3 % moisture, 18.2 % protein, 2.4 % lipid and 1.5 % ash. Heu et al. (2006) reported that skipjack and yellowfin tuna roes contained 71.1–71.2 % moisture, 21.4–21.5 % protein, 2.0–2.1 % lipid and 1.9 % ash, and both tuna roes were no difference in proximate compositions. Intarasirisawat et al. (2011) reported that three different tuna species (skipjack, tongol and bonito) roes contained 72.17–73.03 % moisture, 18.16–20.15 % protein, 3.39–5.68 % lipid and 1.79–2.10 % ash. Iwasaki and Harada (1985) described the proximate composition of raw roe from different species with a wide range in protein content (11.5 % in angler fish to 30.2 % in crab). The chemical composition of roes of 18 fish species was reported and the protein content was estimated to be in the range of 11.5–30.2 % (Iwasaki and Harada 1985). Therefore, yellowfin tuna roe could be source of proteins. The yield of RPCs prepared from YTR were 25.3 % for FDC, 22.2 % for BDC and 22.8 % for SDC, respectively, and uncooked FDC was higher than those of cooked BDC and SDC ($P < 0.05$). It was due to released soluble protein and other compounds to process drip through cook-dried process (boiling or steaming). Moisture and ash content of RPCs ranged from 4.3 to 5.6 %, and 4.3 to 5.7 %, respectively. Lipid content in RPCs was in the range of 10.6–11.3 %. RPCs had a high protein content accounting for 72.3 % for FDC, 77.3 % for BDC and 76.0 % for SDC, but RPCs were lower protein content than 81.4 % egg white as a positive control ($P < 0.05$). From the result and reports, difference in moisture and protein content of RPCs was due to processing condition (Mahmoud et al. 2008). Narsing Rao et al. (2012a) was reported that Channa and Lates roes yielded 20.7 % and 22.5 % of protein concentrates possessing 90.2 % and 82.5 % protein, respectively. Protein content in the range of 39.1–43 % and fat content in the range of 14.1–14.8 % were reported in dried and salted roes of hake (*Merluccius merluccis*) and ling (*Molva molva*) (Rodrigo et al. 1998). Sathivel et al. (2009) was reported that moisture, protein, ash, and fat contents of spray dried soluble protein powder were 4.5 ± 0.5 %, 67.1 ± 0.2 %, 10 ± 1.0 %, and 18.3 ± 0.6 %, respectively. Chalamaiah et al. (2013) was reported that total ash content of dehydrated and defatted protein concentrates was found to be 5.95 % and 1.95 %, respectively.

During the preparation of yellowfin tuna roe concentrate, boiling process drip (BPD) and steaming process drip (SPD) as 2'nd byproduct were generated through cooking process, and data are shown in Table 2. Moisture content of BPD (92.0 %) was higher than that (90.5 %) of SPD ($P < 0.05$). Protein content of BPD and SPD were 5.8 and 6.4 %, respectively with no significant difference. Starting process water (600 mL) for cooking process was concentrated to 29.6 (BPD) and 18.1 mL (SPD), respectively. Total protein in BPD (500.2 mg/100 g roe) and SPD (365.6 mg/100 g roe) were meaningful difference. This means that boiling process could be easier released to soluble protein and other compounds than steaming process. The proximate composition of fish roe depends on the age of fish, seasons and type of processing (Iwasaki and Harada 1985).

Total amino acid

YTR and RPCs contained protein ranged from 81.6 to 83.6 % on a dry base were still a considerable amount of fish protein that could be possible to use as protein resource. To assess protein quality, we determined the content of total amino acid in the YTR and RPCs. Total amino acid composition (g/100 g of protein, %) of RPCs from yellowfin tuna roe and positive control (egg white) is shown in Table 3. The major non-essential amino acids of YTR were glutamic acid (12.3 %), aspartic acid (9.1 %) and arginine (6.6 %), respectively. Leucine (8.8 %) and lysine (8.5 %) were the major essential amino acids in YTR. Intarasirisawat et al. (2011) reported that leucine (8.28–8.64 %) and lysine (8.24–8.30 %) were the

Table 1 Proximate composition of yellowfin tuna roe (YTR) and roe protein concentrates (RPCs) prepared by cook-dried process

Sample	Yield[1] (g)	Protein yield[2] (g)	Moisture (%)	Protein (%)	Lipid (%)	Ash (%)
YTR	100.0	18.2	77.3 ± 0.1	18.2 ± 0.0	2.4 ± 0.1	1.5 ± 0.2
FDC	25.3	18.3	4.3 ± 0.1ab	72.3 ± 0.4c	10.6 ± 0.1a	5.7 ± 0.5a
BDC	22.2	17.2	4.8 ± 0.1a	77.3 ± 1.4b	11.3 ± 0.5a	5.0 ± 0.9a
SDC	22.8	17.3	5.6 ± 1.0a	76.0 ± 1.9b	10.6 ± 0.3a	4.3 ± 0.2a
Egg white			3.0 ± 0.6b	81.4 ± 0.5a		

Data are given as mean values ± SD (*n* = 3)

Values with different letters within the same column are significantly different at *P* < 0.05 by Duncan's multiple range test

FDC freeze-dried concentrate, *BDC* boil-dried concentrate, *SDC* steam-dried concentrate, respectively

[1] Yield is weight (g) of roe protein concentrate obtained from 100 g of raw YTR

[2] Protein yield (g) = yield x protein (%)

Table 2 Moisture and protein content of process drips obtained from yellowfin tuna roe during cook-dried process

Sample	Moisture (%)	Protein (%)	Volume (mL/100 g roe)	Protein[a] (mg/mL)	Total protein (mg)
BPD	92.0 ± 0.1a	5.8 ± 0.0a	29.6	16.9b	500.2
SPD	90.5 ± 0.7b	6.4 ± 0.5a	18.1	20.2a	356.6

Data are means ± standard deviation of triplicate determinations
Values with different letters within the same column are significantly different at $P < 0.05$ by Duncan's multiple range test
BPD boiling process drip, *SPD* steaming process drip, respectively
[a]Based on Lowry's method (1951)

predominant essential amino acids in defatted tuna roe from skipjack, tongol and bonito. YTR had an essential amino acids/non-essential amino acids ratio of 1.11. From these results, total essential amino acids content of YTR (52.6 %) was higher than total non-essential amino acids content of that (47.4 %). The major non-essential amino acids of RPCs were glutamic acid (13.1–13.2 g/100 g of protein), aspartic acid (8.7–9.2 %) and arginine (6.5–6.6 %), respectively. Among RPCs, glutamic acid (13.1–13.2 %) was the predominant essential amino acid followed by aspartic acid (8.7–9.2 %), leucine (8.5–8.6 %)

and lysine (8.5 %). Amino acids namely glutamic acid (13.14 g), aspartic acid (8.08 g) and leucine (7.57 g) were reported per 100 g roe protein of Channa roe protein concentrate (Narsing Rao et al. 2012a).

RPCs had essential amino acid/non-essential amino acid ratio in range of 1.00–1.06. From this result, essential amino acid/non-essential amino acid ratio of PRCs was lower than that (1.10) of egg white, but it was similar to the value reported Heu et al. (2006) for yellowfin tuna roe. Lysine is often considered a first limiting amino acid for cereal food. Therefore, it needs to be

Table 3 Total amino acid composition (g/100 g protein, %) of yellowfin tuna roe (YTR) and roe protein concentrates (RPCs) prepared by cook-dried process

Amino acid	YTR	FDC	BDC	SDC	Egg white
Protein content (%)	82.4	81.6	82.6	83.6	83.9
Asp	9.1b	8.7c	9.0b	9.2b	11.8a
Thr [f]	5.2a	5.0c	5.1b	5.2a	4.7d
Ser	4.9e	5.6b	5.3d	5.4c	5.7a
Glu	12.3e	13.1d	13.2b	13.2c	14.9a
Pro [†]	6.5a	6.1ab	5.8b	5.8b	3.7c
Gly [†]	5.6a	4.9b	5.1b	5.1b	4.0c
Ala [†]	5.6b	6.6a	6.4a	6.4a	6.6a
Cys	0.8a	0.7a	0.7b	0.8a	0.7a
Val [f†]	5.9c	6.3b	6.5b	6.4b	8.2a
Met [f†]	4.0a	2.9b	2.9c	2.8b	2.0c
Ile [f†]	5.9b	5.4c	5.5c	5.4c	6.2a
Leu [f†]	8.8b	8.6c	8.6c	8.5c	9.2a
Tyr	2.5c	3.4a	3.1b	3.0b	0.2d
Phe [f†]	4.3e	4.4d	4.6b	4.6c	6.4a
His [f]	3.4a	3.4a	3.2b	3.2b	2.7c
Lys [f]	8.5a	8.5a	8.5a	8.5a	8.2b
Arg [f]	6.6a	6.6a	6.6a	6.5a	4.8b
TAA	99.9	100.2	100.1	100.0	100.0
EAA	52.6	51.1	51.5	51.1	52.4
EAA/NEAA	1.11	1.04	1.06	1.00	1.10
Hydrophobic amino acid	46.6	45.2	45.4	45.0	46.3

Data are means ± standard deviation of duplicate determination
Values with different letters within the same row are significantly different at $P < 0.05$ by Duncan's multiple range test
TAA total amino acid, *EAA* essential amino acids, *NEAA* non essential amino acids
[f]Essential amino acids for infant
[†]Hydrophobic amino acids

emphasized that the RPCs had a higher content of lysine (8.5 %) than egg white (8.2 %) ($P < 0.05$). The lysine content of RPCs was 8.5 % which was higher than that reported for Channa (6.94 %) and Lates (6.86 %) roe protein concentrate (Narsing Rao et al. 2012a).

YTR and RPCs had hydrophobic amino acids ranged from 45.0 to 46.6 %. Hydrophobicity plays an important positive role in determining emulsifying properties (Chalamaiah et al. 2013). FitzGerald and O'Cuinn (2006) reported that bitterness of protein hydrolysate is associated with the release of peptides containing hydrophobic amino acid residues. Thus, RPCs could possibly be a dietary protein supplement to poorly balanced dietary proteins exhibiting to low bitterness.

Mineral

Mineral (K, S, Na, P, Fe, Mg, Zn, Ca, Fe, Mn, Cu, Pb, Cd, Cr and Ni) composition was determined to identify the nutritional properties as a food compounds. Mineral content of YTR and RPCs is presented in Table 4. K content was the most predominant mineral in YTR (456.0 mg/100 g of sample), followed by P, Na and Mg (437.0, 167.0 and 29.0 mg/100 g, respectively). Minor minerals were Ca and Zn (11.0 and 8.0 mg/100 g, respectively). Heu et al. (2006) reported that predominant mineral in YTR was P (371.5 mg/100 g) followed by K and Ca (325.4 and 61.9 mg/100 g, respectively). Na content of YTR (167.0 mg/100 g) was very lower than that (2473 and 2348 g/100 g, respectively) of dried salted roes of hake *Merluccius merluccius* and ling *Molva molva*

reported Rodrigo et al. (1998). This high Na content is clearly due to the salting process used to obtain the commercial product. FDC had significantly higher K, Na and P content (1179.9, 376.2 and 257.7 mg/100 g of sample, respectively) compared with those of BDC and SDC ($P < 0.05$). Similar K, P and Na contents in defatted-tuna roes were found in salmon roe (Bekhit et al. 2009). P content has been generally associated with the phospholipid content and the presence of phosphoprotein (Mahmoud et al. 2008; Matsubara et al. 2003). The highest S content was found in egg white (1341.3 mg/100 g) as a positive control ($P < 0.05$). Bekhit et al. (2009) reported that salmon roe had S content of 1647–2443 mg/kg (wet basis). Catfish roe protein powder had S content of approximately 0.56 mg/kg (Sathivel et al. 2009). The result implied that tuna roe had a high content of sulfur-containing compounds, which could undergo decomposition during storage, causing off-odor. Fe (9.7–12.9 mg/100 g of sample), Cu (0.4–0.5 mg/100 g of sample), Mn (0.1–0.2 mg/100 g of sample) and Mg (59.1–70.6 mg/100 g of sample) were minor minerals in RPCs. Fe and Cu are classified as the essential trace elements required for physiological and metal metabolic processes of marine organisms (Thanonkaew et al. 2006). Metal ions (Fe, Cu, Mn and Mg) might serve as catalysts for lipid oxidation, and have been shown to exhibit pro-oxidant activity (Thanonkaew et al. 2006). However, Fe and P contents of all samples were lower than those of catla, carp and rohu roes (Balaswamy et al. 2009). The variation in the mineral composition of marine foods is closely related to seasonal and biological differences, area of catch, processing methods, food source and environmental conditions (salinity, temperature

Table 4 Mineral contents (mg/100 g sample) of yellowfin tuna roe (YTR) and roe protein concentrates (RPCs) prepared by cook-dried process

Sample		YTR	FDC	BDC	SDC	Egg white
Moisture (%)		77.3	4.3ab	4.8a	5.6a	3.1b
Mineral	K	456.0 ± 2.0c	1179.9 ± 8.3a	761.6 ± 67.4b	758.9 ± 200.3b	787.0 ± 13.2b
	S	–	992.3 ± 92.6b	858.2 ± 10.3c	818.1 ± 141.8bc	1341.3 ± 1.2a
	Na	167.0 ± 1.0d	376.2 ± 2.1b	287.2 ± 17.5c	268.3 ± 55.5c	1015.8 ± 8.8a
	P	437.0 ± 3.0a	257.7 ± 2.8b	223.4 ± 13.4c	208.5 ± 35.2c	92.5 ± 0.4d
	Mg	29.0 ± 0.0c	66.8 ± 0.4a	70.6 ± 2.5a	59.1 ± 7.7a	–
	Zn	8.0 ± 0.0c	45.0 ± 0.3b	49.9 ± 0.4a	48.4 ± 3.5a	–
	Ca	11.0 ± 0.0c	33.5 ± 0.3b	58.6 ± 12.2a	32.3 ± 3.7b	65.2 ± 0.6a
	Fe	–	9.8 ± 0.1b	12.9 ± 0.2b	9.7 ± 0.8ab	0.6 ± 0.0a
	Mn	–	0.1 ± 0.0a	0.2 ± 0.0b	0.1 ± 0.0b	0.0 ± 0.0b
Heavy metal	Cu	–	0.5 ± 0.0a	0.5 ± 0.1a	0.4 ± 0.1a	0.1 ± 0.0b
	Pb	–	0.2 ± 0.1a	0.1 ± 0.0a	0.3 ± 0.1a	0.0 ± 0.0a
	Cd	–	0.1 ± 0.0a	0.1 ± 0.0a	0.1 ± 0.0a	0.0 ± 0.0b
	Cr	–	0.1 ± 0.0a	0.2 ± 0.1a	0.2 ± 0.1a	0.1 ± 0.0a
	Ni	–	0.0 ± 0.0a	0.1 ± 0.1a	0.0 ± 0.0a	0.0 ± 0.0a

–; not determined
Data are means ± standard deviation of three samples
Values with different letters within the same row are significantly different at $P < 0.05$ by Duncan's multiple range test

and contaminant) (Alasalvar et al. 2002). According to Korean Food Standards Codex, Pb and Cd do not exceed 0.5 and 0.1 mg/kg for fish roe. YTR and RPCs could serve as safety and mineral source.

Hunter color

RPCs were light yellow in color and the L, a, b values are listed in Table 5. FDC was the lightest with L* value of 59.2. L* value of BDC and SDC (55.7 and 55.4, respectively) were no significant ($P > 0.05$). a* value of FDC (6.5) was higher than that of BDC (5.2) and SDC (5.3). FDC had the highest b* value (18.6) than that (17.8 and 17.4, respectively) of BDC and SDC ($P < 0.05$). However, total color change (ΔE) value is not significantly difference among the RPCs. ($P > 0.05$). Whiteness values of FDC (54.7), BDC (52.0) and SDC (51.8) were observed with no significant ($P > 0.05$). Sathivel et al. (2009) reported that the catfish roe soluble protein powder was light and yellow in color (L* = 71.7; a* = 3.9; b* = 28.8).

SDS-PAGE

SDS-PAGE patterns of FDC, BDC and SDC prepared from yellowfin tuna roe and their 2'nd byproduct (BPD and SPD, respectively) are shown in Fig. 2. Actinin band of FDC was observed around 120 K. The major bands of FDC observed at molecular weight around 97 K. FDC exhibited two clear protein bands in the range of 75–100 K. Protein bands at approximately 97 K could be vitellin like protein, which was found in skipjack, tongol, bonito roes (Intarasirisawat et al. 2011), chicken egg yolk (DeMan 1999) and salmon *Oncorhynchus keta* and sturgeon *Acipenser transmontanus* roes (Al-Holy and Rasco 2006). Protein bands in range of 50–37 K and 15 K were observed. They could be the actin, troponin-T and myosin light chain (MLC), respectively. Protein bands of egg white (EW) were observed that in range of 75–100 K, 37–50 K and around 15 K, respectively. More clearance actin band was observed on EW than that of FDC. Whereas, vitellin like protein band of FDC was more clear than that of EW. SDS-PAGE patterns observed for BDC and SDC were comparable. Above 250 K of protein band was clearly observed on BDC and SDC. However, actinin was not

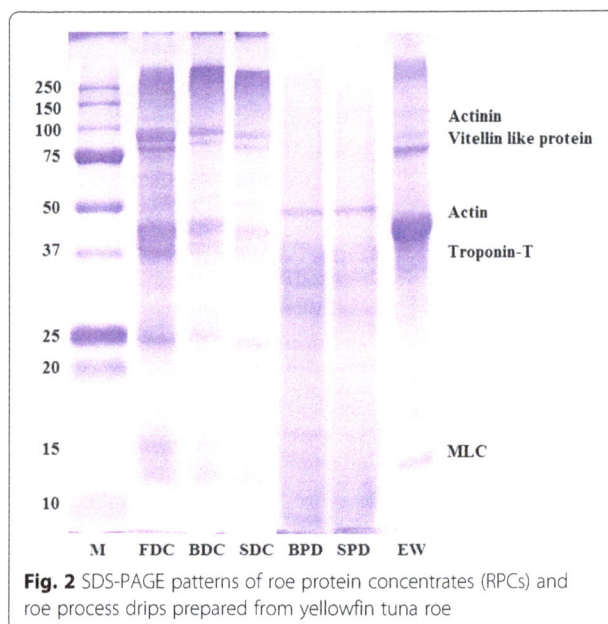

Fig. 2 SDS-PAGE patterns of roe protein concentrates (RPCs) and roe process drips prepared from yellowfin tuna roe

found both of them. Protein bands of 170, 95, and 55 K was reported in Channa and Lates roe protein concentrate (Narsing Rao et al. 2012a). Sathivel et al. (2009) observed the presence of four abundant proteins with molecular weights between 40 and 100 K in spray dried catfish roe protein powder. The major bands of dehydrated and defatted egg protein concentrates from mrigal observed occurred at molecular weight (MW) of 97 K (Chalamaiah et al. 2013). In case of BDC and SDC, band patterns in range of 15–25 K more reduced than those of FDC. Because association and aggregation of protein on low molecular protein was occurred due to cooking process which caused protein degradation and soluble protein like a sarcoplasmic protein was released from YTR. Released soluble protein was found on BPD and SPD. Their major protein band was 50 K. Low molecular soluble protein was observed in range of 25–50 K caused by degradation of high molecular protein.

Conclusion

Fish roe, a by-product in fishery industry, can be utilized as a low cost source of protein for value addition. Production of roe protein concentrate from yellowfin tuna roe is a simple and economical process. This study demonstrated that is feasible to produce concentrates from yellowfin tuna roe by cook-dried process. The roe protein concentrates were found to be rich in protein with essential amino acid. Fish roe protein concentrates may potentially serve as a good source of protein with desirable functional properties. Therefore, these protein concentrates could be used as protein supplements and functional ingredients in human diets.

Table 5 L*, a* and b* color values and whiteness of concentrates (FDC, BDC and SDC) by cook-dried process

Hunter color	FDC	BDC	SDC
L*	59.2 ± 0.1a	55.7 ± 3.7b	55.4 ± 1.9b
a*	6.5 ± 0.1a	5.2 ± 0.6b	5.3 ± 0.4b
b*	18.6 ± 0.0a	17.8 ± 0.1b	17.4 ± 0.1c
ΔE	42.3 ± 0.1a	45.0 ± 3.4a	45.1 ± 1.8a
Whiteness	54.7 ± 0.1a	52.0 ± 3.5a	51.8 ± 1.8a

Data are means ± standard deviation of triplicate determinations
Values with different letters within the same row are significantly different at
$P < 0.05$ by Duncan's multiple range test

Competing interests
The authors declare that they have no competing interests.

Authors' contributions
HJL and SHP carried out the preparation of protein concentrates, participated in the analysis of chemical compositions and drafted the manuscript. ISY, G-WL and YJK participated in searching and screening references and performed the statistical analysis and carried out the SDS-PAGE. J-SK and MSH conceived of the study, and participated in its design and coordination and helped to draft the manuscript. All authors read and approved the final manuscript.

Acknowledgements
This research was supported by Basic Science Research Program through the National Research Foundation of Korea(NRF) funded by the Ministry of Education(NRF-2014R1A1A4A01008620).

Author details
[1]Department of Food and Nutrition/Institute of Marine Industry, Gyeongsang National University, Jinju 52828, Korea. [2]Department of Seafood Science and Technology/Institute of Marine Industry, Gyeongsang National University, Tongyeong 53064, Korea.

References
Alasalvar C, Taylor KDA, Zubcov E, Shahidi F, Alexis M. Differentiation of cultured and wild sea bass (Dicentrarchus labrax): total lipid content, fatty acid and trace mineral composition. Food Chem. 2002;79:145–50.

Al-Holy MA, Rasco BA. Characterisation of salmon (Oncorhynchus keta) and sturgeon (Acipenser transmontanus) caviar proteins. J Food Biochem. 2006;30:422–8.

AOAC (Association of Analytical Chemists). Official Methods of Analysis. 16th ed. Washington DC: Association of Analytical Chemists; 1995. p. 69–74.

Bala BK, Mondol MRA. Experimental investigation on solar drying of fish using solar tunnel dryer. Drying Technol. 2001;19:427–36.

Balaswamy K, Prabhakara Rao P, Narsing Rao G, Govardhana Rao D, Jyothirmayi T. Physicochemical composition and functional properties of roes from some fresh water fish species and their application in some foods. J Environ Agric Food Chem. 2009;8:704–10.

Bekhit A, Morton JD, Dawson CO, Zhao JH, Lee HYY. Impact of maturity on the physicochemical and biochemical properties of chinook salmon roe. Food Chem. 2009;117:318.

Bellagha S, Amami E, Farhat A, Kechaou N. Drying kinetics and characteristic drying curve of lightly salted sardine (Sardinella aurita). Drying Technol. 2002; 20:1527–38.

Chalamaiah M, Narsing Rao G, Rao DG, Jyothirmayi T. Protein hydrolysates from meriga (Cirrhinus mrigala) egg and evaluation of their functional properties. Food Chem. 2010;120:652–7.

Chalamaiah M, Balaswamy K, Narsing Rao G, Prabhakara Rao PG, Jyothirmayi T. Chemical composition and functional properties of mrigal (Cirrhinus mrigala) egg protein concentrates and their application in pasta. J Food Technol. 2013;50:514–20.

Cordero-de-los-Santos MY, Osuna-Castro JA, Borodanenko A, Paredes-López O. Physicochemical and functional characterisation of amaranth (Amaranthus hypochondriacus) protein isolates obtained by isoelectric precipitation and micellisation. Food Sci Technol Int. 2005;11:269–80.

DeMan JM. Proteins. In: DeMan JM, editor. Principles in Food Chemistry. Gaithersburg: Aspen Publishers Inc.; 1999. p. 111–62.

Duan ZH, Zhang M, Tang J. Thin layer hot-air drying of bighead carp. Fish Sci. 2004;23:29–32.

FitzGerald RJ, O'Cuinn GO. Enzymatic debittering of food protein hydrolysates. Biotechnol Adv. 2006;24:234–7.

Heu MS, Kim HS, Jung SC, Park CH, Park HJ, Yeum DM, Park HS, Kim CG and Kim JS. Food component characteristics of skipjack (Katsuwonus pelamis) and yellowfin tuna (Thunnus albacares) roes. J Kor Fish Soc. 2006;39:1–8.

Intarasirisawat R, Benjakul S, Visessanguan W. Chemical compositions of the roes from skipjack, tongol and bonito. Food Chem. 2011;124:1328–34.

Iwasaki M, Harada R. Proximate and amino acid composition of the roe and muscle of selected marine species. J Food Sci. 1985;50:1585–7.

Klomklao S, Benjakul S, Kishimura H. Optimum extraction and recovery of trypsin inhibitor from yellowfin tuna (Thunnus albacores) roe and its biochemical properties. Int J Food Sci Technol. 2013;49:168–73.

Laemmli UK. Cleavage of structural proteins during the assembly of the head of bacteriophage T4. Nature. 1970;227:680–5.

Lowry OH, Rosebrough NJ, Farr AL, Randall RJ. Protein measurement with the Folin phenol reagent. J Biol Chem. 1951;193:265–75.

Mahmoud KA, Linder M, Fanni J, Parmentier M. Characterisation of the lipid fractions obtained by proteolytic and chemical extractions from rainbow trout (Oncorhynchus mykiss) roe. Process Biochem. 2008;43:376–83.

Matsubara T, Nagae M, Ohkubo N, Andoh T, Sawaguchi S, Hiramatsu N. Multiple vitellogenins and their unique roles in marine teleosts. Fish Physiol Biochem. 2003;28:295–9.

Ministry of Ocean and Fisheries. 2014. Yearbook of marine resource. Retrieved from http://www.mof.go.kr/article/view.do?articleKey=5197&boardKey=32&menuKey=396&currentPageNo=1. Accessed 26 Jan 2015.

Narsing Rao G, Balaswamy K, Satyanarayana A and Prabhakara Rao P. Physico-chemical, amino acid composition, functional and antioxidant properties of roe protein concentrates obtained from Channa striatus and Lates calcarifer. Food chem. 2012a; 132:1171–1176

Narsing Rao G, Prabhakara Rao P, Satyanarayana A and Balaswamy K. Functional properties and in vitro antioxidant activity of roe protein hydrolysates of Channa striatus and Labeo rohita. Food Chem. 2012b;135, 1479–1484

Pires C, Costa S, Batista AP, Nunes MC, Raymundo A, Batista I. Properties of protein powder prepared from Cape hake by-products. J Food Eng. 2012; 108:268–75.

Rodrigo J, Ros G, Periago MJ, Lopez C, Ortuiio J. Proximate and mineral composition of dried salted roes of hake (Merluccius merluccius, L.) and ling (Molva molva, L.). Food Chem. 1998;63:221–5.

Sathivel S, Bechtel PJ. Properties of soluble protein powders from Alaska pollock (Theragra chalcogramma). Int J Food Sci Technol. 2006;41:520–9.

Sathivel S, Yin H, Bechtel PJ, King JM. Physical and nutritional properties of catfish roe spray dried protein powder and its application in an emulsion system. J Food Eng. 2009;95:76–81.

Thanonkaew A, Benjakul S, Visessanguan W, Decker EA. The effect of metal ions on lipid oxidation, colour and physicochemical properties of cuttlefish (Sepia pharaonis) subjected to multiple freeze-thaw cycles. Food Chem. 2006;95:591–9.

PERMISSIONS

All chapters in this book were first published in FAS, by BioMed Central; hereby published with permission under the Creative Commons Attribution License or equivalent. Every chapter published in this book has been scrutinized by our experts. Their significance has been extensively debated. The topics covered herein carry significant findings which will fuel the growth of the discipline. They may even be implemented as practical applications or may be referred to as a beginning point for another development.

The contributors of this book come from diverse backgrounds, making this book a truly international effort. This book will bring forth new frontiers with its revolutionizing research information and detailed analysis of the nascent developments around the world.

We would like to thank all the contributing authors for lending their expertise to make the book truly unique. They have played a crucial role in the development of this book. Without their invaluable contributions this book wouldn't have been possible. They have made vital efforts to compile up to date information on the varied aspects of this subject to make this book a valuable addition to the collection of many professionals and students.

This book was conceptualized with the vision of imparting up-to-date information and advanced data in this field. To ensure the same, a matchless editorial board was set up. Every individual on the board went through rigorous rounds of assessment to prove their worth. After which they invested a large part of their time researching and compiling the most relevant data for our readers.

The editorial board has been involved in producing this book since its inception. They have spent rigorous hours researching and exploring the diverse topics which have resulted in the successful publishing of this book. They have passed on their knowledge of decades through this book. To expedite this challenging task, the publisher supported the team at every step. A small team of assistant editors was also appointed to further simplify the editing procedure and attain best results for the readers.

Apart from the editorial board, the designing team has also invested a significant amount of their time in understanding the subject and creating the most relevant covers. They scrutinized every image to scout for the most suitable representation of the subject and create an appropriate cover for the book.

The publishing team has been an ardent support to the editorial, designing and production team. Their endless efforts to recruit the best for this project, has resulted in the accomplishment of this book. They are a veteran in the field of academics and their pool of knowledge is as vast as their experience in printing. Their expertise and guidance has proved useful at every step. Their uncompromising quality standards have made this book an exceptional effort. Their encouragement from time to time has been an inspiration for everyone.

The publisher and the editorial board hope that this book will prove to be a valuable piece of knowledge for researchers, students, practitioners and scholars across the globe.

LIST OF CONTRIBUTORS

Jun Hyung Ryu
Department of Fisheries Biology, Pukyong National University, Busan 608-737, Korea

Hak Jun Kim
Department of Chemistry, Pukyong National University, Busan 608-737, Korea

Seung Seob Bae and Choon Goo Jung
Marine Biodiversity Institute of Korea, Seochun 33662, Korea

Seung Pyo Gong
Department of Fisheries Biology, Pukyong National University, Busan 608-737, Korea
Laboratory of Cell Biotechnology, Department of Marine Biomaterials and Aquaculture, College of Fisheries Science, Pukyong National University, Busan 608-737, Korea

Young Cheol Park, Yeoun Joong Jung, Ka Jeong Lee, Min Seon Kim, Kyeong Ri Go, Sang Gi Park, Soon Jae Kwon, Jong Soo Mok and Ji Hye Yang
Southeast Sea Fisheries Research Institute, National Fisheries Research & Development Institute, 397-68, Sanyang-iljuro, Sanyang-up, Tongyeong 650-943, Republic of Korea

Poong Ho Kim
Food Safety Research Division, National Fisheries Research & Development Institute, 216, Gijang-haeanro, Gijang-up, Gijang-gun, Busan 619-705, Republic of Korea

Md Mostafizur Rahman and Sang-Min Lee
Department of Marine Biotechnology, Gangneung-Wonju National University, 7 Jukheon-gil, Gangneung, Gangwon-do 25457, South Korea

Kyeong-Jun Lee
Department of Marine Life Sciences, Jeju National University, Jeju 63243, South Korea

Tae-Hyun Yoon, Jung-Woo Choi, Changsix Ra and Jeong-Dae Kim
College of Animal Life Sciences, Kangwon National University, Chuncheon 24341, Korea

Seunggun Won
Department of Animal Resources, Daegu University, Gyeongsan 38453, Korea

Dong-Hoon Lee
Gyenoggi Province Maritime and Fisheries Research Institute, 23-2 Sangkwang-Gil, Yangpyeong-Gun 12513, Korea

Eun-Seo Lim
Department of Food Science and Nutrition, Tongmyong University, Busan 608-735, Republic of Korea

Simona Rimoldi, Chiara Ceccotti, Rossana Girardello and Chiara Ascione
Department of Biotechnology and Life Sciences (DBSV), University of Insubria, Via Dunant, 3-21100 Varese, Italy

Giovanna Finzi
Department of Pathology, Ospedale di Circolo, Varese, Italy

Annalisa Grimaldi and Genciana Terova
Inter-University Centre for Research in Protein Biotechnologies "The Protein Factory", Polytechnic University of Milan and University of Insubria, Varese, Italy

Shin-Hu Kim, Jun-Hwan Kim and Ju-Chan Kang
Department of Aquatic Life Medicine, Pukyong National University, Busan 608-737, Republic of Korea

Seong Don Hwang and Ki Won Shin
Aquatic life disease control division, National Fisheries Research and Development Institute, Busan 619-902, Republic of Korea

Young-Sik Choe, Ji-Hoon Lee, Soo-Geun Jo and Kwan Ha Park
Department of Aquatic Life Medicine, College of Ocean Sciences, Kunsan
National University, San-68 Miryong-Dong, Gunsan City, Jeonbuk, South Korea

Jong-Oh Kim, Wi-Sik Kim, Ha-Na Jeong and Myung-Joo Oh
Department of Aqualife Medicine, Jeonnam National University, Yeosu 59626, South Korea

Sung-Je Choi
Algae Research Institute, JeollaNamdo, Wando 59146, South Korea

Jung-Soo Seo and Myoung-Ae Park
Aquatic Life Disease Control Division, Fundamental Research Department, National Fisheries Research and Development Institute, Busan 46083, South Korea

Song-Hun Han
Jeju Fisheries Research Institute, National Institute of Fisheries Science, Jeju 63068, Korea

Joon Sang Kim
Korea Fisheries Resources Agency, Jeju Branch, Jeju 63005, Korea

Choon Bok Song
College of Ocean Sciences, Jeju National University, Jeju 63243, Korea

Nalae Kang, Seo-Young Kim, Ju-Young Ko and You-Jin Jeon
Department of Marine Life Sciences, Jeju National University, Jeju 63243, Korea

Sum Rho
Center of Ornamental Reefs and Aquariums, Jeju 63354, Korea

Jun-Hwan Kim
West Sea Fisheries Research Institute, National Institute of Fisheries Science, Incheon, Korea

Ju-Chan Kang, Hee-Ju Park, In-Ki Hwang, Jae-Min Han and Do-Hyung Kim
Department of Aquatic Life Medicine, Pukyong National University, Busan, Korea

Chul Woong Oh
Department of Marine Biology, Pukyong National University, Busan, Korea

Jung Sick Lee
Department of Aqualife Medicine, Chonnam National University, Yeosu, Korea

So-Hee Son, Jin-Hyeon Jang, Hyeon-Kyeong Jo and Hyung-Ho Lee
Department of Biotechnology, Pukyong National University, Busan 608-737, South Korea

Joon-Ki Chung
Department of Aquatic Life Medicine, Pukyong National University, Busan 608-737, South Korea

Xu Chen, Jun Wang, Yun Wang and Qiang-qiang Liu
South China Sea Fisheries Research Institute, Chinese Academy of Fishery Sciences, Guangzhou 510300, China

Jia-jun Xie
South China Sea Fisheries Research Institute, Chinese Academy of Fishery Sciences, Guangzhou 510300, China
School of Life Science and Technology, Shanghai Ocean University, Shanghai 201306, China

Jin Niu
Institute of Aquatic Economic Animals, School of Life Science, Sun Yat-sen University, NO.135 at Xingang Xi Road, Haizhu District, Guangzhou 510275, Guangdong Province, China

Sang Yoon Lee and Yoon Kwon Nam
Department of Marine Bio-Materials & Aquaculture, Pukyong National University, Busan 48513, South Korea

Jun-Hwan Kim
West Sea Fisheries Research Institute, National Institute of Fisheries Science, Incheon 22383, Korea

Hee-Ju Park, Kyeong-Wook Kim and Ju-Chan Kang
Department of Aquatic Life Medicine, Pukyong National University, Busan 48513, Korea

Joon-Young Jun, Min-Jeong Jung and Byoung-Mok Kim
Division of Strategic Food Industry Research, Korea Food Research Institute, Seongnam 13539, Republic of Korea

Dong-Soo Kim
Jeonbuk Institute for Bioindustry, Jeonju 54810, Republic of Korea

In-Hak Jeong
Department of Marine Food Science & Technology, Gangneung-Wonju National University, Gangneung 25457, Republic of Korea

Ji-Hoon Lee, Ju-Wan Kim, Kyung-Il Park and Kwan Ha Park
Department of Aquatic Life Medicine, College of Ocean Science and Technology, Kunsan National University, 558 Daehak-ro, Miryong-Dong, Gunsan City, Jeonbuk 54150, Republic of Korea

Yun-Kyung Shin
Southeast Sea Fisheries Institute, National Institute of Fisheries Science, Youngwun-Ri 361, Sanyang-Eup, Tongyeong City, Kyungnam, Republic of Korea

Hee-Jeong Lee, Sol-Ji Chae, Periaswamy Sivagnanam Saravana and Byung-Soo Chun
Department of Food Science and Technology, Pukyong National University, 45 Yongso-ro, Nam-Gu, Busan 48513, Republic of Korea

Byeong-Hoon Kim
Marine Science Institute, Jeju National University, Jeju 6333, South Korea

Sang-Woo Hur
Aquafeed Research Center, National Institute of Fisheries Science (NIFS), Pohang 37517, South Korea

Yuki Takeuchi
Department of Electrical Engineering and Bioscience, School of Advanced Science and Engineering, Waseda University, Shinjuku, Tokyo 162-8480, Japan

Akihiro Takemura
Department of Chemistry, Biology and Marine Sciences, Faculty of Science, University of the yukyus, Nishihara, Okinawa 903-0213, Japan

Sung-Pyo Hur
Jeju International Marine Science Research & Logistics Center, Korea Institute of Ocean Science & Technology, Jeju 63349, South Korea

Young-Don Lee
Marine Science Institute, Jeju National University, Jeju 6333, South Korea
19-5, Hamdeok 5(o)-gill, Jocheon, Jeju Special Self-Governing Province 695-965, Republic of South Korea

Alagesan Paari
Department of Marine Bioscience, Gangneung-Wonju National University, Gangneung, Gangwon 25457, South Korea.
East Coast Life Science Institute, Gangneung-Wonju National University, Gangneung, Gangwon 25457, South Korea
Department of Life science, Christ University, Bengaluru, Karnataka, India

Chan-Hyeok Jeon
Department of Marine Bioscience, Gangneung-Wonju National University, Gangneung, Gangwon 25457, South Korea
East Coast Life Science Institute, Gangneung-Wonju National University, Gangneung, Gangwon 25457, South Korea
East Sea Life Resources Center, Korea Fisheries Resources Agency, 36340, Uljin, South Korea

Jeong-Ho Kim
Department of Marine Bioscience, Gangneung-Wonju National University, Gangneung, Gangwon 25457, South Korea
Department of Marine Bioscience, Gangneung-Wonju National University, Gangneung, Gangwon 210-702, South Korea

Sung-Hee Jung
Pathology Division, National Institute of Fisheries Science (NIFS), Busan 46041, South Korea

Hye-Sung Choi
Pathology Division, National Institute of Fisheries Science (NIFS), Busan 46041, South Korea
Inland Aquaculture Research Center, NIFS, Changwon 51688, South Korea

Hyun Ji Lee, Gyoon-Woo Lee, In Seong Yoon, Sung Hwan Park and Min Soo Heu
Department of Food and Nutrition/Institute of Marine Industry, Gyeongsang National University, Jinju 52828, Korea

Sun Young Park and Jin-Soo Kim
Department of Seafood Science and Technology/Institute of Marine Industry, Gyeongsang National University, Tongyeong 53064, Korea

Jun Hyung Ryu
Department of Marine Biomaterials and Aquaculture, Pukyong National University, Busan 608-737, South Korea

Yoon Kwon Nam
Department of Marine Biomaterials and Aquaculture, Pukyong National University, Busan 608-737, South Korea
Department of Fisheries Biology, Pukyong National University, Busan 608-737, South Korea

Seung Pyo Gong
Department of Marine Biomaterials and Aquaculture, Pukyong National University, Busan 608-737, South Korea
Department of Fisheries Biology, Pukyong National University, Busan 608-737, South Korea
Laboratory of Cell Biotechnology, Department of Marine Biomaterials and Aquaculture, College of Fisheries Science, Pukyong National University, Busan 608-737, South Korea

Hyun Ji Lee, Sung Hwan Park, In Seong Yoon, Gyoon-Woo Lee and Min Soo Heu
Department of Food and Nutrition/Institute of Marine Industry, Gyeongsang National University, Jinju 52828, South Korea

Hyung Jun Kim, Jin-Soo Kim and Yong Jung Kim
Department of Seafood Science and Technology/Institute of Marine Industry, Gyeongsang National University, Tongyeong 53064, South Korea

Hyun Ji Lee, Sung Hwan Park, In Seong Yoon, Gyoon-Woo Lee and Min Soo Heu
Department of Food and Nutrition/Institute of Marine Industry, Gyeongsang National University, Jinju 52828, Korea

Yong Jung Kim and Jin-Soo Kim
Department of Seafood Science and Technology/Institute of Marine Industry, Gyeongsang National University, Tongyeong 53064, Korea

Index